International Scientific Conference LOGI 2017

MATEC Web of Conferences Volume 134 (2017)

Ceske Budejovice, Czech Republic
19 October 2017

Editor:

Ondrej Stopka

ISBN: 978-1-5108-5191-7

Printed from e-media with permission by:

Curran Associates, Inc.
57 Morehouse Lane
Red Hook, NY 12571

Some format issues inherent in the e-media version may also appear in this print version.

Printed by Curran Associates, Inc. (2017)

For additional information, please contact EDP Sciences – Web of Conferences
at the address below.

EDP Sciences – Web of Conferences
17, Avenue du Hoggar
Parc d'Activité de Courtabœuf
BP 112
F-91944 Les Ulis Cedex A
France

Phone: +33 (0) 1 69 18 75 75
Fax: +33 (0) 1 69 28 84 91

contact-edps@webofconferences.org

Additional copies of this publication are available from:

Curran Associates, Inc.
57 Morehouse Lane
Red Hook, NY 12571 USA
Phone: 845-758-0400
Fax: 845-758-2633
Email: curran@proceedings.com
Web: www.proceedings.com

TABLE OF CONTENTS

LOGI 2017

The international scientific conference LOGI 2017 deals with the transport and logistics issues especially related to operation of all modes of transport, intermodal transport and terminals construction. The LOGI 2017 conference was organized on the ground of the Clarion Congress Hotel České Budějovice, Czech Republic, 19th October 2017.

VŠTE
VYSOKÁ ŠKOLA TECHNICKÁ A EKONOMICKÁ V ČESKÝCH BUDĚJOVICÍCH

Univerzita Pardubice
Dopravní fakulta Jana Pernera

Vysoká škola logistiky o.p.s.

Department of Transport and Logistics, Institute of Technology and Business in České Budějovice

Jan Perner Transport Faculty, University of Pardubice

College of Logistics – Přerov

in cooperation with

The President of Czech Logistics Association

Václav Cempírek

The Chairman of the Board of Association of Forwarding and Logistics of the Czech Republic

Martin Drábek

and

The Executive Director of Transport Union of the Czech Republic

Petr Kašík

18th International Scientific Conference

LOGI 2017

was held

October 19th 2017

on the ground of the Clarion Congress Hotel České Budějovice, Czech Republic

All manuscripts were supported by the research project *"From horse-drawn railway to intermodal transport"* within Visegrad Fund (no. 21710486).

Visegrad Fund

Edited by:

Ondrej Stopka

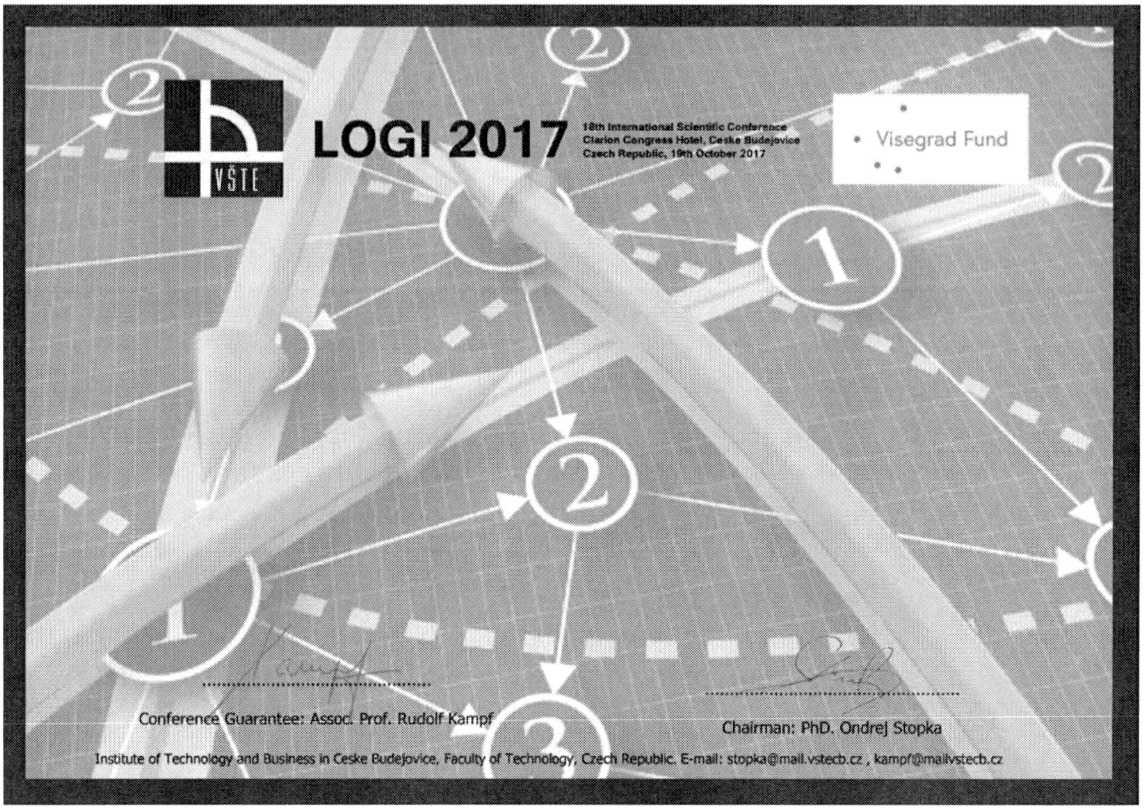

LIST OF TOPICS:

- Aviation, water transport, road transport and railways;
- Intermodal transport and intermodal transport terminals construction;
- Information systems and telecommunications;
- Applied mathematics in terms of transport and logistics;
- Automotive engineering;
- Transport engineering and constructions.

ORGANIZING COMMITTEE:

Conference Guarantee: Assoc. Prof. Rudolf Kampf, Faculty of Technology, Institute of Technology and Business in Ceske Budejovice, Czech Republic
e-mail: kampf@mail.vstecb.cz

Chairman: PhD. Ondrej Stopka, Faculty of Technology, Institute of Technology and Business in Ceske Budejovice, Czech Republic
e-mail: stopka@mail.vstecb.cz

Vice-Chairman: Assoc. Prof. Jan Lizbetin, Faculty of Technology, Institute of Technology and Business in Ceske Budejovice, Czech Republic
e-mail: lizbetin@mail.vstecb.cz

Members: Ing. Ladislav Bartuska, Faculty of Technology, Institute of Technology and Business in Ceske Budejovice, Czech Republic
e-mail: bartuskal@mail.vstecb.cz

Ing. Irena Hvozdova, Faculty of Technology, Institute of Technology and Business in Ceske Budejovice, Czech Republic
e-mail: hvozdova@mail.vstecb.cz

Ing. Martina Hlatka, Faculty of Technology, Institute of Technology and Business in Ceske Budejovice, Czech Republic
e-mail: hlatka@mail.vstecb.cz

Ing. Maria Chovancova, Faculty of Operation and Economics of Transport and Communications, University of Zilina, Slovak Republic
e-mail: maria.chovancova@fpedas.uniza.sk

PhD. Vladimir Luptak, Faculty of Technology, Institute of Technology and Business in Ceske Budejovice, Czech Republic
e-mail: luptak@mail.vstecb.cz

SCIENTIFIC COMMITTEE:

Prof. Jozef Gnap, Faculty of Operation and Economics of Transport and Communications, University of Zilina, Slovak Republic
e-mail: jozef.gnap@fpedas.uniza.sk

Prof. Gabriel Fedorko, Institute of Logistics, Faculty BERG,
Technical University of Kosice, Slovak Republic
e-mail: gabriel.fedorko@tuke.sk

Prof. Vieroslav Molnar, Institute of Logistics, Faculty of Mining, Ecology, Process Control and Geotechnology,
Technical University of Kosice, Slovak Republic
e-mail: vieroslav.molnar@tuke.sk

Prof. Vaclav Cempirek, Jan Perner Transport Faculty, University of Pardubice, Czech Republic
e-mail: vaclav.cempirek@upce.cz

Prof. Gerhard Bahrenberg, Institute of Geography,
University of Bremen, Germany
e-mail: gbha@uni-bremen.de

Prof. Bernd Kortschak, Faculty of Commerce, Logistics, Transport,
University of Applied Sciences Erfurt, Germany
e-mail: kortschak@th-erfurt.de

Prof. Hans-Martin Niemeier, Institute for Transport and Development, School of International Business, Bremen University of Applied Sciences, Germany
Email address: Hans-Martin.Niemeier@hs-bremen.de

Prof. Ulrich Weidmann, Eidgenössische Technische Hochschule Zurich, Swiss Federal Institute of Technology Zurich, Switzerland
e-mail: weidmann@ivt.baug.sthz.ch

Prof. Maria Nadia Postorino, Department of Civil, Energy, Environment and Materials Engineering, Mediterranea University of Reggio Calabria, Italy
e-mail: npostorino@unirc.it

Prof. Alexandre Dolgui, Automation, Production and Computer Sciences Department, Ecole des Mines de Nantes, Nantes, France
e-mail: alexandre.dolgui@mines-nantes.fr

Assoc. Prof. Jozef Gasparik, Faculty of Operation and Economics of Transport and Communications, University of Zilina, Slovak Republic
e-mail: jozef.gasparik@fpedas.uniza.sk

PhD. Vladislav Zitricky, Faculty of Operation and Economics of Transport and Communications, University of Zilina, Slovak Republic
e-mail: vladislav.zitricky@fpedas.uniza.sk

Assoc. Prof. Milos Hitka, Faculty of Wood Sciences and Technology, Technical University in Zvolen, Slovak Republic
e-mail: hitka@tuzvo.sk

Assoc. Prof. Zdenek Cujan, Department of Logistics and Technical Disciplines, College of Logistics, Czech Republic
e-mail: zdenek.cujan@vslg.cz

Assoc. Prof. Petr Prusa, Jan Perner Transport Faculty, University of Pardubice, Czech Republic
e-mail: petr.prusa@upce.cz

Assoc. Prof. Jaromir Siroky, Jan Perner Transport Faculty, University of Pardubice, Czech Republic
e-mail: jaromir.siroky@upce.cz

Assoc. Prof. Borna Abramovic, Faculty of Transport and Traffic Sciences, University of Zagreb, Croatia
e-mail: borna.abramovic@fpz.hr

Prof. Srecko Krile, Department of Electrical Engineering and Computing, University of Dubrovnik, Croatia
e-mail: srecko.krile@unidu.hr

Prof. Fusaomi Nagata, Department of Mechanical Engineering, Faculty of Engineering, Tokyo University of Science, Yamaguchi, Sanyo-Onoda, Japan
e-mail: nagata@rs.tus.ac.jp

Prof. Larisa M. Kapustina, Ural State University of Economics, Yekaterinburg, Russia
e-mail: lakapustina@bk.ru

Prof. Jozef Majercak, Faculty of Operation and Economics of Transport and Communications, University of Zilina, Slovak Republic
e-mail: jozef.majercak@pedas.uniza.sk

PREFACE

The international scientific conference LOGI has been held since 1997, i.e. in the 2017, it was already its 18th Volume.

In the previous volumes of this conference, the issues focused on the establishment, construction and operation of logistics centers, the issues of introduction of co-modality in the transportation and logistics chains, the support of intermodal transport and the reinforcement of the logistics as an independent field within the departments of government executives were discussed.

The topics of the LOGI conferences have been focused on emphasis of the issues of new trends and expectations in transport and logistics, logistics activities and services and progressive technologies utilized within the logistics processes.

The international scientific conference **LOGI 2017** dealt with the basic and applied research in the field of transport, logistics and other related sectors, such as transportation, road transport, railway and intermodal transport, air transport and shipping. The LOGI 2017 conference was organized on the ground of the Clarion Congress Hotel České Budějovice, Czech Republic, 19th October 2017.

A primary objective of Conference was to present the highest quality research results in the field of transport and logistics.

In the invitation sheet, sent half a year before the Conference beginning, the basic information about the Conference as a list of topics, Conference website, important deadlines, members of the committees, instructions for authors, fees, Conference venue as well as the main funder (Visegrad Fund) and partners were indicated (see http://journals.vstecb.cz/publications/logi/introduction/).

All the manuscripts were supported by the *international project "**From Horse-drawn Railway to Intermodal Transport**" - supported by the **Visegrad Fund** (no. 21710486): www.visegradfund.org.*

The individual manuscripts submitted within the LOGI 2017 (the Special issue of Web of Science indexed Journal – „MATEC Web of Conferences", which is online available via platform www.matec-conferences.org) presented the following areas:

- Aviation, water transport, road transport and railways;
- Intermodal transport and intermodal transport terminals construction;
- Information systems and telecommunications;
- Applied mathematics in terms of transport and logistics;
- Automotive engineering;
- Transport engineering and constructions.

All the submissions were directed to the appropriate reviewers (double-blind peer review process was applied). Two reviewers examined each manuscript much them to the expertise according to the interests and subsequently forwarded them to the editor. Selection of manuscripts for the Conference was highly competitive, thus participants had to ensure that their submissions met all the Conference requirements and were of the highest possible quality.

All Conference participants prepared their manuscripts according to the instructions for authors (Template).

edp sciences

MATEC Web of Conferences

Statement of Peer review

In submitting conference proceedings to MATEC Web of Conferences, I certify to the Publisher that I adhere to the **Policy on Publishing Integrity** of the journal in order to safeguard good scientific practice in publishing.

1. All articles have been subjected to peer review administered by the proceedings editors.
2. Reviews have been conducted by expert referees, who have been requested to provide unbiased and constructive comments aimed, whenever possible, at improving the work.
3. Proceedings editors have taken all reasonable steps to ensure the quality of the materials they publish and their decision to accept or reject a paper for publication has been based only on the merits of the work and the relevance to the journal.

Title, date and place of the conference LOGI 2017
19th October 2017, CLARION CONGRESS HOTEL
ČESKÉ BUDĚJOVICE, CZECH REPUBLIC

Proceedings editor(s): ING. ONDREJ STOPKA, PhD.

Date and editor's signature 26.10.2017

Draft Proposal of Determining Charges for Stabling of Rolling Stock within the ŽSR Railway Network

Peter Blaho[1], *Peter* Šulko[2], *Lenka* Černá[3,*], *and Anna* Dolinayová[3]

[1]Železnice Slovenskej republiky (ŽSR), 1. Mája 34, 010 01 Žilina, Slovak Republic
[2]Železnice Slovenskej republiky (ŽSR), Klemensová 8, 813 61 Bratislava, Slovak Republic
[3]University of Žilina, Faculty of Operation and Economics of Transport and Communications, Department of Railway Transport, Univerzitná 1, 0126 Žilina, Slovak Republic

Abstract. The contribution briefly presents analysis of payment definitions for the access to the railway infrastructure in Slovak Republic and in chosen countries of EU and consequently it suggests payment definitions for long-term track occupation by railway vehicles on ŽSR network.

1 Introduction

The essential role of an infrastructure manager is to permanently and safely operate the track as well as to organize a fluid flow of transport on the track. For this activity, the infrastructure manager is entitled to levy carrier's charges for the use of the railway infrastructure, charges for access to service facilities as well as for services provided by the latter. The method of determining the charges as well as the scope of the provided services has been defined in the EP and Council Directive 2001/14/EC **on the allocation of railway infrastructure capacity and the levying of charges for the use of railways** and safety certification which is implemented in the Act No. 513/2009 Coll. of October 2009 on railways and on amendment of some of the acts as well as of the Act No. 514/2009 Coll. on the transport on the railways [1].

The charges for access and use of RI as well as the attitude of the infrastructure manager to requirements of the carriers depend on the scope of provided services. This article will deal with the service of immobilization of rolling stock on the network of ŽSR (Railways of the Slovak Republic), which currently fails to be a part of the offer of services of the infrastructure manager, or the vehicles immobilization on the rails of ŽSR is free of charge. Consequently, this fact causes increased requirements as regards immobilization of complete train sets that takes several days bringing along complications and problems in managing the transport operations.

* Corresponding author: lenka.cerna@fpedas.uniza.sk

2 Charging fees for stabling of rolling stock in selected countries

In spite of the fact that the system of levying charges for the railway infrastructure in different countries is basically the same, based on the Directive of the European Parliament and Council 2001/14/EC, the actual amount of charges for use of the railway infrastructure as well as for services provided by the railway infrastructure based on costs directly incurred for running of the train, nevertheless, logically, has to differ [1].

Similarly, also the fees for provided services in maintenance centres that include the "stabling" track differ. A comparison of the fee for the use of a stabling track, or possibly a fee for parking of vehicles is given in Table 1. Individual fees are first calculated to EUR according to the relevant exchange rate [2].

Slovakia and the Czech Republic, or ŽSR as well as SŽDC respectively, have no special dedicated tracks that serve parking of trains. Parking of vehicles as well as trains is possible in a given agreed station for necessary period of time only upon a mutual agreement between railway companies and the infrastructure manager, as well as subject to given operational conditions. Thus, parking of vehicles is free of charge [3 -7].

Table 1. Comparison of fees for parking of vehicles. Source: authors, based on [3-7]

Country	Charge in the country	Conversion fee
Slovakia	0 €	0 €
Czech Republic	0 €	0 €
Austria	0.2953 € / meter and day	0.2953 € / meter and day
Hungary – MÁV	122 HUF / meter and day	0.39 € / vehicle and day
Hungary – GYSEV	189 HUF / meter and day	0.60 € / vehicle and day
Poland	1.14 PLN / meter and day	0.27 € / vehicle and day

Since Austria charges for "a meter of a vehicle" and Hungary for "the whole vehicle", in order to make a comparison we selected 20 m as an average length of a vehicle. In Poland there is a charge for a track per hour, therefore the charge has to be multiplied by 24 hours.

3 Assumptions of determining the charges for stabling of the rolling stock by ŽSR

The objective of charging the service of parking the wagons and other rolling stock (especially complete trainsets) is, nevertheless, not only to determine a new payment, but rather to incentivize carriers to better utilize the railway vehicles and avoid congesting the tracks of the ŽSR network. Congestion of the running lines of stations has consequently direct negative impact not only on the general management of rail operations, but it even affects in a greater extent the operational transport control in extraordinary transport circumstances. Most of the neighbouring infrastructures charge parking of the rolling stock, and therefore carriers are quite happy to park their rolling stock using the infrastructures that provide the "parking" free of charge [8].

In order to determine the charge for the service of the rolling stock parking appropriately, it is necessary to define the following assumptions:

- Total length of suitable station tracks for stabling.
- Proper knowledge and registration of stabling reasons.
- Stay duration in case of stabling.
- Charge (fee) - payer for stabling.

3.1 Determination of suitable tracks for rolling stock stabling

Within the ŽSR network there are no specially designated stabling tracks, nevertheless parking of vehicles would be possible depending on the operational situation on the available track. According to the code list of the Central Master Data Administration (CSKMD) of the operation information system the ŽSR network registers in total 4 862 station tracks with the usable length of approximately 2 089 km.

In consideration of basic characteristics as well as purpose of individual tracks, pursuant to CSKMD the tracks not suitable for stabling the wagons (vehicles) are the following [8]:

In respect to the handling tracks: joining track, rented track, track with some machinery equipment (washing facility, crane, scales), loading ramps, track for repairing, transhipment track, track for load fitting.

Engine tracks, tour-out tracks, siding tracks, stabling (for HDV, rail equipment, trucks), currently not operated and immobilized tracks

- In respect to the running tracks: main plain tracks, with a platform, joining track, shunting track, siding track, transfer track from and to the siding, rented track, tracks of the siding operator, currently not operated and immobilized (out of operation) track.

In consideration of these specified characteristics from CSKMD there has been formed as the basis for stabling of rolling stock within the ŽSR network approximately 960 running tracks (with the usable length of 564 km) and 1245 handling tracks (with the usable length of 373 km) [9-10].

3.2 Reason for stabling the rolling stock

For each parking of the rolling stock and especially parking of the whole trainset on the track it is necessary to know and register also the reason of the parking. Parking of wagons, vehicles as well as trainsets on the tracks of ŽSR can occur due to reasons on the side of the carrier, ŽSR, third parties, or also due to „force majeure “. CSKMD contains a numbered list of reasons for parking, which include e.g. waiting for HDV, changing of the train crew, instructions, infrastructure restrictions, accidents, unfavourable weather conditions etc. On the basis of a correct and unambiguous definition of the reason it will be possible, or not necessary to determine the charge payer for parking of vehicles [11-12].

3.3 Stay duration in case of stabling

The essential assumptions for determining the charges include relevant information about the stay of a train (stay of wagons and vehicles) at the traffic control point. Due to specific features of the mode of transport, it is necessary to make the analysis of data on passenger train stays and on freight trains stay separately. Presentation of individual values of stay duration and their frequencies already at first sight show evident considerable differences (Fig.1).

 DOI: 10.1051/matecconf/201713400001

Passenger trains already by their nature (transporting passengers) in principle do not have planned long stays in forwarding stations (Fig. 2). A longer stay in a forwarding station (for about 30 minutes) can be caused especially in case of trains with some planned change in its composition (additional coupling and uncoupling of through coaches, setting up and disintegrating the trainsets, direct trains waiting for corresponding trains or possibly activities in case of closure). Long stays of passenger trains are less numerous and they occur only as extraordinary events (traction rolling stock (HDV) or wagon breakdown, accidents) [13].

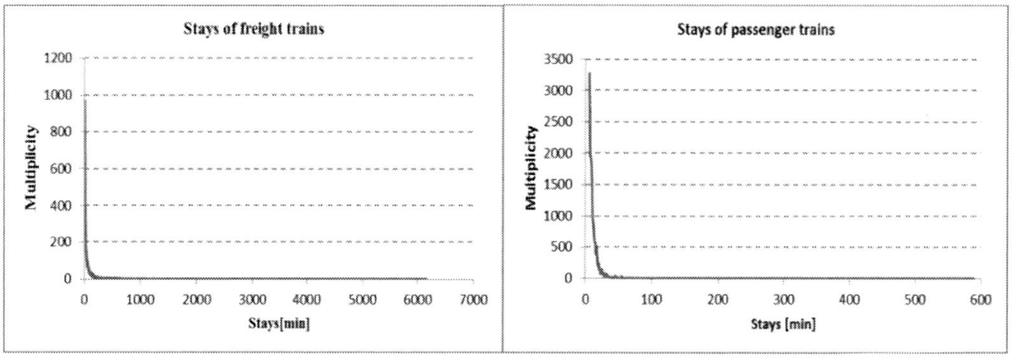

Fig. 1. Stays of trains in the forwarding railway station, total stays. Source: [13]

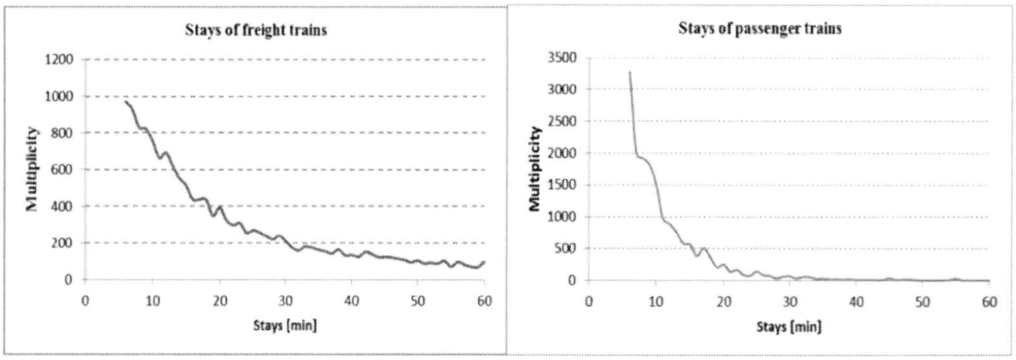

Fig. 2. Stays of trains in the forwarding railway station, stays of up to 60 minutes. Source: [13]

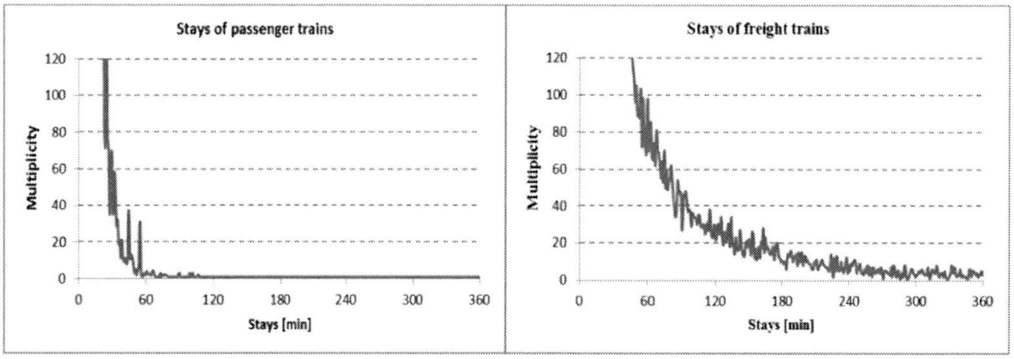

Fig. 3. Stays of trains in the forwarding railway station, stays of up to 360 minutes. Source: [13]

The short stays of freight trains in forwarding stations (about 30 minutes) are mostly due to traffic reasons, longer stays mostly concern transit trains with need of manipulation (change of the trainset composition, HDV change, turnings and changes of train essentials among the trains), but there also closure activities as well as extraordinary situations in traffic. The longest stays are connected especially with stabling of trainsets in the forwarding stations.

Due to the above analysis (approx. 510 000 data on stays per 2015), charging for stabling of wagons, vehicles as well as trainsets was implemented for stays exceeding 360 minutes (6 hours).

3.4 Payer of the charge for vehicles stabling

Within the topic it is also important to determine who will be charged with fees for stabling of vehicles. The charge-payer in this case must be the carrier that stables the given vehicles (wagons, a trainset, train). The reason is the fact that only this given carrier has the possibility to actually start running the given train, the possibility to establish an agreement on handing over the train (wagons) to another carrier in PPS or in the forwarding station or to a siding, this is also just within his competence, or possibly his obligations. In case he fails to meet these obligations or to appropriately agree on their meeting, he will actually have to stable/park the train (trainset, wagons) on the station tracks of the infrastructure manager [14-16].

In the event, that stabling of the train (trainset, wagons) has been caused by the transferee (the carrier taking over the obligations), the burden of proof, agreement and payment would be left only within the competence of the transferor (the carrier handing over the obligations). Thus, charges would be invoiced by the infrastructure manager to the carrier that stables the train, or possibly vehicles, and it will be within the mutual discretion of the transferor and transferee on what they will agree on.

4 Determining the amount of charge for stabling

In consideration of the above the actual amount of the charge for stabling of vehicles should take into account the length of stay of stabling of vehicles (in hours) as well as the length of stabled the current topic in terms of rail transport vehicles (in metres). The amount of charges in total should not represent any amount implying his liquidation (i.e. not too big), but rather incentivizing (i.e. reasonably big), in order to make him use the vehicles more efficiently and to park (stable) the rolling stock on the ŽSR tracks in minimum time periods [17].

The determination of the amount of charges for stabling of the rolling stock should therefore be based especially on the following principles:

- Charging for the railway infrastructure and service facilities.
- Determining the scope and amount of economically eligible costs (EON) related to service provision and a reasonable profit.
- The charge will be determined based on economically eligible costs incurred for service provision in maximum up to the amount of total own costs including a reasonable profit.
- Under the currently applicable „Concept of Charging the Railway Infrastructure in ŽSR" for each budgeted year there is an agreed scope and amount of total economically eligible costs classified as fixed and variable costs.
- Determination of average costs per 1 meter of track has been based on costs for administration, maintenance and operation of tracks within the whole ŽSR network, without taking account of its purpose specification (running, handling etc.).

According to the above given principles the total own costs (total EON in respect to the tracks) then include track depreciation, administrative costs, tracks maintenance and repair costs, internal organization supplies, operating and other general network overhead costs. In the follow-up these elements will be subject to cost-sharing method whereby we will arrive to determined shares of fixed and variable EON.

To determine and specify a unit price we can thus derive from the total costs for the tracks within the ŽSR network per 2015 in the amount of 15 547 580 EUR (variable costs), in the total number of tracks amounting to 4 862 and their total usable length 2 088 805 m. The calculation of average costs per 1 day (every 24 hours) was based on the calendar year, i.e. 364 days. Consequently, pursuant to the above, the amount of charge for stabling of the rolling stock was determined **as 0.0204 EUR per each commenced 1 meter of occupied track and per each commenced 24 hours** [18].

5 Application of the proposed charge to an average train

Due to the above and proposed we can compare what would be the level of levying charges for using stabling tracks, or possibly for parking of the rolling stock by ŽSR. We will thus use the same comparison criterion as mentioned in Chapter 2 (Table 1). Given the average length of vehicle is 20 m, the comparable charge for stabling of one average vehicle per day in individual countries has been illustrated in Figure 4 [19].

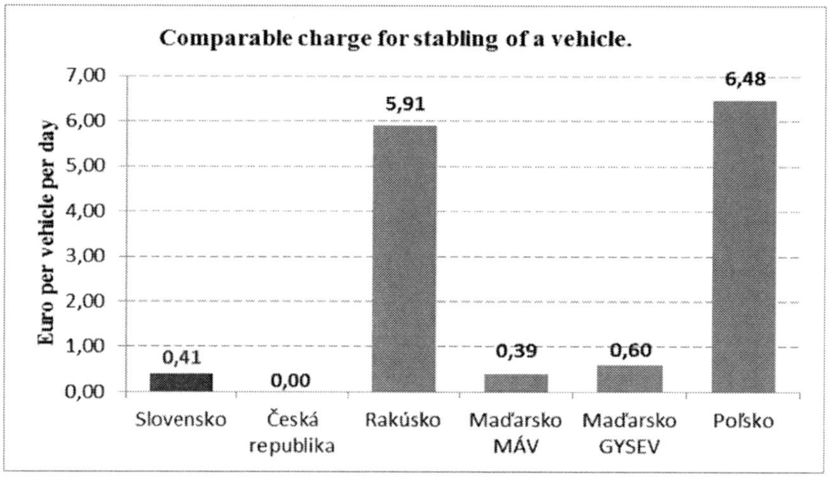

Fig. 4 Comparable charge for stabling of a vehicle. Source: authors, based on [19]

The difference of amounts of charges and their meaningfulness is more evident when they are applied to the whole train. Let us assume an average train of 600 meters comprised of 30 wagons of average length of 20 m each. Then the amount of the charge for stabling this average train in selected countries is as illustrated in Figure 5 [19].

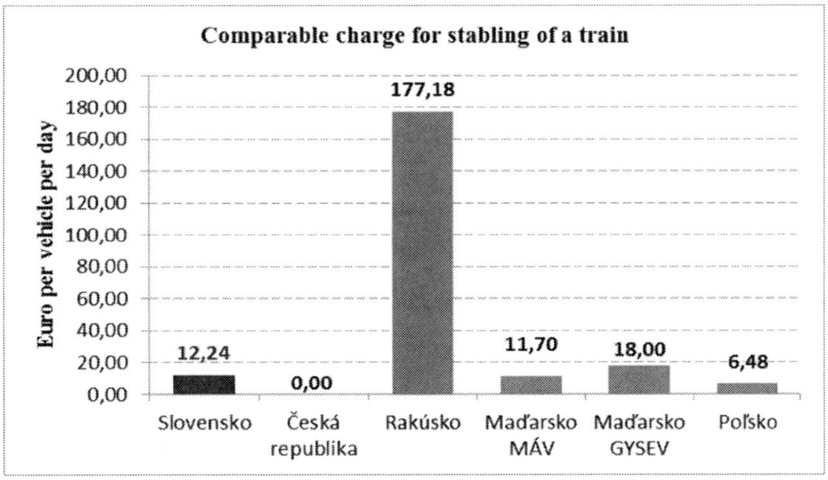

Fig. 5. Comparable charge for stabling of a train. Source: authors, based on [19]

From the above, it can be assumed that stabling of the rolling stock in financial terms would be most convenient in the Czech Railway Infrastructure, whereas the most unfavourable would be the railway infrastructure in Austria.

6 Conclusion

In the conclusion, it can be briefly stated that determining the charge for stabling of the rolling stock is feasible. Its introduction, eventually, would not imply any relatively significant investment. In principle, the estimated reduction in revenues from U_{tp2} a U_{tp3} components of charges for track access to service facilities would be "balanced" by revenues from the new service "stabling track" (i.e. charges for stabling the rolling stock). Thus, there would be a change in distribution of costs as well as revenues, implying that this new charge should not have any impact on the costs, revenues and subsidies from the state budget. Concurrently, the fee U_{tp2} and U_{tp3} applied in respect to carriers would be reduced, but more importantly, this new charge should motivate carriers to more efficient time utilization of the rolling stock and prevent congesting the running tracks of the ŽSR network, which is an undisputable benefit for transport management on the side of the infrastructure manager.

The paper was supported by the VEGA Agency, Grant No. 1/0019/17 "Evaluation of regional rail transport in the context of regional economic potential with a view to effective use of public resources and social costs of transport", at Faculty of Operations and Economics of Transport and Communication, University of Žilina, Slovakia.

This paper is supported by the research project "From horse-drawn railway to intermodal transport" within Visegrad Fund.

References

1. *Directive 2001/14/EC of the European Parliament and of the Council of 26 February 2001 on the allocation of railway infrastructure capacity and the levying of charges for the use of railway infrastructure and safety certification*, Available online: http://data.europa.eu/eli/dir/2001/14/oj (2001)

2. J. Ližbetin, J. Ponický, V. Zitrický, Nase More **63**, 161-169 (2016), DOI: 10.17818/NM/2016/SI16
3. SŽDC, *The Network Statement on nationwide and regional rail networks*, Available online: http://www.szdc.cz/en/soubory/ prohlaseni-o-draze/2017/prohlaseni-2017.pdf (2016)
4. ŽSR, *Network Statement 2016/2017*, Available online: http://www.zsr.sk/buxus/docs/Marketing/SVen/2017/Network_Statement_2016-2017-1.pdf (2017)
5. ÖBB, *Network Statement ÖBB, Infrastruktur AG rail network;* Available online: http://www.oebb.at/infrastruktur/en/_p_Network_Access/Product_ Services__Prices/02_DMS_Dateien/_Train_Path.jsp (2016)
6. MAV, *Network Statement MAV*, Available online: https://www2.vpe.hu/en/network-statement-2016-2017 (2016/2017)
7. *Network Statement, Regulations on train paths allocation and use of allocated train paths by licensed railway undertakings within timetable*, Available online: http://en.plk-sa.pl/files/public/user_upload/pdf/Reg_ przydzielania_tras/Regulamin_2016_2017/13.09.2017/Network_Statement_2016-2017_v22.pdf (2016/2017)
8. P. Droździel, B. Buková, E. Brumerčíková, Transport Problems **10**, 5-13 (2015)
9. O. Stopka, J. Ponický, M. Chovancová, V. Zitrický, Nase More **63**, 200-203 (2016), DOI: 10.17818/NM/2016/SI23
10. J. Ližbetin, P. Vejs, O. Stopka, V. Cempírek. Nase More **63**, 156-160 (2016), DOI: 10.17818/NM/2016/SI15
11. J. Mašek, J. Čamaj, I. Bolešović, *CTTE 2016 – 2nd Conference on Traffic and Transportation Engineering*, 409-416 (Xian, China, 2016)
12. J. Čamaj, J. Mašek, M. Kendra, Procedia Engineering **134**, 224-230 (2016)
13. ŽSR, *Annual report of the* ŽSR, Available online: http://www.zsr.sk/buxus/docs/vyrSpravy/AnnualReport2016.pdf (2016)
14. J. Čamaj, J. Mašek, A. Dolinayová, J. Daniš, *ICTTE 2016 – International conference on Traffic and Transport Engineering,* 955-963 (Belgrade, Serbia, 2016)
15. P. Meško, J. Gašparík, V. Gáborová. *Transport means 2016 – Proceedings of the 20th international scientific conference,* 224-229 (Juodkrante, Lithuania, 2016)
16. B. Buková, E. Brumerčíková, P. Kondek, *ESM 2016 – the 2016 international conference on Engineering science and management* **62,** 249-252 (Zhengzhou, China, 2016)
17. R. Kampf, O. Stopka, I. Kubasáková, V. Zitrický, *WMCAUS 2016 – World Multidisciplinary Civil Engineering-Architecture-Urban Planning Symposium*, 1538-1544 (Procedia Engineering **161**, Prague, Czech Republic, 2016), DOI: 10.1016/j.proeng.2016.08.623. (2016)
18. *Internal materials and regulations of the ŽSR* (2017)
19. I. Šulková. *Draft Proposal of Determining Charges for Stabling of Rolling Stock within the ŽSR Railway Network* (Diploma thesis, University of Zilina, 2016)

The Influence of Free Transport in Rail Passenger Transport on the Elasticity of Demand of Rail Passenger Transport

Eva Brumerčíková[1,*], *Bibiana* Buková[1], and *Pavol* Kondek[1]

[1]University of Žilina, Faculty of Operation and Economics of Transport and Communications Department of Railway Transport, Univerzitná 1, 010 26 Žilina, Slovakia

Abstract. The article focuses on the elasticity of the demand for transport services in rail passenger transport in the Slovak Republic after introducing free transport for students and retirees. The article studies the impact on income and cross elasticity, given that the fare for rail transport is regulated by the state and, therefore, there is no price elasticity. The article characterizes the volume of transport services in rail passenger transport and the impact of the introduction of free transport on the cross and income elasticity.

1 Introduction

The elasticity (flexibility) of demand reflects how the demand for a product changes when its price or the price of substitute goods changes or when revenue changes. The individual coefficients of elasticity measure changes in demand with respect to changing one of the factors, while the other conditions remain unchanged. Demand for rail passenger transport in the Slovak Republic declined and transport became an inferior product. In November 2014, Slovak Republic introduced free rail passenger transport for selected groups of passengers. The article summarizes the changes in elasticity after the introduction of the free rail passenger transport.

2 Demand for transport services

The necessary data for calculating the coefficients of all monitored types of elasticity is the volume of transport services in rail passenger transport in the Slovak Republic for the whole monitored period of 10 years. Demand for transport services is shown in Table 1 and is shown in millions pkm.

* Corresponding author: eva.brumercikova@fpedas.uniza.sk

Table 1. Demand for transport services. Source: [1]

Year	2006	2007	2008	2009	2010	2011	2012	2013	2014	2015
Demand	48 438	46 984	48 655	46 597	46 509	47 453	44 609	45 946	49 098	60 292
Index	-	0,96998	1,03557	0,9577	0,99811	1,0203	0,94007	1,02997	1,0686	1,22799

The table 1, it is clear that Slovakia recorded the largest increase in passenger numbers in 2015, which influenced the introduction of free transport from 17th November 2014. On the contrary, the largest decline in using the rail transport was in 2012. Slovakia does not make significant fluctuations in the use of passenger rail transport except for 2014 and 2015 when introducing free passenger transport. The increase in demand between 2006 and 2015 is 24.47%. The demand for transport services for the year 2016 has not been published yet.

3 Income elasticity

Income elasticity reflects how the demand for rail transport service changes if revenue is changed, i.e. if the real income increases, the demand should increase, too. In the case of rail transport, it is assumed that the increase in the income of the population will increase the number of cars and the demand for rail passenger transport will decrease [2, 3]. Calculation of income elasticity (equation 1):

$$E_p = \frac{\frac{\Delta Q_{ŽD}}{Q_{ŽD}}}{\frac{\Delta I_j}{I_j}} \tag{1}$$

where: E_p- coefficient of income elasticity; $Q_{ŽD}$ - demand for the rail transport services; I_j - unit revenue (income per person).

According to the final value of elasticity we distinguish following results:

good	result	what happens
Normal:	$E_p > 0$	With the increase in income, the amount of purchased goods is increasing.
Luxury:	$E_p > 1$	The amount of purchased goods grows faster than the consumer's income.
Necessity:	$0 < E_p < 1$	The amount of purchased goods is growing more slowly than the consumer's income.
Inferior:	$E_p < 0$	With the growth of income, the demand for inferior goods declines and vice versa [4, 5].

In order to calculate the income elasticity coefficient, the average annual unit income in the Slovak Republic has to be analyzed, as shown in the following Table 2.

Table 2. Average income in Slovak Republic. Source: [1]

Year	2006	2007	2008	2009	2010	2011	2012	2013	2014	2015
Average wage	8 694	9 469	9 991	10 334	10 897	11 218	11 475	11 716	11 932	12 305
Index	-	1,08914	1,05513	1,03433	1,05448	1,02946	1,02291	1,021	1,01844	1,03126

Based on Table no. 2 it is possible to compare the average annual income in the Slovak Republic. The table shows that the average income rises by 3.96% on average in 10 years. Between 2006 and 2015 the average income increased by 41%.

Table 3. Income elasticity coefficients in the Slovak Republic. Source: authors

Year	2006	2007	2008	2009	2010	2011	2012	2013	2014	2015
Income elasticity	-	-0,378	0,657	-1,331	-0,037	0,695	-2,847	1,415	3,546	6,125

According to Table 3, it can be assumed that in 2007, 2009, 2010 and 2012, rail transport in Slovakia was considered an "inferior" good, as the income elasticity coefficient was negative which means that the demand for rail transport decreased with the growth of retirement income. In 2008 and 2011, rail transport can be assessed as a necessity good, so the number of passengers transported is increasing more slowly than the consumer's income. This can be noticed, for example, when comparing the years 2008 and 2007, when average income grew by 5.5%, while the number of passengers transported grew only by 3.5%.

We can denote the rail transport in Slovakia between 2013 and 2015 as a luxury good, when under the influence of the introduction of free transport there was a significant difference between the increase in average income and the increase in the number of passengers transported. In order to be able to state, based on the above mentioned facts, that rail transport is an "inferior" good, Table no. 1 shows the Engel curve for inferior goods [6, 7]. The graph shows that rail transport could be considered a necessity good, as the Engel curve is a concave. In the case of an average income greater than 11500 EUR, rail transport becomes an "exclusive" good when the demand is rising sharply.

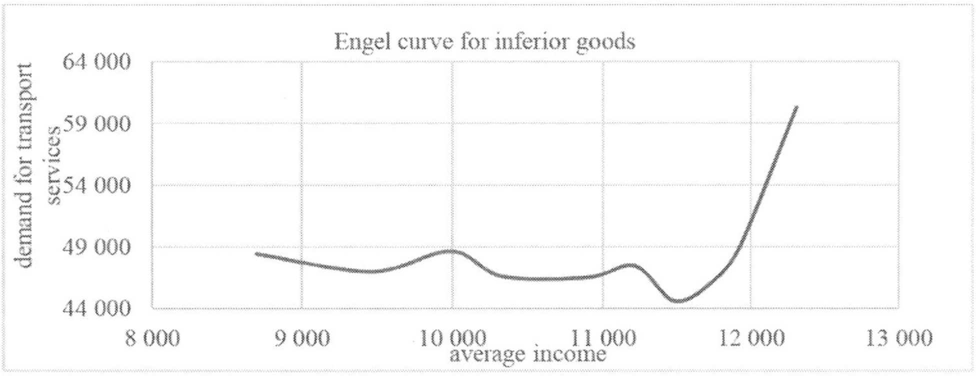

Fig. 1 Engel curve for inferior goods. Source: authors

4 Cross elasticity

Cross elasticity expresses how demand changes if the price of another type of transport changes, provided that the cost of the rail transport service remains unchanged. Cross elasticity gains values:

- $E_k > 0$ - substitution effect, i.e. in the case of a reduction in the price of another type of transport, the demand for rail transport services is reduced, provided that the price of the rail transport service does not change. In this case, the consumer replaces the more expensive good with a cheaper one, hence abandoning the use of rail transport. The higher the cross elasticity of the goods X and Y, the more they are substitutable.

- $E_k < 0$ - a complementary effect, i.e. the service of another type of transport is complementary to the rail transport service. In such a case, the reduction in the prices of rail transport services will cause an increase in demand for rail transport as well as in its supplementary services.
- $E_k = 0$ - cross elasticity does not apply [8, 9].

Cross-elasticity can be expressed as follows:

$$E_k = \frac{\frac{\Delta Q_{ŽD}}{Q_{ŽD}}}{\frac{\Delta P_j}{P_j}} \tag{2}$$

where: E_k - coefficient of cross elasticity; Pj - price of the service of another type of transport.

The demand for rail passenger services is also specific for cross elasticity measurements when compared to individual car traffic. In addition, if the price of individual car transport increases, there is no increase in demand for rail passenger transport [10-12].

Table no. 4 displays the analysis of fuel prices, namely oil fuel, in the Slovak Republic for a time period of 10 years. The given results are the annual average prices of fuel and they are stated in the purchase price.

Table 4. Oil fuel price per 1 l in €. Source: [1]

Year	2006	2007	2008	2009	2010	2011	2012	2013	2014	2015
Average price of diesel	1,318	1,245	1,383	1,096	1,114	1,34	1,441	1,39	1,334	1,135
Index	-	0,9446	1,1108	0,7925	1,0164	1,2029	1,0754	0,9646	0,9597	0,8508

Table 4 shows that the price of fuel, in particular oil fuel, has varied depending on excise duty applied to oil fuel. The lowest decrease can be seen between 2014 and 2015, when the price of oil fell most in the studied period.

The following table no. 5 shows the cross elasticity coefficients for the studied period in the SR.

Table 5. Cross elasticity coefficients. Source: authors

Year	2006	2007	2008	2009	2010	2011	2012	2013	2014	2015
Cross elasticity	-	0,528	0,344	0,169	-0,117	0,118	-0,91	-0,793	-1,529	-1,059

Between 2007 and 2009, substitution effect prevailed in Slovakia, which means that when reducing the price of oil fuel, passengers used rail passenger transport less and used other alternatives instead. For example, the number of passengers decreased by 3% in 2007, and oil fuel prices, which were developing in the same way, decreased by 5.5%. The change occurred in 2012, 2013, 2014 and 2015 when, for example, oil fuel prices decreased by almost 15% in 2015, causing that the number of transported passengers grew by almost 23%. Thus, the monitored goods are already considered complementary during this period because it is no longer possible to reduce oil fuel prices so much that they can compete with free rail passenger transport [13, 14].

5 Conclusion

Two types of elasticity were observed in the elasticity of demand for transport services in rail passenger transport. The first observed parameter was income elasticity, where the impact of changes in the average annual income of the country's inhabitants on the demand for transport by rail was determined. It was discovered that in the past rail transport in the

Slovak Republic was considered to be inferior, later a necessity good, however in the last observed time period it has the characteristics of a luxury good when under the influence of the introduction of free transport there has been a significant difference between the increase in average income and the increase in the number of passengers transported.

The second parameter was cross elasticity, where the effect of the change in the price of oil fuel on the performance of rail passenger transport was determined. In the first observed years in the Slovak Republic, the substitution effect prevailed when oil fuel prices and the number of persons transported in rail transport were developing in the same way. The change occurred only when the free rail passenger transport was introduced, where the reduction in the price of oil fuel did not affect the reduction of the volume of transport by rail.

The paper is supported by the VEGA Agency by the Project 1/0095/16 "Assessment of the quality of connections on the transport network as a tool to enhance the competitiveness of public passenger transport system", that is solved at Faculty of Operations and Economics of Transport and Communication, University of Zilina.

This paper is supported by the research project "From horse-drawn railway to intermodal transport" within Visegrad Fund.

References

1. *Eurostat, statistical data*, Available online: www.ec.europa.eu/eurostat/ (2017)
2. A. Dolinayova, E. Nedeliakova, E. Brumercikova, *Economy of Railway Transport: Monography*, (University of Zilina, Slovakia, 2016)
3. L. Cerna, V. Zitricky, J. Ponicky, *MME 2016*, 126-131 (2016)
4. J. Danis, A. Dolinayova, J. Camaj, AER – Advances in Engineering Research **62**, 187-190 (2016)
5. L. Cerna, V. Zitricky, J. Danis, Open Engineering **7**, 6-13 (2017), DOI: 10.1515/eng-2017-0002
6. J. Masek, M. Kendra, *Regulated and Unregulated Competition on Rails*, 117-129 (2013)
7. Z. Krzysiak, G. Bartnik, W. Samociuk, J. Zarajczyk, K. Plizga, B. Rachwal, S. Wierzbicki, L. Krzywonos, F. Brumercik, Przemysl Chemiczny **96**, 279-282 (2017)
8. L. Jedlinsky, J. Caban, L. Krzywonos, S. Wierzbicki, F. Brumercik, Journal of Vibroengineering **17**, 175-187 (2015)
9. J. Cejka, L. Bartuska, L. Turinska, Open Engineering **7**, 55-59 (2017), DOI: 10.1515/eng-2017-0010
10. J. Lizbetin, O. Stopka, Open Engineering **6**, 441-445 (2016), DOI: 10.1515/eng-2016-0066
11. M. Yarmen, S. Sumaedi, S, Transport Problems **11**, 99-111 (2016)
12. J. Lalinská, J. Gašparík, D. Šipuš, LOGI – Scientific Journal on Transport and Logistics **8**, 1, 74-81 (2017), DOI: 10.1515/logi-2017-0009.
13. P. Stoyanov, P. Gagova, Transport Problems **7**, 37-41 (2012)
14. M. Bak, P. Borkowski, B. Pawlowska, Transport Problems **7**, 27-36 (2012)

Safety of Maritime Transport in the Baltic Sea

Jacek Caban[1,*], *František* Brumerčík[2], *Ján* Vrábel[3], *Piotr* Ignaciuk[4], *Wojciech* Misztal[1], and *Andrzej* Marczuk[1]

[1]University of Life Science in Lublin, Faculty of Production Engineering, 20-612 Lublin, Poland
[2]University of Žilina, Faculty of Mechanical Engineering, Univerzitná 8215/1, 010 26 Žilina, Slovakia
[3]University of Žilina, Faculty of Operation and Economics of Transport and Communications, Univerzitná 8215/1, 010 26 Žilina, Slovakia
[4]Lublin University of Technology, Mechanical Engineering Faculty, 20-618 Lublin, Poland

Abstract. The maritime economy is particularly exposed to danger due to the environment of its operations. These risks result from deliberate actions, as well as incidental (hydro-meteorological conditions, mechanical, etc.). These are the threat dependent and independent of human e.g. environmental, safety health and life of humans or belongings. In spite of attempts to limit the risk of danger, it is impossible to completely eliminate it. May be helpful here all kinds of security systems, alarm, emergency, legislative and educational of ship owners and crews to the defence activities. The article presents some aspects of the state of transportation safety in the Baltic Sea Basin as well as undertaken the attempt to analyse the security state of this sea area.

1 Introduction

Sea and water ways are mainly used for world-wide goods transportation [1]. Maritime transportation plays a significant role in the integrated transportation system, especially in international trade system [2]. Around 90% of world trading is carried out by the shipping industry [3]. Safety is a crucial factor due to the fact that various accidents like ship collisions and groundings often result in great economic loss, fatalities and the environmental contamination [2]. The main directions of movement of goods by sea are borders linking Asia and Europe, Europe and North America, as well as the route connecting Asia with North America.

Shipping is also a highly regulated domain, and regulations have been reinforced in the last two decades [3]. The main principles underlying shipping regulations are harmonized national rules based on international conventions and resolutions given by the International Maritime Organisation (IMO) [4]. At the level of the European Union (EU), the European Commission (EC) translates the regulations, determined at the IMO, into binding laws, with the support of the European Maritime Safety Agency (EMSA) [5]. Further, the rules are adapted to regional conditions by the Baltic Marine Environment Protection Commission (Helsinki Commission, HELCOM), an intergovernmental organization of nine Baltic

* Corresponding author: jacek.caban@up.lublin.pl

coastal countries and the EU [6]. In the globally orientated maritime safety regime, the regional role of the HELCOM in the Baltic Sea is significant [7]. The aim of HELCOM is protect the marine environment of the Baltic Sea from all sources of pollution. It collects data related to maritime traffic, accidents and oil transportation, gives recommendations related to maritime safety and pollution from ships, and coordinates multilateral response in case of major maritime incidents [7].

Since our inception of maritime transport are also present factors causing all kinds of threat in this sector of transport. Therefore, can assume that the dangers accompanying in maritime transport have are continuous in time, impossible to the eliminate entirely, and also amenable to changes in parallel with the development of civilization. The International Convention for the Safety of Life at Sea (SOLAS) of the IMO recommends Vessel Traffic Services (VTS) for areas in which the volume of traffic or the risk of accident is high [8].

Maritime safety is referred to as safety of life, health and property from operating and environmental hazards, and the safety of the marine environment against pollution which are a by-products of human activities at sea. The maritime economy is particularly exposed to danger due to the environment of its operations. These risks result from deliberate actions, as well as random (weather conditions, the mechanical elements state, etc.). At the same time, it is a collection of punks logistics necessary for the functioning of the global economy. Nowadays it is difficult to imagine international trade without the participation of the maritime transport.

The nine bordering countries with Baltic Sea Area and approximate catchment and main sub-basins indicated was shown in Fig. 1.

Fig. 1. Baltic Sea Area with coastal countries, approximate catchment and main sub-basins indicated. KG – Kattegat, RU – Russian enclave of Kaliningrad, G.R. – Gulf of Riga, G. Finland – Gulf of Finland. Note that Gulf of Bothnia consists of two major parts – Bothnian Bay in the north and Bothnian Sea in the south. Source: [9]

This countries jointly suffer from environmental pollution, and are responsible for, the degraded state of the sea. The primary threats caused deliberately by a man is piracy and maritime terrorism. The events caused by factors independent of a man is mainly: the disaster of ships and other objects resulting from the weather conditions, as well as oil spills from ship. These events are often associated with incurring a real losses and the require take the appropriate actions.

The article presents some aspects of transport security on the Baltic Sea and an attempt was made to analyse the security situation of this sea area. One objective of the study is to show the causes of maritime accidents in the Baltic Sea.

2 Factors affecting the safety of sea transport

Dynamic development of maritime transport in the Baltic Sea region is stimulated by a variety of factors [10]. These factors – both external and internal, that is, inter-Baltic, and market and non-market, transport and non-transport – affect with different strength the size and structure of carriage by sea [10]. The Baltic Sea basin is relatively small and closed, and thus particularly sensitive to threats [11]. The Baltic is a shallow sea, its mean depth being 50 meters. While maritime transport is of vital economic importance to the Baltic Sea Area, challenging winter navigation conditions pose additional hazards to ships operating in these waters [12]. The increase in migration of people and the circulation of goods by sea has caused an increase in the volume of traffic on the routes, while technological progress has made it possible to build vessels with a large carrying capacity. According to IMO [13], the Baltic Sea Area has some of the densest maritime traffic in the world [14].

Several maritime accidents tragically demonstrate that an accurate assessment of the situation as well as timely and right decisions both aboard the emergency vessel and by the Search and Rescue (SAR) authorities are of paramount importance when attempting to save the people aboard and for ensuring the safety of the vessel itself [15]. Accidents where inadequate information and communication about the vessel's condition have been identified as factors needing improvement are for instance the flooding and foundering of the Abigail H [16], the grounding and capsizing of the Costa Concordia [17] and the grounding and flooding of the Commodore Clipper [18] or MS 'Jan Heweliusz'.

One area where improvements are considered necessary by especially SAR operators is the communication and information exchange between vessel crew, SAR operators and other relevant parties in maritime distress situations [15]. The importance of shared situational awareness and the quality of information in emergency situations is also stressed by Luokkala and Virrantaus [19] and Seppänen and Virrantaus [20], who highlight the importance of narratives in information sharing.

Every ship, both passenger and cargo is a potential threat to the environment, as fuel and cargo of all kinds can get out of it into the sea. The presence of chemicals in the marine environment contributes to the imbalance of the sea's health and life, as well as reduces the aesthetic of the marine environment [11]. Accidents that happen during transport on shipping routes are difficult to predict. Therefore, sea transport is characterized by a high potential of danger [21]. In the face of these problems appeared the necessity of regulating administrative legislation as well as technical and quality standards, the aim of which is to protect the life and the marine environment. Until recently, there was no necessity, and the ability to assess the level of threats at sea. Only since the 50s of last century, when significantly increased transport of dangerous goods, particularly oil and gas, extremely increased threat to the marine environment [22]. The best measure of safety of navigation on a particular basin are data about accident rates of ships [23]. On the basis of the HELCOM [24] report in the year 2012 44% of the accidents took place in ports, 35% in the open sea, and 14% occurred during the approach to port and 7% of accidents were unidentified. In the years 2009 to 2015 on the Baltic Sea area there were near 439 accidents.

Due to the civilization and structural development of economy, factors threatening the security of maritime transport are also changing. Figure 2 shows the causes of accidents in the maritime transport sector in the years 2009 to 2015.

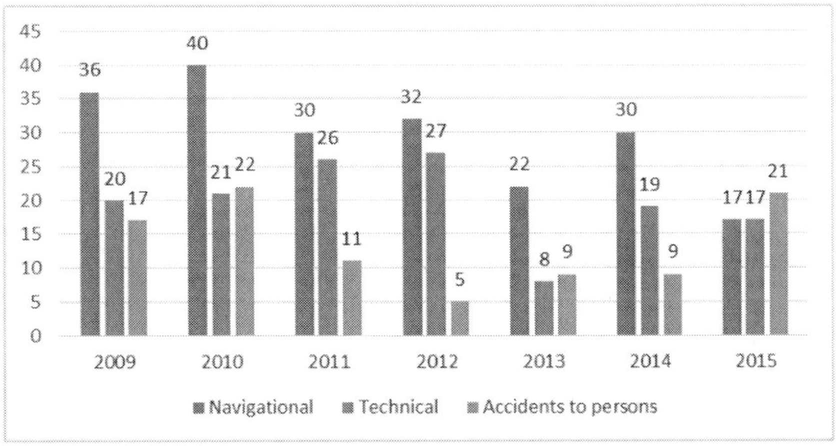

Fig. 2. General classification of types of maritime accidents occurring in the years 2009 to 2015. Source: compiled on the basis of the [25-27]

The largest group of accidents are accidents of the genesis of navigation, the decisive factor in the advantage of this type of accidents are caused by human factors (errors in reading the location, bad training of the crew, etc.), and possible malfunction of the navigation instrumentation. A false sense of security "guarded" by electronics (radars equipped with ARPA, Automatic Identification System AIS, electronic maps, etc.), results in a fairly rapid growth of maritime accidents [23]. Lowering the vigilance of the navigator takes place on the open waters. Since 2009, the number of accidents decreased, except for technical reasons, that may be caused by cutting production costs and lowering standards of components implementation.

Figure 3 shows in detail the number of accidents which occurred for reasons of navigation.

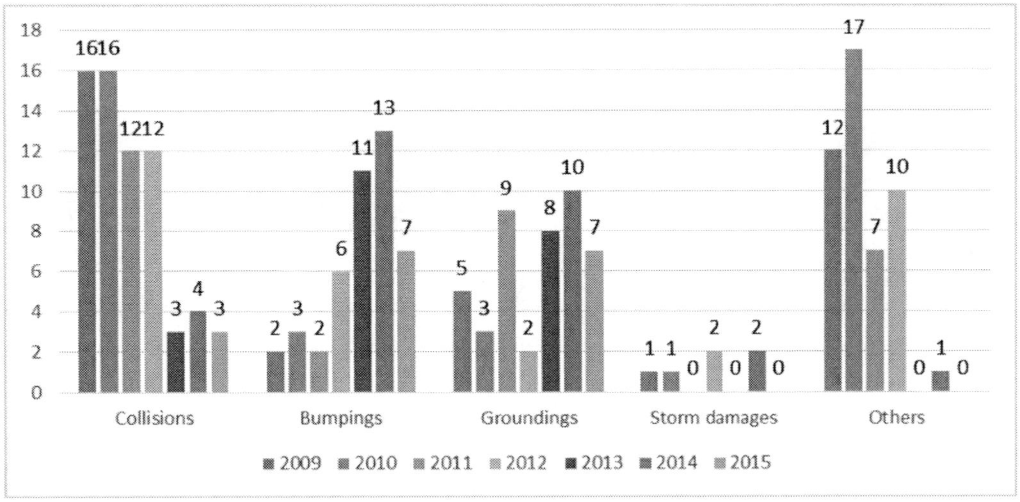

Fig. 3. Detailed classification of accidents due to navigation causes in the years 2009 to 2015. Source: compiled on the basis of the [25-27]

Navigation accidents can be divided into several main groups, the largest ones, consisting of collisions, bumpings, groundings and storm damages. The largest number of collisions took place in 2009 and 2010, bumpings in 2013 and 2014 as well as grounding in the year 2014 and 2011. The largest number of others accidents was in 2010. As you can see in the graph (Fig. 3), the least navigational incidents occurred in 2015.

Figure 4 shows in detail the number of accidents which occurred due to technical reasons.

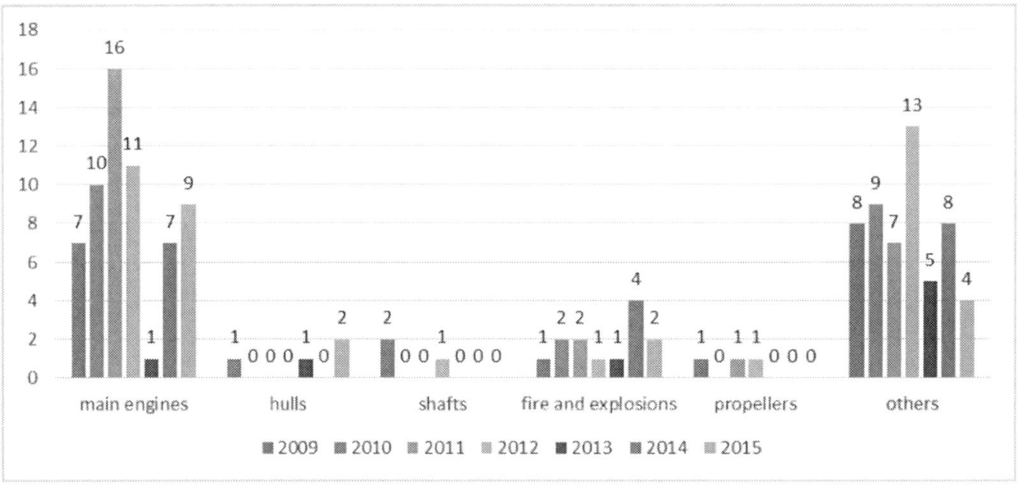

Fig. 4. Detailed classification of accidents due to technical reasons occurred in the years 2009 to 2015. Source: compiled on the basis of the [25-27]

The most common causes of accidents due to technical reasons are main engine failures. The largest number of accidents reported in the year 2011, this number dropped significantly in the year 2014. Accidents due to unsealing hulls, damages to drive shafts and propellers, and fires are relatively rare compared to the damage to engines which are clearly the predominant factor.

Figure 5 shows the detailed classification of maritime accidents caused by human factor.

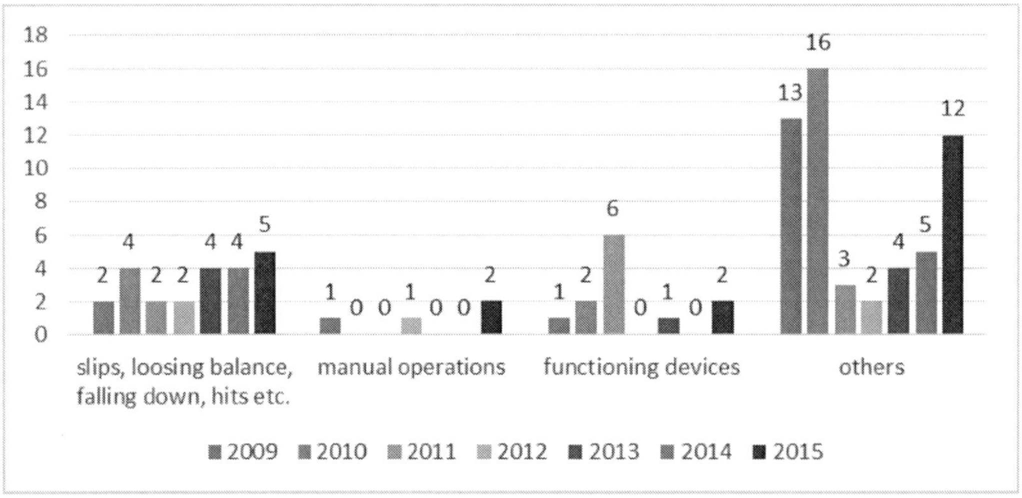

Fig. 5. Detailed classification of accidents caused by human factor in the years 2009 to 2015. Source: compiled on the basis of the [25-27]

The number of marine accidents caused by human factor significantly decreased in the year 2010 – 22 accidents and 2015 – 21 accidents. Random accidents such as slip or loosing balance are at a similar level in view of the difficulty of eliminating these types of events that occur in an uncontrolled way. Most of the accidents caused by devices in services were reported in the year 2011, and in 2012 they were eliminated, but in 2013 was 1 and 2015 – 2, of this type of accidents. The largest number of other reported accidents was in 2010. This type of accident is the highest in all the years analyzed.

3 Conclusion

To assess the value of the operating conditions is used a number of approaches, due to the fact that in practice there are different circumstances allowing (or not) the application of certain methods [28]. Ensuring the safety maritime transport is both a regional and international issue. Particularly important are safety issues during cargo handling in ports and transport terminals [29]. The above mentioned development of transport applications depends on the creation and precision of digital maps [1], a specially for transport safety. All activities protecting the marine environment and human life should be encouraged and supported. The increased number of sailings on maritime routes endanger the environment and causes all sorts of accidents or collisions. An important link in the fight against the threats posed by human activities at sea are maritime safety systems. The latest telecommunication solutions allow you to coordinate the activities undertaken by various institutions, created to ensure safety in the area of environmental protection, all sorts of accidents at sea lanes as well as in the maritime terrorism.

In the analysed period (2009-2015), in the Baltic Sea occurred nearly 439 accidents. On the basis of the analysis, It has been shown that the largest group of accidents are accidents of the genesis of navigation, followed by technical and the last is accidents to persons. It has been shown, that the number of accidents related to people significantly decreased in the year 2012. The lowest number of technical accidents took place in 2013 (22 accidents). Moreover, it has been demonstrated that most of the causes of accidents caused by equipment operation occurred in 2011, which in 2012 they were eliminated. The most common cause of accidents due to technical failures were the main engines. In the category of navigational accidents it has been shown that the greatest number of collisions took place in 2009 and 2010, bumpings in 2013 and 2014 and stranding in 2011 and 2014.

In conclusion, it should be noted that there are a lot of components threatening the safety of sea routes, as well as degrading the environment. By pursuing comprehensive efforts to raise the level of safety in maritime transport it is possible, however, to reduce the risk of these threats. These include a.o. all kinds of alarm systems, rescue, tailored legislative conditions for the preparation of ship-owners and crews of ships for defensive action.

This work was carried out in partnership between University of Life Sciences in Lublin and University of Žilina, as well as Lublin University of Technology and supported by the Slovak Scientific Grant Agency of the Slovak Republic under the project No. VEGA 1/0077/15.

This paper is supported by the research project "From horse-drawn railway to intermodal transport" within Visegrad Fund.

References

1. A. Kalasová, L. Palicka, Promet - Traffic & Transportation **19**, 2, 121-128 (2007)
2. J. Zhang, X. Yan, D. Zhang, S. Haugen, X. Yang, Safety Science **63**, 157-167 (2014)
3. C. Chauvin, S. Lardjane, G. Morel, J.P. Clostermann, B. Langard, Accident Analysis and Prevention **59**, 26-37 (2013)
4. S. Kristiansen, Marine safety – Background, *A. Molland (Ed.), The Maritime Engineering Reference Book, Butterworth Heinemann*, Chapter 11, 786-875 (2008)
5. European Maritime Safety Agency (EMSA), *Quality Shipping, Safer Seas, Cleaner Oceans. Lisbon,* Available online: http://www.roodbovengroen.com/website/managedMedia/mediaItem/159.pdf (2009)
6. B. Hassler, Maritime Policy Management **37**, 489-503 (2010)
7. P. Haapasaari, I. Helle, A. Lehikoinen, J. Lappalainen, S. Kuikka, Marine Policy **60**, 107-118 (2015)
8. International Maritime Organization (IMO), *Vessel Traffic Services*, Available online: http://www.imo.org/OurWork/Safety/Navigation/Pages/VesselTrafficServices.aspx), (2015)
9. H. Backer, J. Leppänen, A.C. Brusendorff, K. Forsius, M. Stankiewicz, J. Mehtonen, M. Pyhäläa, M. Laamanen, H. Paulomäki, N. Vlasov, T. Haaranen, Marine Pollution Bulletin **60**, 5, 642-649 (2010)
10. A.S. Grzelakowski, *Rozwój transportu morskiego w regionie Morza Bałtyckiego, Zeszyty Naukowe Akademii Morskiej w Gdyni*, 67, 73-88 (2010)
11. M. Bogalecka, Journal of Management and Finance **3**, 1, 574-584 (2012)
12. R. Aps, M. Fetissov, F. Goerlandt, J. Helferich, M. Kopti, P. Kujala, *3rd European STAMP Workshop, STAMP UE 2015*, 64-73 (Procedia Engineering **128**, 2015)
13. IMO, *Designation of the Baltic Sea area as a particularly sensitive sea area. Resolution MEPC*, **136**, 53 (2005)
14. R. Aps, M. Fetissov, F. Goerlandt, P. Kujala, A. Piel, *4th European STAMP Workshop 2016*, 2-12 (Procedia Engineering **179**, 2017)
15. J. Nordström, F. Goerlandt, J. Sarsama, P. Leppänen, M. Nissilä, P. Ruponen, T. Lübcke, S. Sonninen, Vessel TRIAGE, Safety Science **85**, 117-129 (2016)
16. MAIB, 2009, *Report on the Investigation into the Flooding and Foundering of the Grab Hopper Dredger Abigail H in the Port of Heysham 2 November 2008* (Marine Accident Investigation Branch, Southampton, UK, 2009)
17. MIT, 2013, *Cruise Ship Costa Concordia Marine Casualty on January 13, 2012 – Report on the Safety Technical Investigation. Ministry of Infrastructures and Transports* (2013)
18. MAIB, 2015, *Report on the Investigation of the Grounding and Flooding of the ro-ro ferry Commodore Clipper in the Approaches to St Peter Port, Guernsey on 14 July 2014* (Marine Accident Investigation Branch, Southampton, UK, 2015)
19. P. Luokkala, K. Virrantaus, Safety Science **63**, 191-203 (2014)
20. H. Seppänen, K. Virrantaus, Safety Science **77**, 112-122 (2015)
21. Z. Kopacz, W. Morgaś, J. Urbański, Zeszyty Naukowe Akademii Marynarki Wojennej w Gdyni **169**, 2, 41-58 (2007)

22. T. Szubrycht, K. Rokiciński, *Gospodarka morska w świetle wybranych zagrożeń współczesnego świata* (AMW, Gdynia, Poland, 2006)
23. K. Marcjan, L. Gucma, *Wykorzystanie analizy incydentów nawigacyjnych w celu oceny bezpieczeństwa nawigacyjnego na obszarach Morza Bałtyckiego o dużym zagęszczeniu ruchu statków, Prace Naukowe Politechniki Warszawskiej – Transport 102*, 77-86 (Politechnika Warszawska, Poland, 2014)
24. Annual report on shipping accidents in the Baltic Sea area during 2012, *Baltic Marine Environment Protection Commission* (HELCOM, 2012)
25. *Statistical Yearbook of Maritime Economy* (Warsaw-Szczecin, Poland, 2015)
26. *Statistical Yearbook of Maritime Economy* (Warsaw-Szczecin, Poland, 2014)
27. *Statistical Yearbook of Maritime Economy* (Warsaw-Szczecin, Poland, 2016)
28. A. Marczuk, J. Caban, P. Savinykh, N. Turubanov, D. Zyryanov, Maintenance Reliability **19**, 1, 121-125 (2017)
29. J. Lizbetin, O. Stopka, *ICTTE – 3rd International Conference on Traffic and Transport Engineering*, 382-387 (Belgrade, Serbia, 2016)

MATEC Web of Conferences **134**, 00004 (2017) DOI: 10.1051/matecconf/201713400004
LOGI 2017

Practical Using Of Network Method to Logistics Chain

Jiří Čejka[1,*]

[1]Institute of Business and Technology in České Budějovice, Faculty of Technology, Okružní 10, 37001 České Budějovice, Czech Republic

Abstract: The article deals with assessing the use of the CPM method (Critical Path Method) to logistics chain of the selected enterprise. The objective is to find the critical path, which is the longest possible path in the project in terms of the time required. The activities on this path are thus called critical, and necessary measures should be taken to avoid delays in these activities, as the delays would have negative impact on the whole project. This method has been developed within the Microsoft Project programme, Microsoft software that is used for creating projects, managing tasks and resources, and tracking the current situation. For the purposes of this article, the MS Project 2016 version was used.

1 Introduction

For a successful project creation and the consequent execution of necessary methods it is necessary to define individual activities needed to carry out the entire process. Activities can be divided into individual levels whereby the better orientation in the project can be achieved. Durations of the activities are then defined, i.e. when the task starts and finishes, dates will be automatically generated [1-5].

An important point is to define continuity of individual tasks (definition of preceding and following tasks, i.e. which activities must be initiated after the completion of the others and which activities can be carried out at same time) and creation of interrelationships. This procedure provides a clear structure upon which the further analyses, such as the network analysis and the critical path resulting from it, can be made.

The project processing along with the determination of interrelationships and following tasks can be characterized as follows:

1. Purchase order from a customer – 1 day
2. Purchase order processing – 3 days
3. Logio
    - 3.1. Production planning – 2 days
    - 3.2. Raw material planning – 2 days
    - 3.3. Distribution planning – 2 days
    - 3.4. Capacity planning – 2 days
4. Collection of milk – 1 day

* Corresponding author: cejka@mail.vstecb.cz

5. Provision of packaging
   5.1. Enough packaging in stock – 1 day
   5.2. Lack of packaging in stock – 30 day
6. Production – 6 days
7. Treatment – 20 days
8. Packing
   8.1. Packing without delay – 1 day
   8.2. Packing with a delay – 1 day
9. Storage
   9.1. Storage without delay – 1 day
   9.2. Storage with a delay – 1 day
10. Distribution
    10.1. Distribution without delay – 1 day
    10.2. Distribution with a delay – 1 day

The main project levels are identical with the main activities in the process – obtaining a purchase order from a customer, purchase order processing, logistic plans in the Logio system, milk collection, provision of necessary materials, production, treatment, packing, storage and distribution. Figure No. 2 shows that a portion of activities contains other partial tasks which need to be done and which represent the process behaviour in certain situations.

Duration in days was defined for all activities and the preceding and following activities were determined. As a result, individual activities amalgamated and the continuity of the entire process was created [4-7].

2 Solutions in graphical representation

The programme allows a graphical representation by means of the Gantt chart, which is a tool showing the sequence and time data. It allows us to carry out simple planning and checks of the partial activities as well as bigger units. This chart ranks among the bar charts, i.e. it represents individual activities on the time line using the bars. Thanks to this representation, the phase and dependency of activities can be easily found out.

The Gantt chart created upon the analysed process can be seen in the right section in Figure No. 3. Each process activity has a corresponding graphical representation on the line. The size of bars in the chart depends on the duration of individual activities and the activities are connected with the links which correspond to their sequence and interrelationship [8-10].

 DOI: 10.1051/mateconf/201713400004

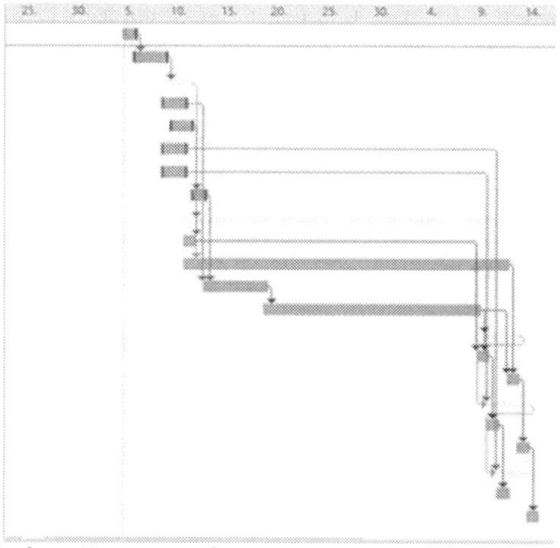

Fig. 1. Gantt chart in MS Project. Source: author

For better overview and time orientation, the flow chart No. 1 is available. This flow chart was generated based on the theoretical basis in Chapter 2 and the processed information.

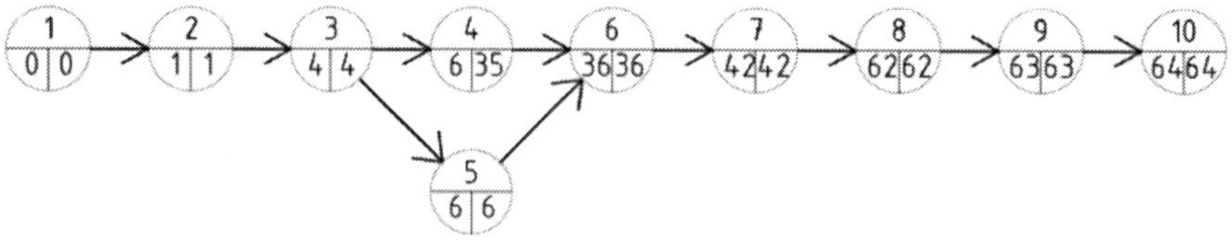

Fig. 2. Flow chart. Source: author

Here, you can see the sequence and interrelationships of individual activities, or more precisely, preceding activities of individual tasks. You can also see the duration of individual activities (numbered edges) and the earliest possible starts (left No. in the node) and ends of individual activities (right No. in the node) [8-11].

3 Critical path

Using the programme and the flow chart tool we can display the critical path upon the Gannt chart. The critical path is defined as the longest possible path in the project in terms of the time required. It is a suitable tool for estimation of the project duration. It leads from the initial (first activity) to the end point (last activity) of the chart, i.e. every project has at least one critical path. Activities lying on this path require more time to do; they do not have any time reserve and need to be focused on if the projects should be completed in time. Their delay would have an adverse effect on the entire project. The critical path is reflected in the time planning and project management of nearly all phases of its life cycle. The critical path of this project is shown in the Graph No. 2. The activities which form this critical path include:

1) Purchase order from a customer
2) Purchase order processing
3) Logio – raw material planning
4) Milk collection
5) Lack of packaging in stock
6) Treatment

 DOI: 10.1051/matecconf/201713400004

7) Packing with a delay
8) Storage with a delay
9) Distribution with a delay

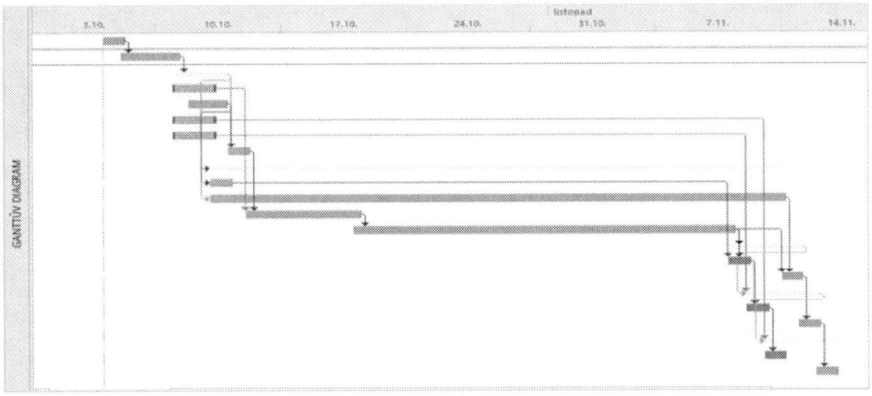

Fig. 3. Critical path of the process in MS Project. Source: author

The critical path always begins with the first activity. The second activity is included in the critical path because no other activities are carried out simultaneously and, hence, the critical path must go through them. The next item is processing in the Logio system which includes the raw material planning as a critical activity, followed by the collection of milk – again included due to a single activity being carried out at a given time. The chart begins to branch out with the next activity – the critical path goes through the lack of packaging in stock [10-13].

The packaging storage room is located in the production plant complex in Český Krumlov. If the packaging is in stock, it can be transported to the production plant during one day and the packing operation can start without any time delays. In case of the lack of packaging, it must be ordered, which is done at the purchase order processing phase, i.e. we can ensure the delivery during the production process which lasts 26 days. Manufacture and delivery of packaging from the supplier which has its central warehouses in Prague takes 30 days.

The Gantt chart helped us to establish that the delay in packing of the products is, based on these assumptions, 3 days when the products are in stock and wait for the packing process. This restraint will cause a delay in the follow-up activities, namely the distribution to the central warehouse in Jesenice and on to the customers. The company has a certain amount of stocks, i.e. an emergency reserve; these, however, are based on the expert estimation of the quantity consumed in previous years. In case of unexpected demand or delay on the supplier´s part, the company may run out of stocks [8-11].

The next activity is the treatment, i.e. an activity when the produced cheese ripens in the ripening cellars. This activity can be neither limited nor eliminated.

Packing with a delay happens at the time when the packaging material is not in stock and the production of the Jihočeská niva cheese is done. Packaging must be ordered – as we know, it is already ordered at the purchase order processing phase. The above figure shows that the delivery of packaging in this situation happens on 14 November, while the treatment ends on 11 November. This implies that the products have to wait in the factory for 3 days until they are packed in the necessary packaging, foil and cardboard boxes. With every other day of delay, the products lose their value. This also results in the delayed entry of the products into the central storage in Jesenice and the next distribution to the customers. The critical path ends with the distribution of the cheese to customers [12-14].

4 Conclusions

The article is concerned with the CPM method, i.e. a critical path method, which is applied on the production process of the selected dairy product of the Madeta company – the Jihočeská niva cheese.

The paper is divided into two parts – the first one focuses on the theoretical determination of the given issue. During this phase I obtained the knowledge which I could apply later. This part also deals with the construction of the flow chart needed later.

The second part focuses on the application of the network analysis method, especially the application of the CPM method on the logistic chain of the selected company. The Microsoft programme, Microsoft Project 2016, was used to solve the problem. Using this programme, the project consisting of individual activities in the chain was created, including determining of the following activities, the entire sequence and durations of individual tasks.

This paper is supported by the research project “From horse-drawn railway to intermodal transport” within Visegrad Fund.

References

1. V. Cempírek, K. Pivoňka, J. Široký, *Fundamentals of Transportation Technology and Management,* 3rd reworked issue, 120 p. (University of Pardubice, Czech Republic, 2002)
2. K. Hrazdilová Bočková, *Project Management: Textbook* (M. Koláček, 2016)
3. J. Jablonský, M. Maňas, P. Fiala, *Multi-criteria Decision-Making*, 1st issue, 316 p. (Prague: University of Economics in Prague, Czech Republic, 1994)
4. J. Kubát, *Lagerstandortoptimierung mit Hilfe des Verfahrens Branch und Bound*, (Ekonomicko matematický obzor **11**, 65-83, Czech Republic, 1975)
5. P. Matejko, L. Černá, J. Majerčák, *International Scientific Conference - Diagnostics Business and Controlling Logistics Slovak Republic* (University of Žilina, Slovak Republic, 2012)
6. O. Pasch, *Microsoft Share Point, Practical Guide of a User* (Brno: Computer Press a.s., Czech Republic, 2011)
7. O. Stopka, M. Chovancová, J. Ližbetin, V. Klapita, Nase More **63**, 3, 195-199 (2016), DOI: 10.17818/NM/2016/SI22
8. L. Unčovský, et al., *Operational Analysis in Company Management* (Alfa, Bratislava, Slovak Republic, 1985)
9. V. Zitrický, J. Gašparík, L. Pečený, Transport Problems **10**, 59-72 (2015)
10. P. Fiala, *Production System Modelling and Analysis,* 1st issue, 260 p. (Prague: Professional Publishing, Czech Republic, 2002)
11. I. Kubasakova, B. Poliakova, L. Krzywonos, Logi - Scientific Journal on Transport and Logistics **4**, 2, 39-46 (2013)
12. J. Jablonský and M. Dlouhý, *Efficiency Evaluation Models of Production Units*, 1st issue, 184 p. (Prague: Professional Publishing, Czech Republic, 2004)
13. R. Hindls et al., *Statistics for Economists*, 7th issue, 420 p. (Prague: Professional Publishing, Czech Republic, 2006)
14. J. Janáček, Doprava **35**, 4, 172-175 (1993)

DOI: 10.1051/matecconf/201713400005

Use of Innofreight Containers for Thermal Coal Transportation

Václav Cempírek[1], *Jaromír* Široký[2,*], *Petr* Nachtigall[2], and *Pavlína* Hlavsová[2]

[1]College of Logistics, Přerov, Palackého 1381/25, 750 02 Přerov, Czech Republic
[2]University of Pardubice, Faculty of Transport Engineering, 532 10 Pardubice, Czech Republic

Abstract. The article discusses the change of siding technology for a fossil fuel power plant using a new reloading system of Innofreight containers. This is a modern system for transporting thermal coal in special transport units – Innofreight containers. The assessment of the new technology takes into account not only technical, but above all technological and economic changes and benefits. There was an analysis of the current situation and it was decided to introduce the new technology of Innofreight system. Its implementation brought additional investment in treatment of industrial tracks for loading, the construction of transshipment facilities for containers, CCTV installation, and a higher degree of security equipment.

1 Introduction

The transportation of thermal coal for thermal power plants is currently provided by unit trains composed of four-axle bogie open hopper cars of the Falls type. The unloading of the cargo takes place in a special unloading hopper by opening the side flaps of the different groups of Falls railcars in either of the two ways [1]:
- manually, by operating the levers of the side flaps from the front platform of the car, or
- automatically, using pneumatic control.

The change of the transportation technology is due to the outdating of Falls railroad cars, which are nearing the end of their service life. For this reason, innovative ways of transporting thermal coal are sought, and the Innofreight container system is one of them.

Thermal coal is loaded into inland Innofreight containers on railroad cars of the Sggrrs series. Every two four-axle bogie cars are connected by a rigid coupling to form an articulated car. The weight of the articulated car is 42.70 tons, the loading weight for the D2 line category (22.5 tons per axle; 6.4 tons per running meter of the car) is 128.2 tons and for the D3/D4 line category (22.5 tons per axle) it is 137.3 tons, the total length of the car over the bumpers is 26.71 meters, the height of the loading area above the top of the rail is 1155 mm [2].

The unloading of coal is carried out using a special tippler, which includes handling skids to gradually remove the individual containers from the railroad cars in a horizontal direction, followed by a 180° rotation to unload them. The unloaded coal is transported by

* Corresponding author: jaromir.siroky@upce.cz

belt conveyors to a storage site and then to a landfill. When introducing the new Innofreight technology, initial investments must be anticipated, both in the Innofreight technology itself and also in the construction and technical adjustments of the siding [3].

2 Technological changes with the introduction of the new Innofreight system

This section describes the current unloading technology of the Falls cars and the new technology of unloading cars with Innofreight containers. The assessment is performed with an emphasis on the technological evaluation of both processes.

2.1 Technological procedures in coal transportation using Falls cars

The siding of the power plant is composed of one shunting circuit. The shunting itself is performed by the head of the shunting crew. The head of the shunting crew must request for a permission to shunt from the dispatcher orally, by public address system, or radio station.

The manually built sets of switches are operated by the head of the shunting crew alone. Centrally built set of switches are operated by dispatchers from the signal box [4].

The head of the shunting crew is responsible for correctly performing the selected procedure when moving the cars on the deep coal storage site, following the operational solution upon agreement with the coal feeding professional and with the dispatcher on shift. Moving the cars on the deep coal storage site may be performed only by shunting.

Table 1. Technological time of terminating train processing. Source: authors

Task	Duration [min]	Technological time
1. securing the train	2	
2. separation of the drive unit	5	
3. receipt of train documentation	2	
4. technical inspection - 32 cars (16 car units), 1.2 min/car + walk	38,4	
5. transportation inspection – the same applies (concurrent)	38,4	
6. notice on termination	2	
Total processing time		49,4 min

The technological processing of a terminating train is shown in Table. 1. The tasks related to the operation of a terminating train are indicated as follows [5]:

1. securing the train,
2. separation of the drive unit,
3. receipt of train documentation,
4. technical inspection,
5. transportation inspection (concurrent with the technical inspection),
6. notice on the termination of processing.

The total processing time of a terminating train is set at 49.4 minutes (see Table 1). The terminating train of the Innofreight set of cars will be processed in a similar manner. This new technology does not have a significant influence on the technological time of processing the terminating train.

The head of the shunting crew is responsible for proper cleaning of the cars and checking them, for proper coupling of the cars into the train, which are reported as ready to be shunted. The cleaning is performed by employees of the power plant. Together with the dispatcher, they are responsible for correct and proper train composition and decommissioning cars with a technical defect [6].

The unloading of thermal coal from the cars is performed on the deep coal storage site in either of the two ways, by manually-operated levers, or automatically using pneumatic operation of the side flaps. Unloading using manual opening of the side flaps is carried out from the car platform. The pneumatic way of unloading coal is performed by compressed air from the central distribution system. Final cleaning of the cars is done by the operator using compressed air also from the central distribution system. In winter when the temperatures are below zero and the coal starts freezing up, a defrosting tunnel with a maximum capacity set for 16 cars is used. The supply and regulation of hot air into the tunnel is controlled from the boiler room, including the operation and setting the output of the main and booster fan.

After the head of the shunting crew shunts and hands over the cars to the unloading personnel, the unloading begins. After unloading the cars, cleaning coal residues, closing the cars, securing the side flaps and cleaning the unloading tracks, the unloading personnel hands the cars over back to the head of the shunting crew. The handover of cars between the unloading personnel and the head of the shunting crew proceeds by means of verbal commands.

2.2 Technological processes during coal transportation in Innofreight containers

The unloading equipment for Innofreight containers must be placed at the unloading handling track. The construction changes require the extension of the handling tracks, which would allow shunting the shunting unit, without restricting the occupancy of running tracks intended for the entry of trains. The shunt will be controlled from the signal box, which will now be placed in close proximity to the Innofreight tippler. This modification will result in the centralization of traffic control on the siding from a single place and will also lead to a greater availability of personnel during reloading the cars with Innofreight containers. Similarly to the existing technology, the train with Innofreight containers will be split into 2 parts with 8 cars each (16 Innofreight containers). The new organization of the shunt requires a modernization of the shunt locomotive and modification for remote control operation, which would produce a saving on the operating personnel (the shunter in particular), as the engine driver will have control over the whole shunting unit, which will not be operated from their footplate.

The reloading of coal by the Innofreight unloading equipment is important in terms of time demands. Unit time for reloading 1 container is 2.5 minutes. The technological chart of the operation of the Innofreight set of cars is shown in Table 2. With regard to the division of the train into two parts, the technological time necessary for operating the remote control is also taken into account and there is also an increase in the time necessary for shunting out and docking with the given group of cars. However, these times do not play a significant role with respect to the total time of unloading one train of Innofreight cars, which is 195 minutes.

Using this method of unloading coal, there is a twofold occupancy of the lead track when shunting the shunting unit on the entering lead track, which may especially in peak demand (supply of thermal coal) have negative effect on the arrival or departure of other trains.

Table 2. Technological chart of unloading the train (split into two parts). Source: authors

Task	Duration [min]	Technological time
1. shunting out the shunting unit	4	
2. shunting in the shunting unit	5	
3. technological time (remote control operation – settings, work)	5	
4. unloading – 2.5 min/container - 32 containers	80	
6. shunting out the shunting unit and shunting it to the siding	4	
7. coupling and shunting out the shunting unit	4	
8. technological time (remote control operation – settings, work)	5	
9. unloading – 2.5 min/container - 32 containers	80	
10. shunting out, docking of the shunting unit	6	
11. shunting the drive unit to the siding	2	
Total processing time		195 min

It is important to bear in mind that especially at the beginning of the operation of this system time reserve needs to be allowed for (due to the lack of experience of the operating personnel). Therefore, the estimated total reloading time of one train may be up to 4 hours.

Table 3. Technological chart of unloading the train (at once). Source: authors

Task	Duration [min]	Technologický čas
1. shunting out the shunting unit	4	
2. shunting in the shunting unit	5	
3. technological time (remote control operation – settings, work)	5	
4. unloading – 2.5 min/container – 64 container	160	
5. shunting out, docking of the shunting unit	6	
6. shunting the drive unit to the siding	2	
Total processing time		182 min

After the track adjustment, it will be possible to unload the whole train at once, without the need for disconnecting the train and further handling. The technological chart will become simpler, as shown in Table 3. This measure will shorten the unloading time by approximately 13 minutes to 182 minutes. In addition, the time of occupying the entering lead track will also be shorter and tasks associated with splitting one train into two shunting units will not need to be performed. This change of technology will produce a saving of workforce necessary for this task.

Similarly to the existing technology of unloading coal from the Falls cars, the same time demands may be assumed for performing a technical inspection (1 hour) and for a technical inspection and testing brakes after the completion of unloading (2 hours).

The total time required for reloading cars with Innofreight containers is approximately 6.5–7 hours [7].

3 Logistics system

The future of the transportation of goods lies in flexibility, speed and ecology. These requirements are fully addressed by the new Innofreight container system, characterized by high degree of flexibility and efficient logistics solution for rail transportation of bulk goods.

Special dump containers of various types are used in combination with an innovative way of unloading them using reloaders with rotary handling systems.

The Innofreight container system is suitable for the transportation of bulk commodities in agriculture, chemical industry, extractive industry, or waste management.

The weight of the MonTainer XXL container designed for thermal coal transportation is 3.4 tons, with a loading weight of 38 tons and loading volume of 38 m3. The container is intended for loose materials with high specific weight and is constructed for maximum useful load and heavy-duty use [8].

Mobile handling equipment with reloading in the horizontal direction is represented by a reloader with a rotary device equipped with a sensor for measuring the weight of unloaded goods with a dumping capacity of 600 m^3/h.

A stationary unloader is used for fast and efficient horizontal unloading. This system is optimal for unloading large volumes of material, or when the handling area for unloaded goods is limited.

The conducted expert study on the change of transportation technology and supply of thermal coal in Innofreight containers concludes that with the annual transportation volume of 2 million tons of coal, the new system will due to its higher loading weight require 129 less unit trains than the current system, i.e. trains composed of Falls cars. The lower number of trains will save 8.8 mil. CZK annually in rail network usage. The calculation is based on the information on the train route from the IS KADR (Route Capacity information system). This substantial saving does not include the discount for combined transport, which is according to the Statetement on the railway system 2017 document 40% [9]. With this saving in rail networks usage, the annual saving would increase by 25.4 million CZK to 34.2 million CZK.

The system using MonTainer XXL containers for the transportation of thermal coal is 13% more efficient in comparison with the currently used Falls cars.

Voestalpine Stahl company in Donawitz, Austria uses containers of the MonTainer XXL type for the transportation of coke. Compared to standard railroad freight cars, these containers have a 30% higher loading weight per unit train.

The Innofreight system is 9-13% more efficient compared to modern freight cars of the Falls type, which have a loading weight of 66.0 or 69.7 tons. The containers are unloaded by one person only.

4 Conclusion

The assessed plan of change in technology includes an independent recommendation that introduces the latest transportation technology for transporting thermal coal. Low-maintenance technological equipment and elimination of the influence of weather conditions combined with low staffing requirements ensures a continuous supply of thermal coal in accordance with the requirements of the end customer (within the agreed time frame).

The new technology requires modifications of the signaling systems, unloading systems, siding tracks and technological procedures, which are necessary not only with regard to the introduction of a new, modern transportation technology, but also with regard to the technical obsolescence of the siding systems. These investment costs have a payback period of maximum 10 years, as the annual saving in rail network usage is around 8.8 million of CZK. The authors conclude that without the proposed investment adjustments it is not feasible to operate the Innofreight system.

This paper is supported by the research project "From horse-drawn railway to intermodal transport" within Visegrad Fund.

References

1. J. Široký, et al., *Transport Technology and Control* (Tribun EU, Brno, Czech Republic, 2014)
2. J. Novák, V. Cempírek, I. Novák, J. Široký, *Intermodal Transport* (University of Pardubice, Czech Republic, 2015)
3. J. Jagelčák, A. Dávid, P. Rožek, *Sea Containers* (EDIS: University of Zilina, Slovak Republic, 2010)
4. S. Bahri, Y. Huseyin, U. Yasin, A. Guneri, G. Bahadir, E. Turan, The Scientific World Journal **2014** (2014), DOI: 10.1155/2014/630320
5. Y.M. Bontekoning, C. Macharis, J.J. Trip, Transportation Research Part A: Policy and Practice **38**, 1-34 (2014)
6. C. Macharis, Y.M. Bontekoning, European Journal of Operational Research **153**, 400-416 (2004)
7. Intermodal terminals. Intermodal terminals in Europe, Available online: http://www.intermodal-terminals.eu/database/terminal/view/id/127 (2013)
8. Multimodal Transport Systems, *CVUT Transport Faculty,* Available online: http://www.fd.cvut.cz/projects/k612x1mp/vn.html (2014)
9. Railway Infrastructure Administration, *Statement on national and regional railway,* Available online: http://www.szdc.cz (2017)

MATEC Web of Conferences **134**, 00006 (2017) DOI: 10.1051/matecconf/201713400006
LOGI 2017

The Possibility of Transferring the Transport Performance on Railway Transport

Václav Cempírek[1,*], *Petra* Vrbová[2], and *Eva* Zákorová[3]

[1]College of Logistics, Přerov, Palackého 1381/25, 750 02 Přerov, Czech Republic and Czech Logistics Association, z.s., president, Pod Výtopnou 367, 186 00 Prague 8, Czech Republic
[2]University of Pardubice, Faculty of Economics and Administration, 532 10 Pardubice, Czech Republic
[3]Invekta, s.r.o., Vlkov 71, 570 01 Litomyšl, Czech Republic

Abstract. The article deals with the transfer of freight from road freight to railway freight transport. In railway transport, individual wagons or groups of wagons with a maximum number of five are used. Combined transport is used for commodity flows. Goods are loaded in ISO 1 containers with an outer length of 20 or 40 feet. Special ramps allow goods to be transported from the ramp to the container and vice versa. The system can be used for storage device, distribution and logistics centres with railway connections to public railway transport.

1. Introduction

The aim of the EU White Paper is to create a so-called "United European Transport Area Plan", which will be created a competitive, resource efficient transport system, namely to move 30% of goods transport over a distance of 300 km from road to other types of transport by 2030. A similar commitment is implemented into the approved Transport Policy of the Czech Republic for the period 2014-2020 with a view to 2050. It is also fully in line with the approved Operational Program Transport for the 2014-2020 period, namely it is a specific objective 1.3 Creating the conditions for greater use of multimodal transport. The main principles set out in the National Emissions Reduction Program will also be implemented, which also concern the transfer of freight transport performance from road to railway freight.

The White Paper - "The way to united European Transport Area, to a Competitive and Effective Transport System" - is a reference document for the creation of a new European transport policy for the period 2012-2020 with a view to 2050, followed by the Trans-European transport network Policy (TEN-T) as the main European instrument for the development of transport infrastructure for long-distance transport flows in order to promote a single European market. The White Paper includes 40 concrete initiatives to build a competitive transport system over the next decade. The main and new objective is to substantially reduce Europe's dependence on imported oil and reduce carbon emissions by 60% by 2050 [2].

* Corresponding author: vaclav.cempirek@vslg.cz

These objectives should be achieved [1, 2]:

a) By moving 50% of medium- and long-distance freight transport from road to railway and waterborne transport and, in the case of passenger transport, significantly increasing the share of railway transport (including air transport over distances of up to 1000 km, while air transport Long distance flights).
b) The introduction of alternative transport energies.
c) The introduction of more efficient engines.
d) The application of ITS in all modes of transport in order to optimize transport and transport processes (as elaborated in EU ITS Policy).

The "New initiatives by the European Commission to future mobility" were published on the last day of May 2015.

The document highlights that the European transport system is undergoing major changes. Moving towards more integrated and more modal mobility systems, which require a more comprehensive and cross-cutting approach and involve a wide range of transport policy areas at EU level.

The aim of the proclaimed documents is to create viable future road systems and intermodal mobility systems, increase competitiveness, strengthen social justice and enforce zero emissions.

Europe on the move includes:

- A policy statement setting out a long-term plan to deliver clean, socially equitable and competitive mobility for the whole of Europe.
- A set of eight legislative initiatives focusing on road transport. Their aim is to improve road transport, in particular the road haulage market; Improving the working and social conditions of workers (truck drivers) and promoting smart road charging across Europe.
- Several non-legislative measures that represent a wide range of EU policy support measures to accelerate the transition to a sustainable, digital and integrated mobility system (investment finance for infrastructure, research and innovation, cooperation platforms, etc.).

2. Way of solving the issue

Meet the implementation of the above mentioned objectives in the introduction of the contribution is very complex because the road freight transport shows a lower total transport cost per ton than that it is in railway transport. By enhancing the level of cooperation between individual carriers in the supply chain, using combined transport systems, a competitive price for transport services can be achieved.

The solution is based on statistical data [3] that are issued for each mode of transport. For the transport of individual consignments, it is possible to rely on the total data for railway transport. The average share of individual wagon consignments from the total data was 35% in individual years.

For a higher degree of cooperation in transport performance, it is important that storage facilities, distribution and logistics centres are connected to railway transport.

In most European countries, with the exception of Switzerland, discussions are being held on the end of the transport of unprofitable individual wagon loads. These shipments are exclusively carried by national freight carriers.

2.1 Development of goods transport in the Czech Republic by transport modes

For the Czech Republic, land transport modes are crucial, which is similar to the share of total traffic volume or traffic performance as the EU countries (see Tab. 1 and Fig. 1).

 DOI: 10.1051/matecconf/201713400006

Since 2010, there has been a decline in the decrease of goods transport in thousands of tons, and also in transport performance in ton-kilometres (ton-km) [3].

Average share of railway freight transport in total transport in thousands of tons was 18.1% and road freight transport 79.5%.

Average share of railway freight transport in overall transport performance in millions ton-kilometres was 21.03% and road freight transport 7.3%.

Table 1. Interdisciplinary comparisons of transport performance. Source: [3]

	2010	2012	2013	2014	2015	2016
Transport of goods in total (thousands of tons)	**451,671**	**435,449**	**447,367**	**491,625**	**549,085**	**539,062**
Railway transport	82,900	82,968	83,957	91,564	97,280	98,033
Road transport	355,911	339,314	351 517	386,243	438,906	431,889
Inland waterway transport	1,642	1,766	1,618	1,780	1,853	1,779
Air transport	14	9	9	9	6	6
Pipelines	11,205	11,392	10,266	12,029	11,040	7,356
Total transport performance (millions ton-kilometres)	**68,495**	**68,087**	**71,509**	**71,421**	**76,613**	**68,172**
Railway transport	13,770	14,266	13,965	14,574	15,261	15,618
Road transport	51,832	51,228	54,893	54,092	58,714	50,315
Inland waterway transport	679	669	693	656	585	620
Air transport	22	17	24	35	31	31

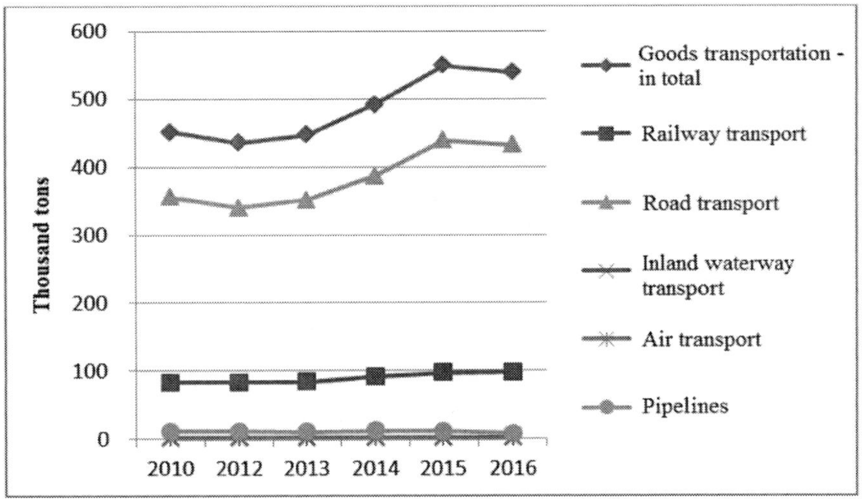

Fig. 1. Graphical representation of transport development in thousands of tons. Source: [3]

2.2 Estimation of individual and group wagon consignments

The carriage of individual and group wagon consignments (maximum of five wagons) reached 34,312 thousand tons in 2016. In the case of greater connection of business entities using railway freight transport, significant savings in CO_2 production and road infrastructure wear would be achieved.

 DOI: 10.1051/matecconf/201713400006

3. Results

The optimal solution for reinforcing individual and group wagon loads in railway transport is the higher degree of use of combined or intermodal transport. The Group of Freight Transports of the Swiss State Railways SBB Cargo has put into operation twenty-foot handling platforms with corner fixing elements that can be placed on a Sgnss railway wagon in a total of 3 platforms. These platforms are manufactured by Invekta, a.s., Vlkov near Litomyšl. The actual weight of one platform is 1.8 t and forms a connection between the railway loading surface of the wagon and the loaded container. Placing the platforms on a railway wagon creates a trolley handling surface that allows devices to handle along the ramp, railcar and container in the logistics chain. The "ramp-ramp" delivery system of goods is created [4, 5].

The platforms can be easily loaded and secured by means of corner fixing elements on a railway carriage, or can be removed and folded into a deck in the combined transport terminal. An example of handling is shown in figure 2 [6].

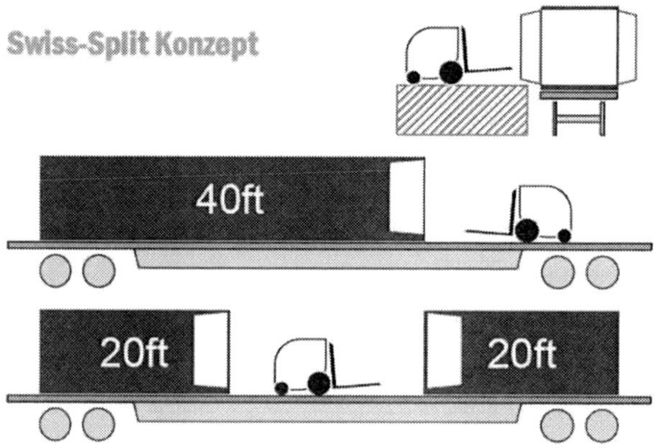

Fig. 2. An example of the use of handling platforms. Source: [6]

4. Discussion

In the case of the completion of the shipment of individual consignments according to the data from 2016, the transfer to the road freight transport would take place. Basic cost calculations can calculate the economic costs of road wear [7].

The semitrailer wears the road communication (infrastructure) about 15.824 CZK per 1 km [8]. The average transport distance by railways for individual consignments is 159 km. The transport volume in tons in 2016 was for individual consignments 34,312 tons.

Such an amount would take them the approximately 1.72 millions road articulated vehicles.

Such an amount would take them approximately 1.72 million of road articulated vehicles according to the relation (equation 1) [9, 10]:

$$N = \frac{Q_{tr}}{N_{loading\ capacity}} \text{ [pcs]} \qquad (1)$$

where: N – number of articulated vehicles [pcs]; Q_{tr} – the volume of carriage by individual wagon consignments [pcs.ton]; $N_{loading\ capacity}$ – loading capacity (payload) of the semi-trailer [t].

$$N = \frac{34.312\text{mil.pcs}}{20\text{ t}} = 1.72 \text{ mil. art. veh.}$$

Considering that, road semi-trailer kits (articulated vehicles) will be used for both directions at 80%, the annual mileage is about 330.24 million km.

To calculate the road infrastructure wear, we use the relation (equation 2) [9-11]:

$$P_{wear} = L_{year} * P_{wear\ art.veh.}\ [\text{CZK}] \tag{2}$$

where: P_{wear} – road infrastructure wear [CZK]; L_{year} – annual traveled km of articulated vehicles [mil. km]; $P_{wear\ art.veh.}$ – road infrastructure wear by the articulated vehicle [CZK/km].

$$P_{wear} = 330.24\ mil.km\ * 16\ \text{CZK/km} = 5.284\ \text{mld. CZK}$$

Calculation of CO_2 production according to equations (3) and (4) in the case that performances of individual consignments are transferred to road freight transport. Is as follows [9, 12]:

$$Q_{CO2} = L_{year} * \emptyset C_D\ [\text{l}] \tag{3}$$

where: Q_{CO2} – total diesel fuel consumption [l]; L_{year} – annual traveled km of articulated vehicles [mil. km]; $\emptyset C_D$ – estimated average diesel fuel consumption [l/km].

$$Q_{CO2} = 330.24\ mil.km * 0.35\frac{l}{km} = 115.6\ mil.l$$

$$TP_{CO2} = C_D * Q_{CO2}\ [\text{t}] \tag{4}$$

where: TP_{CO2} – total production of CO_2 [t]; C_D – total diesel fuel consumption [l]; Q_{CO2} – the amount of CO_2 when burning 1 liter of diesel fuel [t] = 2.64 kg CO_2.

$$CP_{CO2} = 115.6\ mil.l * 0.00264 = 305.184\ t$$

The abovementioned values of road infrastructure wear and calculation of total CO_2 production are only some of the negative impacts of the modal shift of the system of single and group wagon loads on road freight transport. It should also be noted that there is a disproportion between the existing toll system, which is valid only on certain designated networks of roads in the Czech Republic, and the fee for the use of the railway route for railway transport, where the whole railway network of the is charged. High accident rate of road transport is another negative factor in goods transportation on the road [13-17].

5. Conclusion

Intermodal transport represents the transportation of cargo by several transport modes using one and the same intermodal transport unit without handling with its contents during transportation.

The development of transport in the world is characterized by the utmost effort to increase the speed, reliability and accuracy of goods delivery in freight transport using modern technical equipment and information technologies. In Europe, intermodal transport is an indispensable element of transport policy, especially due to reducing the negative effects of road transport on the environment, fuel and energy consumption, maintenance costs on highways and roads, land consumption and road safety.

The paper discusses the possibility of transferring freight transport to railway transport using combined or intermodal transport. For railway transport, it is a complex problem to deliver goods to the final customer's yard, so-called "last mile". This system effectively addresses this problem.

This paper was prepared using the Final Report "Optimization of Transport and Transport Products by ČD Cargo, a.s.", prepared by the Institute of Jan Perner, o.p., Studentská 95, 532 10 Pardubice with the use of the Researchers doc. Ing. Jaromír Široký, Ph.D., Ing. Edvard Březina, CSc., Prof. Ing. Václav Cempirek, Ph.D., Ing. Petr Nachtigall, Ph.D. For Czech Railways Cargo, a.s. in 2013.

This paper is supported by the research project "From horse-drawn railway to intermodal transport" within Visegrad Fund.

References

1. Wascosa, *Modulare Güterwagen für erfolgreichen Einzelwagenverkehr*, Available online: http://www.viwas.eu/wp-content/uploads/2015/10/5-Wascosa.pdf (2016)
2. The European Commission, *European strategies, The White Paper*, Available online: https://ec.europa.eu/transport/themes/strategies/2011_white_paper_en (2011)
3. *Transport Yearbook*, Available online: https://www.sydos.cz/cs/rocenka-2016/index.html (2016)
4. O. Stopka, R. Kampf, Transport **April 2016**, 1-11 (2016), DOI: 10.3846/16484142.2016.1174882
5. M. Tomášiková, M. Tropp, Z. Krzysiak, F. Brumerčík, Logi – Scientific Journal on Transport and Logistics **6**, 2, 40-47 (2015)
6. SBB: The Swiss Way for a Renewed Railfreight System, *ViWaS Final Conference*, Available online: http://www.viwas.eu/wp-content/uploads/2015/10/1-SBB-Cargo.pdf (2015)
7. R. Kampf, J. Gašparík, N. Kudláčková, Periodica Polytechnica Transportation Engineering **40**, 2, 71-75 (2012)
8. T. Brzobohatý, *Analýza vlivů mýtného na intenzitu silniční nákladní dopravy v České republice* (Dopravní federace, Czech Republic, 2008)
9. T. Mirfakhraie, Y. He, R. Liscano, *IMECE - International Mechanical Engineering Congress and Exposition **12*** (ASME, Montreal, Canada, 2015)
10. M. Chovancova, V. Klapita, Open Engineering **7**, 1, 50-54 (2017), DOI: 10.1515/eng-2017-0009
11. T.C. Martin, T.R. Thoresen, Research in Transportation Economics **49**, Special Issue, 55-64 (2015), DOI: 10.1016/j.retrec.2015.04.006
12. M. Prekop, M. Dolejs, Geographia Technica **11**, 2, 78-86 (2016), DOI: 10.21163/GT_2016.112.08
13. B. Sarkan, O. Stopka, J. Gnap, J. Caban, Procedia Engineering **187**, 775-782 (2017), DOI: 10.1016/j.proeng.2017.04.437
14. J. Lizbetin, R. Kampf, K. Jerabek, Z. Caha, *Transport Means – 20th International Scientific Conference on Transport Means*, 1083-1087 (Juodkrante, Lithuania, 2016)
15. H. Neradilova, G. Fedorko, V. Borsosova, *Proceedings of the Third International Conference on traffic and Transport Engineering*, 971-979 (Belgrade, Serbia, 2016)
16. M. Fazekaš, M. Šulgan, Š. Liščák, Transport Problems **8**, 4, 35-44 (2013)
17. R. Kampf, J. Roudná Logi – Scientific Journal on Transport and Logistics **1**, 1, 79-85 (2010)

Minimizing the Risks in the Storage of Metallurgical Material

Zdeněk Čujan[1,*], and *Jiří* Lajtoch[2]

[1]College of Logistics, Přerov, Palackého 1381/25, 750 02 Přerov, Czech Republic
[2]College of Logistics, Přerov, external doctoral student at the University of Žilina, Faculty of Security Engineering, Univerzitná 1, 01026 Žilina, Slovak Republic

Abstract. The paper focuses on minimizing the risks in the storage of metallurgical material in a specific company which needs to cope with a limited space for material storage in general. At present, the metallurgical material, together with other purchased items, is stored in a designated space in the production hall, from which it is directly distributed to the production. The aim of the solution is to propose a more efficient use of the designated storage space while increasing the safety of work and minimizing occupational injuries.

1 Introduction

Storage is an integral part of logistics that involves a fairly wide range of problem areas [1]. It serves as an important link between manufacturers and customers [2, 3]. Closely related topics include stock levels, order cycles, warehouse equipment, warehouse zoning, spatial layout and inventory management [4]. Storage also fulfils several important functions: receiving and storing stock items, creating or improving their utility value and performing the necessary warehousing operations.

2 Current state analysis

The storage facility for metallurgical material is the largest storage space in the company [5]. It is closed, located directly in the production hall and serves for supplying the material to the production [6]. Besides metallurgical material, other materials such as plastics, non-ferrous metals, aluminium and stainless steel materials are stored in the storage facility. The metallurgical material has various profile shapes of different cross-sections (square, rectangular, round, tubular). The supplied length is usually 6 m.

The storage facility is operated by a total of nine employees, including seven operators and two technicians. As for the technological equipment, the facility belongs into the category of mechanized warehouses. In addition to hand manipulation with the material using handling trolleys there are three forklift trucks and two cranes used (one is operated from the ground, the other, which is a bridge crane, is operated from a cabin) [7].

[*]Corresponding author: zdenek.cujan@vslg.cz

The total area of the storage facility is 1,174 m^2, the usable area is 712 m^2. The difference between the total area and the usable area, which is 712 m^2, is used relatively inefficiently as a handling area.

There are a total of six universal racks and one cantilever rack. The material is stored both vertically and horizontally. The cranes are used to stow the material on the cantilever rack.

Purchased materials are usually stored in lengths of 3 m, except for steel profile material, which is stored in lengths of 6 m.

3 Storage systems for metallurgical material

A survey of suppliers offering storage systems for metallurgical material provided a useful overview of solutions which may improve the layout of the storage facility in the given company. This section provides a list of selected products:

Suppliers of cantilever racks

BEG Bohemia spol. s r. o.

The products offered by the company include: pallet racks (drive-in, cable reel, gravity), shelf racks, mobile racks (shelf, pallet, cantilever), cantilever rack, steel mezzanines (heavy mezzanines, steel platforms, warehouse extensions) and other systems [8].

Fig. 1. Cantilever rack Multistong. Source: BEG Bohemia spol. s r. o. Source: [9]

BITO skladovacítechnika CZ s.r.o.

The product range of the company is very wide. It offers various storage systems (several kinds of bins and containers), gravity racking systems for non-palletised loads (shelving, multi-tier installations, mezzanines, automated small parts storage), pallet racking systems (pallet racking, mobile, automated, gravity, drive-in systems), container transport, workshop equipment and storage systems for long loads (cantilever racking) [10].

 DOI: 10.1051/mateconf/201713400007

Fig. 2. BITO cantilever racking. Source: BITO s. r. o. Source: [7]

KREDIT spol. sr.o.

The most used is the UNIKANT cantilever racking, which is primarily intended for storing board or rod-shaped metallurgical material, sheet metal, etc. The material is stowed on the rack girders by handling equipment or by hand individually, as bundles, or other handling units. It is always placed on two or more girders of the same storage level of the rack. Racks are intended for stowing by forklift trucks, special loading machines or cranes with accessories [10].

Fig. 3. UNIKAT cantilever racking. Source: [7]

KASTO Maschinenbau, GmbH

In addition to conventional storage systems such as racks, floor storage and stacking cradles, the company offers the UNITOWER system which has several advantages in terms of space requirements and speed of access to individual items. The storage system can be both single-sided and double-sided. The UNITOWER system features a sliding platform located longitudinally or transversely in the tower system[7].

DOI: 10.1051/matecconf/201713400007

Fig. 4. UNITOWER storage system. Source: [7]

KARDEX s. r. o.
One of the systems supplied by Kardex for the storage of oversized and heavy material is the Towermat vertical lift system. Its design makes it suitable for fitting to a given space and for storing long profile materials, sheets or pallets [7].

Fig. 5. Towermat automated storage system. Source: [7]

The Towermat system enables efficient material storage up to a total weight of 5,000 kg per tray.

3 Design of a more efficient layout of the storage facility

In order to design a more efficient layout of the storage facility, the following requirements were established after consulting the company management:

- maximizing storage space using vertical storage,
- optimizing shelf layout based on size so that shorter or residual material can be stored in longer shelves as well,
- minimal load bearing capacity of one shelf 1,000 kg,
- rack height 7 m,
- shelf width 3 to 6 m,
- maximal diameter of stored material at least 270 mm,
- storing material in the handling direction,

- using semi-automated or fully automated system,
- minimizing necessary operator presence,
- minimizing the number of operators [1].

These requirements were processed into a request for quotation which was sent to selected suppliers of storage systems.

Based on the subsequent analysis, the LS (LongSpan) Tower storage system supplied by KREDIT was chosen for storing all profile material.

Compared to similar systems provided by other companies, LS Tower offers greater usable depth of its storage shelves and its load bearing capacity is greater by 1,0 kg. It also has 13 shelves more than other comparable systems. This particular storage system is able to store metallurgical material with a total weight of up to 58,0 kg.

It will replace the two universal racks which are currently located at the front of the storage facility. Free usable area will be created which can be used as a central section where the material is received. In addition, the system will also replace racks located in the middle part of the storage area where 19,000 kg of material is stored and provide storage capacity for 21,0 kg of material which is currently stored on the floor [10].

Fig.6. LS Tower automated vertical storage system. Source: [10]

Using the proposed LS Tower vertical automated storage system, the original storage racks can be replaced obtaining more free space of 198 m^2 in the storage facility which can be efficiently used for storing material on the floor, as a handling area, etc.

4 Evaluation of the proposed solution

To evaluate the proposed solution, logistics indicators of its economy and logistics indicators of its productivity were used.

4.1 Logistics indicators of the proposed solution economy

Using the indicators of economy, the level of the logistics systems efficiency will be determined before and after the implementation of the proposed LS Tower vertical automatic storage system.

- Indicator of equipment costs per operator:

$$K = \frac{E_m}{Z_m} \tag{1}$$

where: E_m is the sum of all the purchase costs of the handling equipment in the evaluated logistics system; Z_m is the number of operators employed in the evaluated logistics system for material handling.

Indicator of equipment costs after the implementation of the new storage system:

$$K = \frac{2{,}822{,}650}{9}$$

$$K = 313{,}628\ CZK$$

The calculation shows that the equipment costs per operator after the implementation of the new storage system will be 313,628 CZK per operator.

4.2 Logistics indicators of the proposed solution productivity

Using the indicators of productivity, the level of the logistics systems efficiency will be determined before and after the implementation of the proposed LS Tower vertical automatic storage system.

- Indicator of usable storage area:

Indicator of usable storage area before the implementation of the new storage system:

$$K_s = \frac{S}{S_c} \cdot 100\ [\%] \tag{2}$$

where: S is the area used for storing the material [m^2]; S_c is the total area of the storage facility [m^2].

Indicator of equipment costs after the implementation of the new storage system:

$$K_s = \frac{712}{1{,}174} \cdot 100$$

$$K_s = 60.6\%$$

Indicator of usable storage area after the implementation of the new storage system:

$$K_s = \frac{514}{1{,}174} \cdot 100\ [\%]$$

$$K_s = 43.8\%$$

The implementation of the new storage system will save 17% of the storage area.

- Indicator of material picking:

Indicator of material picking before the implementation of the new storage system:

$$V_p = 20\,\text{min}(33\%)$$
$$M_n = 130\ CZK/hour$$

$$C_n = V_p \cdot M_n \tag{3}$$

$$C_{n1} = \frac{33 \cdot 130}{100}$$

$$C_{n1} = 43\ CZK$$

where: V_p is the time of picking one item requested for production; M_n is the labour costs per operator; C_{nl} is the total costs.

Indicator of material picking after the implementation of the new storage system:

$$V_p = 15\,\text{min}(24.75\%)$$
$$M_n = 130\ CZK/hour$$

$$C_n = V_p \cdot M_n$$

$$C_{n2} = \frac{24.75 \cdot 130}{100}$$

$$C_{n2} = 32\ CZK$$

The calculation shows that the implementation of the new automated storage system will save 11 CZK per picking one requested item.

- Indicator of financial savings:

Indicator of financial savings before and after the implementation of the new storage system:

$V_p = 8249/year\ 2016$

$$C_{n1} = 43CZK$$
$$C_{n2} = 32CZK$$
$$U_{mn1} = V_p . C_{n1} \tag{4}$$
$$U_{mn2} = V_p . C_{n2} \tag{5}$$

$$U_{mn} = U_{mn1} - U_{mn2} \tag{6}$$

where: V_p is the time of picking one item requested for production per year; C_{n1} is the total costs before the implementation; C_{n2} is the total costs after the implementation; U_{mn1} is the labour costs savings before the implementation; U_{mn2} is the labour costs savings after the implementation.

$$U_{mn1} = 8249.43 = 354{,}707\ CZK/year$$

$$U_{mn2} = 8249.32 = 263{,}968\ CZK/year$$

$$U_{mn} = 354{,}707 - 263{,}986 = 90{,}739\ CZK/year$$

The evaluation shows that the total financial savings are 90,739 CZK.

5 Conclusion

Most companies aim to use their storage facilities more effectively. This holds for the storage of metallurgical material as well. Unsuitable storage racks, which cannot accommodate larger quantities of material, and low to zero identification of stored items – these are the most common shortcomings that lead not only to time losses but also to increased administration demands and hence to greater error rates [10].

These risks can be reduced by the implementation of automated storage systems which also increase safety, productivity and business competitiveness.

This paper is supported by the research project "From horse-drawn railway to intermodal transport" within Visegrad Fund.

References

1. Z. Cujan, Open Engineering **6**, 470-475 (2016)
2. G. Fedorko, N. Husáková, G. Dudáš, Acta Montanistica Slovac*a* **15**, 14-22 (2010)
3. V. Molnar, K. Pacutova, in *Proceedings of the third international conference on traffic and transport engineering (ICTTE)*, 570-575 (Belgrade, Serbia, 2016)
4. J. Fabianova, P. Kacmary, V. Molnar, P. Michalik, Open Engineering **6**, 270-279 (2016)
5. N. Husakova, G. Fedorko, V. Molnar, S. Honus, in *Ecology, economics, education and legislation conference proceedings, SGEM 2016,* **II**, 187-194 (2016)
6. H. Neradilova, G. Fedorko, Open Engineering **6,** 700-710 (2016)
7. B. Martináková, *Skladování hutního materiálu* (Přerov, Czech Republic, 2017)
8. R. Rohan, *Hutní Materiál* (Prague, Czech Republic, 1987)
9. BEG Bohemia, Available online: http://www.beg-regaly.cz/en (2017)
10. Z. Čujan, Z. Málek, *Výrobní a obchodní Logistika* (Zlín, Czech Republic, 2008)

Simulation as an Instrument for Research of Driver-vehicle Interaction

Kristián Čulík[1,*], *Alica* Kalašová[1], and *Simona* Kubíková[1]

[1]University of Žilina, Faculty of Operation and Economics of Transport and Communications, Department of Road and Urban Transport, Univerzitná 8215/1, 010 26 Žilina, Slovakia

Abstract. This report describes opportunities for using driving simulators for research of driver-vehicle interaction. We have focused on examples of the using of the current equipment of the University Science Park of University of Žilina. Laboratory was built for research of driver-vehicle interaction. It's also important to give examples of evaluation after several measurements. The SNA - 211 REN driving simulator was used for each experimental measurement. It was necessary to use some additional equipment, because the driving simulator was used only for simulation. During simulations, drivers have been performing tasks connected with vehicle driving. Driver's eye tracking in specific conditions was also important part of our research. We have tested the behaviour of drivers when they have used mobile phones for texting and GPS device with touchscreen.

1 Introduction

The rising pace of life raises the claim for psychological resistance of a person and that affects human thinking and behaviour. This is always reflected during driving the vehicle. From the view of road transport, it is necessary to research the driver-vehicle interaction, because its results can help to increase safety of road traffic.

Driving simulators are devices, which can be used as a safe and practical instrument for research of driver-vehicle interaction under laboratory conditions. Nowadays the driver is still fully responsible for his vehicle and its safety systems, but subjects for research are not only the parts of the vehicle. We can also simulated different influences of road infrastructure and environment.

Therefore, we have done series of measurements with many drivers with using training driving simulator SNA – 211 REN. This device isn't research driving simulator as we need, but it can simulate some basic traffic situations. Measurements were done thanks to external recording and using additional equipment. Software of the simulator is not editable. Drivers had predefined specific tasks in all parts of research. This task has included driving through the narrow areas, texting during freeway ride, setting the GPS navigation device etc.

[*] Corresponding author: kristian.culik@fpedas.uniza.sk

 DOI: 10.1051/matecconf/201713400008

2 Driver-vehicle interaction

Road traffic, especially motor transport as its most important part, can be defined as a complex system of many partial events and phenomena. They together affect driving individual vehicles and all signs that characterize transport. Significant part of partial transport events is very variable and their appearance is random. Therefore, the transport can be attributed to stochastic character and its each partial events can be considered as random variables in mathematical meaning. If we evaluate the transport from the view of fluency, safety and economy, its complex effect is depended on the quality of certain factors, which can be incorporated into three basic groups in system human - vehicle - environment [1, 2].

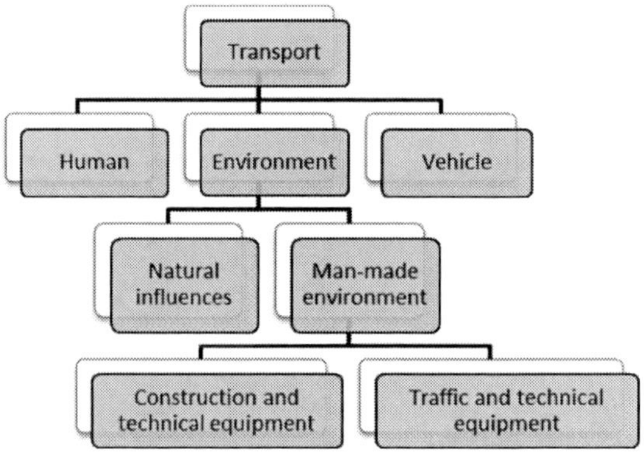

Fig. 1. Scheme of the transport system. Source: [1]

2.1 Driver's behaviour

Driver's behaviour is a process with two levels. First, it is unconscious level, which includes automatic activities during driving. Second level is conscious, when driver starts automatic activities after changing the driving conditions and also decides about the most suitable action. Vehicle driving is very difficult and complex activity and it requires permanent attention and reactions to stimuli [1-3].

From the view of traffic psychology, it's possible to express the driver behaviour during ride with following equation (1) respective function:

$$R = f(S\text{-}O) \tag{1}$$

where: R – driver's behaviour (reactions)
S – perceived and acting stimuli on the driver
O – driver's personal characteristics

2.2 Significance of driver-vehicle interaction

Pyramid in the figure 2 shows approximate ratios between specific levels of cases. This cases result from interaction between road users – it's C. Hydén's pyramid of interaction. From the picture is obvious, that a small portion of the conflicts may end up in an accident. Specifically regarding the consideration of the potential conflicts or near accidents, complementary sources of accident records are not considered, but this task is quite important [4].

DOI: 10.1051/matecconf/201713400008

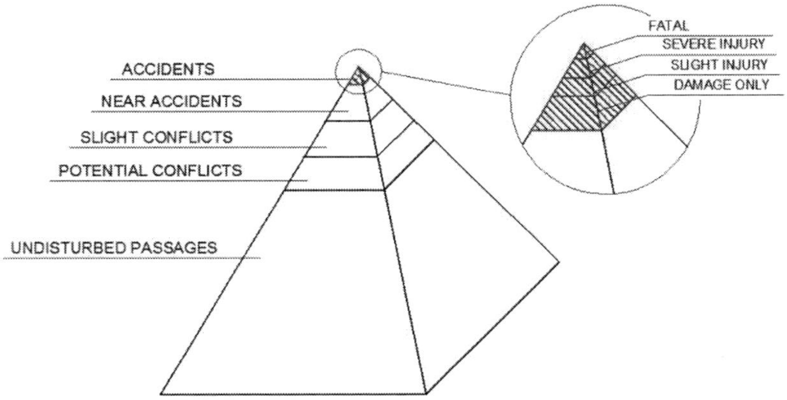

Fig. 2. Pyramid of interaction among road users. Source: [4]

From the picture is obvious, that is only possible to measure conflicts on the top of the pyramid. Under the peak of the pyramid, there is a lot of different kind of conflicts between vehicles respectively road users. These conflicts are not simply defined, because a physical accident isn't happen. Number of these conflicts we can estimate with for example from subjective opinions expressed in questionnaire. Better method for research of driver's behaviour is to measure their actions and trajectory of their vehicles in safe conditions. Ideal device for this kind of research is driving simulator. With high accuracy, we can record distance between vehicles, distance between vehicle and obstacle and so on [3-5].

3 Simulations for research of driver-vehicle interaction

The driving simulator is simply defined as a device which is used for simulation of driving a road vehicle in environment, which is similar to real conditions in road traffic. Light driving simulators (devices, that do not include the whole car, but only some parts of cockpit) are effective instrument for driver training, but they usage is more important in research of human-machine (driver-vehicle) interaction [5, 6].

Nowadays only a little attention is given to the research of driver-vehicle interaction in the Slovak republic. University of Žilina also does not have a full-valued research driving simulator, which could be used for measurements and also capable to automatically evaluate the impact of the vehicle and its components to the driver.

At the University Science Park of University of Žilina, there is a training driving simulator SNA – 211 REN (fig. 3) for research in the field of road transport. It is a device, which simulates driving a lorry (Renault Midlum) and a truck (Renault Magnum) with trailer or tank trailer.

Fig. 3. The cabin of training driving simulator SNA - 211 REN. Source: authors

3.1 Equipment, drivers and preparation

The training driving simulator SNA - 211 REN can only simulate the real conditions on the roads, but it's necessary to use additional devices to record the process of driving in virtual environment and also tools, which are necessary for particular driver's task. It is very important to find the ways for easy measurements and collecting the output data. The training driving simulator, which was used for our experiments, cannot provide recording of input or output data. For driver's eye tracking, we have used SMI Eye Tracking Glasses 2 connected with recording device – smartphone Samsung. Other parts of research have needed digital video camera.

For specific experiments, we have chosen people from the same age category 23-26 years. We hadn't established special requirements to the drivers before the drivers were chosen. It was very important to provide the simulation in undisturbed conditions, so the instructions for the driver were always very simple. It guaranteed that the results were minimally affected.

3.2 Quantification of driver's skills during driving in slalom

First task for drivers included driving the vehicle in slalom. This task is evaluated in complex graph in figure 4. Black line in the graph connects the absolute sums of time, which was needed to accomplish the tasks. The values in seconds are for all drivers together. It's obvious that the value of time is rapidly improving with experience and accustoming the driver to the virtual vehicle. Histograms of driver's mistakes are irregular and also decreasing. Left columns show the conflicts with traffic cones, right are numbers of unwanted exits from asphalt lane. Virtual environment of the simulator had these settings during the simulation: adhesion 90% asphalt, load 0%, visibility - day.

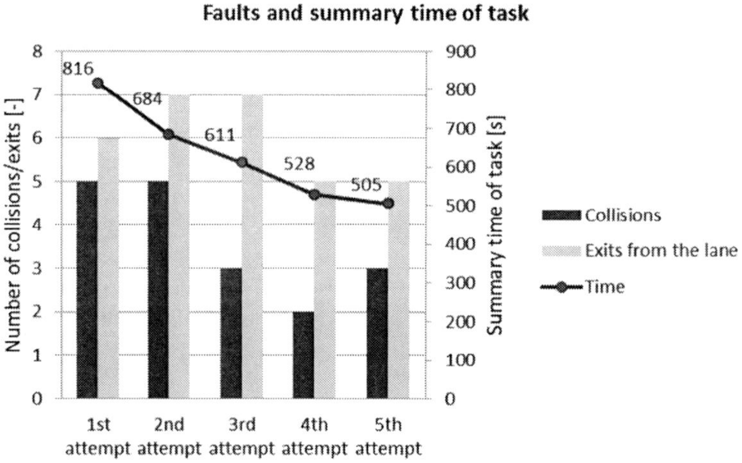

Fig. 4. Summary graph of driver's mistakes and slalom time. Source: authors

During the data processing, we made two groups of drivers. In the first group was faster half of drivers and in the second the remaining slower drivers. From the more detailed evaluation of changes in average speed, it's interesting that the growth of the average speed of the faster group has a concave shape and the slower group has a convex graph. More experienced drivers can drive with higher speed and their accustoming to the vehicle is faster.

3.3 Driving through the narrow areas

In the next part of our research, we have decided to test how drivers can drive through the narrow areas, because the SNA – 211 REN simulates the truck driving, so the vehicle is quite bigger than personal car. Simulator has one specific scene – "Narrow areas", which is ideal for this task. It includes road circuit with five points – at each are two walls and between them is narrow area only for one vehicle with only little allowance. The vehicle, which is used in this simulation, is Renault Midlum.

In this task, every driver has drove through the narrow areas as fast as he can, but without contact with any wall. Driver has driven the circuit five times, so every driver has passed 25 narrow points.

For the recording time and vehicle speed, we have used external video camera recording. Outputs of record are these data:
- start time, time in half of circuit, finish time
- distances
- minimal speed in narrow area (for every point)

Minimal speed in narrow area was read with accuracy 1 km/h and it is minimal value from all showed values during driving in the narrow area. Accuracy of times is 0,04 second, because the frame rate is 25 frames per second. For the evaluation, we have made partially automatic form, which include driving time, travelled distance, calculated section speed, as well as number of fails, which were made by drivers during this task.

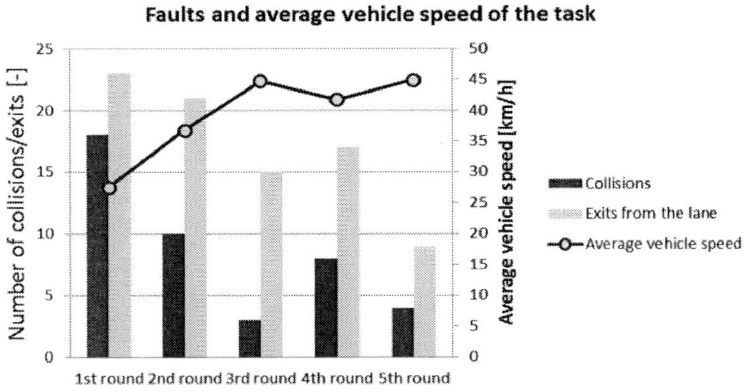

Fig. 5. Summary graph of driver fails and vehicle speed. Source: authors

Graph in the figure 5 shows that the sum of failures has decrease tendency, with the exception of the fourth attempt. With the successful passing the narrow areas drivers has increased their vehicle speed and they was also less concentrated. It turned out during four rounds, where faults have increased.

Other option for evaluation of driver's behaviour during ride with lorry is comparison of their achieved results. Columns in the next graph (fig. 6) show all faults during this task – collisions with walls and also number of escapes from the lane. Vehicle speed is average speed of all drivers and vehicle speed in narrow area is an average value from all minimal speeds.

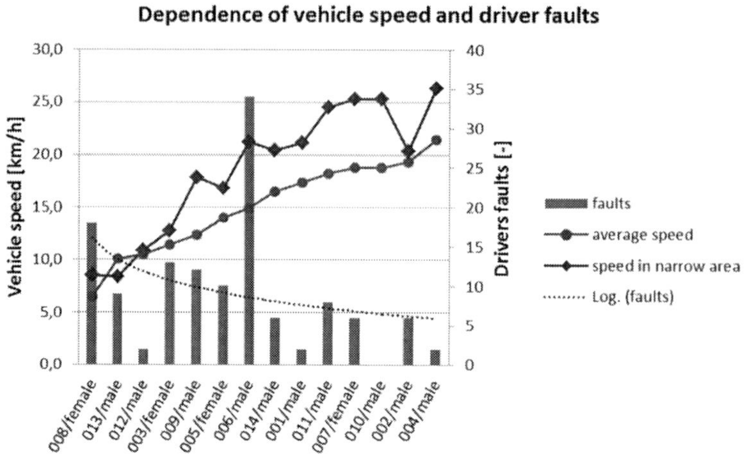

Fig. 6. Graph of vehicle speed, speed in narrow areas and driver's faults. Source: authors

3.4 Reaction time

Simulation for this task had a predetermined set of obstacles, which are generated in certain distance from the vehicle during driving. SNA – 211 REN has summer and winter alternative of this scene. We have chosen the winter scene, because the visibility of obstacles is lower, so the task can be harder for driver. Obstacles imitate the fall of tree, wild animal running across the road etc.

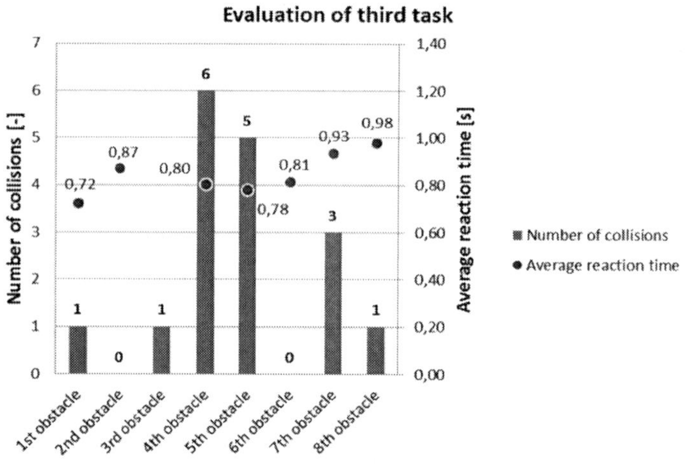

Fig. 7. Graph of average reaction times and numbers of collisions. Source: authors

From the graphical representation of task results in the figure 7 is clear that six drivers couldn't prevent collision with sixth obstacle. Average reaction time was only 0,80 second, so the drivers were concentrated, but only one driver has stopped in front of the obstacle. Video analysis showed, that this obstacle in this scene (falling tree) is starting relatively late. The position of the vehicle is so close to the obstacle, that it is impossible to stop the vehicle from higher speed. We wanted to measure reaction times, so bypassing the obstacle was forbidden.

3.5 Eye tracking

Simulation of driving has many advantages. One of the biggest is possibility to research behaviour, which is forbidden in road traffic by law, for example driving under the influence of alcohol, phoning at the wheel etc. Driving simulators give us the opportunity to drive virtual vehicle in safe conditions, so other disturbing elements are not dangerous. So it's also possible to test the influence of radio, air condition, on board systems and so on.

For this part of our research, we have chosen freeway simulation scene, because shifting the eight gears on the hilly terrain can be difficult and it requires a lot of attention. On the freeway, driver must only check only position on the lane and vehicle speed. During the eye tracking every driver had a certain distance for free ride and then they have been doing two specific tasks during driving:

- first task was writing an SMS on the smartphone
- second task was setting target destination on a GPS device with touchscreen.

Output of the eye tracking glasses is video format .avi, which is necessary to next analysis with accuracy 10 points per second. We have used time stamps function in video editor. Marking times, when the driver change the point of view, is the most effective and accurate method of evaluation. After creating all time stamps, we have exported them to the table editor.

Table 1. Eye tracking results of specific driver. Source: authors

Measurement:	E-04	Scene:	freeway	
Date:	12.4.2017	Load	100%	
Driver:	009/male	Adhesion:	90% asphalt	
Driv. licence:	B	Visibility	mist	
Value	Value meaning		Time	Ratio
1	view in field of view		164.8	61.8%
2	view of dashboard and cabin		44.1	16.5%
3	view of mobile phone/GPS		57.9	21.7%
		∑	266.9	100.0%

Complex graphical evaluation of the tasks is more important than individual numerical results. Figure 8 shows pie chart with average total time of viewing the specific areas during free ride. During this part, the drivers were fully concentrated. They have looked to the road and environment 81% of the time and to the cabin and its equipment remaining time (18%)

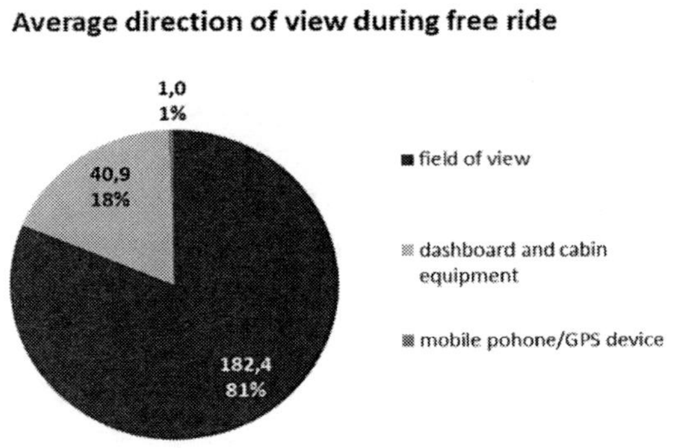

Fig. 8. Pie chart of results from eye tracking during free ride. Source: authors

However, eye tracking results are more important, when tested driver has been doing some other task during his drive. Average percentage ratios of driver's view are showed in pie charts in the figure 9.

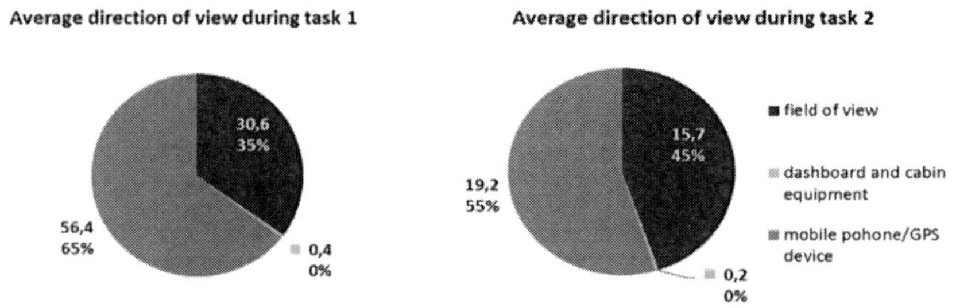

Fig. 9. Pie chart of results from eye tracking during the task 1 and 2. Source: authors

Pie charts show only specific part of the experiment, time between starting and ending the task. It's obvious, that the first task was done by average driver in 1 minute and 27.4 seconds, the second one in 31.1 seconds. Results of first task showed that drivers have given 65% of their attention to the mobile phone. When drivers have been setting the target on the GPS device, they have been looking to its touchscreen for 19.2 s (approximate value) and 157 seconds they were concentrated on driving.

3 Conclusion

The main aim of our paper was to determine the behaviour of the driver while he is driving. All tasks were done in virtually simulated environment and in real vehicle cabin. Training driving simulator SNA – 211 REN located at University Science Park of University of Žilina is a device with only few predefined simulations and also small variability of settings. It seems that it will be necessary to use different simulation device in future. One of our aims was to make sample measurement methodology and also realise measurements. The most effective method for the research of driver-vehicle interaction is combination of driving simulator and eye tracking device, which has given us the relevant results. Thanks to the simulation, it is also possible to do tasks that are not legal in real conditions.

This paper is supported by the research project "From horse-drawn railway to intermodal transport" within Visegrad Fund.

References,

1. A. Kalašová, M. Mikušová, *Bezpečnosť cestnej dopravy a dopravná psychológia*, 224, (University of Žilina, Slovak Republic, 2017)
2. M. Korfant, M. Gogola, *TRANSCOM 2017 – International scientific conference on sustainable, modern and safe transport* **192**, 433-438 (2017)
3. J. Ondruš, Ľ. Černický, Communications: scientific letters of the University of Žilina **18**, 118-123 (2016)
4. C. Hydén, *The development of a method for traffic safety evaluation: The Swedish Traffic Conflicts Technique*, 229, Available online: http://www.tft.lth.se/fileadmin/tft/video_in_traffic/Swedish_conflict_technique.pdf (1987)
5. M. Mikušová, *TST 2011 – Modern transport telematics: 11th international conference on transport systems telematics*, 70-77 (Katowice-Ustron, Poland, 2011)
6. M. Miko, J. Ondruš, *Dopravné inžinierstvo 2016: International scientific conference* **1**, 165-174 (University of Žilina, Slovak Republic, 2016)

Quality Assessment in the Logistics of Rail Passenger Transport

Milan Dedík[1,*], *Jozef* Gašparík[1], and *Zdenka* Záhumenská[1]

[1]University of Žilina, Faculty of Operation and Economics of Transport and Communications, Department of Railway Transport, Univerzitná 8215/1, 010 26 Žilina, Slovakia

Abstract. The article is focused on analysis and rating of the quality in the passenger railway transport logistics. At first is described the relationships between logistics and transport, logistical demands for transport and functions of logistics in rail transport. In the third chapter is explained quality assessment including the description of the quality indicators in passenger rail transport and making a proposal of the new methodology for this rating. Consequently this methodology is tested for chosen transport route and is concerned with monitoring quality of the transport routes in the railway network.

1 Introduction

The relationship between logistics and transport is very narrow as transport ensures the physical relocation of the product from the place of production to the point of consumption. Transport is a significant factor in time benefit, it is the carrier of speed and reliability of product relocation, it is one of the most important elements of the logistical system. In the logistical chain from material supplier to customer, it is a very important and irreplaceable place [1, 2].

When choosing the optimum kind of transport and the mode of transport, it is very important to take into account, in particular, the type and characteristics of the goods to be transported, the method of packing, the volume, the weight, the legislative rules and other conditions. In this selection, account must be taken of the characteristics of the various modes of transport:

1. Ability to maximize transport safety, including the elimination of shocks and possible damage to the goods.
2. Ability to transport networks, the ability to provide traffic to any destination.
3. Ability to transport any quantity of materials and goods.
4. Ability to provide the optimal degree of time security and reliability.
5. Ability to ensure the optimal level of home-to-house speed.
6. Ability to provide the maximum degree of comfort
7. Ability to minimize the cost of shipping

* Corresponding author: milan.dedik@fpedas.uniza.sk

Several types of conventional and unconventional transport can be used to implement transport services. Each type has some advantages and disadvantages and has an important place in the transport system.

Conventional modes of transport are:

- railway transport
- road transport
- water transport
- air transport

Unconventional modes of transport include, for example, pipeline transport, cableways and so on [3-6].

2 Functions of logistics in rail transport

Rail transport is by no means a guarantee of such high-quality availability, manoeuvrability and flexibility as road transport, but in logistics, its function is very important. It can provide transport in much larger quantities, in the case of freight transport offers the possibility to transport also heavy loads, shipments with specific properties, etc. Thus, it can offer a more favourable transport price per unit of consignment (e.g., ton).

In passenger rail transport, logistics deals in particular with the planning, organization, management and control of all activities between operators entering the transport process, thereby creating the conditions for ensuring safe, reliable, sufficiently fast and convenient passenger transport at the optimum price level in cooperation with other types of transport. The basic strategic objective is to ensure the maximum speed, quality and comfort of the territory with a positive impact on the optimal interconnection of the regions and their economic development. To achieve this goal, the introduction of an integrated transport system, integrating transport and tariff conditions, increasing the coordination of transport modes and co-operation between individual carriers, is the most appropriate solution. The more integrated the transport system is in place, the better for the traveling public, and the logistics of public passenger transport work at a higher level [3, 6].

The main tasks of logistics in passenger rail transport are:

- Review of passenger traffic flows at certain times in transport sessions using different survey / marketing methods.
- Optimization and control of these transport streams
- Improving the various factors of the transport process
- Improving the relationship with customers
- Ensuring quality services and optimal travel culture - cleanliness and comfort of the means of transport, seating, technical condition of the vehicle
- Ensuring optimal cost and ease of customer clearance
- Solving the complex logistics chain - road from house to house [1, 7, 8].

3 System for quality assessment of rail passenger transport logistics

Logistics always strives to find the optimal solution for the provision of passenger transport services, with a strong emphasis on quality to the fullest extent possible, as at present the passenger wants not only to dislocate, but he is interested in traveling and demanding a certain level of service. The assessment of the quality of passenger rail transport logistics is a very broad topic and can be viewed from several angles. The most relevant way of assessing quality is the concept examining selected quality indicators in passenger transport, which are a very important and decisive factor for the traveling public. They can be divided into quantitative, qualitative and others. These quality indicators form the basis of the proposed methodology for assessing the quality of rail passenger transport logistics. Within the proposed methodology, a scoring scale will be developed for these indicators, which will provide a comprehensive assessment of the quality of passenger transport logistics. In practice, this methodology will be applied and tested for a selected transport routes [2, 9, 10].

The supporting and decisive indicators are broken down as follows:

- number of connections
- average speed of achievement
- time interval between links
- types of train
- availability of railway stations and stops
- transport distance - coefficient γ
- transportation services and comfort [11, 12].

3.1 Methodology of scoring and objective evaluation of quality indicators

In the case of the evaluation of selected quality indicators (factors), the point method of multi-criteria analysis will be used. Indicators with a slightly higher weight can be rated at most by ten points; lower weight factors can rise to eight points. This system of classification on points and the subsequent scale for the evaluation of individual factors was determined by expert estimation and in this way carried out in consultation with selected representatives of a broad professional public. Consequently, an analysis of the evaluation of a certain transport routes will be carried out, which will use the biasing of these indicators. The Bratislava – Sereď route will be chosen transport report for testing the new methodology. Table 1 contains a set of all factors with the maximum possible number of points achieved in the complex evaluation of the transport routes [11]:

Table 1. System for point assessment of quality indicators. Source: [11]

Factors influencing on the quality assessment of a particular transport session :	Maximum achieved point number:
Number of connection per day	10
Average speed of achievement	10
Average time interval between links	8
Transport distance coefficient γ	8
Types of train	8
Starting and finishing station availability	8 (4+4)
Detailed information about the services provided in the home, destination and transit stations	8
SUMMARY:	**60**

From the above data it follows that the maximum possible rating of the selected transport route can be up to 60 points. The final quality evaluation with a point scale, percentage expression and verbal evaluations, according to which the final evaluation of the selected transport routes will be carried out is mentioned in table number 2 [5, 8].

 DOI: 10.1051/matecconf/201713400009

Table 2. Final assessment for transport routes. Source: [11]

The amount of achieved points:	Percentage expression:	Verbal assessment:
60 – 49	100 – 81 %	Green field - the quality of the transport session is excellent, strict criteria are met
48 – 37	80 – 61 %	Yellow-green field - the quality of the transport session is fairly good, most of the criteria are met
36 – 25	60 – 41 %	Yellow field - the quality of the transport session is average, about half of all criteria are met
24 – 13	40 – 21 %	Orange field - The quality of the transport session is below average, most of the criteria are not met
12 – 0	20 – 0 %	Red field - The quality of the transport session is very low, only the minimum criteria are met

3.2 Point assessment for transport routes evaluation

However, for the implementation of the above-mentioned final evaluation, it is necessary to exactly define a scale for the classification of individual factors for routes evaluation. The more favourable the value of a particular factor, the final score will be higher. This evaluation is shown in Tables 3 and 4.

For the factor, the number of connections per day is considered only with logically and geographically meaningful connections, taking into account that no two connections have the same time to leave the home station or time to arrive at the finishing station. In case of starting and finishing destination availability factor, it is very important to differ starting and finishing station availability separately. In both cases, a maximum of four points can be obtained. For this factor, the scale for city and village is different. However, it is also possible to obtain the "Bonus points" in the case when the high quality and heavy urban mass transportation that runs between the railway station and the centre of the individual settlements is ensured [13-15].

Table 3. Point assessment of selected quality indicators. Source: [11]

Number of connection per day:	Point assessment:	Starting and finishing destination availability in villages(km) :	Point assessment:	Starting and finishing destination availability in cities (km) :	Point assessment:
0	0	0.4 and less	4	0.6 and less	4
1	1	0.41 – 0.80	3	0.61 – 1.20	3
2	2	0.81 – 1.20	2	1.21 – 1.80	2
3	3	1.21 – 1.60	1	1.81 – 2.50	1
4 – 5	4	1.60 and more	0	2.51 and more	0
6 – 7	5				
8 – 9	6				
10 – 11	7				
12 – 14	8				
15 – 17	9				
18 and more	10				

Table 4. Point assessment of further selected quality indicators. Source: [11]

Average time interval between links (min):	**Point assessment:**	**Average speed of achievement (km/h):**	**Point assessment:**	**Transport distance – coefficient γ :**	**Point assessment:**
0 – 30	8	100 and more	10	1 – 0.91	8
31 – 60	7	99 – 90	9	0.90 – 0.81	7
61 – 90	6	89 – 80	8	0.80 – 0.71	6
91 – 120	5	79 – 70	7	0.70 – 0.61	5
121 – 180	4	69 – 60	6	0.60 – 0.51	4
181 – 240	3	59 - 50	5	0.50 – 0.41	3
241 – 300	2	49 – 40	4	0.40 – 0.31	2
300 – 900	1	39 – 30	3	0.30 – 0.21	1
900 and more	0	29 – 20	2	0.20 - 0	0
		19 – 10	1		
		9 – 0	0		

In the case of the transport distance factor, a new coefficient γ is proposed. This coefficient describes the transport distance deflection, which expresses the advantage of the realization of the selected transport route by rail, is considered. This coefficient calculates the ratio of the air distance of the session and the distance of the transport session by rail. The higher the value of the route, the greater the distance of the transport route along the railway line will be more direct and more suitable for the passengers [16, 17].

For factors train types and detailed information on the services provided at the destination, finish stations and transit stations, it is not possible to set an exact scale for the point assessment. For train types, this rating is based on the combination and number of trains of higher and lower quality. The factor expressing information on services at stations is based on their specific quality and quantity [11].

3.3 An example of application at transport route

A practical example of the use of this methodology will be the Bratislava - Sereď transport route (on a working day).

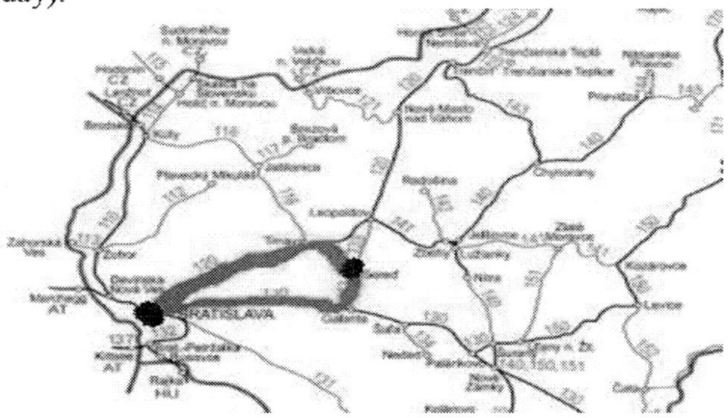

Fig. 1. Illustration of Bratislava – Sereď transport route. Source: [11]

In the case of the train types factor, it can be stated that the transport session can be reached by train with categories R (fast train), RR (regional fast train) and Os (suburban train). These are middle and lower category trains and based on this statement, the factor will be assigned a half point. In case of availability of station Bratislava hl. st. a bonus point for the above reason, which represents the heavy urban mass transportation in Bratislava, will be added to the point score resulting from the scale. For detailed information about the services provided at the Bratislava hl. st., Trnava, Galanta and Sereď can be considered with six points from the eight possible. In the case of the newly introduced coefficient of the transport distance deflection γ, the value of the air transport distance is 47.5 km and the distance by rail 61.5 km. The coefficient γ will be 47.5 / 61.5 = 0.77.

The total score for the transport route will be the same as it is listed in table 5.

Table 5. Complex point assessment of Bratislava – Sereď route. Source: [11]

Factors influencing on the quality of a particular transport relations:	**The value adjacent to the factor:**	**Achieved point number according to scale :**
Number of connection per day	23	**10**
Average speed of achievement	64.06 km/h	6
Average time interval between links	49.09 min	7
Transport distance - coefficient γ	0.77	6
Types of train	-	4
Availability – railway station Bratislava hl. st.	1.6 km	3
Availability – railway station Sereď	1.5 km	2
Detailed information about the services provided in the home, destination and transit stations	-	6
SUMMARY:		**44**
Percentage expression:		**73.33%**

This complex assessment shows that the transport session received 44 points out of 60 possible, representing 73.33 percent of the quality. It means that it can be included in the so-yellow-green field - the quality of the transport route is fairly good, most of the criteria are met [11].

4 Conclusion

The proposed methodology for assessing the quality of passenger transport logistics should, in practice, help the infrastructure manager or the transport coordinator in the Integrated Transport System to simplify, more accurately and more readily identify the selected quality indicators and evaluate the transport routes. This will make it easier to discover in the problematic spots in train traffic diagram, which should help to make it better in future years.

At present, quality assessment in the logistics rail passenger transport is not implemented through a comprehensive methodology and the results often do not have the required reporting value, the suggested facts would certainly help to better gauge train traffic diagram quality and thus to help to avoid the introduction of poor quality variants. On this basis, there is a prerequisite for improving the train traffic diagram, linking the different types of transport, increasing the number of passengers, increasing customer satisfaction, as well as a comprehensive improvement and improvement in rail transport that has stagnated in recent years. From the point of view of logistics, this methodology should help to find the optimal alternative for moving people with the maximum synergy effect of selected quality indicators. The main result should be a high-quality logistical chain with satisfied customers [13, 18].

The paper is supported by the VEGA Agency by the Project 1/0095/16 "Assessment of the quality of connections on the transport network as a tool to enhance the competitiveness of public passenger transport system", that is solved at Faculty of Operations and Economics of Transport and Communication, University of Žilina.

This paper is supported by the research project "From horse-drawn railway to intermodal transport" within Visegrad Fund.

References

1. M. Kendra, M. Babin, P. Šulko, *BulTrans-2013 – Anniversary scientific conference on aeroanutics, automative and railway engineering and technologies*, 95-97 (Sofia, Bulgaria, 2013)
2. T. Skrúcaný, M. Kendra, M. Škorupa, Procedia Engineering **192**, 806-811 (2017)
3. I. Gros, I. Barančík, Z. Čujan, *Velká kniha logisitky* (Prague, Czech Republic, 2016)
4. M. Šulgan, J. Gnap, J. Majerčák, *Postavenie dopravy v logistike* (University of Zilina, Slovak Republic, 2008)
5. M. Škorupa, M. Kendra, Procedia Engineering **192**, 800-805 (2017)
6. G. Fedorko, V. Molnár, J. Strohmandl, M. Vasiľ, *Transport Means 2015 – 19th international scientific conference*, 466-469 (2015)
7. L. Černá, A. Dolinayová, J. Daniš, *Transport means 2016 – 20th international scientific conference*, 414-419 (Juodkrante, Lithuania, 2016)
8. L. Černá, V. Zitrický, J. Daniš, Open Engineering **7**, 6-13 (2017)
9. J. Mašek, M. Kendra, J. Čamaj, *Transport means 2016 – 20th international scientific conference*, 584-588 (Juodkrante, Lithuania, 2016)
10. E. Majerčáková, P. Majerčák, Logi – Scientific Journal on Transport and Logistics **6**, 1, 90-99 (2015)
11. M. Dedik, *Connection quality assessment in the railway transport from the railway network aspect* (Diploma thesis, University of Zilina, Slovak Republic, 2016)
12. G. Fedorko, H. Neradilova, M. Sutak, V. Molnar, *Transport means 2016 – 20th international scientific conference*, 169-174 (Juodkrante, Lithuania, 2016)
13. M. Chovancova, V. Klapita, *Transport means 2016 – 20th international scientific conference*, 112-116 (Juodkrante, Lithuania, 2016)
14. R. Kampf, L. Ližbetinová, K. Tislerová, Open Engineering **7**, 26-30 (2017)
15. P. Majerčák, E. Majerčáková, *Transport means 2016 – 20th international scientific conference*, 688-692 (Juodkrante, Lithuania, 2016)
16. *Euroekonom.sk – Diplomova_teoria_obchod*, Available online: http://www.euroekonom.sk/download2/diplomovka-teoria-obchod/Teoria-Diplomova-praca-Logistika-predmet-a-obsah-logistiky.pdf (2017)
17. V. Lendel, M. Varmus, Logi – Scientific Journal on Transport and Logistics **3**, 2, 21-34 (2012)
18. O. Stopka, I. Simkova, V. Konecny, Nase More **62**, 3, 126-130 (2015), DOI: 10.17818/NM/2015/SI7

MATEC Web of Conferences **134**, 00010 (2017) DOI: 10.1051/matecconf/201713400010
LOGI 2017

Structural Analysis of a Modified Freight Wagon Bogie Frame

Ján Dižo[1,*], *Jozef* Harušinec[1], and *Miroslav* Blatnický[1]

[1]University of Žilina, Faculty of Mechanical Engineering, Department of Transport and Handling Machines, Univerzitná 8215/1, 010 26 Žilina, Slovak Republic

Abstract. This paper presents results of selected load cases for the modified freight wagon bogie frame. It comprises of two main parts. The first part is addressed to the introduction of a modified freight wagon bogie. There are described conditions for rail vehicles approval and mainly load cases which every bogie must meet for commissioning. The next part deals with computer modelling and analyses of a frame of this modified freight wagon bogie. It was analysed using the FE method and calculations were focused on the assessment of stresses of the frame structure under exceptional load. Calculation conditions were based on valid standards.

1 Introduction

The transport of goods by railways represents an important element of a transport service. Given than it is an environmentally friendly kind of goods transport, its importance is seen in the international and intermodal transport. Therefore, railway transport is nowadays an inseparable part of the transport system.

The rail transport in this kind of transport allows an efficient way to move large quantities of goods over longer distances [1]. Nowadays the design of rail vehicles meets conflicting requirements. On one hand there is the rail vehicles weight reducing and on the other hand all railway transport means must comply strict safety criteria. Every new designed railway vehicles and also construction units, e.g. bogies must before commissioning satisfy the terms set out in codes [2, 3].

2 Description of the solved freight wagon bogie

The largest development of freight wagon bogies design in Europe began after 1950. The main reasons for the design of new freight wagon bogie types have been constantly increasing requirements for higher speed transport and higher axle loads while reducing the risk of accidents. The effort of engineers has resulted in the design and production of the Y25 bogie [4].

This bogie is the most commonly used bogie for freight wagons mainly in the Central Europe region. The suspension of the Y25 bogie consists of a duplex coil springs with

* Corresponding author: jan.dizo@fstroj.uniza.sk

bilinear characteristic and friction dampers. The wagon body is connected with bogies by high stiffness centre pivot. The yaw movement of the bogie is damped by sidebearer pair. The friction forces between friction damper surfaces are created through the inclined Lenoir link and they are approximately proportional to the vertical load [5-7]. The Y25 bogie is most commonly designed for freight wagons running at the speed up to 100 km/h and axle load of 22.5 t (alt. 25 t) or at the speeds up to 120 km/h and axle load of 20 t.

Since demands of the best ratio of the unloaded weight and the capacity weight of transport means are constantly required, engineers try to modify their design within their capabilities and maintaining its all positive operational properties and reliability [8-10]. Developers of rail vehicles have therefore come to the possibility to modify a frame of the original Y25 bogie by removing buffer beams [11, 13]. Such a frame is lighter, which is beneficial for the capacity weight increasing. Figure 1 shows design of the original and the modified freight wagon bogie [13].

Fig. 1. Design of the original (left) and modified (right) freight wagon bogie. Source: [13]

If we compare the unloaded weight of the original bogie and the modified bogie with comparable equipment (brake type, bogie determination, etc.) we can find out that one bogie saves approx. 250 kg [13] and since standard wagons (i.e. the most often used wagons with Y25 bogies) are usually equipped with two such bogies, the wagon weight is no longer negligible.

From the operation point of view reducing the wagon unloaded weight means increasing the wagon capacity weight, lower operating costs when a wagon is transported in an unloaded state by saving energy consumption [14] and from the production point of view we can save material costs and partly to reduce needed production time.

Nevertheless every intervention in the load-bearing construction requires a through strength calculations supported by experimental measurements on test benches [15, 16]. The bogie frame is the main carrying part of the bogie. Therefore it is not possible to perform such the essential modification of the frame without adequate analyses [17, 18]. They come out and rely on strict requirements embedded in relevant standards and regulations.

In European standard EN 13749 the considered analysed bogie is classified in the category B-V: bogies freight rolling stock with single-stage suspensions [2, 19].

In the next part of the article, the requirements for wagon bogie as well as the results of the strength calculations and analyses of the modified bogie frame for chosen load cases are presented.

3 Rail vehicle bogies loads

A real bogie and its frame are loaded in real operational conditions by a wide spectrum of a load which depends on actual level of loading, quality of a track expressed by variations form its geometric ideal position load or an eventual buckling of the track in the given track section and various others factors. These result in the variable operational load which is practically not possible to describe by load dependencies. Therefore for the load of bogie frames the replacement load spectrum was determined according to which new developed bogies are tested [20]. For this testing load bogie frames are designed and optimised in terms of strength and virtual testes. This testing load is intended in legislation, such as UIC edict, European and national norms or internal directives of a vehicle operator [2, 3, 19]. In this contribution the bogie frame is tested using virtual models while the calculation conditions are based on the European standard, in which requirements on bogie frame are defined.

Generally, various load cases of bogies frames can be divided into external and internal.

For all categories in all applications the internal load is defined as well as the external load:

- Exceptional load, which should not produce permanent deformation or excessive deflections.
- Normal service loads, which should not induce fatigue cracks.

European standard regulates the calculation of an internal load of bogies by accelerations determination, which the appropriate component must withstand without defects or loss of function.

3.1 Combinations of individual loads of bogie

Loads defined above are in reality not individual, but they act concurrently in various combinations and in various intensity. For needs of the strength and dynamic design of a bogie frame the European standards submit several selected load combinations.

They are compiled to meet the bogie frame design to withstand possible combined operational loads.

In the European standard two combinations of a load are defined static load combination and dynamic load combination.

Prescribed loads combinations represent the base not for design and optimisation of bogie frames but also subject for verification tests on test benches. They are distinguished static and dynamic load cases.

Bogie structures are loaded by a very high number of dynamic load of a variable size. Effects of such load are concentrated in critical locations of a bogie frame (e.g. points of action forces, geometry changes, welded joints etc.). The objective of fatigue tests is verification, if a bogie frame has a sufficient fatigue strength, i.e. if a cyclic operational load does not result to initiation of fatigue cracks or fractures. Fatigue stresses is possible to determine by two methods, i.e. fatigue resistance, if the fatigue lifetime is known (stress value, in which under load by determined number of cycles the fatigue damage not occur) and cumulative damage, if the stress is constantly under of the fatigue lifetime level for all determined load combinations.

In our work we have considered only the static load combination.

4 Computer analysis of the bogie frame

Ride properties significantly influence on rail vehicles mechanical systems dynamic behaviour [21]. We can theoretically predict the movement of the wheelset on a track by means of the wheelset and track geometric characteristics analysis [22]. Geometric characteristics define the rail/wheel profiles contact couple geometrical relationship. The shape of the contact couple crucial influences the size of the contact patch and contact stress between wheel and rail value [23]. This creates loading and excitation forces acting inside vehicle and track systems [24]. The analysis of the mechanical systems dynamics may be analysed using various methods [25, 26].

The analysed frame is component of a freight wagon bogie. According to standard this bogie belongs the category B-V namely Freight bogies with a central pivot and two sidebearers [2, 19]. We have analysed this bogie frame in such way that we proceeded in accordance with a valid standard.

In boundary conditions and loads definition of the bogie frame we have come out of the standards.

4.1 Bogie frame load definition

The total weight of the bogie was $m = 5.0$ t, wheelbase $b = 1.8$ m. The bogie is determined for a freight wagon with the total weight of $M_W = 90$ t.

The bogie is loaded in the vertical direction by the force:

$$F_Y = \left(\frac{M_W}{2} - m\right) \cdot g \tag{1}$$

For the vertical direction the value of the exceptional load is given by these formulas:
- if vertical forces act only in the centre pivot:

$$F_{YP\max} = 2 \cdot F_Y \tag{2}$$

- if vertical forces act in the centre pivot and on the one sidebearer:

$$F_{Y1\max}\left(or\ F_{Y2\max}\right) = 1.5 \cdot F_Y \cdot \alpha \tag{3}$$

$$F_{YP} = 1.5 \cdot F_Y\left(1 - \alpha\right) \tag{4}$$

where F_Y is the total vertical load the a bogie, F_{YP} is the vertical force acts in the centre pivot, F_{Y1} and F_{Y2} are vertical forces act on sidebearers, α is coefficient for the body swinging. In our case we have considered with the value of $\alpha = 0.3$.

For the lateral force of the exceptional load acting on the every wheelset there is applied following:

$$F_{Z1\max} = F_{Z2\max} = \frac{F_{Z\max}}{2} = 10^4 + \frac{F_Y + m \cdot g}{6} \tag{5}$$

The longitudinal force straining bogie frame is given:

$$F_{X1\max} = 0.1 \cdot \left(F_Y + m \cdot g\right) \tag{6}$$

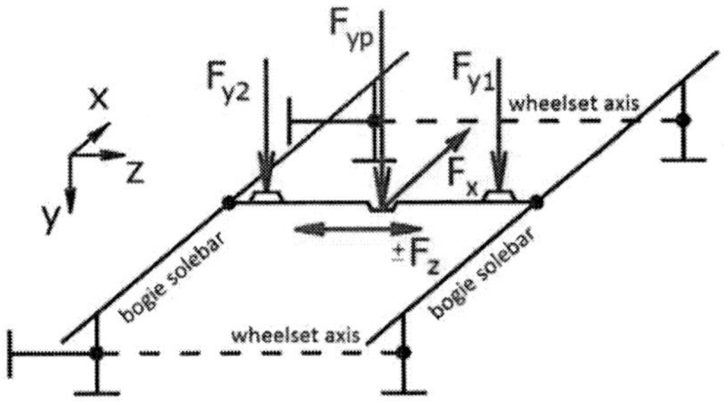

Fig. 2. Scheme of load point applications. Source: [2, 19]

The load involved in case of the wagon impact is possible to substitute for the static longitudinal force acting in places, where the equipment is connected to the bogie. Its value is determined from the mass of individual elements and maximum acceleration acting on them during wagon impact [19].

4.2 Calculations and results

For strength analyses using FEM we have used Ansys software. It allows engineers to create computer models of structures, machine components or whole systems, apply operating load and other design criteria and study physical responses [27, 28], such as stress levels, pressure, deformations, etc.

Process of the bogie frame computer analysis includes several steps, which are needed to perform for successful calculation. We can it describe as follows:

- Creation of a CAD model of the considered body, appropriate model files generation.
- Setting up a FEM model in Ansys.
- Model - generation of 8-nodes shell elements and 10-nodes quadratic tetrahedron elements (axle guides and centre pivot).

The bogie frame is generally made of combination S235 and S355 steels, which minimal yield stress is 340 - 440 MPa and 520 - 630 MPa, respectively. The S355 steel is used for the centre pivot and axle guides. Therefore, we have defined these parameters of material model:

- Material was considered homogenous, isotropic, linear and elastic.
- Mechanical properties - Young's modulus of elasticity $E = 2.1{\cdot}10^{11}$ Pa, Poisson's ratio $\mu = 0.3$.

Next there are presented results from computer analyses of the modified bogie frame, which was subjected to four load cases corresponded to the exceptional load described in chapter 4.1.

Table 1 contains calculated data, which are needed to apply as loads on the bogie. The third column indicates, which formula from chapter 4.1 was used for calculation.

Table 1. Calculated loads for bogie frame. Source: authors

Load	Calculated value [kN]	Formula
F_Y	392.4	(1)
F_{YPmax}	784.8	(2)
F_{YImax}	176.58	(3)
F_{YP}	412.02	(4)
F_{ZImax}	83.575	(5)
F_{XImax}	44.145	(6)

As mentioned above, in this work we have focused on four most unfavourable load combination cases. These load combinations are embedded in the described standard. They practically represent combinations of calculated loads from Table 1 and Figure 2 shows points of application and directions of individual loads.

Load combinations are listed in Table 2.

Table 2. Calculated load combinations. Source: authors

Load case	F_{YPmax} [kN]	F_{YP} [kN]	F_{YImax} [kN]	F_{ZImax} [kN]	F_{XImax} [kN]
1st	784.8	0	0	0	0
2nd	0	412.02	176.58	0	0
3rd	0	412.02	176.58	83.575	0
4th	0	412.02	176.58	0	44.145

Figure 3 shows results from strength numerical analyses of the modified freight wagon bogie frame under four load cases according to the table. We have evaluated stresses according to the HMH hypothesis.

The frame structure is satisfactory, if in all load cases evaluated stresses are less than the yield stress of the used material.

Let's look at Fig. 3. The first load case is the state when the highest value of the exceptional load acts on the frame ($F_{YP\max}$ = 784.8 kN), but only in the centre pivot. Under influence of this load the highest values of the stress arise in the centre pivot region. The maximum is 460.71 MPa. This would indicate, that the frame structure in this region is at the limit of acceptability. But we have to consider important facts.

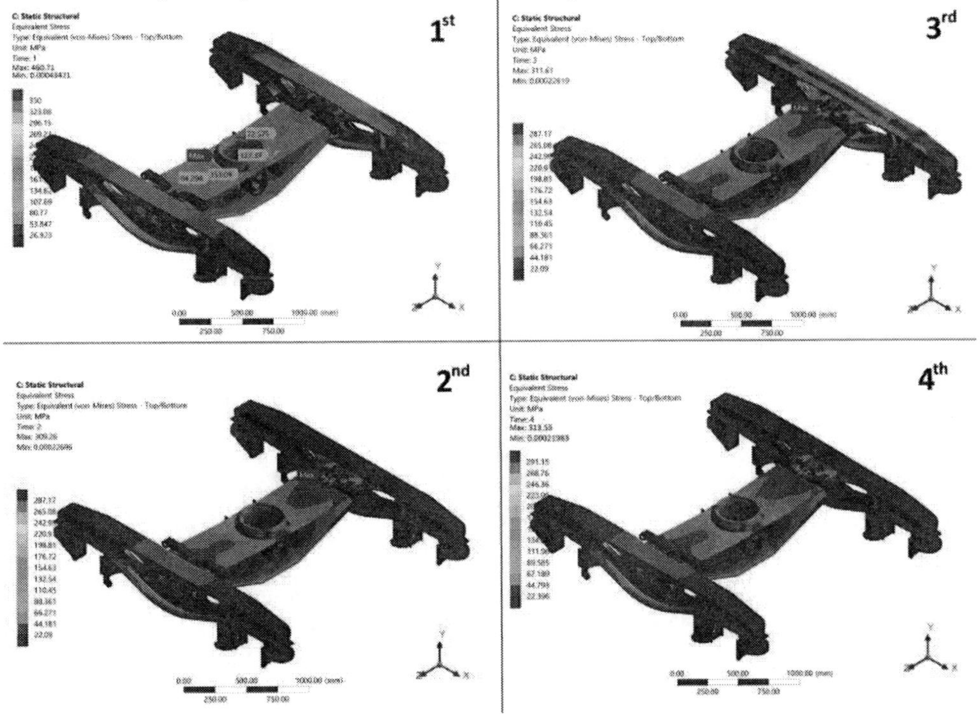

Fig. 3. Results of strength analyses. Source: authors

The centre pivot is made of steel with the higher yield stress, therefore for this part the minimal yield stress value of 520 MPa is valid. Moreover during calculations more variants of FE mesh size were used and compared. Based on results we have found out that this extreme value is obviously caused by calculation errors. Other values are lower and comply given limits (see Fig. 3, 1st).

In the second exceptional load case the frame was loaded by forces acting in the centre pivot and on the one sidebearer (namely F_{YP} = 412.02 kN, $F_{Y1\max}$ = 176.58 kN). This load simulates the body swinging. From results we can see (Fig. 3, 2nd), that the maximum stress is in the loaded sidebearer area and it reaches value of 309.26 MPa. It is even safely below the yield stress of the S235 steel.

In the third load case the exceptional load is formed in addition to vertical forces (F_{YP} = 412.02 kN, $F_{Y1\max}$ = 176.58 kN) also by the lateral force ($F_{Z1\max}$ = 83.575 kN). As we can see in Fig. 2, two forces act in the centre pivot and one loads frame on the sidebearer. The maximum stress is again in the loaded sidebearer area (Fig. 3, 3rd). The value of 311.61 MPa is higher than in the previous load case, but still within permitted limits.

Finally, let's move to the fourth load case. When we compare this case with the previous one, we can notice, the vertical load is the same (F_{YP} = 412.02 kN, $F_{Y1\max}$ = 176.58 kN), but the horizontal force acts in the longitudinal direction ($F_{X1\max}$ = 44.145 kN). This force simulates the frame load resulting from a running wagon dynamics. The calculated maximum stress value (Fig. 3, 4th) is slightly higher in comparison with the third load case and that is 313.55 MPa. It is still well below the yield stress.

In the future, we will research mechanical properties and dynamic behaviour of an entire modified bogie. We will create a mechanical system of a modified bogie in a multibody software and our just analysed FE model will serve as an important input for setting up a multibody system with a flexible body in order to study its dynamic properties and to compare with the original bogie for the detection of possible problems in terms of long-term operation [29, 30].

5 Conclusion

To summarising, we have created the three-dimensional model of a modified freight wagon bogie frame. Because it is a fairly new design of a freight wagon bogie, we wanted to know, if this new frame structure is able to meet strict criteria for all new designed rail vehicles and their components, mainly bogies. From all scale of load cases we have chosen four cases included as exceptional loads. We have performed numerical analyses of the bogie frame structure and based on results we have found out that the structure meets requirements prescribed in the standard. For making clear conclusions it would be necessary to compare results from numerical calculations and measurements to each other. This is the intention of further research in this area.

This article was supported by the project VEGA 1/0927/15 “Research of the use of alternative fuels and hybrid drives on traction vehicles with aim to reduce fuel consumption and air pollutants production”.
The work was supported by the Cultural and Educational Grant Agency of the Ministry of Education of the Slovak Republic in project No. KEGA 077ŽU-4/2017: Modernization of the Vehicles and engines study program.
This publication is the result of the project implementation: Modern methods of teaching control and diagnostic systems of engine vehicles ITMS 26110230107 supported by the Operational Programme Education funded by the ESF.

VÝSKUMNÁ AGENTÚRA Európska únia Štrukturálne fondy

This paper is supported by the research project “From horse-drawn railway to intermodal transport” within Visegrad Fund.

References

1. M. Kendra, M. Babin, D. Barta, Transport Research Arena 2012 **48**, 743-752 (2012), DOI: 10.1016/j.sbspro.2012.06.1052
2. EN 13749, *Railway Applications - Wheelsets and bogies - Method of specifying the structural requirements of bogie frames, European Committee for Standardization* (Brussels, Belgium, 2011)
3. UIC 566, *Loading of coaches bodies and their components* (1994)
4. J. Dižo, J. Harušinec, M. Blatnický, Soucasne problemy v kolejovych vozidlech, 29-36 (2017)
5. J. Gerlici, T. Lack, Soucasne problemy v kolejovych vozidlech, 117-125 (2017)
6. T. Lack, J. Gerlici, Y25 freight car bogie models properties analysis by means of simulation computations (to be published)
7. M. Maňurová, A. Suchánek, Manufacturing Technology **16**, 390-396 (2016)

8. O. Lunys, S. Dailydka, S. Steišunas, G. Bureika, Procedia Engineering **134**, 64-71 (2016), DOI : 10.1016/j.proeng.2016.01.040
9. J. Gerlici, O. Nozhenko, G. Chernyak, T. Lack, M. Gorbunov, R. Domin, The development of diagnostics methodological principles of the railway rolling stock on the basis of the analysis of dynamic vibration processes of the rail (to be published)
10. J. Grenčík, R. Poprocký, P. Volna, J. Galliková, Use of risk assessment methods in maintenance for a more reliable rolling stock operation. (to be published)
11. P. Štastniak, L. Smetanka, M. Moravčík, Manufacturing Technology **17**, 250-256 (2017)
12. J. Gerlici, M. Gorbunov, O. Nozhenko, T. Lack, S. Kara, K. Kravchenko, Soucasne problemy v kolejovych vozidlech, 109-116 (2017)
13. Tatravagónka Inc., *Products, Bogies* (2017)
14. Ľ. Kašiar, P. Zvolenský, D. Barta, L. Bavlna, M. Mikolajčík, P. Droždziel, Diagnostyka **17**, 3, 95-101 (2016)
15. J. Gerlici, T. Lack, J. Harušinec, Applied mechanics and materials **486**, 387-395 (2014), DOI: 10.4028/www.scientific.net/AMM.486.387
16. J. Gerlici, T. Lack, Applied mechanics and materials **486**, 379-386 (2014), DOI : 10.4028/www.scientific.net/AMM.486.379man
17. V. Hauser, O. Nozhenko, K. Kravchenko, M. Loulová, J. Gerlici, T. Lack, Car body and bogie connection modification for track curves passability improvement (to be published)
18. V. Hauser, O. Nozhenko, K. Kravchenko, M. Loulová, J. Gerlici, T. Lack, Innovative wheel tread design aimed to tramcar-track interaction improvement when passing curves of a small radius (to be published)
19. D. Kalinčák, *Rail vehicles I* (In Slovak) (University of Žilina, Faculty of Mechanical Engineering, Slovak Republic, 2008)
20. R. Mełnik, M. Kostrzewski, Structural Health Monitoring II **518**, 281-288 (2012), DOI : 10.4028/www.scientific.net/KEM.518.281
21. J. Gerlici, T. Lack, Applied mechanics and materials **486**, 396-405 (2014), DOI: 10.4028/www.scientific.net/AMM.486.396
22. J. Gerlici, T. Lack, Wear **271**(1-2), 246-258 (2011), DOI: 10.1016/j.wear.2010.10.052
23. J. Gerlici, M. Gorbunov, K. Kravchenko, O. Prosvirova, T. Lack, Noise and temperature reduction in the contact of tribological elements during braking (to be published)
24. M. Maňurová, A. Suchánek, Manufacturing Technology **16**, 1020-1027 (2016)
25. P. Hejma, M. Svoboda, J. Kampo, J. Soukup, Procedia Engineering **177**, 3-10 (2017), DOI: 10.1016/j.proeng.2017.02.175
26. P. Baran, M. Brezáni, P. Kukuča, P. Šťastniak, Procedia Engineering **192**, 34-39 (2017), DOI: 10.1016/j.proeng.2017.06.006
27. F. Klimenda, J. Soukup, Procedia Engineering **177**, 11-16 (2017), DOI: 10.1016/j.proeng.2017.02.176
28. J. Soukup, B. Skočilasová, J. Skočilas, Vibration of the steel frame of the primary sand classifier (to be published)
29. J. Dižo, J. Harušinec, M. Blatnický, Manufacturing Technology **15**, 781-788 (2015)
30. M. Kostrzewski, R. Mełnik, Procedia Engineering **192**, 439-444 (2017), DOI: 10.1016/j.proeng.2017.06.076

The Impact of Delays of Freight Trains in JIT System on Example One Track Line

Matúš Dlugoš[1,*], *Lumír* Pečený[1], *Pavol* Meško[1], and *Peter* Šulko[2]

[1]University of Žilina, Faculty of Operation and Economics of Transport and Communications, Department of Railway Transport, Univerzitná 8215/1, 010 26 Žilina, Slovakia
[2]Železnice Slovenskej republiky (ŽSR), Klemensová 8, 813 61 Bratislava, Slovak Republic

Abstract. The paper describes the problem between freight trains operated by Metrans Danube, a.s. and passenger trains operated by Regiojet, a.s on the line Bratislava – Dunajská Streda - Komárno. It is very important to emphasize this problem because, nowadays, mostly freight trains across the Europe drive in just in time system, it means, that shipments are delivered at exact time and place on time. At this line there are many problems not only this track line is single track, so it is necessary to cross the trains very often in traffic points which maximum 2 traffic tracks but also organizing of transport is more complicated which is provided by Railway infrastructure manager – this case is very important example when railway traffic technology slows down all processes in the logistic chain.

1 Introduction

Railway line Bratislava – Dunajská Streda – Komárno, marked in timetable 131, connected cities Bratislava – Dunajská Streda – Komárno. It is 100 km long single track with gauged 1435 mm. The maximum speed limit is 80 kph. The object of the track is situated in an unpretentious terrain in region Žitný ostrov. Maximum inclination is 5‰ and category of line class is D4. Organizing and transport service is provided from the center of operating point in Dunajská Streda. They operated transport from Komárno st. 4 (without) to Bratislava-Nové Mesto (without) [1].

There was an extensive reconstruction between Podunajské Biskupice – Dunajská Streda in 2011 [2].

* Corresponding author: matus.dlugos@fpedas.uniza.sk

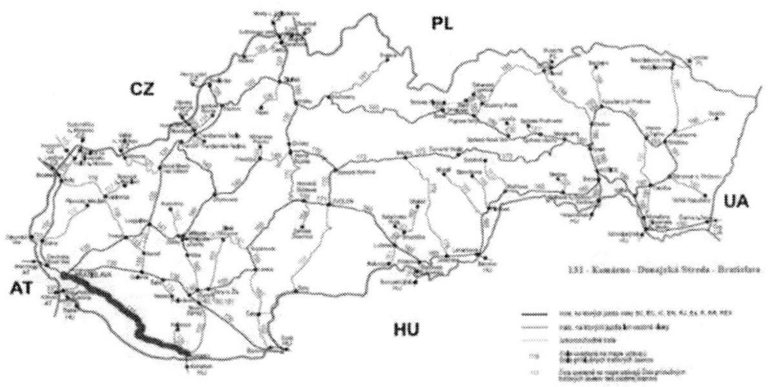

Fig. 1. Track 131 marked in the scheme of ŽSR railways. Source [2]

2 Contemporary operators

At present, both – the passenger trains and freight trains are operated at this line. By the end of 2011, personal transport was provided by ZSSK which transported about 790 000 passengers a year. From the March of 2012, the private passenger operator RegioJet started to operate passenger transport. The act was signed in December 2010 but valid from March 2012 to December 2020. Until this day, the number of transported passengers is about 1,412 mil. from the beginning 2017. The number of passenger trains is growing and nowadays is the operated periodic timetable with half an hour period in the peak hour.

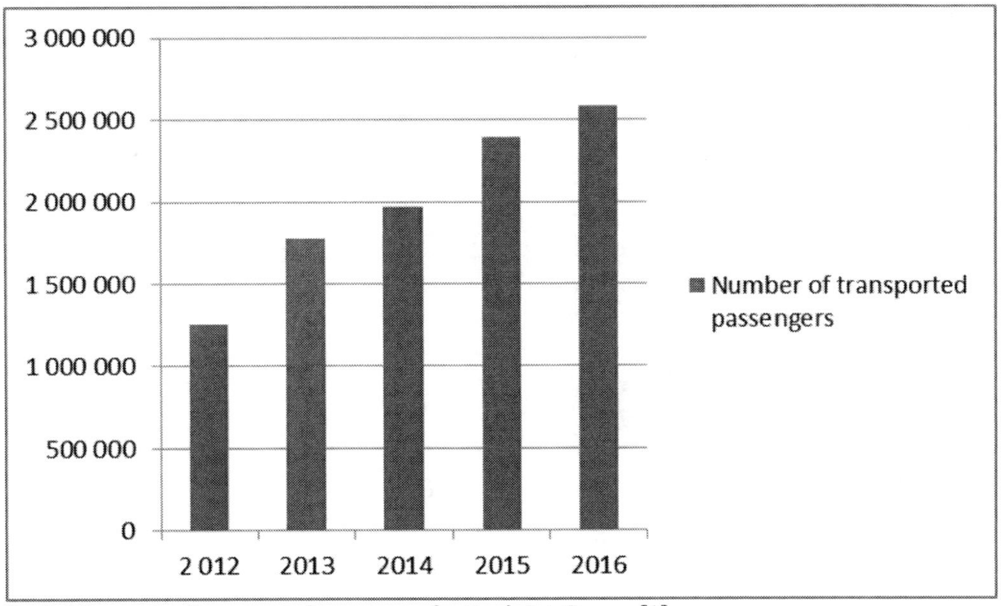

Fig. 2. The graph of transported passengers by RegioJet. Source [1]

Freight operator Metrans Danubia provides and operates terminal of intermodal transport in Dunajská Streda. They transport containers and exchange superstructures from the terminal to the other terminals around the Europe, for example to Koper or Trieste. [3].

Their trains drive in just in time system, so it is very important to keep the time of arriving in terminals. On avarage 8-9 trains daily leave the terminal from Dunajská Streda to the others.

The last one, state freight operator ZSSK CARGO, provides mainly transport cereals. They load the wagons in stations on the track to Komárno. They only provide handling trains, there [4].

3 JIT system

As we mentioned, the Just In Time technology or Just In Time system is the most widespread logistics system in the field of supplies, manufacturing and distribution. In the field of transport, JIT system means to distribute or deliver shipments from one terminal, stations or ports to the other on time, to reduce storage costs, efectivelly save storage place, decrease the number of train requirements, which includes number of employees – locomotive drivers, shifters and in the end increase efficient use of them [5].

In addition, JIT technology is required:

- shorter and more reliable shipping time,
- more sophisticated communication,
- fewer carriers with long-term relationships,
- efficiently designed conveying means and material handling equipment,
- quality decision models on the use of means of transport (own, public, contract) [6].

4 The reasons that affect the origination of the train delays

Due to the fact the mentioned track is single track line, the number of trains is from year to year higher and the capacity is already exhausted. It is mainly caused by passenger trains which are ordered by Ministry of Transportation of Slovak Republic, however the package of vlkm is every year the same. Not only passenger trains are the biggest problem. There are also the other reasons why it is complicated to provide freight trains of Metrans company don´t drive on time but usually delay.

We can divide reasons of delay due their origination:

- First group is formed by the effects caused by infrastructure,
- secondone is caused by problems on the side of operators,
- lastone is formed by problems caused by third side [7].

4.1 Effects caused by infrastructure

There are many reasons and problems, why it is complicated to provide trains on this track. At first, there is a problem with the capacity. As a single track with no trackside security equipment, trains drive from one station to the next station. Between some stations there is an automatic block signal, that divides interstationary section into 2 block sections, it is situated mainly between stations which are in a large distance [8].

Another reason is that some of the railway stations have only 2 traffic rails to cross the trains, for example station Lehnice. It isn´t possible to cross more than 2 trains. Closely linked with the stations are self-resetting switches. It means that trains from the one direction drive only to the branch by 40 kph. The lenght of traffic rails is also problem, mainly for freight trains of Metrans, because usually it is a problem to cross passenger train with freight train [9].

The speed limit is also problem, track is built in a very flat terrain and they drive only 80 kph [1].

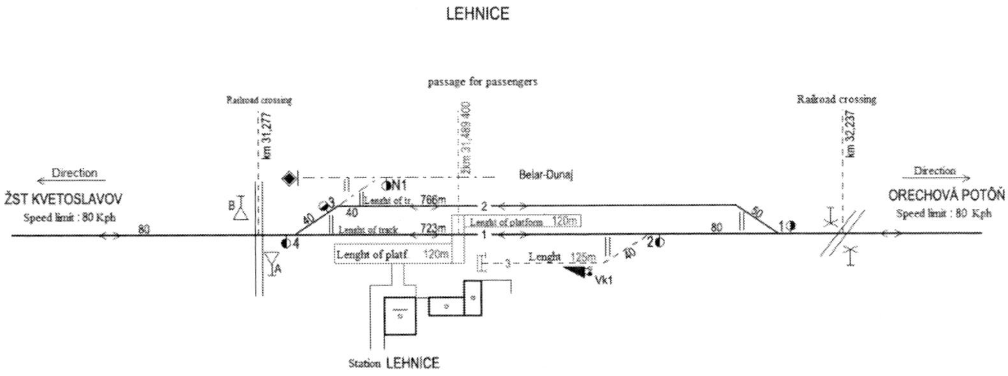

Fig. 3. Scheme of railway station Lehnice. Source [2]

4.2. Effects on the side of operators

In the actual train traffic diagram 2016/2017 there are 27 passenger trains from Bratislava to Dunajská Streda (Komárno) and 30 trains to Bratislava operated by Regiojet. Except of this 2 pairs of trains are operated as regional express situated during the rush hours, when many trains is necessary to cross very often [10].

For example in the morning hours, there is a passenger train 4315 from Bratislava to Komárno, it leaves at 8:05 and arrives 10:13 and during the journey it crosses 7 times. The intersection interval is too much short, so if the opposite train is delayed for more than 5 minutes, it disrupts the whole traffic diagram and passenger trains are delayed the whole day, because the time to turnover the rolling stock is short in final station, so final delay time may change [11, 12].

The another problem seems to be in connections. Some of the passenger trains from Komárno or Dunajská Streda are connections to Regiojet trains to Prague and in reverse.

The most common problem is train failure, it is cause mainly by the old types of rolling stocks or insufficient quality of service.

Operator Metrans Danubia provides intermodal transport of containers across Europe. The most common destinations are port Koper, Trieste, Marien St. Lorenzen, Budapest, Česká Třebová or Haniska pri Košiciach (Slovakia) [12, 13].

The majority of these places are so far, so there is a very high risk that it could delay why it is necessary to communicate with other international operators and make a deal how to solve the problems to avoid a delay. The most common problems are damaged wagons or broken trains and then If train is delay, it will affect the turnovers of wagons and locomotives and furthermore it can cause that train will stop there longer, because it didn´t come on time and actually other operator can manipulate in terminal [14-16]

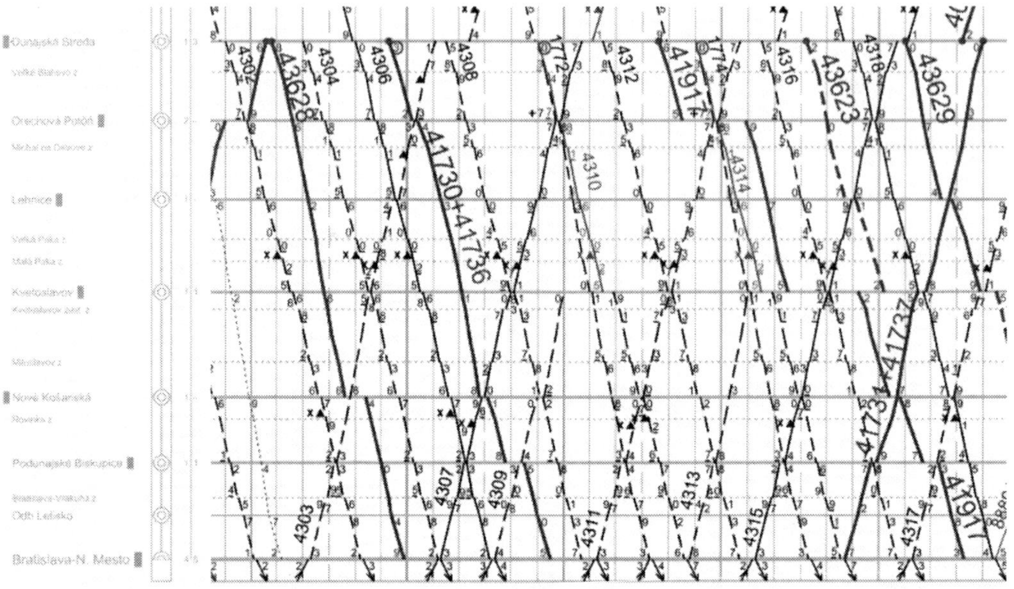

Fig. 4. Part of the train traffic diagram with passenger and freight trains between Bratislava – Dunajská Streda. Source [2]

4.3. Effects caused by third side

Effects caused by third side are connected mainly with things which operators can´t influence, for example, bad weather, traffic accident on the road when the trucks drive to terminal with containers or custom officer stop the trains and check the papers [17, 18].

5 Conclusion

The basic problem of this line is that it is singletrack. There isn´t enough capacity so there will be more problems in the future how to provide personal and freight trains, because year to year the amount of passengers is higher and the operator Regiojet wants to add other trains. Also Metrans is forced to make a shorter trainsets because of lenght of rails in stations what causes that they break a trainset in two pieces and drive more than it´s necessary and it causes delay.
Another option is to build second track but this responsibility is for manager of railway infrastructure – ŽSR. This aim should be a part of the state policy. Nowadays the railway infrastructure manager also Ministry of Transport don´t have an action plan how solve this situation.

This paper is supported by the VEGA Project 1/0095/16 ´Assessment of the quality of connections on the transport network as a tool to enhance the competitiveness of public passenger transport system, that is solved at Faculty of Operations and Economics of Transport and Communication, University of Žilina.

This paper is supported by the research project "From horse-drawn railway to intermodal transport" within Visegrad Fund.

References

1. G. Fedorko, V. Molnár, J. Strohmandl, M. Vasiľ, *Transport Means 2015 – 19th international scientific conference*, 466-469 (Kaunas University of Technology, Lithuania, 2015)
2. H. Neradilova, G. Fedorko, Open Engineering **6**, 700-710 (2016), DOI:10.1515/eng-2016-0085
3. I. Kubaskova, M. Šulgan, Logistics for forwarders and road transport (in slovak), 229 p. (University of Zilina, Slovak Republic, 2013)
4. *Internal documents of Manager of Infrastructure – ŽSR* (2017)
5. J. Danis, A. Dolinayova, J. Camaj, PRTDSR, *ESM 2016 – the 2016 international conference on Engineering science and management*, 187-190 (Advances in engineering research **62**, Zhengzhou, Henan, China, 2016)
6. J. Gasparik, B. Abramovic, M. Halas, Promet-Traffic and Transportation, 283-290 (2015)
7. J. Gasparik, V. Zitricky, Transport **25**, 4, 387-393 (2010), DOI: 10.3846/transport.2010.48
8. L. Černá, A. Dolinayová, J. Daniš, *Transport means 2016 – 20th international scientific conference*, 414-419 (Juodkrante, Lithuania, 2016)
9. L. Černá, V. Zitrický, J.Daniš, Open Engineering **7**, 1, 6-13 (2017)
10. M. Dlugos, *Optimalizácia obehu vlakových súprav RegioJet, a.s. na trati Komárno – Dunajská Streda – Bratislava hl. st.* (Diploma thesis, University of Zilina, Slovak Republic, 2017)
11. M. Kendra, P. Šulko, M. Babin, *BulTrans-2013 – Anniversary scientific conference on aeroanutics, automative and railway engineering and technologies*, 95-97 (Sofia, Bulgaria, 2013)
12. V. Gaborova, L. Peceny, J. Gasparik, *ICTTE – 3rd International Conference on Traffic and Transport Engineering*, 462-466 (Belgrade, Serbia, 2016)
13. V. Zitricky, L. Černá, B. Abramovic, *12th International Scientific Conference of Young Scientists on Sustainable, Modern and Safe Transport*, 994-999 (Procedia Engineering, High Tatras, Slovakia, 2017)
14. O. Stopka, J. Ponický, M. Chovancová, V. Zitrický, Nase More **63**, 3, 200-203 (2016), DOI: 10.17818/NM/2016/SI23
15. J. Lalinská, J. Gašparík, D. Šipuš, Logi – Scientific Journal on Transport and Logistics **8**, 1, 74-81 (2017), DOI:10.1515/logi-2017-0009
16. Z. Zahumenska, J. Gasparik, *12th International scientific conference of young scientists on sustainable, modern and safe transport*, 976-981 (Procedia engineering, High Tatras, Slovakia, 2017)
17. V. Zitricky, P. Šulko, *LOGI 2014 – 15th international scientific conference*, 377-384 (Pardubice, Czech Republic, 2014)
18. K. Košťálová, Logi – Scientific Journal on Transport and Logistics **2**, 2, 38-47 (2011)

MATEC Web of Conferences **134**, 00012 (2017) DOI: 10.1051/matecconf/201713400012
LOGI 2017

Analysis of the Utilization of Machinery in the Production Process Using Computer Simulation

Gabriel Fedorko[1,*], *Stanislav* Honus[2], and *Soňa* Badiarová[1]

[1]Technical University of Košice, BERG Faculty, Logistics Department, Park Komenského 14, 040 01 Košice, Slovakia
[2]Technical University of Ostrava, Centre ENET, 17. listopadu 15, 708 33 Ostrava-Poruba, Czech Republic

Abstract. For the efficient operation of each production process, individual machines and equipment must be used in the maximal possible measure. For this reason, it is necessary to know their utilization and to take measures to ensure their effective use. For performing such an analysis and to design subsequent measures, the use of the computer simulation method is very effective, e.g. simulation in Tecnomatix Plant Simulation program.

1 Introduction

The production process has a major impact on the efficiency of each industrial enterprise. It affects its market position, its ability to compete, satisfying the requirements of individual customers, and, last but not least, influences the overall economic situation and logistics [1, 2]. For the proper functioning of the transformation process, it is necessary to make effective use of available production capacities, machinery, and equipment [3, 4]. This is a problem that many people still underestimate and do not pay adequate attention to it. A frequent argument why they neglect this fact is the difficulty of collecting and evaluating information. However, current production logistics recognizes several methods that can be efficiently used to solve the problem, such as the use of Petri Nets [5] and the method of computer simulation [6].

Precisely the method of computer simulation is a very effective means. More specifically described by Chramcov et al. [7]. In their contribution, the specific use of heuristic algorithms is discussed. Heuristic algorithms can be used, for example, for reducing economic costs [8]. This issue is more closely dealt with by Bucki et al. [9], which focuses on automated manufacturing systems in his research.

However, heuristic optimization can also be used in the creation of production plans as described [10]. This is specifically about the use of discreetly oriented simulation for the need to efficiently optimize the planning of the production process [11]. A similar issue was dealt with also Bensmaine et al. [12].

As follows from mentioned, the computer simulation method has a wide spectrum of applications in logistics. For this reason, it was also used in the analysis and subsequent

* Corresponding author: gabriel.fedorko@tuke.sk

DOI: 10.1051/matecconf/201713400012

adjustment of the production process of the selected part from the assembly of welding pliers.

2 Characteristics of the production process

The production process of the selected part from the welding pliers set, which is shown in Fig. 1, begins by aligning the clamping surfaces on a conventional milling machine, which takes about 2 minutes. At the same time, the program is created by the 3D model and drawing documentation that lasts 30 minutes. Meanwhile, the trained operator clamps the part into Machine 1. The generated program is inserted into the machine and the NC production is started.

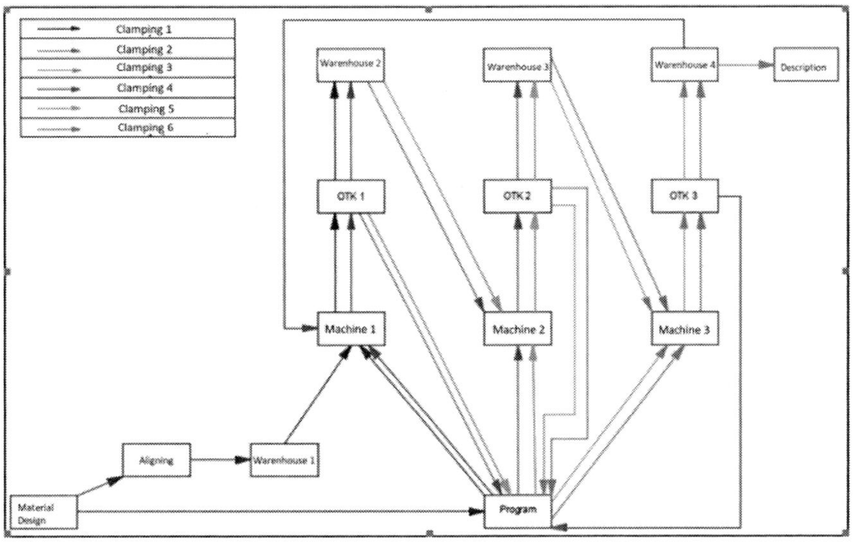

Fig. 1. Diagram of the production process. Source: authors

After the first clamping (Fig. 2), which lasts for 30 minutes, a part goes through the control (OTK). If the checked part is in good order, production continues and every fifth piece is checked. The parts are stored while the programmer generates the program for the second clamping to be run on Machine 2. In parallel with the creation of the program, the operator clamps the part and arranges the tools according to the instructions of the programmer. This operation ends after 25 minutes (Fig. 2).

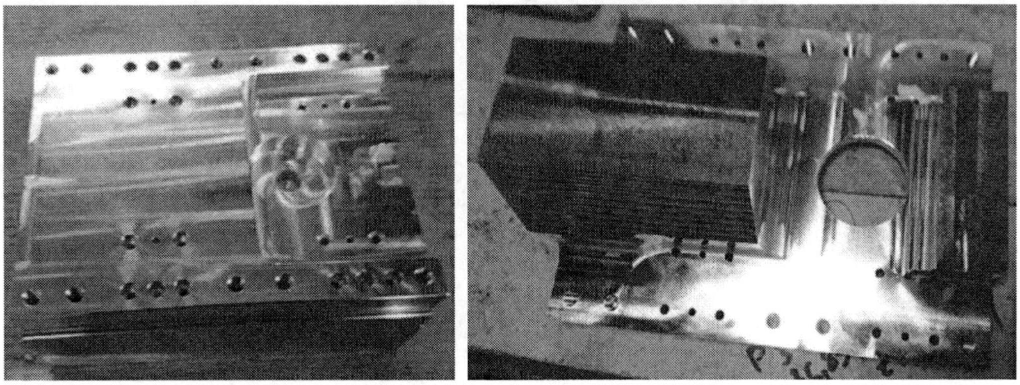

Fig. 2. A machined part after the first clamping and the second clamping. Source: authors

DOI: 10.1051/matecconf/201713400012

If a worker at control approves the quality of the part manufactured, production continues and checked is every tenth piece because the correctness of the manufactured part can be easily visually checked by the servicing worker in the machining center.

The parts are moved to the warehouse and the programmer generates the third clamping program (Fig. 3), which takes 45 minutes. The production in the third machine is running for 30 minutes. The difficulty of this process requires a check of every third part. After the third clamping is waiting for Machine 1, in which the fourth clamping will take place (Fig. 3). Production proceeds the same procedure till the sixth clamping (Fig. 4), working times are variable and are listed in Tab. 1.

Table 1. Working times for individual operations and clamping. Source: authors

Program	Programming [min]	Machining [min]	Control [min]
1. clamping	30	30	2
2. clamping	20	25	1
3. clamping	45	30	4
4. clamping	10	6	0,5
5. clamping	15	5	1
6. clamping	20	10	10

The first clamping (Fig. 2) serves as a priority for creating the surfaces on which the next clamping is performed. At the same time, tolerated holes are drilled in the component, which could be created even with the third clamping, but it would be technologically more demanding. Milling resulted in surface areas that serve to clamp the blank in the following operations. When the first piece passes through the check, a stop is installed in the machine, making it easier for the operator to replace the blanks and eliminating the need to focus each workpiece separately.

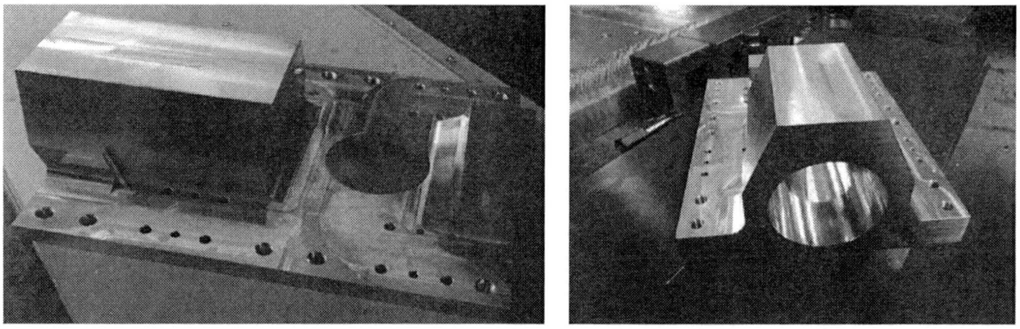

Fig. 3. A machined part after the third clamping and the fourth clamping. Source: authors

There is a part after the finishing operations in Fig. 3. Its dimensions are identical to the drawing documentation, which is achieved by the smaller forces needed to hold the material in the machine. After this machining, increased attention is paid to the control as the resulting elements are geometrically tolerated. The check is carried out on every third piece, since the elements that are created fit exactly into the welding pliers' assembly.

After the third clamping, the blanks are stored on the pallets until the Machine 1 is released from working in the first position. After it has finished, it is necessary to clean the machine, arrange the tools and set the clamping device that is specific for this kind of machining. The operation performed here consists of milling an exact hole (Fig. 3). Zero point focusing as well as the measurement to control adherence to tolerated dimensions is automatic and performed by the machining center with a workpiece probe. The check is simplified and therefore only every fifth piece is checked outside the machine.

After the machine 2 is released, the machining center is ready for fifth clamping. This clamping consists of a T-shaped groove milling (Fig. 4) into which the metal segment of the assembly fits. The focus is also performed automatically. Check measurements are performed by the operator outside the machine.

In the final phase of the production process (Fig. 4), the welding joint is formed on the part together with screw holes ensuring its functionality. The parts in this form head to the output control, where the trained worker checks them using the 3D measuring device. Products that have passed the final control are marked with a labeling device. The description consists of the assembly number, position number, and date of manufacture. The expedition worker is responsible for correctly packing products in boxes, which includes preventing damage during shipment to the customer and marking the shipment for easier identification of packaged products.

Fig. 4. A machined part after the fifth clamping and the sixth clamping. Source: authors

3 Simulation model

The entire production process as described has already been realized inefficiently from the point of view of utilizing the capacities of individual machinery. However, there was no information available that would accurately quantify the percentage utilization of available machinery capacities. For this reason, it was necessary to analyze the whole process in detail, using the method of computer simulation.

The outputs of the individual simulation experiments are the time values of the whole process, the percentage expression of the machine utilization, their downtimes and the storage space utilization (Fig. 5). After finishing the simulation of the existing production process, the total production time (Fig. 6) of 500 pieces with 3 machines was determined.

DOI: 10.1051/matecconf/201713400012

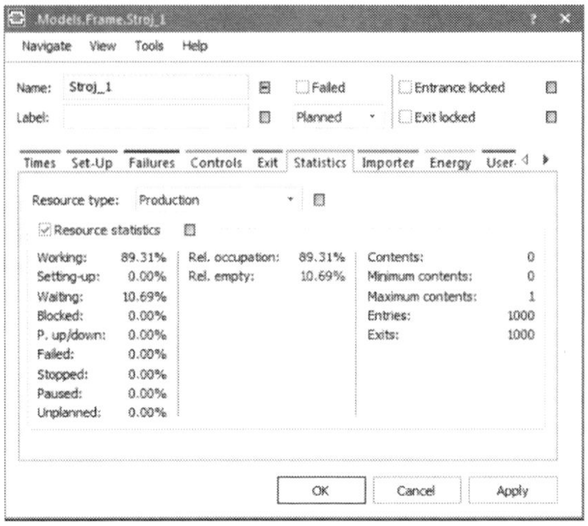

Fig. 5. Output values of the Machine 1 after the production process simulation. Source: authors

Simulation time: 13:23:54:00.0000

Cumulated Statistics of the Parts which the Drain Deleted

Object	Name	Mean Life Time	Throughput	TPH	Production	Transport	Storage	Value added	Portion
Popisovanie	Vstup	12:01:39:34.5600	500	1	0.64%	0.00%	99.36%	0.64%	

Fig. 6. Output values in Tecnomatix program after the production process simulation.

Percentage utilization of machines and their downtimes are shown in Tab. 2. As obvious from results, the production process used in the company for the work order processing is not ideal. Machine 2, for example, shows downtime of 25.57 % of the total working time. The task was, therefore, to try to optimize the process to achieve higher efficiency of work equipment.

Table 2. Results of original production process simulation. Source: authors

Machine	Working time [%]	Downtimes [%]
Machine 1	89.31	10.69
Machine 2	74.43	25.57
Machine 3	99.24	0.76.

4 Proposal for efficient machines utilization

On the basis of the information obtained from the simulation experiments and the material flow analysis, several variants were proposed to redistribute the activities of the machines so that the most time-consuming clamping in the production process were processed by several machines at once (Tab. 3). This has been supposed to shorten the total time needed to produce the chosen component, increase machine utilization and eliminate downtime that is undesirable in existing production.

Based on suggestions for possible solutions, the first clamping was divided into the devices - Machine_1 and Machine_2. This operation ended at both machining centers at the same time. Meanwhile, Machine_3 would work on the second clamping, and also Machine_1 after the work there was finished.

Machine_2 then processes the blank with a third clamping. By analogy, after the processes that run on the remaining two machines are completed, they are connected to the second machining center, so all available devices work on the third clamping together until they complete it.

The fourth clamping was again assigned to machines - Machine_1 and Machine_2. Machine_3 then performs the fifth component manufacturing process in sequence.
The final operation is run on Machines 1 and 2 again.

Table 3. Proposal for reallocation of processes between individual machines. Source: authors

Clamping	1.	2.	3.	4.	5.	6.
Machines	1	3	2	1	3	1
	2	1	3	2		2
			1			

At the end of this experiment, the overall simulation time was found to be shorter compared to the production process currently used by the company. By using an optimized production process, 500 units on 3 machines are manufactured in 12 days 15 hours and 41 minutes (Fig. 7). The difference between these two processes is 1 day 8 hours and 13 minutes.

Simulation time: 12:15:41:00.0000

Cumulated Statistics of the Parts which the Drain Deleted

Object	Name	Mean Life Time	Throughput	TPH	Production	Transport	Storage	Value added	Portion
Drain	Vstup	11:14:18:17.1600	500	2	0.66%	0.00%	99.34%	0.66%	

Fig. 7. Output values in Tecnomatix program after work process optimization. Source: authors

Percentage of machines downtime has decreased significantly in this production process and the production on the machines has increased. The production on all machines and their downtime before and after optimization is listed in Tab. 4.

Table 4. Results of a modified production process experiment. Source: authors

Machine	Productivity [%] after/before the change	Downtimes [%] after/before optimization
Machine 1	99.69 / 89.31	0.31 / 10.69
Machine 2	99.83 / 74.43	0.17 / 25.57
Machine 3	91.35 / 99.24	8.65 / 0.76

5 Conclusion

The problem of determining the use of machinery in the production process with a focus on optimizing and subsequently increasing machine productivity is relatively simple but very effective when using computer simulation [13]. Several available simulation software can be used to verify proposals. In the case described in the paper, the simulation program Tecnomatix Plant Simulation was used, in which the basic model of the production process was created.

The model corresponds to the actual process used by the company to produce the specified component. This process at first glance showed signs of inefficiency, which was confirmed by the results of the simulation experiments.

The adjustments that could be made in the production were limited by the company's conditions. Only three machining centers could be used for the production, and only improvements were made that were attainable with the equipment that the company has. Therefore, the adjustments concerned mainly logistics or more efficient division of production among machines.

Prior to optimization, the 90 % productivity limit exceeded only one machine. Machining center number 2 showed a high value of downtime. Of the total processing time, it was more than 25%. The total production time of the 500 products was 13 days, 23 hours and 54 minutes.

The next step was to create an optimization proposal that was applied to the previous model. The result has increased the productivity of all machines beyond 90% and reduced downtime below 10%. The total production time was significantly shortened, which can be considered a great result.

However, this procedure should also be applied to the production of the remaining parts of the assembly ordered by the customer. This application of computer simulation and subsequent optimization is evidence of the importance of paying attention to factors influencing the transformation of inputs to outputs of the production process. In a relatively simple way, by using software simulation, it is possible to reduce costs, increase production and improve the ability of an enterprise to compete in a particular market.

Regarding Tecnomatix program, program basic functionality is quite simple at first glance, but possibilities of this program are almost unlimited thanks to the SimTalk language [14]. This is also related to the area for which this software is intended [15]. Using this program, it was possible to model the production process and immediately perform the simulation. But fine-tuning the correct behavior of the entire process was achieved by logic programming that was relatively complicated and time-consuming.

This work is a part of these projects VEGA 1/0063/16, KEGA 018TUKE-4/2016.

This paper is supported by the research project “From horse-drawn railway to intermodal transport” within Visegrad Fund.

References

1. P. Ficzere, Z. Ultmann, Á. Török, Transport **29**, 278-284 (2014)
2. E. Tomková, N. Husáková, J. Strohmandl, *Transport Means - Proceedings of the 19th International Conference on Transport Means*, 694-697 (Kaunas University of Technology, Lithuania, 2015)
3. J. Fabianova, P. Kacmary, V. Molnar, P. Michalik, Open Engineering **6**, 270-279 (2016)

4. R. Kampf, L. Lizbetinova, K. Tislerova, Open Engineering **7**, 26-30 (2017)
5. E.J. Macías, M.P. de la Parte, Simulation **80**, 143-152 (2004)
6. R. Jašek, M. Sedláček, B. Chramcov, J. Dvořák, *ICNAAM 2015 - International Conference of Numerical Analysis and Applied Mathematics*, (S. T.E., S. T.E., T. C., and S. T.E., American Institute of Physics Inc., Tomas Bata University in Zlin, 1738, 2016)
7. B. Chramcov, R. Bucki, S. Marusza, Adv. Intell. Syst. Comput. **210**, 423-434 (2013)
8. V. Molnar, K. Pacutova, *ICTTE 2016 - Proceedings of the third international conference on traffic and transport engineering*, 570-575 (Belgrade, Serbia, 2016)
9. R. Bucki, B. Chramcov, P. Suchánek, J. Univers. Comput. Sci. **21**, 503-525 (2015)
10. S. Horn, G. Weigert, E. Beier, *ISSE 2006 - 29th International Spring Seminar on Electronics Technology: Nano Technologies for Electronics Packaging*, 422-427 (Dresden University of Technology, Department of Electrical Engineering, Electronics Packaging Laboratory, Dresden, Germany, 2006)
11. N. Brnjac, B. Abramovic, M. Maslaric, PROMET-TRAFFIC Transp. **22**, 303-307 (2010)
12. A. Bensmaine, M. Dahane, L. Benyoucef, Int. J. Prod. Res. **52**, 3583-3594 (2014)
13. I. Kubasakova, R. Kampf, O. Stopka, Communications **16**, 9-13 (2014)
14. H. Neradilova, V. Laskovsky, *16th International Multidisciplinary Scientific Conference SGEM2016, Book 1*, **2**, 929-934 (2016)
15. J. Strohmandl, UPB Sci. Bull. Ser. D Mech. Eng. **76**, 223-230 (2014)

MATEC Web of Conferences **134**, 00013 (2017) DOI: 10.1051/matecconf/201713400013
LOGI 2017

Comparison of the Traditional and Autonomous AGV Systems

Gabriel Fedorko[1,*] *Stanislav* Honus[2], and *Roland* Salai[1]

[1]Technical University of Košice, BERG Faculty, Logistics Department, Park Komenského 14, 040 01 Košice, Slovakia
[2]Technical University of Ostrava, Centre ENET, 17. listopadu 15, 708 33 Ostrava-Poruba, Czech Republic

Abstract. Automation is now an indispensable part of our world, which helps us produce more efficient, precisely and with lower costs. One of the major automation inventions were AGV vehicles, an automated guided vehicles, which meant a breakthrough in the internal logistics of many businesses. Today, these robots are nothing special, but their advanced functionality moves them to a higher level than before. This article describes one of their functionalities, which is an efficient and safe navigation while preserving the basic idea of their autonomy.

1 Introduction

The beginnings of AGV systems, in other words Automated Guided Vehicles, were in the 50 years of the 20th century [1]. AGV is automatically guided and drive vehicle or equipment which is replacing the transport and part of the handling of the material, part of the internal logistics activities [2].

Expansion of AGV's has continuously covered almost all sectors of business, production and industry. AGV`s have eliminated errors caused by human activity, because they work at a predefined standard, more precisely. Basic advantage is still cost reduction connected with personal cost [3, 4]. The AGVs replaced initially simple logistics activities, such a transfer of material between the production halls, later between the lines, later were these vehicles able even replace the material handling by moving the material directly to the place of its consumption [5].

Basis of each AGV is control unit, security and guiding equipment, power supply and power unit [6]. The shape and accessories of each AGV are constructed according to the specialized activity for which the model is designed [7, 8]. The simplest solutions replace function of classic logistics trains [9]. More specified AGVs, completely replace material handling from its autonomous picking and loading, through transportation and final delivery at place of consumption.

By increasing number of AGVs grow the necessity for their coordination, control and dispatch management. The usage of autonomous vehicles is moving out of the basic concept of autonomy.

* Corresponding author: gabriel.fedorko@tuke.sk

2 The principle of traditional AGV system

The basic principle of the operation of traditional AGV system is in a pre-defined route placed on the floor or located in the floor, which is followed by reading device, which is a part of each traditional AGV.

In addition, other support devices are moving toward the basic functionality "tracking pre-defined route "to a higher level. We speak about safety bumpers, optical scanning devices, obstacle sensors, audio and visual warning devices, Wi-Fi connection, GPS modules, remote controls, communication via smartphone, camera systems and others.

The first AGV routes operated on an induction principle, which was created by the induced magnetic field in the floor. Later versions came to the color lines principle, so the AGVs tracked the color footprint on the floor, as well as the version where the route is made by magnetic tape on floor.

Newer versions use GPS navigation modules to navigate and are able to communicate with other ones and keep simple traffic rules. A very popular and affordable version is still a version using a magnetic tape.

2.1 Magnetic guidance with obstacle sensor

We consider this solution as a traditional and most advanced because of the easy application and use. The principle is as was mentioned above in the physical placement of the magnetic tape on the floor that the AGV reader will read. On the sides of the guiding magnetic tape are placed marks "tags" with opposite magnetism such as the basic magnetic tape route. These tags are also read by the AGV reading unit and processed by the control unit (Fig. 1).

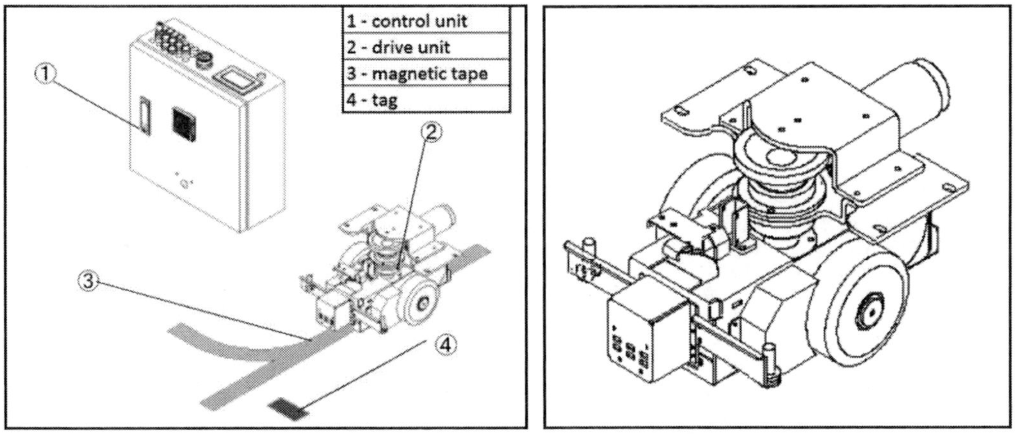

Fig. 1. Basic elements of AGV systems and AGV drive unit with magnetic tape and tags reader. Source: [10]

AGV works in a written program where each tag assigns parameters to which the AGV is to be set. These are parameters such as speed, turn right – left, warning signals, usage of the handling devices, obstacle sensor field settings, and others. Obstacle sensors play an important role in the proper autonomy of these AGVs.

2.1.1 PBS Obstacle sensor

Chosen model for comparison is the HOKUYOU obstacle sensor: PBS - 03JN. Sensor scans the space in the horizontal plane of its location, with 180 degrees radius and max scanned distance 20 m. Mentioned PSB sensor communicates via PBS software configurator with a PC where it is possible to see an online scan of the space around area with free space and obstacles [11].

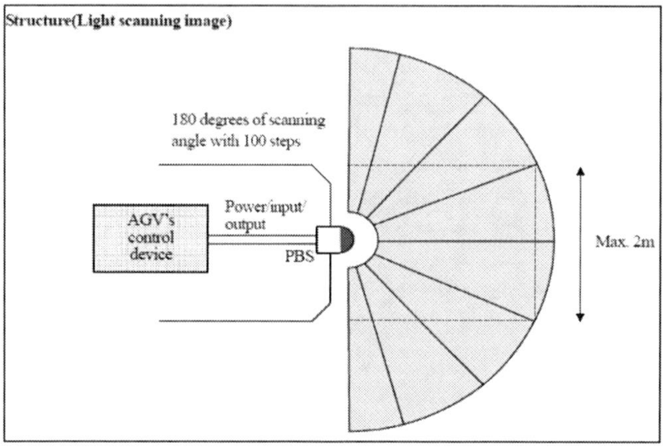

Fig. 2. PBS obstacle sensor detection guarantee range [12] and PBS obstacle sensor. Source: [13]

Using the PBS configurator, it is possible to create several areas of different shapes. These areas can be divided into fields where the pre-defined rules, how will AGV react when there will be presence of obstacle in the area and field. Colors fields are represented in this case with the yellow field farthest from the sensor, where visual and audio signaling occurs. The other is the orange field, which forces the AGV to slow down. There are also sound and visual alarms. The last field is a red field where the obstacle presence will cause AGV immediate stop.

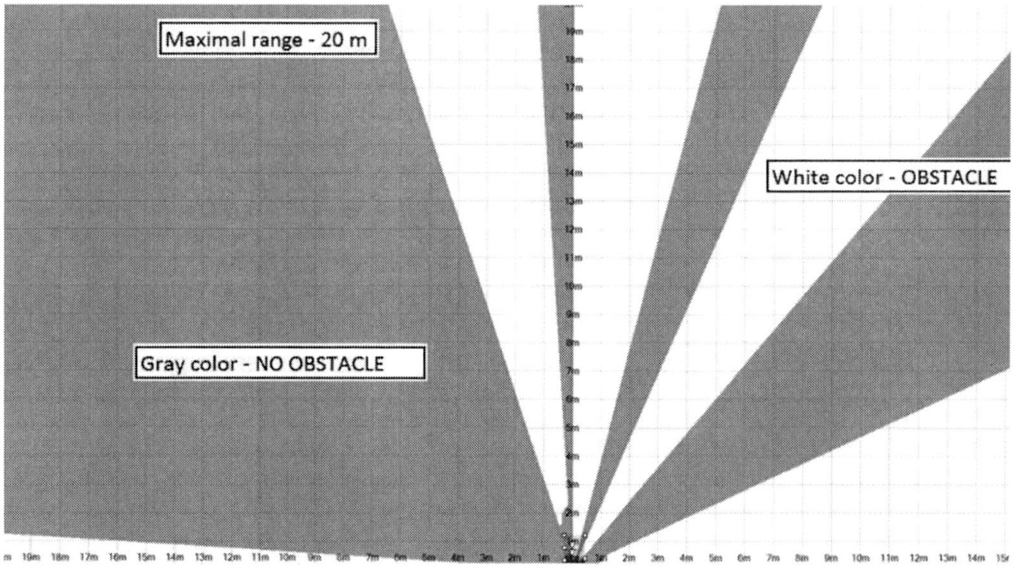

Fig. 3. PBS obstacle sensor maximal range and color indication – online data snapshot. Source: authors

AGV activity is restarted after removing the obstacle out from the red field of the area. This also the main disadvantage of these traditional AGV systems where the vehicle is strictly connected to fixed routes, and don't move toward until the obstacle is removed. If there is a higher number of vehicles AGV and their frequency on the same route, there may be a delay of several delivery units.

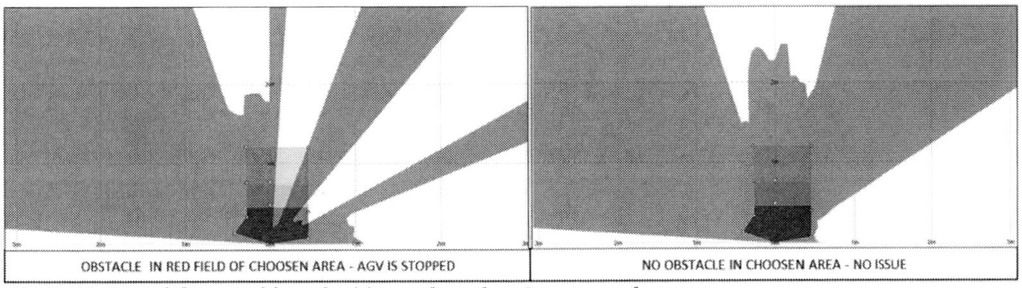

Fig. 4. PBS model area with and without obstacles. Source: authors

Another negative factor of magnetic tape is maintenance and replacement in case of damage, complicated replacement in case of new layout, and especially the risk of mismanagement in the case of a missing tag. In such cases when the tag is not present, AGV works as if it were in another part of the predefined route, and a collision or injury could easily happen.

3 A modern solution through the Virtual Map

The newest guidance systems are laser sensors, reading the tags placed on the pillars and walls in the production, from which AGV may identify its location, and also GPS guided vehicles. Compared and tested solution, which was built in cooperation with the Technical University in Košice, (project of Mr. Bačík).

The AGV was guided by a virtual map where the scanner scanned the space around and created a virtual map of the environment. In the map, with the simple click, we create the route on which we want AGV to move. Thus, the route realization time takes only a couple of minutes compared to the traditional magnetic tape, which in the first phase requires hours of work and settings. In the map are defined points that represent physical tags of the traditional system, this action in virtual map is also very fast. Points replace their function with magnetic tape.

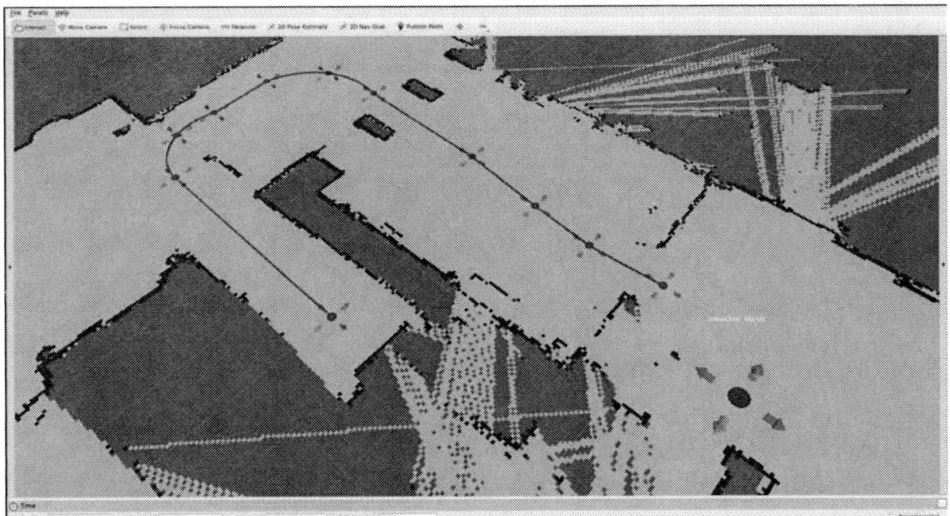

Fig. 5. Snapshot of virtual map with defined points. Source: authors

In the environment where the vehicle movement is allowed, it is necessary to physically mark the starting point, or put the starting tags. By them can AGV vehicle identify its position in map like starting line and be prepared for round. During the movement, control unit via scanner compares the actual spare around with the virtual map, i.e., the movement is not so precise and fixed, with accuracy few centimeters as compared traditional system.

Some of the advantages are almost zero costs when new route is going to be created or updated after plant layout change, and no maintenance costs compare magnetic solution. Solution creates a virtual environment where it is possible to maneuver, and don't be fixed to route. It means that in the virtual map can be defined maximum maneuvering space in which the AGV can move in the moment when an obstacle is in the area of predefined route. AGV control unit create an alternative route using maximum manoeuvring space and if possible vehicle will go around the obstacle, and returns to the originally defined route, where it continues the assigned task.

Table 1. Basic factors comparison of traditional and autonomous systems. Source: [14]

COMPARISON	TRADITIONAL VEHICLE AGV	AUTONOMOUS VEHICLE AGV
MANEUVERING	movement acc. to guiding elements , tag, color line, magnetic tape etc.	free movement
INTEGRATION	expensive and lengthy	easy and simple
OBSTACLE DETECTION	stop and waiting	move around

4 Conclusion

AGV vehicles are part of enterprise logistics for more than 60 years and still expanding and developing very quickly. However, the the basic idea for which they were created as well as other robots is to simplify the work of man. Knowledge and experience gained from the development of autonomous AGVs also play an important role in the development of autonomous cars that will be part of our life in next year as AGVs are today.

This work is a part of these projects VEGA 1/0063/16, KEGA 018TUKE-4/2016.

This paper is supported by the research project "From horse-drawn railway to intermodal transport" within Visegrad Fund.

References

1. G. Ullrich, *Automated Guided Vehicle Systems* (Springer Berlin Heidelberg, Berlin, Heidelberg, Germany, 2015)
2. H. Neradilova, G. Fedorko, *12th international scientific conference of young scientists on sustainable, modern and safe transport*, 638-643 (Elsevier science BV, Amsterdam **192**, Netherland, 2017)
3. V. Molnar, K. Pacutova, *ICTTE 2016 - Proceedings of the third international conference on traffic and transport engineering*, 570-575 (Belgrade, Serbia, 2016)

4. N. Brnjac, B. Abramovic, M. Maslaric, PROMET-TRAFFIC Transportation **22**, 303-307 (2010)
5. H. Neradilova, G. Fedorko, Open Engineering **6**, 700-710 (2016)
6. R. Kampf, L. Lizbetinova, K. Tislerova, Open Engineering **7**, 26-30 (2017)
7. L. Bartuska, O. Stopka, Chovancova, J. Lizbetin, *Transport Means - Proceedings of the 19th International Conference on Transport Means*, 107-111 (Kaunas University of Technology, Lithuania, 2016)
8. R. Kampf, O. Stopka, I. Kubasakova, V. Zitricky, *WMCAUS 2016 – World Multidisciplinary Civil engineering-Architecture-Urban planning Symposium 2016*, 1538-1544 (Procedia Engineering **161**, Elsevier science BV, Amsterdam, Netherlands, 2016)
9. J. Fabianova, P. Kacmary, V. Molnar, P. Michalik, Open Engineering **6**, 270-279 (2016)
10. T.C. S.r.o.: *Automaticky Navádený Vozík, Návod na údržbu* (2017)
11. L. Aichikikai, *Techno Systems Co: CarryBee K Type 1 AGV Construction Manual*, ver 10.07.06, Available online: http://www.aiki-tcs.co.jp/carrybee?lang=en (2016)
12. L. Hokuyo, *Automatic CO: Measuring Distance Type, Obstacle Detection Sensor PBS-03JN, Instruction Manual* (2002)
13. *M. Systems: Manu Systems*, Available online: https://en.manu-systems.com/HOK-PBS-03JN.shtml (2017)
14. *Systechgroup: Systechgroup presentation on " Fórum praktickej logistiky, IPA, 20.9.2016 Gbeľany*, Available online: http://systechgroup.eu/sk/2016/07/29/testicek/ (2016)

Analysis of Work Accidents in Timber Transport in Slovakia

Miloš Gejdoš[1,*], and *Mária* Vlčková[2]

[1]Technical University in Zvolen, Faculty of Forestry, Department of Forest Harvesting, Logistics and Ameliorations, T. G. Masaryka 24, Zvolen 960 53, Slovakia

Abstract. From the aspect of safety at work, we consider the timber transport under the risky works. Data about accidents at work in forestry for the period 1996 – 2014 were summarized. The database contained more than 1500 work accidents in forestry. Laboratory Accident Coding Methodology was used, according to European Statistics on Accidents at Work (ESAW). In total, 105 registered work accidents in timber transport, during the investigated period, were recorded. This correspond to 7% share of registered work accidents. From the anthropological aspect, the legs and whole body were most frequently injured. The highest number of accidents on Tuesday were recorded, with the most risk month of February. The highest accident rate in 2002 were recorded. In general, forestry is classified as a risky activity. The timber harvesting and transport process belongs to the most risky works, from the aspect of the risk formation on the work accident, which are in forestry conducted. From all phases, which are in this process involved, the phase of timber transport is the least risky, in terms of the risk rate on accident formation as well as the absolute abundance of accidents.

1 Introduction

Timber transport was in the past one of the main production phases of the timber harvesting-production and transport process in the Slovak forestry sector. With the growth of timber sales directly on the site, respectively on the forest depot, gradually decrease the importance of the main wood storages, and thus also the share of timber transport realized directly by forestry enterprises. The cost on timber transport in Slovakian forestry was highest in the last 7 years in 2010 (over € 38 million) and the lowest in 2013 (almost € 25 million). In the point of view the transported timber volume since 1990, it was in 2005 carried out (almost 9.3 million m^3), which is also related to the treatment of calamities in the High Tatras. From 2012, the volume of transport is from 5 to about 5.5 million m^3 [1]. In 1990, the volume of wood harvesting was 4.7 mil. m^3. It is clear that, despite the almost doubled harvested volume in the last 30 years, the volume of timber transported by forestry enterprises has not risen significantly, which confirms the decrease in its share in the individual phases of timber harvesting.

* Corresponding author: gejdos@tuzvo.sk

From the aspect of work safety, we include the timber transport also as hazardous work, even though from the aspect of work accidents share in the production phases of timber harvesting a transport process, here is the lowest risk occurrence of work accident [2]. The most risky phase in this process is the timber harvesting phase. The timber transport is most important phase for the costs calculation and optimization. The risk of work accidents in this phase can partly influence the optimization processes [12, 13].

In the presented paper, we analyze the risk of work accident occurrence in the timber transport phase of State forestry entities and self-employed persons. At the same time, we analyze which part of the body from the anthropological aspect is the most vulnerable to potential occupational accidents in this type of transport.

2 Methodology

Data about work accidents in timber transport from records from the National Labor Inspectorate, from the General Directorate of Lesy SR, state enterprise in Banská Bystrica were drawn. The existing database of work accidents has been supplemented by accidents involving the self-employed persons, as well as accidents in the Military forests and properties, state enterprise. Accident questionnaires were sent to several forestry enterprises (to obtain the necessary information on the circumstances of the accident). From the 30 sent questionnaires, half returned.

In the creation of a database of bad work accidents registered in forestry entities, the methodology for the coding of work accidents according to European Statistics on Accidents at Work (ESAW) [3] was used. The consequences of an injury characterize the type of injury and the injured part of the body. Several parts of the body were injured only with the most serious injuries. In order to determine the source and cause of work accident, the Slovakian national classification was used which is governed by a national Decree [4].

The obtained data include a period of 18 years (from 1996 to 2014). The total database of identified work accidents over the reference period contained approximately 1500 work accidents in forestry.

3 Results

In total, 105 registered work accidents in timber transporting were recorded during the period. What represent approximately 7% of registered accidents in forestry activities. Most accidents were recorded in 2003 (totally 17). Any work accident was fatal. In terms of the time of the accident, the most hazardous time is between 10 am - 11 am (17 accidents) and midday between 1 pm - 4 pm (16 accidents). Figure 1 shows the number of accidents by individual sources that were caused according to Decree 500/2006 as amended. According to this classification, the very most important source of accidents was the single transport mean (38 accidents). In terms of the cause of the accident, the ESAW methodology [3] classifies a total 14 causes of an accident:

1. Incorrect or unfavorable source of injury (not workplace).
2. Missing or insufficient protective equipment and security.
3. Unprotected, insufficient or unsuitable personal protective equipment.
4. Unfavorable state or misconception of the workplace or communication (even if the workplace is the source of the injury).
5. Deficiency in illumination and visibility, adversive effects of noise, shocks and harmful air on the workplace or on communication.
6. Incorrect organization of work.

7. Guiltlessness with the conditions of safe work and lack of necessary qualification (e.g. theoretical knowledge, skill, training, adaptation).
8. Using of unsafe practices or methods of work, including unlicensed, anti-burglary, prohibition and instruction procedures, stay in a restricted area.
9. Removal or non-use of prescribed safety devices and protective measures.
10. Do not use or misuse the prescribed and assigned personal protective equipment.
11. Endangered by others, e.g. Distraction at work, jitters, quarrels, other wrong and dangerous actions.
12. Deficiencies in personal assumptions for proper work performance (e.g. lacking physical assumptions, sensory shortcomings, unfavorable personal qualities and immediate psychophysiological states).
13. Endangered by animals and natural elements
14. Uncertain causes

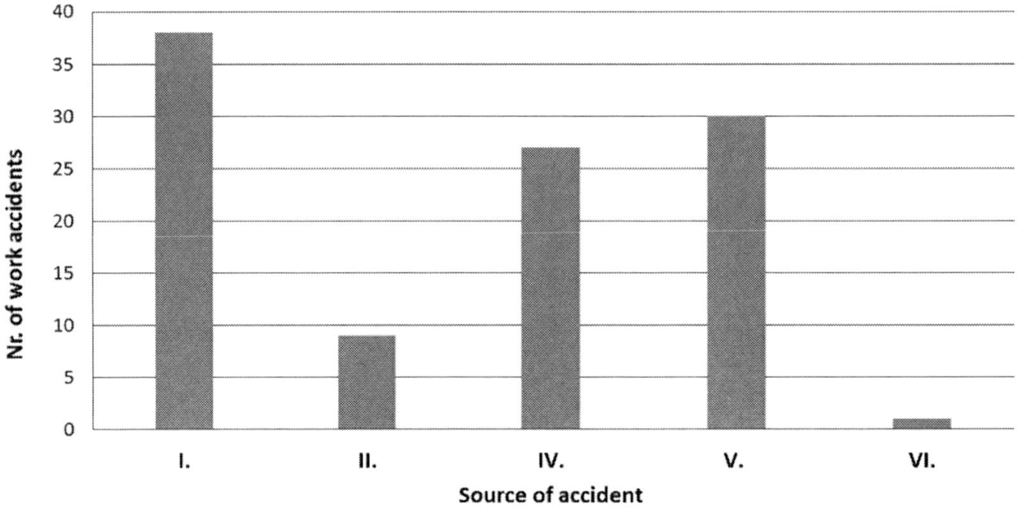

Fig. 1. Number of work accidents according to accident. Source: [4]

The most common causes of work accidents were the causes of groups 11-14 (72 accidents). Causes of groups 7 to 10 caused 26 accidents, and causes of groups 1-6 caused 7 accidents. Nearly half of all accidents have become workers aged 46 to 55 (48% of accidents). The average age of the affected worker was 44 years.

Figure 2 shows the accident classification according to the injured part of the body. The report shows that the most risky part of the body in the phase of hauling and timber transport are legs (35 accidents). From the anthropometric point of view, the legs are one of the parts of the body that the biggest increase in the past 20 years were recorded [5]. The objectively highest level of foot and hand injuries is understandable in terms of the location of the controls and the work performed in the timber transport.

The most common type of injury was open wounds (38 injuries) and fractures (33 injuries). In eight cases, the brain was shaken, in 5 cases multiple injuries, in 15 cases of foot wrinkling. Three injuries to the trunk did not indicate the type of injury. On average, the most hazardous days for the accident were Monday and Tuesday (Figure 3). In months, the most risky month was February with 16 accidents, then the summer months of June and August with 12 accidents. It is especially interesting from the aspect of the fact that major part of the principal felling (the most volume trees) and their transport is realized mainly in the winter months. The average length of practice in accidents was 14.8 years, so it can be said, that it was not workers who would not have enough work experience in the job.

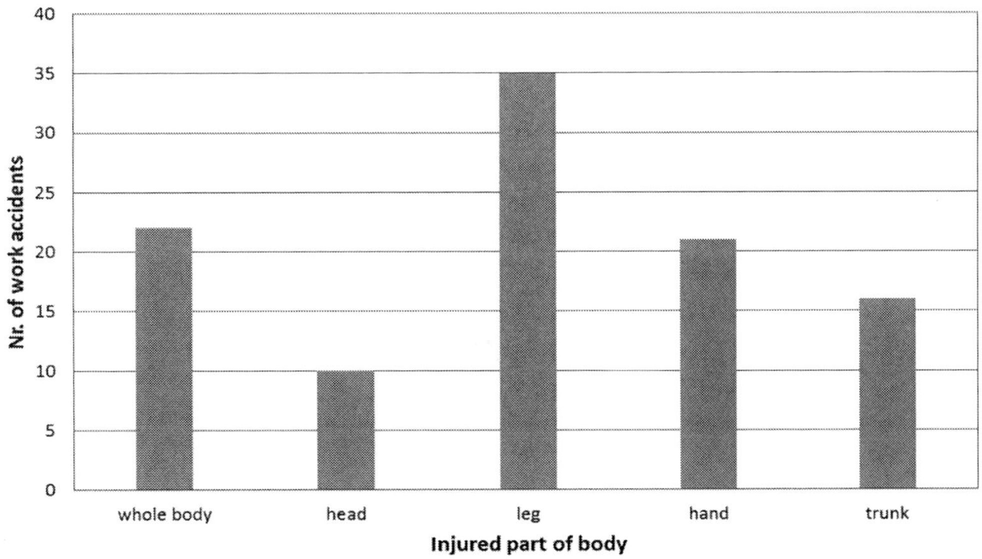

Fig. 2. Number of work accidents according to injured part of body in timber transport. Source: authors

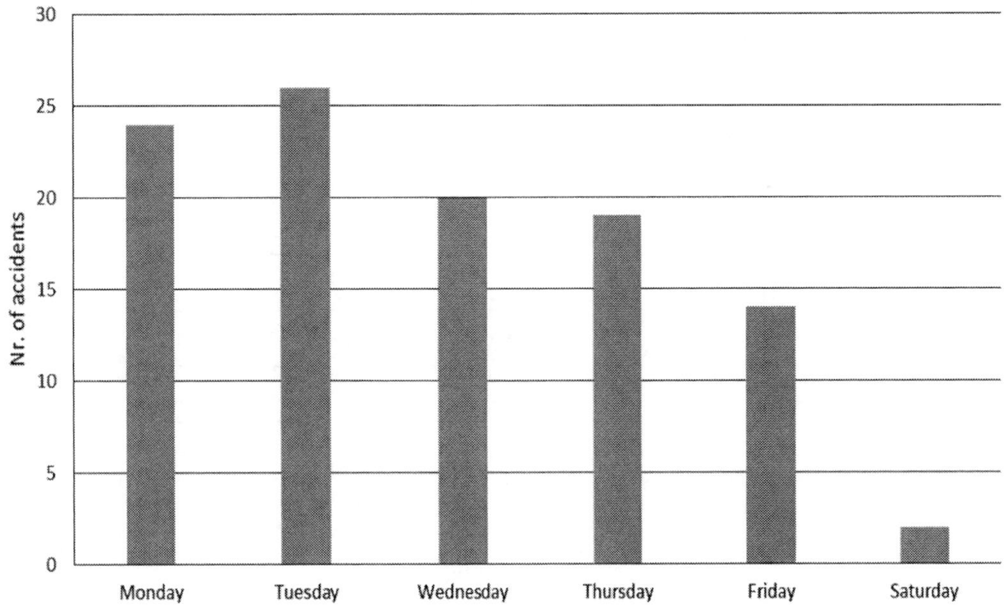

Fig. 3. Number of work accidents according to day in week. Source: authors

The rates of accidents are estimated on the number of accidents (d) divided with the volume of transported timber in mil. m^3 (N) [6].

$$\text{Rate} = (\text{Nr. of accidents}) \,/\, (\text{transported timber in mil. } m^3) \qquad (1)$$

In assessment of accidents rate, it was confirmed, that the timber transport is not a major risk phase in the harvesting and transportation process. In 2000, the level of work accidents in timber transport was 2.85 % and in 2012 at 0.8 % (Table 1). The highest accident rate was in 2002. Almost exclusively (up to one case), all accidents became male and only one reported accident came from the self-employed person.

For a small number of identified cases in timber transport, would be methodologically known statistical analyzes (regression analysis, contingency tables, classification trees, Mann-Kendall test) have not provided any relevant results. The results of these analyzes to the risks of the whole harvesting and timber transport process are available in [2].

Table 1. Rate of accidents in the period 2000 – 2012. Source: authors

Year	Transported timber [in mil. m^3]	Rate of accidents
2000	3512	2,85
2001	3763	3,19
2002	3713	4,04
2003	4750	3,58
2004	5041	1,39
2005	9287	1,18
2006	6882	1,02
2007	6744	0,15
2008	7160	0,28
2009	9006	0,33
2010	8294	0,24
2011	5965	0,84
2012	5028	0,80
2013	5103	0,00
2014	5352	0,37

4 Conclusion

The philosophy of sustainable development is basically oriented to improve the quality of people´s life, especially the working part. Work safety and health in a comprehensive understanding integrates technical safety, work environment and psychosocial factors and creates and influences the overall quality of life for the individual and society. The number of work accidents worldwide is steadily increasing. According to International Labor Organization data, more than two million people die annually from the consequences of occupational accidents and diseases. Each year, an average of 270 million occupational accidents will occur and more than 160 million people occupational illnesses gets sick to some occupational disease [7].

In general, forestry is classified as a risky activity. The harvesting and timber transport process belongs to the most risky works in terms of the risk of at work accident creation in the forestry industry. From all phases involved in this process, is the phase of timber transport, the least risky, in terms of both the risk of accident and the absolute abundance of accidents. Timber transport is largely made by trucks, which are optionally equipped with a hydraulic crane. However, it still belongs to the risky works. The current system of registration and evidence of work accidents, especially among service providers, self-employed persons (on this basis provides the main activities of about 90% in forestry subjects) focuses on severe work accidents with serious health consequences and serious occupational accidents resulting to death. For other work accidents is only a registration requirement.

The proportion of accidents in the timber transport phase was around 7%. In another work [8] realized in the state enterprise Lesy SR in 1984-1993, the share of accidents in the following phases was as follows: wood harvesting phase - 32%, wood skidding - 29%, works on forest depot - 9%, timber transport - 8% , work on main wood depots - 6%, repair and maintenance - 15% and control activity - 1%. Analyzes confirm that historically the proportion of accidents in the timber transport phase does not change fundamentally.

The most injured part of the body is the leg (33% of all accidents) and then the whole body (21% of all accidents). The wounds and fractures were the most common type of injury, which is related to the transport of heavy loads and falls from inattention.

The very dangerous factors of the working environment and the working process of private forest entrepreneurs are unfavorable microclimates, increased working load, noise and bad lighting [9, 10, 11]. Inappropriately high work length during the day, insufficient work breaks, inappropriate conditions for a balanced diet coupled with harmful manufacturing factors create fatigue and a considerable vulnerability to worker health damage. Employers and private entrepreneurs, who do not employ anyone, should therefore be in their own interest to ensure compliance with the safety and safety requirements in the forestry business [2].

The research described in this paper was financed jointly by the: APVV-16-0297 Updating of anthropometric database of Slovak population.

This paper is supported by the research project "From horse-drawn railway to intermodal transport" within Visegrad Fund.

References

1. M. Moravčík et al., *Ministry of Agriculture and Rural development of the Slovak Republic* (2016)
2. M. Vlčková, *Analýza vývoja pracovných úrazov a chorôb z povolania vo vybraných lesníckych subjektoch SR [Development analysis of work accidents and occupational diseases in selected Slovakian forestry subjects]*, 149 p. (Dissertation thesis, Technical University in Zvolen, Slovak Republic, 2014)
3. *European Statistics on Accidents at Work (ESAW)* (European Union, ISBN 978-92-79-28419-9, 2013)
4. *Vyhláška Ministerstva práce, sociálnych vecí a rodiny SR č. 500/2006, ktorou sa ustanovuje vzor záznamu o registrovanom pracovnom úraze [Decree of the Ministry of Labor, Social Affairs and Family of the Slovak Republic No. 500/2006, which establishes a pattern of registration of registered work accident]* (2006)
5. M. Hitka, E. Mračková, R. Sedmák, Advanced Materials Research **1001**, 401-406 (2014)

6. H. Checkoway, N.E. Pearce, D.J. Crawford-Brown, *Research methods in occupational epidemiology* (New York: Oxford University Press., 1989)
7. J. Suchomel, K. Belanová, M. Vlčková, Ľ. Ivan, *Analýza pracovných úrazov v Lesoch SR š.p. [Analysis of work accidents in Forestry state enterprise of Slovak republic]* (Zvolen: Technická univerzita vo Zvolene, Slovak Republic, 2008)
8. J. Suchomel, *Analýza vybraných kritérií a metód pre modelovanie a optimalizáciu ťažbového procesu [Analysis of selected criteria and methods for harvesting process modelling and optimization]* 148 p. (Zvolen: Technická univerzita vo Zvolene, Slovak Republic, 1999)
9. M. Němec, M. Dado, Z. Danihelová, Akustika **21**, 1, 31-34 (2014)
10. T. Grladinović, L. Oblak, M. Hitka, Drvna industrija **58**, 3, 141-146 (2007)
11. M. Danilovič, S. Antonič, Z. Dordevič, P. Vojvodič, Forestry work-related injuries in forest estate „Sremska Mitrovica" in Serbia. Sumar. List. **140**, (2016)
12. M.F. Costa, N.C. Fiedler, G.R. Mauri, Scientia Forestalis **41**, 99, 299-305 (2013)
13. P. Lijewski, J. Merkisz, P. Fuc, A. Ziolkowski, L. Rymaniak, W. Kusiak, Fuel consumption and exhaust emissions in the process of mechanized timber extraction and transport. Eur. J. For. Res. **136**, 1, 153-160 (2017), DOI: 10.1007/s10342-016-1015-2

Selected Options for Modelling of Transport Processes Particularly in Relation to Intermodal Transport

Jozef Gnap[1], and *Jaroslava* Kubáňová[1,*]

[1]University of Žilina, Faculty of Operation and Economics of Transport and Communications, Department of Road and Urban Transport, Univerzitná 1, 010 26, Žilina, Slovak Republic

Abstract. The first specialized road-rail combined transport operators were founded by road hauliers, who experienced road capacity bottlenecks and collaborating railways in the late 1960s in Europe. Combined transport is thus the optimal embodiment of the European concept of co-modality, whereby the strengths of both participating modes of transport – road and rail – can best be exploited: the flexibility of road transport in pre- and post- haulage, while the low energy need, safety, speed and reliability of rail transport.

1 Introduction

Multimodal transport – is the movement of cargo from origin to destination by several modes of transport where each of these modes have a different transport provider or entity responsible, but under a single contract. A Single carrier contracted to fulfil a single journey [1].

Intermodal transport - is the movement of goods in one and the same loading unit or road vehicle, which uses successively two or more modes of transport without handling the goods themselves in changing modes. By extension, the term intermodality has been used to describe a system of transport whereby two or more modes of transport are used to transport the same loading unit or truck in an integrated manner, without loading or unloading, in a door to door transport chain [1].

Combined transport - is a form of intermodal transport, which is the movement of goods in one and the same loading unit or road vehicle, using successively two or more modes of transport without handling the goods themselves in changing modes. Combined transport is intermodal transport where the major part of the journey is by rail, inland waterways or sea, and any initial and/or final legs carried out by road are as short as possible [1].

The negative consequences of transport such as pollution, climate change, noise, congestion and accidents pose problems to the economy, health and well-being of European citizens. Freight transport continues to grow and road freight transport, in particular, is projected to increase by around 40% by 2030 and by little over 80% by 2050. The EU

* Corresponding author: jaroslava.kubanova@fpedas.uniza.sk

 DOI: 10.1051/matecconf/201713400015

transport policy aims therefore at reducing road transport towards less polluting and more energy efficient modes of transport [2].

Four types of actions support greater use of intermodal solutions.

1. The internalisation of external costs in all modes of transport, with a view to send appropriate pricing signals to users, operators and investors. The social and environmental costs of transport should be paid in line with the polluter pays principle.
2. More targeted investments into physical infrastructure, aimed at better interconnections between the single modal networks.
3. Better used of information (on traffic, capacities, availability of infrastructure, cargo and vehicle positioning).
4. Direct support for intermodal transport, as provided by the Combined Transport Directive (Council Directive 92/106/EEC), which aims to increase the competitiveness of the combined transport (defined as intermodal transport with a strictly limited road leg). The EU also provides financial support to multimodal/intermodal transport [3].

The following definitions are given by the United Nations Economic Commission for Europe: Combined transport is intermodal transport where the major part of the European journey is by rail, inland waterways or sea and any initial and/or final legs carried out by road are as short as possible.

2. Selected techniques of combined transport

2.1 Rail - Road

Lorry

Lorries make it possible to provide a door-to-door service, as they cover the short distances separating factories and terminals. They enable the major advantages of road haulage to be tapped, i.e. a network that reaches further and is denser.

Container

Containers lead to better logistical management of the areas used for loading and unloading goods, since their rigid structure enables up to six of them to be stacked in one pile. Container lengths have been standardised at between 20 and 53 feet.

Swap body

Swap bodies are standardised loading units equally suitable for carriage on road vehicles or railway wagons. As they can be used in a broad range of situations, are simple in design and inexpensive, and this form of conveyance has been highly successful and is currently one of the most widely used transport systems on the market.

Semi-trailer

While semi-trailers are more costly and heavier, their advantage is that they can be coupled directly to a tractor and do not require a road chassis, unlike containers and swap bodies.

Terminal

Terminals are interchange hubs between rail and road traffic. They are fitted with all the equipment required to handle and tranship loading units from one transport mode to the next in a rapid and efficient manner: gantries and mobile cranes, modern computer systems integrating tracks, storage areas, transhipment areas and connections to roads and motorways [4].

2.2 Benefits of combined transport

Combined Transport presents a range of benefits. It contributes towards a better quality of life and proposes a seamless transport solution in order to improve the productivity of the entire chain.

The main benefits of combined transport can be summarised as follows:

- Friendly towards the environment: on routes where volumes can be bundled and distance is appropriate, combined transport provides substantial energy gains and lower CO2 emissions.
- Friendly towards society: reduces accidents, road congestion and dependency on energy reserves.
- It allows for a better use of existing capacity.
- It combines the flexibility of road with the economies of rail on long journeys for large volumes.
- It is well integrated in the logistics chain.
- It is safe and secure: reduce risk for goods during the transport journey.

3 Rail/road combined transport in Europe

Combined transport (CT) represents an important cornerstone of the European freight market (see Fig. 1).

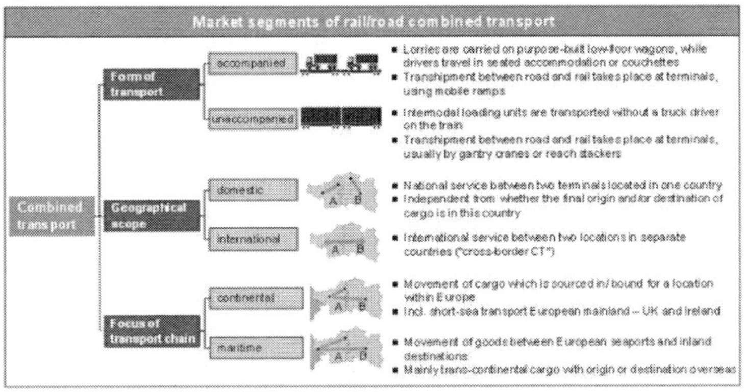

Fig. 1. Overview of market segments in rail/road combined transport. Source: [5]

The total volume of combined transport in Europe, including unaccompanied and accompanied combined transport, adds up to 21,0m TEU in 2015. Compared to 2013, the total combined transport volume recorded a slight increase of +1% (see Fig. 2).

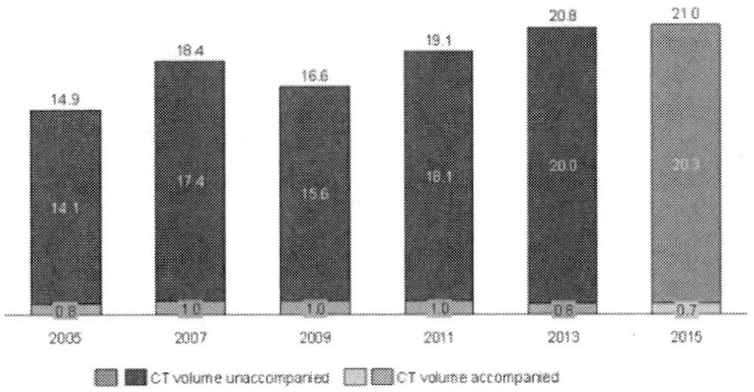

Fig. 2. Development of total combined transport volumes 2005 to 2015 in million TEU. Source: [5]

Based on the total CT tonnage transported, the increase amounts to approx. +8.5% from 2013 to 2015 and is therefore considerably higher than the growth in TEU. The positive development of the overall CT market is the same for both measurements in TEU and in tonnes. However, the stronger growth of tonnes in contrast to TEU indicates a trend towards transporting heavier shipments than some years ago. The following table depicts the total CT market development from 2005 to 2015.

Table 1. Development of total combined transport volumes 2005 - 2015 in million tonnes. Source: [5]

Segment	2005	2007	2009	2011	2013	2015
CT volume unaccompanied	145.5	181.5	164.6	191.8	203	218
CT volume accompanied	10.2	13.6	15.1	14.9	10.8	13
Total	155.7	195.1	179.7	206.7	213.8	231

The total CT market increase is mainly driven by volumes of the unaccompanied segment, while the accompanied CT volume increased in tonnage and is approaching the level of 2011. In 2015, the unaccompanied CT segment's market share amounts to approx. 94% of the total CT market. Consequently, nearly 90% of the CT providers offer unaccompanied CT services, while only 1% is focused on accompanied CT and approx. 10% provide both unaccompanied and accompanied CT services [6].

3.1 Transportation of intermodal units by Bohemiakombi transport operator

Bohemiakombi is a key Czech combined transport operator. Their aim is to move goods flows from the roads to the railways, which provides users with marked benefits, such as various legislative exceptions.

When planning transport, what is very important for them are the interests of the road haulier, which they try to harmonise with the options provided by railways, thereby ensuring reliable, fast and environmentally-friendly transport. There are several reasons to work with them:

- They coordinate mutually beneficial co-operation between road and rail transport companies.
- The goods are never handled — they load the whole unit onto a railway truck.
- Rail transport is not as limited as road transport (e.g. the ban on lorries on roads at weekends and on public holidays, higher weights, etc.).

Handling system in the terminal of Bohemiakombi is depicted in following Fig. 3.

Fig. 3. Handling system in the terminal of Bohemiakombi. Source: [7]

Through their system of complete combined transport trains they offer fast, reliable and regular links to industrial areas in the Ruhr, Baden, northern Germany, the Benelux,

Scandinavia, Italy and other places. In the Czech Republic they arrange follow-on rail transport to Czech and Slovak combined transport transhipment points, Poland and Austria. They also transport freight directly to the customer's industrial spur line. At the current time their company operates on the following transport lines:

- Lovosice – Duisburg/Rotterdam v. v.
- Lovosice – Hamburg v. v.
- Lovosice – Antwerpen v.v.
- Ostrava (Paskov) – Duisburg/Hamburg/Antwerpen/Rotterdam v.v.
- Brno – Duisburg/Hamburg/Antwerpen/Rotterdam v.v.
- Přerov - Duisburg/Hamburg/Antwerpen/Rotterdam v.v.
- Brno – Rostock v. v.

Transport condition

Orders are accepted by our sales department. Orders for transport on the regular lines have to be delivered by 10 a.m. one day before the end of loading. The orders must contain:

- the loading unit no.,
- requested relation,
- date of transport,
- type and length of the loading unit,
- gross and net weight of the unit,
- information about RID (class, UN number, name of the goods, packing group).

Line: **Lovosice CD – DUSS - Duisburg-Ruhrort Hafen Ubf DUSS**

Timetable:

From Monday to Thursday
Departure: 7:45 p.m.
Arrival: 12:40 p.m.
Duration: 16:55

Saturday:
Departure: 9:30 a.m.
Arrival: 5:40 a.m.
Duration: 1day 20hours

4 Comparison of road freight transport and intermodal transport

In this article, we compare just delivery of the load in a purely time perspective. And we design a procedure for calculating the number of trailers required to operate one complete train with intermodal semitrailers. In the future, it would be interesting to compare the costs of each type of transport.

Intermodal transport

If we check Bohemiacombi's timetable (from Monday to Thursday), we can see that the transfer from the Lovosice CD-DUSS to Duisburg-Ruhrort Hafen Ubf DUSS line runs about **17 hours.**

Road freight transport

Through the planner program Viamichelin we find the recommended route (see Fig.4).

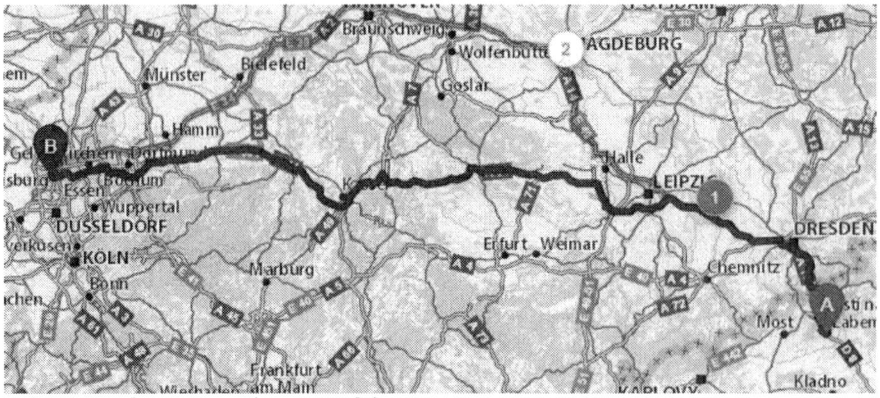

Fig. 4. The route from Lovosice to Duisburg. Source: [8]

This route includes motorways, rather *644 km* on motorways of total approx. 690km. We schedule daily driving periods and breaks for one driver taking into account REGULATION (EC) No 561/2006 OF THE EUROPEAN PARLIAMENT AND OF THE COUNCIL of 15 March 2006 on the harmonisation of certain social legislation relating to road transport.

Distance: 690 km
Average speed: 60km/h
Crew: monomial crew
Start: 7:45 p.m.

Daily driving periods and breaks for one driver with regular daily rest period are summarized in following Table 2 and 3.

Table 2. Daily driving periods and breaks for one driver with regular daily rest period. Source: authors

Start	Finish	Mode	Duration
7:45 p.m.	0:15 a.m.	driving	4.5 hour
0:15 a.m.	1:00 a.m.	break time	45 min
1:00 a.m.	5:30 a.m.	driving	4.5 hour
5:30 a.m.	4:30 p.m.	rest period	11 hour
4:30 p.m.	7:00 a.m.	driving	2.5 hour

Total time: 23 hour 15 minute

Table 3. Daily driving periods and breaks for one driver with reduced daily rest period. Source: authors

Start	Finish	Mode	Duration
7:45 p.m.	0:15 a.m.	driving	4.5 hour
0:15 a.m.	1:00 a.m.	break time	45 min
1:00 a.m.	5:30 a.m.	driving	4.5 hour
5:30 a.m.	2:30 p.m.	rest period	9 hour
2:30 p.m.	5:00 a.m.	driving	2.5 hour

Total time: 21 hour 15 minute

If we compare delivery times, it is better to use intermodal transport. And that's just a time reason.

If we looked at other benefits, they are:

- increases the available working time of drivers,
- minimizing the risks associated with road transport (congestion, stealing of cargo, atc.),
- avoid driving bans,
- to keep the minimum wage paid to the driver in Germany,

- exemption from road tax and toll payment,
- higher delivery speed,
- transfer of road loads to the railways,
- impact on the environment, etc.

However, the continued growth of road transport presents challenges when it comes to traffic congestion, road safety, and air pollution. Though trucks can provide door-to-door convenience, it is imperative that suppliers and governing bodies facilitate intermodal approaches to freight. Intermodal transport, sometimes referred to as combined transport, is the movement of cargo using two or more means – including road, rail, air, and sea. The term is also used interchangeably with intermodal transport, in which goods are transported in a single intermodal container using multiple transport modes [9].

Transport operators have to understand that public infrastructure sets a general limit to the size of vehicles and combined transport loading units, and that any developments in this area must comply with rules for this type of infrastructure. The road networks seems to pose the fewest constraints, mainly because road vehicle manufactures are able to offer chassis and trucks with relatively low platforms if the operator is willing to pay extra for the special design and small-diameter tires. Carriers tend to use combined transport. They think this is the future of long-distance transport within Europe. Although after-sales services are still cheap, they see more advantages than disadvantages. The main problem is to convince the customer that his goods will be partly transported by road and partly by rail.

4.1 Designing a procedure for calculating the number of trailers required to operate one complete train with intermodal semitrailers

A complete train departs from the Lovosice terminal; it means the maximum number of intermodal semi-trailers per line is 20. The terminal operates 9 hours per day. The attractive area of customers, from which intermodal trailers are imported, has a range of 30 km. The trailers run at an average speed of 55 km/h. The handling time with the semi-trailer in customers is 25min, and the handling time in terminal takes 20min per one semitrailer on average.

Calculation of trailers turn-over time:

$$Tto = [\,(2*lz)\,/\,vt\,] + thc + tht \qquad (1)$$
$$Tto = [\,(2*30)\,/\,55\,] + 0.42 + 0.33 = 1.84h$$

where: Tto – trailer´s turn-over time; thc – handling time in customers; tht – handling time in the terminal; lz – transport distance; vt – the technical speed of the trailer [10]

Calculation of number of trailers turnarounds per day:

$$Po = DWT\,/\,Tto \qquad (2)$$
$$Po = 9\,/\,1.84 = 4.89 \doteq 5 \text{ turnarounds}$$

where: Po – number of trailers turnarounds per day; DWT – daily working time

Calculation of required number of trailers:

$$TR = PN\,/\,Po \qquad (3)$$
$$TR = 20\,/\,5 = 4 \text{ trailers}$$

where: TR – required number of trailers; PN – number of semi-trailers

Thus, we need four trailers to operate a complete train from Lovosice terminal.

5 Conclusion

For in-line hauls, combined transport will use the installations specific to the mode on each leg. In additional, it needs facilities to change from one mode to the other. These will be found mainly in the terminals, combined transport cannot thus do without terminal equipment. However, as is the case with freight transport, poor connectivity can create a barrier to intermodalism [11].

Creating hubs or terminals is perhaps the most crucial step for combined transport. Intermodal transport is essential for removing passenger and freight vehicles from road networks worldwide. As many nations strive to meet carbon emissions targets and make their roads safer, it is crucial that commuters and suppliers discover alternate modes of movement. If public and private sector bodies combine capital and resources to streamline shifts from mode to mode, it will undoubtedly result in safer and more sustainable roads. Intermodal transport can best provide sustainable transport solutions once the right infrastructure and management solutions are in place. But cargo isn't the only thing that moves more sustainably using multiple modes of transport.

This paper is supported by the research project "From horse-drawn railway to intermodal transport" within Visegrad Fund.

References

1. J. Gnap, *Modelovanie dopravného a prepravného procesu v cestnej nákladnej doprave*, 1st ed., 120 p. (University of Zilina, Slovak Republic, 2013)
2. L. Moravcik, Transport and communications **1** (2010)
3. I. Kubasakova, R. Kampf, O. Stopka, Communications: scientific letters of the University of Žilina **16**, 2, 9-13 (2014)
4. O. Stopka, R. Kampf, J. Kolar, I. Kubasakova, Ch. Savage, Communications: scientific letters of the University of Žilina **16**, 2, 14-19 (2014)
5. L. Wattignies, *Combined transport in Europe*, Available online: http://uic.org/IMG/pdf/2016_report_on_combined_transport_in_europe_cles.pdf (2017)
6. M. Poliak, Journal of Economics, 206-220 (2013)
7. J Weise, Available online: https://www.kombiverkehr.de/en/transport/#fahrplan (2017)
8. I. Fišer, Available online: http://bohemiakombi.cz (2017)
9. J. Kapusta, A. Kalašová, Scientific journal of Silesian University of technology – series: Transport **93**, 49-58 (2016)
10. I. Kubasakova, J. Jagelcak, Communications: scientific letters of the University of Žilina **18**, 2, 109-112 (2016)
11. S. Kubíková, A. Kalašová, Ľ. Černický, *TST 2014 – Telematics - support of transport*, 414-423 (Katowice/Kraków/Ustroń, Poland, 2014)

Logistics of Entry and Parking of Vehicles at Large Production Companies

Jozef Gnap[1,*], *Pavol* Varjan[1], and *Štefánia* Semanová[1]

[1]University of Žilina, Faculty of Operation and Economics of Transport and Communications, Department of Road and Urban Transport, Univerzitná 8215/1, 010 26 Žilina, Slovakia

Abstract. The paper deals with the logistics of entry of freight vehicles to be unloaded or loaded in logistics and industrial parks. This problem is not systematically addressed when designing new logistics parks and industrial companies in the Slovak Republic neither in Europe nor in the world. Based on the research applied to the production company, a methodology for calculating the number of parking spaces was proposed. As it is possible to use simulation methods for solving this issue, the authors also deal with this possibility. The aim is to verify the accuracy of the proposed coefficients as well as to propose changes in the Slovak technical standard.

1 Introduction

The logistics of entry of freight vehicles to be unloaded or loaded within logistics and industrial parks is an issue which must be addressed systematically already when designing new logistics or industrial parks. Unfortunately, many newly built industrial companies as well as logistics parks in the Slovak Republic do not have any parking lots for freight vehicles in front of entries to their premises. However, these companies often require adherence to loading and unloading time windows (e.g. two hours), even in the case of long distance transport of over 1 000 km. Furthermore, the companies assume that carriers plan the transport in compliance with compulsory safety breaks, daily and weekly rest periods with the use of public parking lots or carriers try to avoid complying with the social legislation (e.g. using magnets to disable tachographs).

The capacity of public parking lots for road freight vehicles is insufficient in Europe [1, 2]. Also, some cities restrict parking or entry of freight vehicles into wider city centres.

By using specific cases, the authors of the paper point out that it is necessary to deal with the logistics of entry of freight vehicles into logistics and industrial areas, also with regard to introducing intelligent tachographs which will be connected to GNSS and thus breaking the social legislation will not be possible. Therefore, logistics managers in logistics parks and industrial companies should have an overview of the requirements of Regulation (EC) No. 561/2006 on the harmonisation of certain social legislation relating to road transport and they should plan schedules for transport including loading and unloading in respect of this regulation [3]. The designers of logistics and industrial parks should have data on the number of parking spaces to be designed for freight vehicles. Based on own

* Corresponding author: jozef.gnap@fpedas.uniza.sk

research, the authors proposed a procedure for calculating the necessary number of parking spaces for articulated freight vehicles.

2 Problems of the logistics of entry and parking of vehicles at large production companies

2.1 Approach to the problem solving for the production company

Department of Road and Urban Transport elaborated an expert study for the large company MONDI SCP, a.s. The expert study dealt with the issue of entry of freight vehicles into the company premises and its main objective was to shorten the waiting time before entering the company and time of staying inside the company [4].

The first step of this study was to identify the flows of incoming and outgoing vehicles. It was necessary to carry out a survey in order to verify the data provided by the company. The survey objective was to obtain input data for a deeper analysis of times of freight vehicle movements, duration of individual activities related to loading as well as workload of the gatehouse at the entrance to the company.

During the survey, the following time intervals were measured:

• **Vehicle turnover** – time interval between stopping a vehicle onto the scale when entering the company and its departure from the scale when leaving the company.

• **Time spent in warehouses** – time interval between the vehicle arrival at the parking lot before loading and its departure from the parking lot after completed loading (this applies to those vehicles which are loaded only in one warehouse). In the case that a vehicle is loaded in several warehouses, this indicator represents a time interval between the arrival at the parking lot of the first warehouse and departure from the parking lot of the last warehouse.

• **Vehicle turnover in the warehouse** – time interval between the vehicle arrival at the parking lot of the particular warehouse before loading and its departure after completed loading.

• **Waiting time before loading** – time measured from the moment when the vehicle arrives at the parking lot till it is moved to the loading ramp and ready to be loaded.

• **Loading time** – time between arrival at/departure from the loading ramp.

• **Waiting time after loading** – time interval between leaving the loading ramp and departure from the parking lot after document processing.

Based on the data obtained from the survey, the following indicators were evaluated:

• vehicle turnover,

• overall time that the vehicle spends in warehouses,

• time that vehicle spends in individual warehouses as well as duration of individual activities performed in those warehouses (waiting time before loading, loading time, document processing after loading),

• load of the entry road by incoming and outgoing freight vehicles.

As an example, Figure 1 depicts the processed results of the survey that was carried out between 5:38 and 11:40.

The turnover of all vehicles arriving for loading and unloading was measured. In overall, 51 vehicles were observed to enter and also leave the company. Within the observed vehicles, the minimum time of turnover represented 39 minutes and the maximum turnover time was 4 hours and 9 minutes. The average turnover time was 2 hours and 5 minutes. Figure 2 depicts the measured turnover times of individual vehicles.

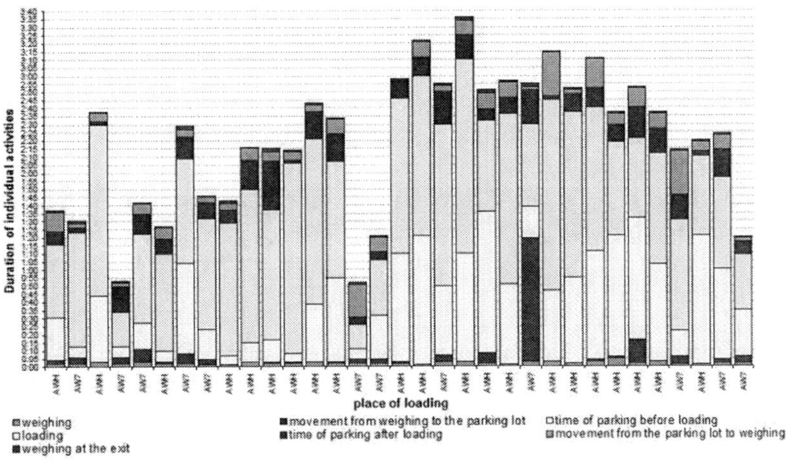

Fig. 1. The measured duration of individual activities of vehicles during their stay in the company. Source: [4]

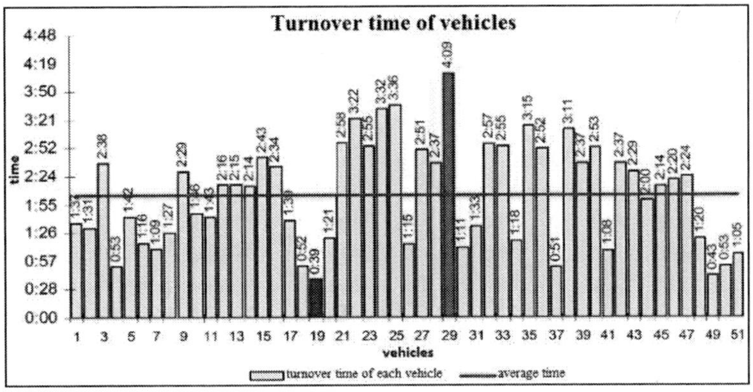

Fig. 2. Turnover time of vehicles. Source: [4]

If large companies do not address the logistics of entry and parking of freight vehicles incoming into the company, problems with traffic congestion or violation of the social legislation mainly in terms of adherence to daily and weekly rest periods may occur. The problems with violation of the social legislation arise mostly as a result of waiting time before loading, blocking public roads and staying in the company premises. The output from the expert study was a proposal to change the organization of entry and parking in front of the company. Other proposals included placing the second scale for weighing vehicles at the exit and changing the system of vehicle routing inside the company campus. One of the proposals was successfully implemented and it contributed to improving the logistics of entry of freight vehicles into the company.

2.2 Calculation of the necessary number of parking spaces at logistics and industrial parks

At present, a 24-hour operation is very common in large production companies. Therefore, high intensity of freight transport can be assumed for such companies. These companies often require carriers to adhere to loading and unloading time windows (e.g. two hours). The issue of parking of freight vehicles is, of course, related to this fact. If a company does not ensure a sufficient number of parking spaces already within constructing new production buildings or does not build additional parking capacity when needed, the following complications may arise:

- crowded parking areas around public roads near the companies; this may result in "dangerous parking" in some cases,
- driving even during compulsory breaks and rest periods may have negative impact on the safety of other road users,
- inadequate or no services provided to drivers at parking areas or pollution of the road infrastructure surroundings.

2.2.1 The issue of freight vehicle parking at KIA Motors Slovakia

KIA Motors Slovakia launched its production in 2006. In 2016, the company produced more than 339 500 cars and 612 000 engines [5].

The company campus can be used as a good example for solving the problem of the logistics of entry and parking of freight vehicles. This production company has a parking lot which is, however, designed only for passenger cars and buses. There is no parking lot determined for freight vehicles. As already mentioned in the paper, production companies often require adherence to loading and unloading time windows. However, some companies do not allow to park freight vehicles inside its premises mainly due to limited space. As KIA Motors Slovakia does not have a parking lot for freight vehicles, problems with parking arise.

The second problem related to freight transport is adherence to the requirements of Regulation (EC) No. 561/2006 on the harmonisation of certain social legislation relating to road transport. Examination within the expert study mentioned in Chapter 2.1 showed that the average time of a vehicle turnover was 2 hours and 9 minutes with the assigned time window (e.g. 2 hours). Oftentimes, drivers must take compulsory breaks, daily or weekly rest periods after unloading (or loading). As they do not have possibility to stop and park their vehicle within the company campus, they have to look for a solution not to break the social legislation. One of possibilities is to apply the exception set out in Chapter IV, Article 12 of Regulation (EC) No. 561/2006: *"Provided that road safety is not thereby jeopardised and to enable the vehicle to reach a suitable stopping place, the driver may depart from Articles 6 to 9 to the extent necessary to ensure the safety of persons, of the vehicle or its load. The driver shall indicate the reason for such departure manually on the record sheet of the recording equipment or on a printout from the recording equipment or in the duty roster, at the latest on arrival at the suitable stopping place"* [6]. The driver can manually indicates the exception and reason on the record sheet or the printout from the recording device; however, additional driving time may take only a few minutes unless the driver finds a safe place to stop. Such a place should be a proper parking lot designed for freight vehicles and not the edge of public roads.

In case of KIA Motors Slovakia, the problem is the absence of a freight vehicle parking lot that is designed only for this company. For example, such parking lots can be seen in the car factory PSA Trnava in the Slovak Republic. It is also necessary to note that there is insufficient number of public parking lots near KIA e.g. at petrol stations. The nearest petrol station, which is about 2.5 km away from KIA depending on the entrance to the company, has available only 12 parking spaces for freight vehicles. This petrol station is very often used by carriers serving KIA production company. Moreover, nearby the city of Žilina, there is no rest area planned yet that could be used for temporary parking of freight vehicles before arriving at the company for loading or unloading in accordance with the time windows. Therefore, freight vehicles park on public roads near the company premises (Fig. 3).

Fig. 3. Parking of freight vehicles near KIA. Source: authors, based on Google maps

2.2.2 Calculation of the number of parking spaces at KIA Motors Slovakia

In terms of the theoretical calculation of the number of parking spaces intended for freight transport, it was first necessary to obtain data on the number of freight vehicles incoming into the company. The survey was carried out on 20th and 21st October 2016 (Thursday and Friday) at predetermined locations near individual entrances to the company (three entrances intended for freight transport). A 24-hour survey was employed to obtain the most accurate information about the number of freight vehicles of N3 category. As the time window before entering the company was set at the level of 2 hours, it was necessary to divide the transport intensity into 2-hour intervals. Tab. 1 contains two-hour intensity intervals of freight vehicles of N3 category incoming into the company campus. Data are provided separately for each entrance as well as in overall. For the theoretical number of parking spaces calculated for whole production company, however, it is sufficient to work further only with the total number of observed incoming vehicles.

Table 1. Intensity of entry of N3 category freight vehicles. Source: authors

	12:00-14:00	**14:00-16:00**	**16:00-18:00**	**18:00-20:00**	**20:00-22:00**	**22:00-24:00**	**00:00-02:00**	**02:00-04:00**	**04:00-06:00**	**06:00-08:00**	**08:00-10:00**	**10:00-12:00**
Entrance 1	14	23	26	10	11	11	15	10	14	22	16	9
Entrance 2	32	36	24	26	32	24	23	19	18	39	36	51
Entrance 3	12	14	5	5	12	6	7	14	10	7	16	16
Total	**58**	**73**	**55**	**41**	**55**	**41**	**45**	**43**	**42**	**68**	**68**	**76**

Based on Tab. 1, it can be concluded that the highest intensity of freight transport was between **10:00 and 12:00 h**. During this time interval, **76** freight vehicles of N3 category entered into the company. This means that 76 freight vehicles would have to wait 2 hours before entering the company and therefore at least 76 parking spaces would be needed. This value is then the basis for next calculation. However, this value should be increased by a share of vehicles which must draw daily or weekly rest period before loading or unloading. The calculation of the number of parking spaces for the vehicles drawing a daily or weekly rest period was based on the total number of vehicles entering the company. During the survey, **655** freight vehicles entered and **645** vehicles exited the company campus. The following formula was used for the calculation:

$$PPM = I_{max} + (PV_{DO} \div k_{pd}) + (PV_{TO} \,.\, k_{pt}) \tag{1}$$

Where: PPM – necessary number of parking spaces; I_{max} – maximum intensity of entry of vehicles in 2-hour intervals [vehicles]; PV_{DO} – number of vehicles, drivers of which draw a daily rest period [vehicles]; k_{pd} – coefficient of the use of a parking space by vehicles, the drivers of which take daily rest periods; k_{pt} – coefficient of the use of a parking space by vehicles, the drivers of which take weekly rest periods, PV_{TO} – number of vehicles, drivers of which draw a weekly rest period [vehicles].

Further, it was necessary to determine a share of those freight vehicles, drivers of which draw daily and weekly rest periods before/after loading or unloading. To determine this share, further surveys were carried out to find out the number of vehicles parked at observed locations for at least 11 hours (daily rest period) and 45 hours (regular weekly rest period). The number of freight vehicles parked at the petrol station near the company (mentioned in Chapter 2.2.1) and on the road sections (marked with black and white lines in Fig. 3) was surveyed. A part of the first survey was carried out on 14th August 2017 (Monday) at 19:00. The vehicle registration numbers (VRN) of N3 category vehicles parked at observed locations were recorded. The second part of the survey was then carried out on 15th August 2017 (Tuesday) at 6:00 in the morning and vehicle registration numbers were again recorded and matched with those from the previous evening. The same procedure of the survey was repeated on 16th and 17th August 2017 (Wednesday and Thursday). The results are processed in Tab. 2.

The second survey was focused on determination of the number of vehicles parked due to drawing a regular weekly rest period (at least 45 h). The procedure was the same as in the previous survey (dealing with the share of vehicles drawing a daily rest period). The survey started on 19th August 2017 (Saturday) at 7:00 and it lasted till 8:30. The next observation was then carried out on 21st August 2017 (Monday) at 6:30. The results are shown in Tab. 2.

Table 2. The average number of parked vehicles (daily/regular weekly rest period). Source: authors

Daily	**Evening (19:00_19:30)**	**Morning (6:00_6:30)**	**Match**	**Share [%]***
No. of vehicle	34	45	22	3.23 %
Weekly	**Saturday (07:00_8:30)**	**Monday (6:30_7:15)**	**Match**	**Share [%]***
No. of vehicle	33	39	23	3.46 %

* *calculated from the total number of incoming vehicles during the survey mentioned in Chapter 2.2.2 (665 vehicles = 100 %)*

When calculating the number of parking spaces for vehicles, the drivers of which draw a daily rest period, it is necessary to take into account duration of the regular rest period which is 11 hours. Therefore, the survey result of 22 vehicles, which on average draw the daily rest period, is to be further divided by the coefficient – k_{pd}. This coefficient expresses how many times one parking space can be used by alternating vehicles, the drivers of which draw the daily rest period. Taking into account the daily rest period of 11 hours, the coefficient can be calculated as:

$$k_{pd} = 24\ hours \div 11\ hours = 2.182 \tag{2}$$

To calculate the number of parking spaces for vehicles, the drivers of which draw a weekly rest period, the results of the survey focused on the rest period lasting at least 45 hours are used. The result represented 23 vehicles. As the calculation is based on the 24-hour survey and the regular weekly rest period is at least 45 hours, the value of 23 vehicles is to be multiplied by the coefficient – k_{pt}. This coefficient takes into account that a vehicle drawing the 45-hour weekly rest period is, of course, parked longer than 24 hours.

$$k_{pt} = 45\ hours \div 24\ hours = 1.875 \tag{3}$$

The calculated values can be then fitted into the formula (1):

$$PPM = 76 + (22 \div 2.182) + (23 \cdot 1.875)$$
$$PPM = 76\ veh. + 10.083\ veh. + 43.125\ veh.$$
$$PPM = 129.208\ vehicles$$

Under the given conditions and regarding the intensity of entry of N3 category freight vehicles, it is can be concluded that there is a need to ensure 130 parking spaces.

2.2.3 Proposal for a calculation of parking spaces at logistics and industrial parks

In Chapter 2.2.2, we calculated the number of parking spaces at the production company of KIA Motors Slovakia on the basis of data obtained from the surveys conducted. The following formula (1) can be used as a basis to propose general procedure for the calculation which can be used when constructing new logistics and industrial parks.

$$PPM = I_{max} + (PV_{DO} \div k_{pd}) + (PVTO \, . \, k_{pt})$$

Where: $k_{pd} = 2.182$ and $k_{pt} = 1.875$

To refine the formula (coefficient adjustment), it would be appropriate to carry out more surveys. This calculation proposal addresses only the issue of number of parking spaces; however, it is necessary to note that parking lots should be also equipped with social facilities (toilets, showers or refreshments).

2.3 Possibilities of using simulation methods to refine the calculation of the number of parking spaces

Based on examining open and closed queueing theory systems applied for simulating operation time of freight vehicles in practice, the Erlang distribution seems to be the most appropriate to be employed [7]. As mentioned in [8] and [9], the case of so-called shifted probability distribution is very common in transport technology.

The distribution function of a random variable is:

$$F(y) = \int_{y_0}^{y} \frac{b^a \cdot (y - y_0)^{a-1}}{\Gamma(a)} e^{[-b(y-y_0)]} dy \tag{4}$$

If parameter a is an integer, this indicates that a random variable has the shifted Erlang distribution with the following distribution function:

$$Fe(y) = 1 - e^{-b(y-y_0)} \cdot \sum_{j=1}^{a-1} \frac{[b \cdot (y - y_0)]^j}{j!} \tag{5}$$

Parameter y_0 (shifting) is calculated in practice as a minimum technological time necessary for loading or unloading of a freight vehicle with certain cargo.

In simulation models, we can use the theoretical probability distribution to obtain random variables if we acquire the input data of an appropriate range and quality and the **goodness of fit test** (Chi-squared test, Kolmogorov–Smirnov test) does not show deviations at the selected significance level. In this particular case, the shifted Erlang distribution seems to be the most appropriate to be used.

Based on the above mentioned characteristics, the flowchart of the simulation model for open and closed queueing theory system for operation of dumpers was created [7]. After adjustment, this model can be also used when examining the logistics of entry of freight vehicles into logistics and industrial parks as well as intermodal transport terminals [10].

3 Conclusion

The objective of the paper was to point out the selected aspects of the issue of logistics of entry and parking at large logistics and industrial parks. The authors provided the outputs of the research which dealt with the issue of entry of freight vehicles into the particular company campus. The main goal of the survey was to shorten waiting time before entering the company as well as time of staying inside the company premises. The paper also pointed out the problems associated with no or insufficient capacity of parking spaces. This problem is also related to the social legislation of drivers' work time.

As in many cases the parking capacity or logistics of entry is not addressed in the Slovak Republic when constructing new logistics and industrial parks, it was necessary to propose a general procedure or formula to calculate the number of parking spaces. The proposed procedure was verified at the production company of KIA Motors Slovakia. Based on the surveys carried out, it was possible to obtain input data for calculation of number of parking spaces for freight vehicles of N3 category.

The proposed calculation formula can be also used to calculate necessary number of parking spaces for other companies. However, it is necessary to carry out further surveys for a given logistics or industrial park in order to refine the formula in terms of the accuracy of coefficients for calculation of number of parking spaces for vehicles, the drivers of which draw daily or weekly rest periods. The authors suggest to include their proposal for calculation of parking spaces for freight vehicles into Slovak technical standards as it already exists in case of parking lots for passenger cars under the construction proceedings according to STN 73 6110/Z2 Design of urban roads. In comparison with Czech technical standard ČSN 73 6056 Parking areas for road vehicles from 2011, the issue related to parking lots for freight vehicles is elaborated more in detail, however, the calculation methodology misses.

This paper is supported by the research project "From horse-drawn railway to intermodal transport" within Visegrad Fund.

References

1. *Dalnice, parkoviste, kamiony*, Available online: http://www.denik.cz/ekonomika/dalnice_parkoviste_kamiony2007.html (2007)
2. J. Čársky, *Parkování vozidel nákladní dopravy,* Available online: http://www.silnice-zeleznice.cz/clanek/parkovani-vozidel-tezke-nakladni-dopravy/ (2010)
3. M. Poliak, J. Gnap, *Práca vodičov nákladných automobilov a autobusov a používanie tachografov* (University of Zilina, Slovakia, 2015)
4. J. Gnap et al., *Logistika vstupu č. 2 a č. 3 MONDI SCP, a.s.* (University of Zilina, Department of Road and Urban Transport, Žilina, Slovakia, 2008)
5. Kia-reports, Available online: http://www.kia.sk/sk/o-nas/vyrocne-spravy (2016)
6. Regulation (EC) No. 561/2006
7. J. Gnap, *Modelovanie dopravného a prepravného procesu v cestnej nákladnej doprave* (University of Zilina, EDIS, Slovakia, 2013)
8. J. Černý, P. Kluvánek, *Základy matematickej teórie dopravy,* 278 p. (VEDA, Bratislava, Slovakia, 1991)
9. D. Malindžák et al., *Aplikácia modelovania a simulácie v logistike podniku*, (Technical University of Kosice, Slovakia, 2009)
10. J. Lizbetin, O. Stopka, *ICTTE – Proceedings of the third international conference on traffic and transport engineering* (Belgrade, Serbia, 2016)
11. M. Rieser, *Adding Transit to an Agent-Based Transportation Simulation,* (Technische Universität Berlin, Germany, 2010)
12. M. Al-Azzawi, *Freight and Logistics Transport Modelling and Planning* (LAP LAMBERT Academic Publishing, Saarbrücken, Germany, 2012)

Model of the Heavy Freight Vehicle's Running Speed Used for Diversion Traffic Routes Determination

Jiří Hanzl[1,*], and *Ladislav* Bartuška[2]

[1]Institute of Technology and Business in České Budějovice, Faculty of Technology, Okružní 517/10, 37001 České Budějovice, Czech Republic
[2]Czech Technical University in Prague, Faculty of Transportation Sciences, Department of Logistics and Management of Transport, Horská 2040/3, 12800 Prague, Czech Republic

Abstract. This paper describes the construction process of a discrete model of a heavy truck running speed. The running speed calculation is based on the technical parameters of the road (geometry, gradient, curves, etc.) and valid Czech legislation for road designing. In this article, road factors influencing the final heavy freight vehicle's running speed are presented as well. The main assumption is the fact that every heavy vehicle moves on the road by speed allowed by the geometry of this road, road signs and traffic flow characteristics (intensity, density or road capacity).

1 Introduction

A heavy freight vehicle (truck) is the most limiting factor in road traffic planning and when designing roads. That's why we're talking about this vehicle as a "design (determinant) vehicle". In the case of increased traffic volume, the traffic flow of vehicles on the road without the possibility of overtaking or on the road with the limited overtaking ability is at just the same speed as the speed of the heavy freight vehicle. If we want to determine the current speed of this vehicle at each point(s) of the road, we must take into account, in addition to the valid Czech legislation, also its own road geometry and traffic constraints on the road.

In order to describe the issue, the authors created a professional search detailing the individual factors on the road influencing the speed of the heavy freight vehicle. In addition to the maximum permitted driving speed, additional constraints are considered, such as railway crossings, road geometry parameters, etc. The text of the paper is based mainly on the thesis entitled "Determination of alternative routes of main roads in the event of extraordinary and planned traffic constraints", where the whole issue is described in more detail and the outcome of which is the analytical model for valuation of diversion traffic routes according to their impedance [1]. This discrete model also uses, in addition to the driving dynamics of a selected type of vehicle, the road geometry parameters (directional

* Corresponding author: hanzl@mail.vstecb.cz

ratios, curve radius, width arrangement, etc.) and maximum permitted speed on the road section supplemented by other decision criteria.

2 Running speed calculation model

For the determination of the vehicle's running speed on the road, the so-called determinant slow vehicle is considered because, as already described in the introductory part of the paper, it is the most limiting factor of the entire traffic flow. According to the Czech technical standards as a design determinant vehicle is considered the vehicle with a total weight of 44 tonnes, with a totally variable ratio continuously variable in the range from 2.070 to 49.680 (with regard to the optimal working conditions of the engine unit), with a radius of 0.526 m, the efficiency of the transmission unit 0.8888, air resistance coefficient 0.7, coefficient of rolling resistance 0.006, and engine driving torque 1600 Nm (for ascending route), 0 Nm (free run) and -550 Nm (for braking) [2-4].

The input parameters of the calculation, which have a significant influence on the resulting running speed of the determinant slow vehicle on a given segment of the road, are given in Table 1. Segmentation of a road means its division into separate sections with a similar character of traffic, whereby the calculation model selects a minimum speed on the given road segment (or maximum speed limit). This velocity is the standard rate at which the entire traffic flow will move [5].

Table 1. Input parameters of the calculation based on the geometry of the road, the legislative regulations and the characteristics of the traffic flow. Source: [1]

a) width of road traffic lane	a [m]
b) directional conditions of the road (curve radius)	R [m]
c) vertical alignment of road (longitudinal slope)	s [%]
d) maximal allowed speed on the road	v_{vk} [km/h]
e) speed restrictions on the road	v_{om} [km/h]
f) traffic restrictions for heavy freight vehicles on the road	
g) estimated traffic intensity on the road	I [cars/h]

The selected parameters in Table 1 are described in more detail in the following sections of the paper.

3 Parameters that influence the driving speed

3.1 Width of the road traffic lane *a [m]*

The wide-lane arrangement of the road is an important parameter for the determination of design (determinant) speed in road design. To determine the running speed depending on the width of the lane of the road, the table 2 is used [2].

Table 2. Design speed dependence on the road driving lane width. Source: [2]

Design category of road				Driving lane
Designation	*b [m]*	Design speed v_n *[km/h]*	Road class	*a [m]*
S	6.5	60	III.	2.75
S	7.5	70	II., III.	3.00
S	9.5	80	I., II.	3.50
S	11.5	90	I.	3.50

3.2 Directional conditions of the road (curve radius) *R [m]*

Directional curve radius is one of the most important parameters of the infrastructure, which significantly affects the vehicle speed passing vehicles the road turning. In simplified calculations, the authors consider a simple circular arc without transitions, for which the following formula can be used to calculate the limit speed v_m which is still possible to pass safely the road bend with vehicle [6]. Together with the size of the radius R, the gravitational acceleration g and the cross slope of the road p are added to the calculation [5].

The general formula for limiting speed calculation has the following form:

$$v_m = 3.6 \cdot \sqrt{g \cdot R \cdot (f + 0.01 \cdot p)}, \qquad (1)$$

where: v_m - limit velocity *[km/m]*; R - directional curve radius *[m]*; f - adhesion coefficient value (for the calculation of the achievable speed, the value f = 0.25) *[m]*; p - cross slope (if there is an opposite cross slope in the directional curve, the negative value is set) [%] ; g - gravitational acceleration 9.81 [m/s^2] [6].

3.3 Vertical alignment of road (longitudinal slope) *s [%]*

The calculation of the running speed of the design (determinant) vehicle is based on the pictorial annex "K2 - Relationship between the longitudinal tilt of the vehicle and the speed of the design vehicle for the other road network (with initial speed 70 km / h)" of standard [2]. Since the change in velocity depending onthe size of the longitudinal slope is not characterized by a linear curve, it is necessary to linearize these curves (Figure 1) to express the rate of change of velocity depending on the length of the section at which this change occurred. By setting the initial velocity v_0 *[km/h]*, the convergence velocity v_k *[km/h]*, the velocity change coefficient Δv *[s^{-1}]* and the length of the road section l_k *[m]* into the uniformly accelerated motion equation, it is possible to calculate the average velocity of the slow vehicle due to the longitudinal slope of the road v_s *[km/h]* [8].

DOI: 10.1051/matecconf/201713400017

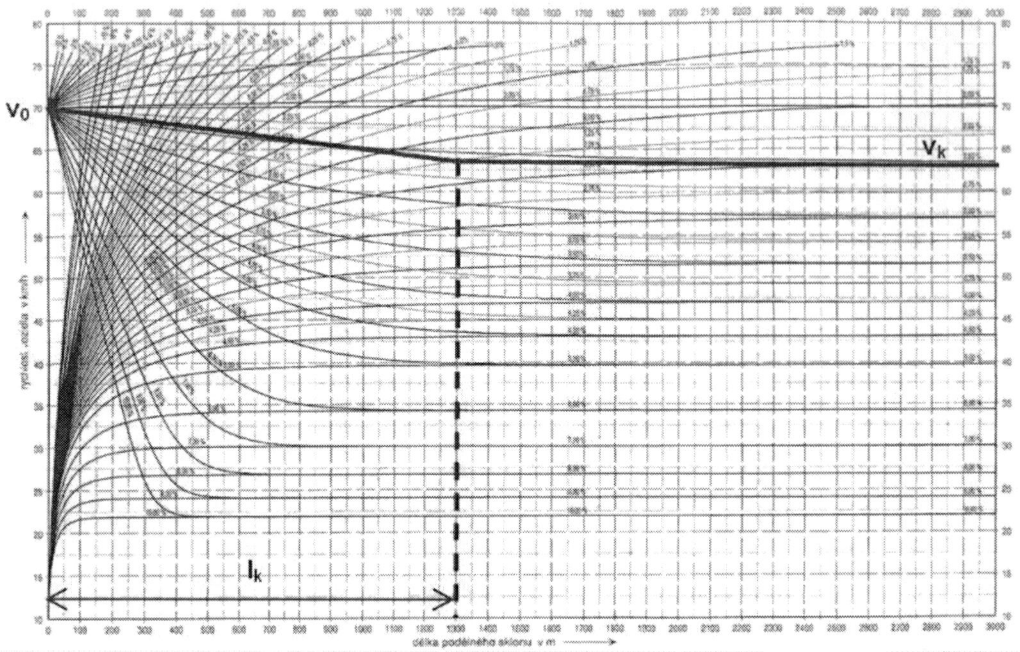

Fig. 1. Linearization of curves expressing the relationship of the longitudinal slope of the road and the vehicle speed. Source: [1]

3.4 Maximal allowed speed on the road v_{vk} *[km/h]*

In the Czech Republic, according to [7], the maximum permitted speed of heavy freight vehicles on free road is determined without speed restrictions according to Table 3:

Table 3. The highest allowed speed of heavy freight vehicles on roads in the Czech Republic. Source: [2]

designation of the road	v_{vk} *[km/h]* - extravilan	v_{vk} *[km/h]* – urban area
D	80 km/h	50 – 80 km/h
S I, II, III	80 km/h	50 – 80 km/h
MK	80 km/h	50 – 80 km/h

3.5 Speed limit on road section v_{om} *[km/h]*

Speed restrictions on the road may be represented by vertical traffic signs or railway crossings. For example, in the case of a railway-road crossing, it is necessary to base the limit speed on valid legislation on road traffic. If there is a railway-road crossing with a light signal system (if the intermittent white signal of the signalling device is flashing), the driver of vehicle has to drive in the distance of 50 m from the railway crossing and when passing the crossing at speed of not more than 50 km/h. In the case of no controlled railway-road crossing without light signal devices or the device is off, the driver may drive in the distance 50 m from the railway crossing and when passing the railway crossing at a speed not exceeding 30 km/h.

3.6 Other road traffic restrictions for heavy freight vehicles

To calculate the driving speed of heavy freight vehicles, it is important that the driving of such vehicles on the road section is not prohibited or otherwise restricted. If there is any such traffic constraint, mostly determined by traffic signs, there is no point in continuing to calculate the driving speed. For example traffic signs "No entry for trucks" or "No entry for vehicles which immediate weight exceeds the limit indicated", and others [9].

3.7 Estimated traffic intensity on the road *I [cars/hour]*

The last input parameter of the calculation, based on the traffic characteristics of the traffic flow, is the traffic intensity *I*. With increasing intensities (traffic volumes), individual vehicles will interact with each other during the ride, which will affect the resulting speeds of the entire traffic flow. Not only the geometric parameters of the road and the traffic constraints on the route (and the calculated maximum limit speeds), but also the total traffic flow intensity, will influence the resulting driving speed [10].

Level of Service (LOS, UKD in Czech) is a qualitative measure used to relate the quality of traffic service. LOS (UKD) is used to analyse highways and main roads by categorizing traffic flow and assigning quality levels of traffic based on performance measure like speed, density, etc. For Level of Service in the "E" category, with the minimum longitudinal slope and the minimum share of slow-moving vehicles, the following values apply to the two-lane road according to [2]:

Table 4. Traffic flow characteristics due to critical density for 2-lane road. Source: [1]

Type of road	Traffic density at LOS (UKD) in category "E" H_{opt} *[cars/km]*	Road capacity I_{max} *[cars/hour]*	Optimal speed until reaching road capacity v_{opt} *[km/h]*
2 - lane road	20 cars/km	1250 cars/hour	62.5 km/h

From the fundamental diagram (Figure 2) expressing the dependence of the velocity of the traffic flow on the intensity, it is obvious that this relation is parabolic. As the velocity increases, the traffic intensity increases until its maximum (road capacity I_{max}) value is reached. Subsequently, despite the increase in velocity, the traffic intensity decreases again [11].

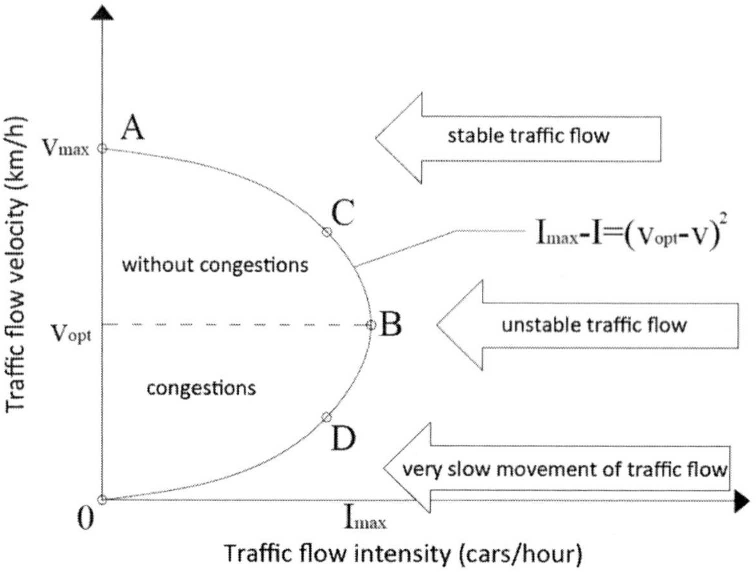

Fig. 2. Parabolic course of traffic flow velocity dependence on traffic intensity. Source: [11]

The parabola curve has the equation:

$$I_{max} - I = (v_{opt} - v)^2, \qquad (2)$$

Where I_{max} - maximal traffic intensity value (road/lane capacity) *[cars/hour]*,
I - current traffic intensity *[cars/hour]*,
v_{opt} - the optimal speed at which the traffic flow reaches its maximum intensity I_{max} *[km/h]*,
v - current traffic flow velocity [km/h]. [1]

By adjusting the equation and setting up the variable values I_{max} a v_{opt} from Table 4, we obtain the equation for calculating the velocity of the traffic flow on the selected road section as a function of the current traffic intensity:

$$v_I = v_{opt} \pm \sqrt{I_{max} - I} \qquad (3)$$

Where v_I - the traffic flow velocity on a given road section in one driving lane depending on traffic intensity *[km/h]*,
v_{opt} - the optimal velocity at which the traffic flow reaches the maximum intensity I_{max} in one driving lane *[km/h]*,
I_{max} - maximal traffic intensity value (road/lane capacity) *[cars/hour]*, I_{max}*=1250 cars/hour,*
I - current traffic intensity on given road section *[cars/hour]*.

Based on the actual traffic density H_j *[voz/km]* on the road in one driving lane and according to Fig. 2, the appropriate formula of calculation is selected according to the following:

1) $H_j < H_{opt}$ $\quad v_{I,j} = 62.5 + \sqrt{1250 - I}$,

2) $H_j = H_{opt}$ $\quad v_{I,j} = 62.5$,

3) $H_j > H_{opt}$ $\quad v_{I,j} = 62.5 - \sqrt{1250 - I}$. [1]

4 The calculation model output of the running speed of the heavy freight vehicle

As mentioned above, the calculation model selects a minimum (maximum limiting) speed v_{min} on a given road section. Its schematic representation is given in Fig. 3. Therefore, the following will apply [12, 15]:

$$v_{\min} = \min\{v_n; v_m; v_s; v_{vk}; v_{om}; v_I\} \quad (4)$$

Where v_{min} - minimum speed (or maximum limiting) on a given road section *[km/h]*,
v_n – design speed (for road designing) on a given road section *[km/h]*,
v_m – limit speed on a given road section *[km/h]*,
v_s – average speed due to the longitudinal slope of the road section *[km/h]*,
v_{vk} – the highest allowed speed on a given road section *[km/h]*,
v_{om} – speed restrictions a given road section *[km/h]*,
v_I – the traffic flow speed on a given road section in one driving lane depending on traffic intensity *[km/h]*.

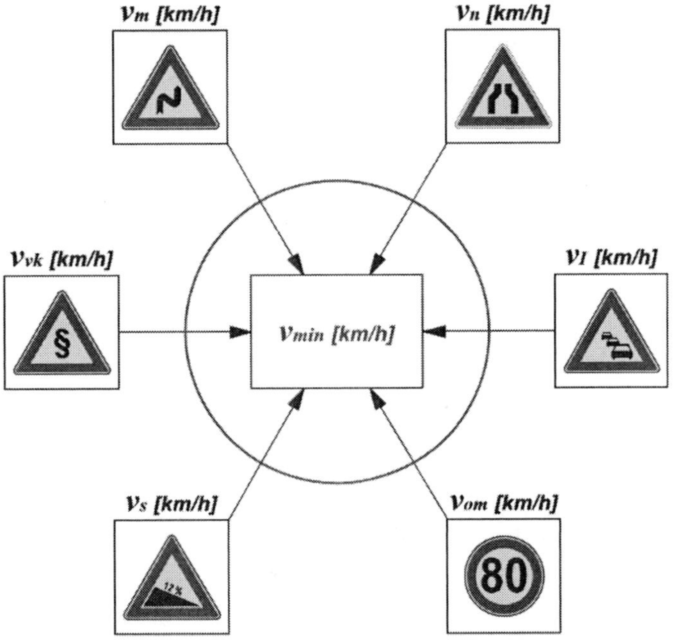

Fig. 3. Schematic representation of the maximum limiting speeds on a given road section. Source: authors

5 Conclusion

The main objective of this paper was to familiarize readers with the main factors influencing the resulting velocity of the heavy freight vehicle (determinant vehicle for road designing) as the most limiting factor of the entire traffic flow moving on the road section. The contribution describes the individual parameters (characteristics) directly affecting the freight vehicle driving speed. It is not only a legislative measure, but also a design parameters of the road and the characteristics of the traffic flow, which are described. As mentioned in chapters above, the text of the paper is based largely on [1], in which the whole issue is described in more detail as part of the so-called analytical model for peer assessment of roads, expressed in financial equivalence.

This paper is supported by the research project "From horse-drawn railway to intermodal transport" within Visegrad Fund.

References

1. J. Hanzl. *Determination of alternative routes of main roads in the event of extraordinary and planned traffic constraints: dissertation* (Czech Technical University in Prague, Faculty of Transportation Engineering, Czech Republic, 2017)
2. Czech Technical Standards CSN 73 6101, *Designing of roads and highways*
3. J. Vrabel, O. Stopka, V. Rievaj, B. Sarkan, K. Pruskova, P. Michalk, Communications: Scientific Letters of the University of Zilina **18**, 2, 124-128 (2016)
4. F. Brumercik, M. Lukac, Logi – Scientific Journal on Transport and Logistics **3**, 1, 45-52 (2012)
5. J. Hanzl, et al. *Transport Means – 20th International Scientific Conference on Transport Means*, 354-360, (Juodkrante, Lithuania, 2016)
6. Czech Technical Standards CSN 73 6102, Z1: 2011, *Design of intersections on roads.*
7. (CZ) Act No. 361/2000 Coll. *Road Traffic*, as amended
8. M. Tomasikova, F. Brumercik, A. Nieoczym, Logi – Scientific Journal on Transport and Logistics **6**, 1, 130-135 (2016)
9. R. Kampf, S. Lorincová, M. Hitka, O. Stopka, Sustainability **9**, 9, 1561, (2017), DOI: 10.3390/su9091561
10. M. Hitka, L. Ližbetinová, Z. Caha, Y. Xu, Communications: Scientific Letters of the University of Zilina **18**, 2 (2016)
11. A. Csikos,T. Charalambous, H. Farhadi, B. Kulcsar, H. Wymeersch, Transportation Research Part C: Emerging Technologies **83**, 120-133 (2017)
12. L. Bartuška, J. Čejka, Z. Caha, Nase More **62**, 91-96 (2015), DOI:10.17818/NM/2015/SI1
13. J. Jagelčák, I. Kubasáková, Naše more **61**, 5-6, 106-116 (2014)
14. O. Stopka, M. Chovancová, J. Ližbetin, V. Klapita, Nase More **63**, 3, 195-199 (2016), DOI: 10.17818/NM/2016/SI22.
15. G. Fedorko, V. Molnár, J. Strohmandl, M. Vasil', *Transport Means – 19th International Scientific Conference on Transport Means*, 466-469 (Kaunas University of Technology, Lithuania, 2015)

The Design of Integrated Logistics Management System of an Industrial Company

Martin Hart[1,*]

[1]Tomas Bata University in Zlin, Faculty of Logistics and Crisis Management, Logistics Department, T.G. Masaryka 5555, 760 01 Zlin, Czech Republic

Abstract. In the contemporary global business markets environment, when the business markets are getting more and more commercial, there are growing demand for effective management of material flows. The effectivity and effectiveness of planning, management and control the material flows across an industrial company and its distribution networks, represents one of the main pillar regarding the high level of competitive advantage within the frame of supply chains. Thus, the company information management system design should have also included a module of integrated logistics management system to ensure required level of material flow management effectivity and effectiveness. The article deals with brief description of the issues on company management, company information management systems and logistics management. Further it's stated the methodology to created integrated logistics management system, which is containing the methodics to design logistics management sub-systems of purchasing, manufacturing, distribution and reverse material flows. The essential methodics of the stated methodology is the methodics to create independent demand forecasting sub-system.

1 Introduction

Global business environment and business markets commercialisation enhance by distinctive manner the requirements putting on industrial companies within the context of competitive advantage high level achievement and living environment protection. Wide range of the products and services in the industrial or business markets is the factor to seek the ways for costs reduction, which are connected with their production or distribution. The one of the ways, how to reach mentioned costs reduction, which are related to the products manufacturing, it's the design or the optimization of the company's management systems, which embody the high level of effectivity and effectiveness. That's the one of the prospective ways, how to increase competitive advantage level in the current markets. The one of the main modules of the company's information management systems in the contemporary turbulent global business environment is the integrated logistics management system. The effectivity and the effectiveness of the planning, management and control, the material flows of the industrial companies or the entire supply chains, represents important

[*] Corresponding author: hart@utb.cz

potential source to reach the high level of competitive strength in the up-to-date commercialized business markets.

The integrated logistics management system of an industrial company should be formed by the sub-systems of the planning, management and control, the particular functional areas. Those functional levels, namely purchasing, manufacturing, distribution and reverse material flows, are crosscut by the material flows. Well, the design or optimization of the company's information management systems or more precisely the design of the integrated logistics management systems is the field, which should be taken into account by all the nodes or the elements of the contemporary supply chains.

2 Enterprise management and enterprise information system

The success and the prosperity of the current enterprises are given by the high-quality company's management. The enterprise management is constituted by the effective design and the cross connection of the particular company's functional fields. The essential functional fields of the company are:

- research and development,
- purchasing,
- manufacturing,
- logistics – distribution,
- finance & accounting,
- sales & marketing,
- human resources.

The companies or the organisations of the particular industries, at present time of the global dynamic market environment, are still more realising the importance of the company management quality – stability and flexibility. This fact is not just from the company's IS/IT point of view, but it's especially from the first rate managerial decisions making process point of view [1].

The single functional levels of the company it's necessary, in current dynamic market environment, to take as the systems [2]. These systems should have been composed in the sense to reach required level of stability, flexibility, resilience and living environment protection. Thus, firstly there are created particular functional areas – the systems and the processes which are running in them. Subsequently the functional areas are integrated thru the process management principles including the planning, management and control of the all flows, which are streaming in the internal or the external company's supply chains [3].

The company's management includes the application of the exact methods developed for the analysis and synthesis given processes, which are running in the particular functional fields or across the functional fields. Further it's important to apply the methods for the decision-making process purposes, within the scope of the decision support systems [4].

The company management should represent the process of the effective treatment (management) relevant knowledge and information from the given company functional levels, which should cooperate at the in advance defined level – basis [5].

Enterprise information management system constitutes wholly important part of the successful company. Effective transmission, analysis, synthesis and the storing of the relevant data and information is the essential function of the enterprise information management systems. That fact can represent considerable competitive advantage, if it's respected the principle of the data – information effectivity and effectiveness [6].

3 Logistics management

The logistics management represents the one of the most important enterprise management fields. The effective management of the material flows in the internal and external company's supply chains is the area, where there are just applied the logistics management principles and methods. Thus, the logistics management plays the key role during enterprise strategies and plans creation – e.g. the issues, which are concerning with the required customer service setting, the suppliers relationships, manufacturing plants and warehouses locations, the distribution, inventory management or demand forecasting, and others [7].

The one of the requirements of the contemporary industrial or business markets, which is putting on the organizations, is the achievement of the high level responsiveness. Such requirement should be met at the lowest possible costs and the created added value should be on the other hand as the highest as possible. This matter of fact is the challenge to design new approaches or principles of the planning, management and control in the logistics management – the creation of the new methodologies or methodics [8].

4 Methodology of integrated logistics management system creation

The integrated logistics management system (ILMS) should be based, in up-to-dates companies, on the process management principles applied across particular enterprise functional fields. The design or the application of the ILMS should be realised continuously together with the actualisation or the implementation of the enterprise information system. The enterprise management system should have, in context of the enterprise management systems integration across the supply chains, a module for collaborative planning, forecasting and replenishment (CPFR) – as a foundation for other management modules.

ILMS should be designed for the management as the internal so the external material flows of the given company and that in context with the information and data gained with the CPFR module utilisation.

4.1 Methodology of the integrated logistics management system design

The methodology, which is still under development, is now composing by 5 methodics to design forecasting or logistics management sub-systems, as it is stated in the following points:

1. Methodics of independent demand forecasting sub-system design
2. Methodics of purchasing management sub-system design
3. Methodics of manufacturing management sub-system design
4. Methodics of distribution management sub-system design
5. Methodics of reverse material flow management sub-system design

Each of the above stated methodics can be further worked out into comprehensive methodology to design given forecasting or management information sub-system.

4.2 Methodics to design independent demand forecasting sub-system

The methodics or the steps succession, which are necessary to realise for design or for optimization of the enterprise forecasting sub-system, it's possible to sum up into 9 following points:

1. an enterprise description or a description of its plants for which about to be created independent demand forecasting system (sales or consumption),
2. a supply chain visualization – internal and external parts,
3. the foundation of a point or the points of the supply chain dividing – the order penetration point theory,
4. an extensive portfolio segmentation of the accumulated material items – an inventory of e.g. the final products, semi-finished products or raw materials; in the order penetration point or points of the supply chain,
5. the analysis of the existing systems for independent demand forecasting,
6. a determination and statistical processing of the input data for demand forecasting (sales or consumption),
7. a basic analysis of demand pattern – a trend, a seasonality, an increase or a decrease,
8. a creation process of the most suitable models for independent demand forecasting of the particular material items segments – the inventory which is accumulated in the order penetration point of the supply chain; their application to forecast, [9]
9. a creation of CPFR module (collaborative planning forecasting and replenishment) for the purpose to integrate the company's forecasting system with the forecasting systems of the other companies, organizations or the retailers within the scope of given supply chains.

Above stated methodics of the independent demand forecasting system is universally applicable in any industry or in any tertiary sphere organization.

4.3 Methodics to design purchasing logistics management sub-system

The methodics to design purchasing logistics management sub-system, it's possible to sum up into 5 following points:

1. the current state analysis of the purchasing management system in the company,
2. a creation or an up-dating of the raw material or purchased items warehousing spaces layouts,
3. a creation of the particular management sub-systems:
   - a. data sub-system
   - b. forecasting sub-system
   - c. planning sub-system
   - d. inventory management sub-system
   - e. a selection and an evaluation suppliers sub-system, supply quantity determination
   - f. a setting particular logistics processes sub-system
   - g. personal sub-system
   - h. economic sub-system
   - ch. purchasing administration sub-system
   - i. control sub-system – a setting of performance measurements
   - j. sub-system of risk analysis and living environment protection,
4. an integration of created sub-systems,
5. a putting into operation [10, 11].

Stated methodics to design purchasing logistics management sub-system is universally applicable in any industry or tertiary sphere organization.

4.4 Methodics to design manufacturing logistics management sub-system

The methodics to design manufacturing logistics management sub-system, it's possible, as well as the methodics regarding a purchasing, to sum up into 5 following points:

1. the current state analysis of the manufacturing management system in the company,
2. a creation or an up-dating of the manufacturing spaces layouts including associated warehouses,
3. a creation of the particular management sub-systems:
    a. data sub-system
    b. forecasting sub-system
    c. planning sub-system
    d. an optimal production batches determination sub-system
    e. inventory management sub-system
    f. a particular logistics processes determination sub-system
    g. personal sub-system
    h. economic sub-system
    ch. manufacturing administration sub-system
    i. control sub-system – a setting of performance measurements
    j. sub-system of risk analysis and living environment protection
    k. sub-system of maintenance planning and management,
4. an integration of created sub-systems,
5. a putting into operation. [12]

Stated methodics to design manufacturing logistics management sub-system is universally applicable in any industry or tertiary sphere organization, as well as 2 methodics briefly described above.

4.5 Methodics to design distribution logistics management sub-system

The methodics to design distribution logistics management sub-system, it's possible, as well as 3 previous methodics to sum up into 5 following points:

1. the current state analysis of the distribution management system in the company,
2. a creation or an up-dating of the distribution spaces layouts including associated warehouses,
3. a creation of the particular management sub-systems:
    a. data sub-system
    b. forecasting sub-system
    c. planning sub-system
    d. sub-system to vehicle full loading and an optimal transportation routes determination
    e. inventory management sub-system
    f. a setting of particular logistics processes sub-system
    g. personal sub-system
    h. economic sub-system
    ch. distribution administration sub-system
    i. control sub-system – a setting of performance measurements
    j. sub-system of risk analysis and living environment protection
    k. sub-system of fleet management (administration, planning and management),
4. an integration of created sub-systems,

5. a putting into operation. [13]

The methodics to design distribution logistics management sub-system is universally applicable in any industry or tertiary sphere organization as the previous 3 methodics briefly described above.

4.6 Methodics to design logistics management sub-system of the reverse material flows

The methodics to design logistics management sub-system of reverse material flows, it's possible, as well as 4 previous methodics to briefly describe by the 5 following points:

1. the current state analysis of the reverse material flows management system in the company,
2. a creation or an up-dating of the warehouse layouts which are intended for reverse material flow items,
3. a creation of the particular management sub-systems:
   a. data sub-system
   b. forecasting sub-system
   c. planning sub-system
   d. sub-system to vehicle full loading and an optimal transportation routes determination
   e. inventory management sub-system
   f. a setting of particular logistics processes sub-system
   g. personal sub-system
   h. economic sub-system
   ch. administration sub-system
   i. control sub-system – a setting of performance measurements
   j. sub-system of risk analysis and living environment protection
   k. sub-system of fleet management (administration, planning and management)
4. an integration of created sub-systems,
5. a putting into operation [10, 11].

The methodics to design reverse material flows management sub-system is also universally applicable as the previous methodics in any industry or tertiary sphere organization.

5 Conclusion

The supply chains of given industrial branches are getting more and more complex and the material flows, which are streaming in them are getting more and more bulky [14, 15]. Due to markets globalization, the industrial and business markets are getting also the markets more or less with commercial character. Thus, there are seeking out the new ways, as to gain or to retain the customers. The one of the way is to reduce the costs thru the progressive information enterprise management systems, and that within the scope of long-term socio-economic sustainability and living environment protection. These systems should meet the requirements of flexibility, stability, resilience and security with connection to contemporary supply chain development.

The important part of the information enterprise management systems is, in context of up-to-date supply chain development, the integrated logistics management system.

The article dealt with brief explanation of the essential notions: enterprise management, enterprise information management system and logistics management. The major article part has been the presentation of the created methodology to design integrated logistics management system, which has been created on the basis of former author's research. The methodology is universally applicable in any industry or organization of tertiary sphere. Thru methodology application, it's possible to create the integrated logistics management system in the practice – the management sub-systems of purchasing, manufacturing, distribution and reverse material flows.

The created methodology should contribute to the development of the issues regarding the enterprise information management systems in the theory and also in the practice. The article has presented new-fashioned approach of enterprise management with the emphasis on the effectivity and effectiveness connected with the planning, management and control of the internal and external company's or supply chain's material flows for which is the methodology also applicable.

This paper is supported by the research project "From horse-drawn railway to intermodal transport" within Visegrad Fund.

References

1. R.L. Tammineedi, Info. Sec. J. **19**, 1, 36-50 (2010), DOI: 10.1080/19393550903551843
2. D. Malindzak, *Proceedings of the 6th Carpathian Logistics Congress*, 20-28 (Zakopane, Poland, 2016)
3. M. Kavan, *Production and Operational Management*, 424 p. (Prague: Grada, Czech Republic, 2002)
4. W.M.P. van der Aalst, A.H.M. ter Hofstede, M. Weske, Lect. Not. in Com. Sci. **2678**, 1-12 (2003), DOI: 10.1007/3-540-44895-0_1
5. T.S. Raghu, A. Vinze, Dec. Sup. Sys. **43**, 3, 1062-1079 (2007)
6. S. Salah, A. Rahim, & J. A. Carretero, Inter. Jour. of Bus. Exc. **6**, 2, 161-191 (2013), DOI: 10.1504/IJBEX.2013.052478
7. J. Gattorna, A. Day, J. Hargreaves, Log. Inf. Man., **4**, 2, 2-86 (1991), DOI: 10.1108/09576059110143603
8. F. Nilsson, The Inter. J. of Log. Man. **17**, 1, 38-54 (2006), DOI: 10.1108/09574090610663428
9. M. Hart, *The Approaches to Independent Demand Forecasting in an Industrial Company*, 203 p. (Dissertation Thesis, VŠB-TU Ostrava, Czech Republic, 2010)
10. M. Hart, P. Taraba, J. Konecny, *Proceedings of the 5th Carpathian Logistics Congress*, 110-115 (Priessnitz Spa, Jeseník, Czech Republic, 2015)
11. M. Hart, P. Taraba, M. Tomastik, *Proceedings of the 6th Carpathian Logistics Congress*, 99-104 (Zakopane, Poland, 2016)
12. M. Hart, P. Taraba, J. Konecny, *Proceedings of the 3rd International Conference on Sustainable Design and Manufacturing*, 191-202 (Chania, Crete, Greece, 2016)
13. M. Hart, P. Taraba, M. Tomastik, *Proceedings of the 28th Annual Nordic Logistics Research Network Conference* (Turku, Finland, 2016)
14. P. Kacmary, D. Malindzak, Ac. Mon. Slo. **15**, 1, 53-60 (2010)
15. P. Kacmary, J. Saderova, A. Rosova, M. Straka, App. Me. and Mat. **683**, 61-65. (2014), DOI: 10.4028/www.scientific.net/AMM.683.61

Application of the Vogel Approximation Method to Reduce Transport-logistics Processes

Martina Hlatká[1,*], *Ladislav* Bartuška[1], and *Ján* Ližbetin[1]

[1]Institute of Technology and Business in České Budějovice, Faculty of Technology, 37001 České Budějovice, Czech Republic

Abstract. The article deals with logistics processes in companies. Logistics aims to optimize material, finance and information flows to optimize overall costs in companies. The first part of the article describes the methods by means of which the processes in companies can be optimized, namely the Vogel approximation method. In the second part of the article, the method is implemented in a particular manufacturing company with the result of distribution route optimization and reduction of operating costs of transport.

1 Introduction

Logistics has undergone a major change in recent decades. Unlike in previous years, today it deals with the flow of material, information and financial resources in all areas of a company. This implies that logistics does not only deal with material flow in individual businesses, but also assesses individual influences from time, place and space point of view in order to satisfy customers and achieve optimal costs associated with these activities. It is a comprehensive science which improves corporate processes and enables the company to respond more quickly to market and customer demands. Nowadays, the emphasis is placed on the quality and high level of services provided; thus, the use of logistics is a necessity.

In general, optimizations are increasingly used as their results bring proposals for improving processes in companies. The advantage of optimization analyses is cost reduction. In addition to economic improvement, there is also shortening of travelled routes, which is also associated with a reduction in the time required to carry out activities. The aim of such an optimization is the most efficient use of means of transport, technology, and human resources with regard to reducing economic costs [1-4]. Currently, there are many ways to optimize processes in a company. The most commonly used are those using operational research methods or graphical representations [4-6].

2 Methods

The Vogel approximation method deals with finding an optimal solution taking into account the relationships of price indices. This method compares always the two lowest

* Corresponding author: hlatka@mail.vstecb.cz

price indices in both a column and a row. Using these comparisons, which are called differences, the task is to evaluate a field whose "inoccupation" by the cheapest option would have the worst impact on the overall result.

In the procedure of the Vogel approximation method, the differences between two smallest rates for each row and column are first counted. The individual differences are written outside the table. For each row, the value is written to the left of the table and for each column the value is written below the table. Subsequently, the highest of the calculated differences is selected, which now indicates the area for the next calculation step. The next step is to find a field with the lowest rate in the selected area, i.e. column or row. We put the highest value in this cell. The maximum value is derived from the restrictive conditions normally written in the last column and row of the assembled matrix.

In this research, the modified version of the Vogel approximation method is fundamental. This method does not take into account restrictive capacities and rates. The purpose of the modified version calculation is to find the shortest distance which enables completing the transport task.

The first step in this research is to adjust the distance matrix that was assembled during the previous method. The change takes place on the main diagonal, where zero values are located. These zero values will be changed to X. Since there is no capacity limitation in the specified delivery routes, the two lowest values of the distance are searched in each column and row. The difference is calculated from the two smallest values. This difference is then written outside the table the same way as in the normal method. The steps are performed for all columns and rows, and the difference with the highest value is selected from the calculated differences. In contrast to the usual Vogel approximation method, where the lowest rate would now be selected and the maximum under the restrictive conditions would be added, a cell with the lowest value of distance is chosen in the selected area. This procedure is repeated until the entire transport task is complete [5-8].

2.1 Input values

The application of the Vogel approximation method will be carried out in a meat and sausage company. It is a small family business, where the optimization of the transport-logistic processes has not yet been dealt with. Based on the interview with the owner of the company it was possible to analyse essential data necessary for the compilation of a detailed list of current stopping points which the company serves by its own internal transport. The first task is to compile a detailed list of stops that will include the exact addresses where the goods will be delivered. Subsequently, the task is to optimize the original routes. Deliveries of goods are always on weekdays, i.e. from Monday to Friday. Each day of the week has more or less modified route. The Friday delivery route was selected for the application of the method. On Friday, the goods are unloaded only six times. This short route is 64.9 kilometres.

Table 1. Places of unloading on Friday. Source: authors

	Town	**Address**
V_1	Volduchy	Volduchy 63; 338 22; Volduchy
V_2	Chrást	Dlouhá 481; 330 03; Chrást
V_3	Chrást	Tř. Čs. Odboje 111; 330 03; Chrást
V_4	Plzeň	Sladkovského 660/34; 326 00; Plzeň 2
V_5	Plzeň	Křimická 927/48; 318 00; Plzeň 3
V_6	Plzeň	Domažlická 182/13; 318 00; Plzeň 3
V_7	Plzeň	U letiště 1074/2; 301 00; Plzeň 3

The normal version of the Vogel approximation method takes into account limited capacity. Each cell typically contains a certain rate. A difference is made between the two smallest rates in the row and it is then written at the edge of the row; the same is done in the columns. Then the highest value from the rate differences is found. According to the value, the row or column in which the lowest rate is found is determined. This procedure is repeated until the transport task reaches the optimum result [9].

The modified Vogel approximation method does not include rates and limited capacity. The matrix only uses the distances of individual points. Thus, in doing the transport task, the difference between the two smallest distances in the row is first sought [10, 11]. The same procedure is repeated for the other rows and columns, where the difference is always written behind the prepared matrix. The highest value is selected from the differences and it determines the selected column or row, from which the smallest value determining the distance is selected. This value is highlighted in light green in the table.

Now, there is another change compared to the normal Vogel approximation method. Instead of occupying only one cell, it is necessary to cross out the entire row and column in which the calculation procedure is located. This step is important in order to effectively calculate such a route that will serve all scheduled stop points. The unloading place, where the transport task is located, is marked in yellow in the table. In addition, it is necessary to eliminate the possibility of more routes. It is necessary to remove such a point, which would early complete the transport task. If two identical values are found, it is purely a matter of choice which option will be chosen for the next calculation step. Subsequently, the procedure is repeated-from the recalculation of differences to crossing out unnecessary cells [12-17].

3 Results

3.1 Vogel approximation method – Friday distribution route

To clarify the process of the modified version of the Vogel approximation method, the task begins with calculating the smallest matrix of distances, and thus the shortest calculation of the whole week as well. The smallest distance matrix occurs for day delivery on Friday. This matrix has the dimensions of 7x7 cells, so the clear explanation of the procedure will be the most transparent on this matrix [16, 17].

Table 2. Friday - VAM - Modification 1. Source: authors

	V_1	V_2	V_3	V_4	V_5	V_6	V_7	
V_1	X	~~10.7~~	11.5	22.1	24.4	29.2	25.9	0.8
V_2	10.7	X	~~1.0~~	12.0	14.3	19.0	15.7	9.7
V_3	~~11.5~~	1.0	X	~~11.0~~	~~13.3~~	~~18.0~~	~~14.8~~	10.0
V_4	22.2	~~11.7~~	10.7	X	4.0	7.9	4.7	0.7
V_5	24.9	~~14.4~~	13.4	3.9	X	7.3	4.7	0.8
V_6	24.7	~~14.2~~	13.3	3.8	1.9	X	4.6	1.9
V_7	26.3	~~15.8~~	14.8	4.8	4.8	4.2	X	0.6
	0.8	9.7	9.7	0.1	2.1	3.1	0.1	

In the first step of the calculation, it is necessary to find the differences in the smallest distance values in each row and in each column. The values of individual differences are then written outside the table. In the first row V_1 we calculate the difference of values for cells [V_1; V_3] = 11.5 km and [V_1; V_2] = 10.7 km. The difference of these values is 0.8 kilometres, which is written in the right-hand outer part of the table [5, 18].

The second step is to select the highest value of all the differences. In the table, this value is highlighted in yellow. The highest distance difference is 10.0 kilometres, located in row V_3. As a result of this step it is determined that the selection of the smallest distance in row V_3 follows. This value is based in cell [V_3; V_2] = 1.0 km and is highlighted in yellow inside the table.

This is followed by crossing out values that would untimely complete the transport task or would pose a multi-ring problem. It begins by crossing out all the values in the row and the column in which the calculation is located. A provisional route is $\mathbf{V_3 \rightarrow V_2}$ for the time being. It is necessary to avoid cycling of this route. Therefore, the value located in cell [V_2; V_3] is crossed out as well [5, 6, 19].

Table 3. Friday - VAM - Modification 2. Source: authors

	V_1	V_2	V_3	V_4	V_5	V_6	V_7		
V_1	X	~~10.7~~	~~11.5~~	22.1	24.4	29.2	25.9	0.8	10.6
V_2	10.7	X	~~1.0~~	~~12.0~~	~~14.3~~	~~19.0~~	~~15.7~~	9.7	1.3
V_3	~~11.5~~	1.0	X	~~11.0~~	~~13.3~~	~~18.0~~	~~14.8~~	10.0	X
V_4	~~22.2~~	~~11.7~~	10.7	X	4.0	7.9	4.7	0.7	0.7
V_5	~~24.9~~	~~14.4~~	13.4	3.9	X	7.3	4.7	0.8	0.8
V_6	~~24.7~~	~~14.2~~	13.3	3.8	1.9	X	4.6	1.9	1.9
V_7	~~26.3~~	~~15.8~~	14.8	4.8	4.8	4.2	X	0.6	0.6
	0.8	9.7	9.7	0.1	2.1	3.1	0.1		
	11.5	X	0.8	0.1	2.1	3.1	0.1		

The procedure is repeated in the second step. First, differences of the smallest values for the rows and columns are calculated. There is nothing to subtract for row V_3 row and columnV_2 anymore, so the position outside the table is replaced by X. After calculating the differences, the highest value is selected, i.e. 11.5 below column V_1. This determines the column from which the lowest possible value is selected. This value is 10.7 kilometres located in the cell with coordinates [V_2; V_1].

Then the entire column and row of the above mentioned cell are crossed out. Due to the possibility of early completion of the transport task, it is necessary to check such cell values that would untimely return the calculation to the previous places of unloading without servicing all the specified delivery unloading places. The provisional route is as follows: $\mathbf{V_3 \rightarrow V_2 \rightarrow V_1}$. It is necessary to avoid early closure of the route from V_1 to V_3. Therefore, it is necessary to have the value in cell [V_1; V_3] also crossed out [5, 6, 20].

Table 4. Friday - VAM - Modification 3. Source: [authors]

	V_1	V_2	V_3	V_4	V_5	V_6	V_7			
V_1	X	~~10.7~~	~~11.5~~	~~22.1~~	24.4	29.2	25.9	0.8	10.6	2.3
V_2	10.7	X	~~1.0~~	~~12.0~~	~~14.3~~	~~19.0~~	~~15.7~~	9.7	1.3	X
V_3	~~11.5~~	1.0	X	~~11.0~~	~~13.3~~	~~18.0~~	~~14.8~~	10.0	X	X
V_4	~~22.2~~	~~11.7~~	10.7	X	~~4.0~~	~~7.9~~	~~4.7~~	0.7	0.7	0.7
V_5	~~24.9~~	~~14.4~~	~~13.4~~	3.9	X	7.3	4.7	0.8	0.8	0.8
V_6	~~24.7~~	~~14.2~~	~~13.3~~	3.8	1.9	X	4.6	1.9	1.9	1.9
V_7	~~26.3~~	~~15.8~~	~~14.8~~	4.8	4.8	4.2	X	0.6	0.6	0.6
	0.8	9.7	9.7	0.1	2.1	3.1	0.1			
	11.5	X	0.8	0.1	2.1	3.1	0.1			
	X	X	2.6	0.1	2.1	3.1	0.1			

After the second modification, it is necessary to evaluate the differences in the rows and columns. The completely crossed out rows V_2 and V_3 are omitted. In addition, the differences of the smallest values in columns V_1 and V_2 are not counted. The highest value of all calculated differences is 2.6, which is located in column V_3. On account of this, the

 DOI: 10.1051/matecconf/201713400019

lowest distance in column V_3 is sought. The result is 10.7 km. This cell is yellow in the table again. The provisional route calculated by the modified version of the Vogel approximation method is as follows: $\mathbf{V_4 \rightarrow V_3 \rightarrow V_2 \rightarrow V_1}$. Now again, it is also necessary to remove cells that would early terminate the transport task. Therefore, the cell with coordinates [V_1; V_4] with the value of 22.1 kilometres is crossed out [6, 7, 20, 21].

Table 5. Friday - VAM - Modification 4. Source: authors

	V_1	V_2	V_3	V_4	V_5	V_6	V_7				
V_1	X	~~10.7~~	~~11.5~~	~~22.1~~	24.4	~~29.2~~	25.9	0.8	10.6	2.3	2.3
V_2	10.7	X	~~1.0~~	~~12.0~~	~~14.3~~	~~19.0~~	~~15.7~~	9.7	1.3	X	X
V_3	~~11.5~~	1.0	X	~~11.0~~	~~13.3~~	~~18.0~~	~~14.8~~	10.0	X	X	X
V_4	~~22.2~~	~~11.7~~	10.7	X	~~4.0~~	~~7.9~~	~~4.7~~	0.7	0.7	0.7	X
V_5	~~24.9~~	~~14.4~~	~~13.4~~	3.9	X	~~7.3~~	4.7	0.8	0.8	0.8	0.8
V_6	~~24.7~~	~~14.2~~	~~13.3~~	3.8	1.9	X	~~4.6~~	1.9	1.9	1.9	1.9
V_7	~~26.3~~	~~15.8~~	~~14.8~~	~~4.8~~	~~4.8~~	4.2	X	0.6	0.6	0.6	0.6
	0.8	9.7	9.7	0.1	2.1	3.1	0.1				
	11.5	X	0.8	0.1	2.1	3.1	0.1				
	X	X	2.6	0.1	2.1	3.1	0.1				
	X	X	X	0.1	2.9	3.1	0.1				

The step which follows after further modification of the table is the calculation of the individual differences for the rows and columns of the table. The completely crossed out columns V_1, V_2, V_3, as well as the rows V_2, V_3 and V_4 are omitted. Then, the highest value is taken from all the differences in the outer part of the table. This value is 3.1 km and is located below column V_6. The resulting value is 4.2 km, which is the lowest value of the determined column V_6. This value is highlighted in yellow in the cell with coordinates [V_7; V_6].

After finding the minimum value, it is necessary to cross out all the values in both the column and the row. Furthermore, the cell with the value of 4.6 km and with coordinates [V_6; V_7] is crossed out. This point has to be eliminated as the early closure of the distribution stations circuit could occur. The provisional route calculated using the Vogel approximation method is: $\mathbf{V_4 \rightarrow V_3 \rightarrow V_2 \rightarrow V_1}$ and $\mathbf{V_7 \rightarrow V_6}$.

Table 6. Friday - VAM - Modification 5. Source: authors

	V_1	V_2	V_3	V_4	V_5	V_6	V_7					
V_1	X	~~10.7~~	~~11.5~~	~~22.1~~	~~24.4~~	~~29.2~~	25.9	0.8	10.6	2.3	2.3	2.3
V_2	10.7	X	~~1.0~~	~~12.0~~	~~14.3~~	~~19.0~~	~~15.7~~	9.7	1.3	X	X	X
V_3	~~11.5~~	1.0	X	~~11.0~~	~~13.3~~	~~18.0~~	~~14.8~~	10.0	X	X	X	X
V_4	~~22.2~~	~~11.7~~	10.7	X	~~4.0~~	~~7.9~~	~~4.7~~	0.7	0.7	0.7	X	X
V_5	~~24.9~~	~~14.4~~	~~13.4~~	3.9	X	~~7.3~~	~~4.7~~	0.8	0.8	0.8	0.8	0.8
V_6	~~24.7~~	~~14.2~~	~~13.3~~	~~3.8~~	1.9	X	~~4.6~~	1.9	1.9	1.9	1.9	1.9
V_7	~~26.3~~	~~15.8~~	~~14.8~~	~~4.8~~	~~4.8~~	4.2	X	0.6	0.6	0.6	0.6	X
	0.8	9.7	9.7	0.1	2.1	3.1	0.1					
	11.5	X	0.8	0.1	2.1	3.1	0.1					
	X	X	2.6	0.1	2.1	3.1	0.1					
	X	X	X	0.1	2.9	3.1	0.1					
	X	X	X	0.1	22.5	X	21.2					

The calculation is almost at the end of the transport task calculated by means of the modified Vogel approximation method. The difference between the two smallest distances in each column and row is counted, which is then written outside the table. The differences for rows V_2, V_3, V_4 and V_5 are no longer counted since the rows are completely crossed out. This is also true for columns V_1, V_2, V_3 and V_6.

The highest value is sought of all the calculated differences from the outside of the table. Such a cell is in yellow and refers to column V_5, from which the lowest value that is not crossed out is then selected. This is a cell with coordinates [V_6; V_5] with the value of 1.9 km. The provisional sections of the route are as follows: $\mathbf{V_4 \rightarrow V_3 \rightarrow V_2 \rightarrow V_1}$ and $\mathbf{V_7 \rightarrow V_6 \rightarrow V_5}$. In order to avoid the early termination of the transport task, we also cross out cell [V_5; V_7].

Table 7. Friday - VAM - Modification 6. Source: authors

	V_1	V_2	V_3	V_4	V_5	V_6	V_7					
V_1	X	~~10.7~~	~~11.5~~	~~22.1~~	~~24.4~~	~~29.2~~	25.9	0.8	10.6	2.3	2.3	2.3
V_2	10.7	X	~~1.0~~	~~12.0~~	~~14.3~~	~~19.0~~	~~15.7~~	9.7	1.3	X	X	X
V_3	~~11.5~~	1.0	X	~~11.0~~	~~13.3~~	~~18.0~~	~~14.8~~	10.0	X	X	X	X
V_4	~~22.2~~	~~11.7~~	10.7	X	~~4.0~~	~~7.9~~	~~4.7~~	0.7	0.7	0.7	X	X
V_5	~~24.9~~	~~14.4~~	~~13.4~~	3.9	X	~~7.3~~	~~4.7~~	0.8	0.8	0.8	0.8	0.8
V_6	~~24.7~~	~~14.2~~	~~13.3~~	~~3.8~~	1.9	X	~~4.6~~	1.9	1.9	1.9	1.9	1.9
V_7	~~26.3~~	~~15.8~~	~~14.8~~	~~4.8~~	~~4.8~~	4.2	X	0.6	0.6	0.6	0.6	X
	0.8	9.7	9.7	0.1	2.1	3.1	0.1					
	11.5	X	0.8	0.1	2.1	3.1	0.1					
	X	X	2.6	0.1	2.1	3.1	0.1					
	X	X	X	0.1	2.9	3.1	0.1					
	X	X	X	0.1	22.5	X	21.2					

In the previous step, the last unnecessary cells with distance values in Friday matrix were eliminated. The last two values that were not marked or crossed out remain. In this step, these last cells are highlighted in yellow. These distance values logically complete the whole circular issue of this transport task.

It is clear from the last modification of the calculation in Table X that the final route of the optimized Friday route using the modified version of the Vogel approximation method is: $\mathbf{V_1 \rightarrow V_7 \rightarrow V_6 \rightarrow V_5 \rightarrow V_4 \rightarrow V_3 \rightarrow V_2 \rightarrow V_1}$. The resulting length of the route is 25.9 + 4.2 + 1.9 + 3.9 + 10.7 + 1.0 + 10.37 = 58.3 km.

The original length of Friday route was 64.9 km. The newly created route optimized by the Vogel approximation method is 58.3km. This will shorten the route by 6.6 km. The original and optimized route composition is shown in Table 8 [22, 23].

Table 8. Friday – Measure proposals. Source: authors

Current:	V_1	V_2	V_3	V_4	V_5	V_6	V_7	V_1	**64.9 km**
Optimized:	V_1	V_7	V_6	V_5	V_4	V_3	V_2	V_1	**58.3 km**

As the number of travelled kilometres is reduced, transport-logistic costs will also be reduced as shown in Table 9. The table shows an annual cost savings of 983 CZK.

Table 9. Friday – Economic indicators. Source: authors

Route	**Costs per 1 travel [CZK]**	**Annual costs [CZK]**	**Difference of annual costs versus current route [CZK]**
Current	193.19	9,660	
Opti. – VAM	173.54	8,677	- 983

For a better visual idea, the new route is shown on the map in Figure 1. The light blue route represents the original route. The new distribution circuit is pink. In places where the original and new routes overlap the colour is purple.

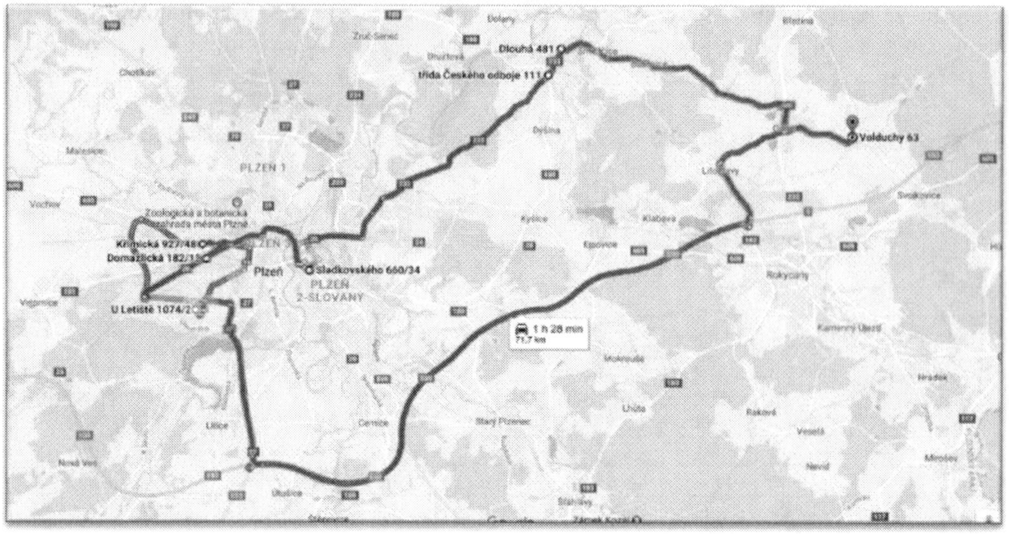

Fig. 1. Comparison – Friday route. Source: authors

4 Conclusion

The goal of the Vogel approximation method was to reduce the operating costs of transport in a selected company. Since it is a small family business, the transport is carried out by Volkswagen Transporter. In order to achieve desired results, it was necessary to create a list of delivery routes for each working day of the week with the company owner. Subsequently, Friday route was selected for implementation of the method due to the least number of stops. On the basis of Friday delivery results, the Vogel approximation method will apply to all delivery routes.

It has been confirmed that using operational research can really improve transport-logistic processes in the selected company. According to the economic evaluation, annual costs may be reduced and the total number of travelled kilometres has been decreased.

This paper is supported by the research project "From horse-drawn railway to intermodal transport" within Visegrad Fund.

References

1. N. Ganikhodjaev, S.F. Zakaria, *International Conference on Operations Research and Statistics*, 111-116 (Singapore: Global Science and Technology Forum, 2011)
2. S. Kurian, H.N. Ramanathan, P.S. Chacko, Journal of Supply Chain Management Systems **2**, 4, 13-19 (2013)
3. J. Fotr, *Tvorba strategie a strategické plánování: teorie a praxe* (Grada, Prague, Czech Republic, 2012)
4. J. Široký, et al., *Technologie dopravy* (University of Pardubice, Czech Republic, 2014)
5. O. Stopka, M. Chovancová, J. Ližbetin, V. Klapita, Nase More **63**, 3, 195-199 (2016), DOI: 10.17818/NM/2016/SI22

6. M. Weiszer, G. Fedorko, V. Molnar, *CLC 2013 – Carpathian Logistics Congress*, 330-335 (Krakow, Poland, 2104)
7. L. Bartuška, J. Čejka, Z. Caha, Nase More **62**, 91-96 (2015)
8. J. Jablonský, *Operační výzkum: kvantitativní modely pro ekonomické rozhodování* (Professional Publishing, Prague, Czech Republic, 2007)
9. R. Sen, P. Rathindra, *Operations research: Algorithms and Applications* (Rekha Printing Pvt, New Delhi, India, 2012)
10. P. Fiala, et al., *Operační výzkum: nové trendy* (Professional Publishing, Prague, Czech Republic, 2010)
11. L. Bartuška, O. Stopka, J. Ližbetin, *Transport Means – 19th International Scientific Conference on Transport Means*, 215-218 (Kaunas: Kaunas University of Technology, Lithuania, 2015)
12. F.S. Hillier, G.J. Lieberman, N. Bodhibrata, B. Preetam, *Introduction to operations research* (Mathematics in science and engineering, New Delhi, India, 2014)
13. N.K. Tiwari, K. Shandilya Shishir, *Operations research* (PHI Learning, New Delhi, India, 2006)
14. O. Stopka, I. Simkova, Transport and Telecommunication Journal **16**, 4, 320-329 (2015), DOI: 10.1515/ttj-2015-0029
15. J. Štůsek, *Řízení provozu v logistických řetězcích* (C.H. Beck, Prague, Czech Republic, 2007)
16. T. Klieštik, *Transport Means - 17th International Conference on Transport Means*, 308-311 (Kaunas: Kaunas University of Technology, Lithuania, 2013)
17. V. Klapita, Logi – Scientific Journal on Transport and Logistics, **3**, 2, 57-64 (2012)
18. O. Stopka, B. Sarkan, M. Chovancova, L.M. Kapustina, Communications: Scientific Letters of the University of Zilina **19**, 2, 18-22 (2017)
19. J. Široký, *Transport technology and control* (Librix.eu, Brno, Czech Republic, 2014)
20. R. Kampf, J. Cejka, M. Telecky, Communications: Scientific Letters of the University of Zilina **18**, 2, 129-132 (2016)
21. G. Fedorko, H. Neradilova, M. Sutak, V. Molnar, *Transport Means – 20th International Scientific Conference on Transport Means*, 169-174 (Juodkrante, Lithuania, 2016)
22. M. Chovancová, V. Klapita, Open Engineering **7**, 1, 50-54 (2017), DOI: 10.1515/eng-2017-0009
23. F. Meszaros, R. Markovits-Somogyi, Z. Bokor, Logi – Scientific Journal on Transport and Logistics **3**, 1, 70-82 (2012)

Legislative Conditions for Intermodal Transport in Slovakia

Juraj Jagelčák[1,*], *Monika* Kiktová[1], and *Jaroslava* Kubáňová[1]

[1]University of Žilina, Faculty of Operation and Economics of Transport and Communications Department of Road and Urban transport, Univerzitná 1, 010 26 Žilina, Slovak Republic

Abstract. This article deals with conditions for intermodal transport in Slovakia. The first part of the article describes definition "combined transport" and "intermodal transport". Maximum allowed lengths, axle loads and masses of vehicles and vehicle combinations effective in Slovakia are given in summary figures. In the second part of the paper is described abnormal transport up to 60 tonnes in relation to a combined transport.

1 Introduction

In Slovakia, a term "combined transport" is defined in the Law no.514/2009 Z. z. on a railway transport where §2 section 4 defines the combined transport using the definition from the Council Directive 92/106/EHS (identical text is not applied in Slovak definition):
"Combined transport means the transport of goods between Member States where the lorry, trailer, semi-trailer, with or without tractor unit, swap body or transport container of 20 feet or more uses the road on the initial or final leg of the journey and, on the other leg, rail or inland waterway or maritime services where this section exceeds 100 km as the crow flies and make the initial or final road transport leg of the journey; — between the point where the goods are loaded and the nearest suitable rail loading station terminal of loading for the initial leg, and between the nearest suitable rail unloading station terminal of unloading and the point where the goods are unloaded for the final leg, or; — within a radius not exceeding 150 km as the crow flies from the inland waterway port or seaport of loading or unloading" [1].

The Council Directive 2015/719/ES states, that the limit of 150 km may be exceeded with intention to reach the nearest terminal suitable to provide intended service; the nearest suitable terminal may be placed in other Member State. The „intermodal transport operation" is defined in the Government Regulation no. 349/2009 Z. z. on a maximum permissible dimensions and mass of vehicles and combinations (in relation with exceptions for these transports) as follows:
The intermodal transport operation is:

- the combined transport operation (As defined in Law no.514/2009 Z. z. on a railway transport); the nearest suitable transport terminal for the envisaged service may be placed in

* Corresponding author: juraj.jagelcak@fpedas.uniza.sk

other Member state than the Member State in which the shipment was loaded or unloaded, or

- transport operations engaged in the transport of one or more containers or swap bodies, up to a total maximum length of 45 feet/13.716 m, using waterborne transport, provided that the length of the initial or the final road leg does not exceed 150 km in the territory of the Union; the distance of 150 km referred to above may be exceeded in order to reach the nearest suitable transport terminal for the envisaged service by the defined vehicles ; the nearest suitable transport terminal providing a service may be located in a Member State other than the Member State in which the shipment was loaded or unloaded.

The Government Regulation no. 349/2009 Z. z. defines also the term „Loader", similarly as it is defined in the ADR Convention. The Government Directive defines Loader as a person which loads goods into or onto a vehicle or into an intermodal loading unit or loads an intermodal loading unit onto a vehicle or vehicle combination.

Intermodal loading unit is defined as a container, swap body, semi-trailer or other similar loading unit suitable for use within intermodal transport.

2 Driving restrictions in road freight transport in Slovakia

The Law no. 8/2009 Z. z. on road traffic regulations sets restrictions of operation of some types of vehicles in § 39. Operation of motor vehicles and combinations of maximum allowed mass exceeding 7 500 kg and motor vehicles of maximum allowed mass exceeding 3 500 kg articulated with a trailer on highways, motorways and 1st class roads is forbidden on the first rest day and on Saturday following after working day from 1st July to 31st August from 07:00 to 19:00 and on Sunday and the last rest day from 00:00 to 22:00. If only one rest day follows after a working day, the restriction is effective from 00:00 to 22:00. These restrictions are totally effective during about 75 days a year [1-4]

The vehicles used within a combined transport, for loading and discharge of ships, railway wagons or aircraft are not subject to these restrictions. However, also vehicles carrying dangerous goods (according to the ADR Convention) are not subject to these restrictions in general as well as carriages related to the use of such vehicles.

In addition to this, driving a freight vehicles or vehicle combinations with a maximum total allowed mass over 12 000 kg is forbidden on the 3rd class roads, with the exception to serve specific area (mainly for vehicles being loaded or unloaded in a place accessible only by using such roads). Similar local restrictions are also on selected 2nd class roads (parallel with the toll roads). No specific exception from these restrictions applies to vehicles used within intermodal transports.

3 Maximum allowed vehicle dimensions and masses in road and intermodal transport

Maximum allowed vehicle dimensions and masses in Slovakia are based on the European Council Directives (96/53/ES, revised by 2015/719/EU), but some specific differences applies, some of them are applicable and important for intermodal transport.

A tandem axle means two axles in a row, the distance between the midpoints of which is not more than 1.8 m (partial wheelbase). A tridem axle means three axles in a row, the sum of partial wheelbases of which is not more than 2.8 m. A load on an individual axle of a tandem or tridem axle of semi-trailers and trailers must not exceed 10 t.

Maximum allowed lengths, axle loads and masses of vehicles and vehicle combinations effective in Slovakia are given in following figures (CT = combined transport).

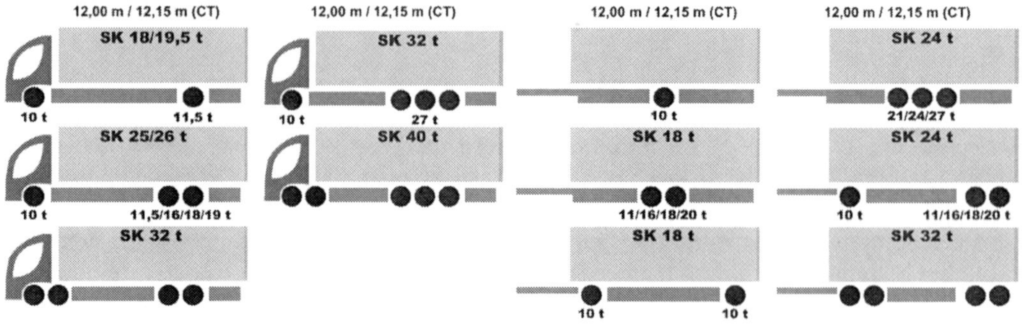

Fig. 1. Maximum parameters of typical vehicles and trailers. Source: [5]

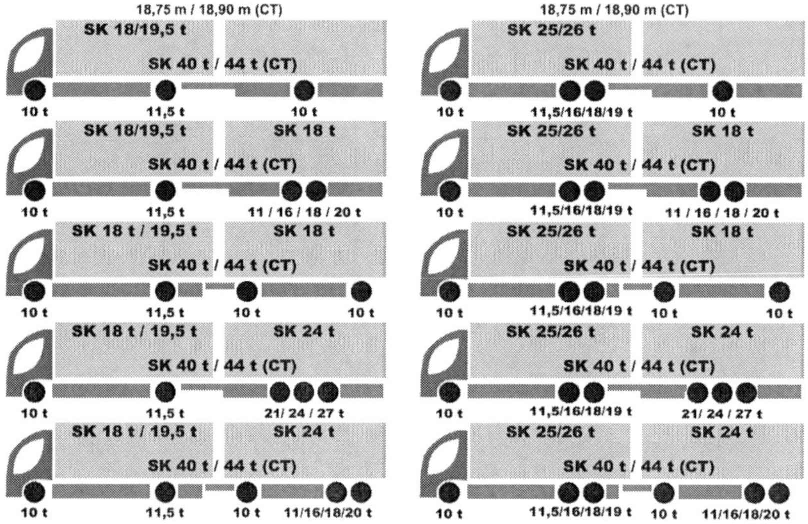

Fig. 2. Maximum parameters of typical road trains. Source: [5]

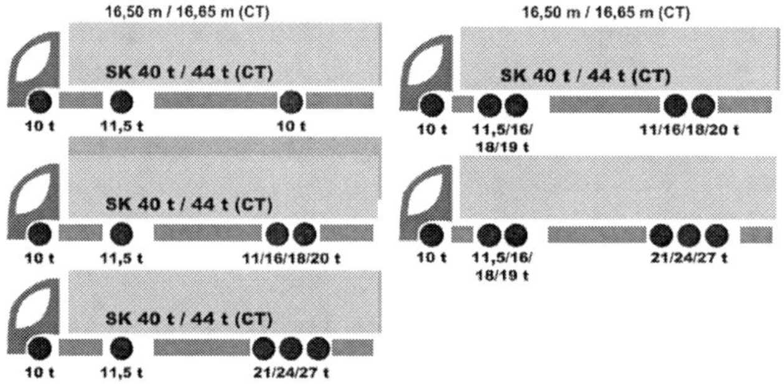

Fig. 3. Maximum parameters of typical semi-trailer combinations. Source: [5]

Following figure shows vehicle combinations consisting of a motor vehicle with two semi-trailers or with two trailers or with a semi-trailer and a trailer allowed in Slovakia (examples given, because legislation does not specify number of axles only the length and gross combination mass).

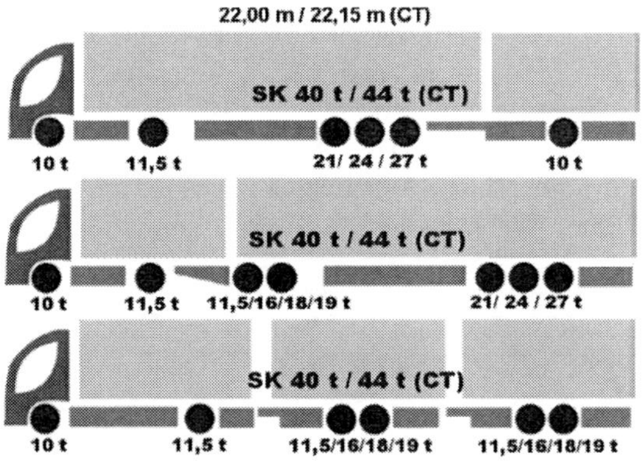

Fig. 4. Maximum parameters of vehicle combinations longer than 18.75 m. Source: [5]

The following figure present semi-trailer combinations (2-axle tractor with 3-axle semi-trailer or 3-axle tractor with 3-axle semi-trailer) typically used for intermodal transport of containers, tank containers, swap bodies and cranable semi-trailers. Vehicle combination 2+3 is mostly used.

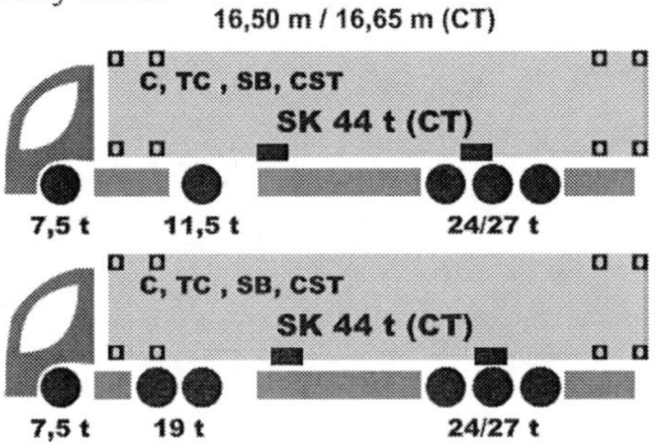

Key:
C = container
TC = tank container
SB = swap body
CST = cranable semi-trailer
CT = combined transport

Fig. 5. Typical semi-trailer combinations used for transport of various intermodal loading units. Source: [5]

The distance measured horizontally between the axis of the fifth-wheel king pin and any point at the front of the semi-trailer must not exceed 2.04 m. Maximum distance between the axis of the fifth-wheel king pin and the rear of a semi-trailer is 12.00 m / 12.15 m (CT).

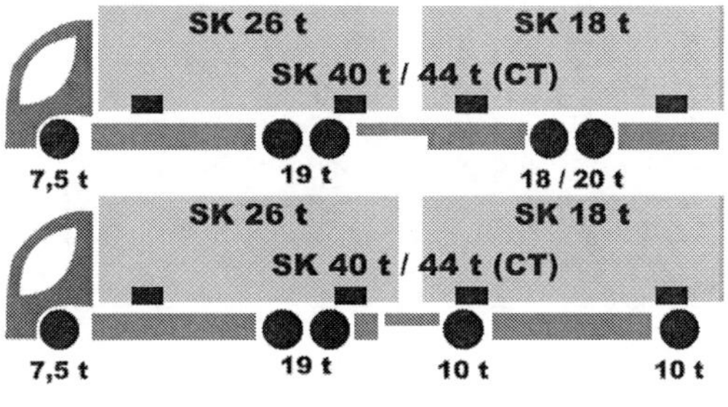

Fig. 6 Typical road trains used for a transport of C-swap bodies 22'-24' and 20' containers. Source: [5]

For road train the following condition (based on the requirement in the Directive 96/53/ES) has to be met: Maximum distance measured parallel to the longitudinal axis of the road train from the foremost external point of the loading area behind the cabin to the rearmost external point of the trailer of the combination, minus the distance between the rear of the drawing vehicle and the front of the trailer is 15.65 m. It corresponds with length of two swap bodies C782 carried on a road train.

Maximum distance measured parallel to the longitudinal axis of the road train from the foremost external point of the loading area behind the cabin to the rearmost external point of the trailer of the combination is 16.40 m. The distance between the rear axle of a motor vehicle and the front axle of a trailer must not be less than 3.00 m.
This requirement is not applicable for road trains with two trailers or with both a semi-trailer and a trailer.

3.1 Comparison of selected specific parameters of maximum allowed dimensions and mass of vehicles and vehicle combinations relevant for combined transport

Table 1. Values valid in SLOVAKIA in comparison to EU directive are presented in following table. Source: [5]

Parameter	Maximum limit Slovakia	Maximum limit EU (2015/719/EU)
Max. allowed length of a semi-trailer combination	16.50 m[1]	16.50 m
Max. allowed length of a road train	18.75 m	18.75 m[2]
Max. allowed length of a combination consisting of a motor vehicle with two semi-trailers or with two trailers or with a semi-trailer and a trailer	22.00 m	16.50 m or 18.75 m (such category is not specifically defined)
Max. allowed gross mass of a semi-trailer combination used in a combined transport – two-axle tractor + three-axle semi-trailer	44 t	42 t for transport of containers/swap bodies; 40 t for transport of cranable semi-trailers
Max. allowed gross mass of a semi-trailer combination used in a combined transport – three-axle tractor + three-axle semi-trailer	44 t	44 t for transport of containers/swap bodies; 40 t for transport of cranable semi-trailers
Max. allowed gross mass of a road train used in a combined transport	44 t[3]	40 t
Max. allowed gross mass of a vehicle combination used otherwise than in a combined transport	40 t	40 t
Max. allowed gross mass of a road train with four axles consisting of a two-axle motor vehicle and a two-axle trailer	40 t not specifically defined but regarding particular vehicle masses only 18 +18 = 36 t 19.5 +18 = 37.5 t	36 t
Max. allowed mass of a semi-trailer	40 t	36 t

[1,3] These lengths may be exceeded by 15 cm for vehicles/vehicle combinations operating in a combined transport in general.
[2,4] These lengths may be exceeded by 15 cm for vehicles/vehicle combinations engaged in the transport of 45-foot containers or 45-foot swap bodies, empty or loaded, provided that the road transport of the container or swap body in question is part of an intermodal transport operation in vehicles or vehicle combinations.
[3] If it is allowable for such type of combination regarding a maximum allowable axle load/mass of particular vehicles articulated in a combination; e.g. a combination of a two-axle motor vehicle and a three-axle trailer, has the maximum allowable gross mass of 42/43.5 t due to max. gross masses of particular vehicles (18 t +24 t or 19.5 t + 24 t).

Parameter	Maximum limit Slovakia	Maximum limit EU (2015/719/EU)
combination with four axles consisting of a two-axle motor vehicle and a two-axle semi-trailer, if the distance between the axles of the tandem axle of semi-trailer is 1.3 m ≤ d ≤ 1.8 m	but regarding particular axle loads as a theoretical maximum 10+11.5+18 = 39.5 t	
Max. allowed mass of a semi-trailer combination with four axles consisting of a two-axle motor vehicle and a two-axle semi-trailer, if the distance between the axles of the semi-trailer is greater than 1.8 m	40 t / but in combined transport may be reached as a theoretical maximum regarding particular axle loads 10+11.5+20 = 41.5 t	38 t when the maximum authorized weight (MAW) of the motor vehicle (18 tonnes) and the MAW of the tandem axle of the semi-trailer (20 tonnes) are respected and the driving axle is fitted with twin tyres and air suspension or suspension recognized as being equivalent
Maximum allowed load on tridem axle of motor vehicles		
Sum of distances between the axles ≤ 2 m	27 t	Not defined

Related types of vehicles in term of allowed gross combination mass 44 tonnes:

- vehicle combination transporting containers within the combined transport,
- combination of a tractor and a cranable semi-trailer within the combined transport,
- vehicle combinations carrying a tank container or a swap body within the combined transport.

Therefore this allowed gross mass is applicable for transport of containers, swap bodies, as well as cranable semi-trailers within a combined transport regardless of the type of vehicle/vehicle combination used (however, requirements for vehicle gross mass or axle loads remain a limiting factor). Gross mass of 44 t is not applicable to the transports within the RoLa system, where a standard gross combination mass of 40 t is applicable [1, 6, 7].

European Directive 2015/719/EU prescribes increased maximum allowed gross mass in intermodal transports for two-axle motor vehicle with three-axle semi-trailer carrying, in intermodal transport operations, one or more containers or swap bodies, up to a total maximum length of 45 feet gross combination mass 42 tonnes and for three-axle motor vehicle with two- or three-axle semi-trailer carrying, in intermodal transport operations, one or more containers or swap bodies, up to a total maximum length of 45 feet gross combinations mass 44 tonnes.

So the increased mass limits are not valid for transport of cranable semi-trailers because they are defined only for transport of containers or swap bodies according EU directive. Higher payload of cranable semi-trailers in intermodal transport can be used only when both a state of origin and a state of destination apply higher limits for such transports in their national legislation as Slovakia, Czech Republic or Italy does. Maximum mass of 44 tonnes may be applied only for combinations with a three-axle tractor, in other cases a maximum mass of 42 tonnes is applicable (so the increased limits in some cases do not cover even a tare mass of ILU + higher mass of three-axle tractors). Both above stated mass limits are defined only for semi-trailer combinations in EU directive. They are not applicable neither for trailer combinations (road trains), typical used e.g. for transport of two C-swap bodies (those tare mass of a about 4 tonnes lowers a payload of such road train), nor for modular road trains with both a semi-trailer and a trailer, allowing a transport

DOI: 10.1051/matecconf/201713400020

of a container and a swap body/transport of two containers together (however, also a problem of length of such road trains has to be considered). [8, 9]

Particular Member States can define the mass limits differently and do not have to prescribe higher limits than those stated in the Directive 2015/719/EU. On the contrary, they can prescribe some specific legislative rules.

In case of semi-trailer combinations, only gross combination mass (GCM) and an axle loads are considered in Slovakia. Gross mass of particular vehicles in the vehicle combination (tractor, semi-trailer) is not considered. European Directive does not use such approach, so it may be inapplicable in legislation of other countries. If maximum gross mass of a tractor and a tridem axle of semi-trailer is considered, the GCM of a combination consisting of a 2-axle tractor and a 3-axle semi-trailer must not exceed 42 tonnes (18 t + 24 t) according EU directive.

In Slovakia, maximum allowed load on an extended tridem axle (sum of distances between axles maximum 2.8 m) of semi-trailers/trailers is 27 t (it means, when one of distances between the axles is over 1,4 m). This is not defined in the EU directive but it is broadly used by manufacturers of containers chassis. The combination of particular wheelbases 131 + 141 cm or 141 + 131 cm (instead of a standard combination of wheelbases 131 + 131 cm) are usually used for such purpose. The 27 t load on tridem axle is limited by a maximum usually used technical load of axles (9 t/axle) and tyres of standard semi-trailers (load index 4 500 kg per wheel). Use of such semi-trailers allows to reach 44 t GCM in combined transport using 2-axle tractor and a 3-axle semi-trailer equipped with the extended tridem axle (18 t + 27 t).

Slovak legislation specifically defines vehicle combinations with two semi-trailers or two trailers or with a semi-trailer and a trailer. Maximum allowed length/GCM of such combinations is 22 m / 40 t (or 22.15 m / 44 t when they are used within combined transport). However, such length is not sufficient for the European modular system (EMS). A maximum length of 25.25 m and a maximum mass up to 60 t used for EMS in Nordic countries is subject to permission for abnormal transport in Slovakia. Following requirement applies for such road train with two trailers: a mass of the middle trailer has to be higher or equal to a mass of the last trailer in the combination.

A tolerance for axle loads when bulk loads are carried can be mentioned as another specific aspect of Slovak legislation. When bulk loads are carried, maximum allowed load on a single axle, tandem axle and tridem axle may be exceeded by up to 6 %, while a maximum allowed gross vehicle/combination mass is not exceeded (349/2009 Z. z., §4). This is applicable for a transport of bulk, unpacked solid substrates or items in vehicles, vehicle combinations, swap bodies, containers or tanks, where it is unable to secure the substrates/items to prevent movement and the cargo may move during standard transport operation, mainly due to emergency braking. This is not applicable for packed cargo, liquid and gaseous substances carried in tanks [10, 11].

4 Conclusion

Vehicles and vehicle combinations used in intermodal transport in Slovakia have to be accompanied with a declaration issued by a Loader to a road carrier. Gross mass of carried intermodal loading unit as well as a gross mass of the vehicle/vehicle combination carrying the intermodal loading unit must be presented in the declaration. The declaration shall be carried on the vehicle/vehicle combination operating on public roads (349/2009 Z. z., §5, sec. 8).

It is recommended that a declared mass should be as accurate as possible and verified by weighing (using processes according to a Methodical instruction 39/2016 on the VGM in accordance with the SOLAS Convention).

In intermodal transports, a way of loading of intermodal loading units has a strong influence on a load distribution on particular axles. Containers should be loaded in such way, that eccentricity of a centre of gravity should not exceed 5 % (not more than 60 % of cargo mass in a half of a container) [6, 10, 11].

It is very difficult to obtain accurate information about the position of a centre of gravity of ILU. Such information is usually not provided within standard business relations in comparison to the information about ILU gross mass and cargo mass.

Results of the analysis of a maximum total container gross mass in dependence on various position of a centre of gravity point at the fact that a position of containers on semi-trailer container chassis is more suitable for higher allowed load of a kingpin together with the use of 3-axle tractors, because a centre of gravity may be located in the middle or slightly forwards for a maximum payload [6, 12]. Using of three-axle tractors is advantageous if higher gross mass of vehicle combinations (more than 44 tonnes) is allowed. Use of 3-axle tractors is not efficient in Slovakia because they increase vehicle combination tare mass and lower the payload with GCM of 44 t. Heavier vehicle combinations can be used as abnormal transports with the permission.

This paper is supported by the research project "From horse-drawn railway to intermodal transport" within Visegrad Fund.

References

1. J. Kubanova, C. Schmidt, Communications: Scientific letters of the University of Žilina **18**, 2, 104-108 (2016)
2. O. Stopka, M. Chovancová, J. Ližbetin, V. Klapita, Naše more **63**, 3, 195-199 (2016)
3. T. Skrúcaný et al., *DYN-WIND'2017* (MATEC web of conferences **107**, 00072, 2017)
4. J. Kubáňová, B. Poliaková. *Transport means 2016 – 20th international scientific conference*, 22-26 (Juodkrante, Lithuania, 2016)
5. J. Jagelčík, *Nakladanie a upevňovanie nákladu v cestnej doprave* (University of Žilina, Slovak Republic, 2015)
6. J. Jagelčák, I. Kubasáková, Naše more **61**, 5-6, 106-116 (2014)
7. J. Ližbetin, O. Stopka, V. Cempírek, Naše more **63**, 3, 156-160 (2016)
8. J. Zamecnik, J. Jagelcak, Communications: scientific letters of the University of Žilina **17**, 4, 21-27 (2015)
9. J. Ližbetin, J. Ponický, V. Zitrický, Naše more **63**, 3, 161-169 (2016)
10. J. Vrabel, J. Jagelcak, J. Zamecnik, J. Caban, Procedia Engineering **187**, 89-99 (2017)
11. J. Zamecnik, J. Jagelcak, S.S. Peterssen, Communications: scientific letters of the University of Žilina **18**, 2, 76-82 (2016)
12. I. Kubasakova, J. Jagelcak. Communications: scientific letters of the University of Žilina **18**, 2, 109-112 (2016)

Potential for Intermodal Transport of Chemical Goods in Slovakia

Juraj Jagelčák[1,*], *Ján* Zámečník[1], and *Monika* Kiktová[1]

[1]University of Žilina, Faculty of Operation and Economics of Transport and Communications Department of Road and Urban transport, Univerzitná 1, 010 26 Žilina, Slovak Republic

Abstract. This article deals with intermodal transport of chemical goods in Slovak republic. Analysis is based on information from interviews with companies and logistics service providers. The first part of the article describes importance of Intermodal transport and basic transport routes for intermodal transport. Respondents considered advantages and disadvantages of intermodal transport. Possible improvements inside companies and improvements of external framework conditions to promote modal shift are described in the second part of the paper.

1 Introduction

This paper was completed with information from interviews with 5 manufacturing companies and 6 logistics service providers (LSP). It is necessary to say that many companies are not willing to provide related information because they consider them as confidential business data.

Bigger industrial enterprises evaluate the importance of intermodal transport as significant. Intermodal transport is also a perspective solution for future due to actual situation of transport market. There are several problems in the field of road transport – such as lack of drivers with worsening trend (less attractive of the profession due to its characteristics – long time away from home, safety and security risks [1-4], high costs for driver licence and qualification card,...) or increase of costs (toll, minimum wages in some of the European countries,...). As an additional problem (as is resulting from the statistics mentioned above) can be mentioned the differences between export and import cargo flows [5] from/to particular European countries, resulting to problematic backhaul. Therefore, a lack of free vehicles for the cargo may occur as a result of these factors (worse availability of some types of vehicles, e.g. tank vehicles). Using of intermodal transport may cover intended transport amount growth in future. Intermodal transport is important mainly for long-distance transports [6-8] and for oversea transports. The individual relations where the intermodal transport using brings an advantage depend on offer of intermodal connections [6-9]. Existing intermodal services are mainly focused on transport to the important European maritime ports – Hamburg, Rotterdam, Koper, Bremerhaven, so the overseas transport over these ports is mostly done intermodal [6] and on the other hand, the overseas

* Corresponding author: juraj.jagelcak@fpedas.uniza.sk

transport play an important role in performances of intermodal transport. The lower transport costs are reached by using the intermodal transport in these relations and transit time is also favourable. One of the respondents expressed that some of the production sites depend on the intermodal lines for their productions, as they are 100% supplied by intermodal transport [9]. Among further important benefits stated by the respondents can be mentioned moving storage forward in the chain = having a warehouse closer to production sites allowing for safety stock and better lead times between order and delivery [9] and intermodal transports are still done during holidays, such as the x-mas / new year period which is a dangerous period for the supply chain of some sites. Ships and trains will still go, so shipments by container are a huge benefit in that period [6].

2 Important Transport Routes

An offer of intermodal lines is an important factor of potential of using intermodal transport. The most important intermodal operators in Slovakia are Metrans, a. s. and Rail Cargo Operator – CSKD, s. r. o. They operate mainly a container trains to/from maritime ports, although some continental lines are also available. The respondents – industrial companies named following intermodal transport routes which they use or consider as a potential for intermodal transports to ports Hamburg, Koper, Rotterdam, for transport of 45ft units [10] to England, to south France/north Spain, from north Slovakia to north Germany – other than ports, from Slovakia to Turkey (save 2 days – min. risk), North West Slovakia – mid UK, Šal'a (SK) – North UK, North Portugal and North Spain, Belgium – Malacky – Terminal Dunajska Streda. However, significant part of transports are the transports to neighbouring countries. Using of intermodal transport to these countries is not so advantageous due to shorter distances [11] and not well-developed network of intermodal terminals. If intermodal transports to neighbouring countries with a transport distance about 500 km would be used, using of cranable semitrailers should be more suitable than system based on containers – but terminals in Slovakia are not equipped to handling them. If we talk about Balkan countries (Romania, Bulgaria, Serbia), offer of intermodal rail lines to these countries is very limited yet [6]. But the amount of international trade with these countries is less than that with West-European countries.

3 Potential for modal shift

Respondents – industrial companies and distributors of chemicals reported such share of intermodal/intermodal transport on total transport performances is 8 %; 12 %; 32 %; approx. 50 % of inbound deliveries. One respondent reported the share of intermodal transport as very low within Slovakia (but having higher level in other countries). In presence, the respondents use intermodal transport in these routes from/to marine ports – 5 - 7% of transport volume, marine containers from/to company by truck/single wagons, transport of raw materials – import from places out of Europe, transport of semitrailers of ferries from ports in Germany to Scandinavian countries (Norway, Sweden, Finland) as well as between ports in Denmark and the United Kingdom (used in approx. 15 % of transports performed by the respondent company), intermodal transport of semitrailers by rail in relations Denmark – Austria/Italy, Sweden – Norway, Norway – the Netherlands.

Overseas transports of chemical commodities are already performed mostly intermodal using containers (existing intermodal services to important European ports are well-developed), alternatively using block trains and reloading in port terminals in case of large-quantity bulk substrates. New potential of growth is therefore mainly in continental intermodal transports. Main motivation for modal shift from road to rail transport by the enterprises´ point of view is in these fields in the field of transport to/from marine ports, Nordic countries road-rail-sea (cost and transit time advantages), for transport to/from Turkey, higher efficiency for heavy goods. Maximum allowed weight of road combination used in intermodal transport is 44 tonnes (versus 40 tonnes in direct road transport) according to Slovak legislation. This becomes an advantage mainly when cranable semitrailers are used as ILU, because a difference of tare weight (curb weight) between cranable and standard semitrailer is much lower than a tare weight of container/swap body, therefore the payload of cranable semitrailer is higher (up to 29 tonnes). But following limiting factors have to be considered as load distribution on axles of tractor and semitrailer (this has to be considered even before loading a container/swap body, because a distribution of cargo in ILU affects position of ILU´s centre of gravity and subsequently distribution of load on particular axles of road vehicle carrying the ILU) [10] and maximum mass of road combination allowed also in the destination (origin) country on the other side of transport chain. The Council Directive 2015/719/EU prescribes mass 42 tonnes in intermodal transport of containers and swap bodies by standard combinations with 2-axle tractors. The Council Directive does not prescribe any higher weight limit than usual 40-tonne limit for transport of cranable semitrailers within intermodal transport).

Another aspects are better situation with free capacities of intermodal transport in comparison with road transport (which is affected by lack of drivers, tank vehicles...). Companies consider intermodal transport as a perspective solution which can cover intended production growth in future. However, there is a deficit of intermodal transport units (containers) in presence yet.

The respondents consider mainly the following routes as perspective to shift transport from road to intermodal for trains from Zilina to north Germany – one time per week, Ruzomberok to Germany, Czech republic, Italy – one train per week, from west Slovakia/Bratislava to France, from Sala to South France / North Spain, from southwest Slovakia to Ludwigshafen (DE), Colmar-Berg (LUX).

One of the respondents see a potential also in transport of cargo from Ukraine to Europe due to advantage - elimination of delays of trucks on border crossing. A problem is, that amount of international trade between Slovakia and Ukraine (as well as other east-European countries) is low (if we do not take imports of bulk raw materials such as crude oil, natural gas or iron ore into account – such commodities are unsuitable for intermodal transports). Moreover, east-European countries of former USSR have different railway gauge (1 520 mm). Intermodal services can be efficiently operated only when a sufficient amount of cargo flow (optimally – sufficient for operating several trains per week due to flexibility) exists to/from particular region in some of these countries (area surrounding a terminal up to approx. 150 km). We have also take into account that travelling speed in railways in east-European countries are lower than those reached in west-European railways [12]. The same situation is in Balkan countries.

One of the respondents see also a potential for perspective extension of intermodal transport using as a possibility of deliveries to customers – motivated by lack of tank vehicles.

A respondent – intermodal operator see a potential of intermodal transport for dangerous goods transport due to safety improvement (lower risk of accidents, better safety of road traffic) and simplified handling with dangerous goods transported in tank containers.

4 Benefits and advantages of Intermodal transport

The respondents reported mainly lower costs, sufficient free capacities also shorter transit time in specific routes as key advantages of intermodal transport. They considered environmental and social responsibility aspects as less important factors. In general, the

 DOI: 10.1051/matecconf/201713400021

advantages of intermodal transport were considered as lower transport costs, regular connection, short transit time in specific relations or periods (e.g. around Christmas), good transport capacity, possibility of transport 45ft containers, ability to supply sites despite of holiday restrictions, allowing to setup safety stock closer to production sites, possibility of using of additional services such as issuing VGM declaration in terminals, it is the only one possibility if vendors/consignees are located outside Europe, diversification of transport modes, higher efficiency for heavy goods in relation with higher allowed weight of trucks. This is an advantage mainly in intermodal transport using portable semitrailers with higher payload, better safety of dangerous goods transport, better quality of transport services to localities covered by well-developed and advantageous intermodal services. One of the respondents referred that intermodal transport is more flexible in comparison with road transport for its purposes due to lack of tank vehicles. Intermodal transport does not have to be arranged several weeks in advance. In general, it can be said that companies in Slovakia focus on the lowest prices much more than on conceptual solutions.

5 Barriers and disadvantages of Intermodal transport

As barriers and disadvantages of multimodal transport shall be considered lack f ILU's for modal shift; small motivation of road hauliers for modal shift; small motivation to buy cranable semitrailers; lack of ILU's for packed load in Europe; strong difference between export and import flows – therefore missing backhaul of intermodal transport units; longer transport duration in comparison with the road transport in many cases. It relates mainly with transports to shorter distances.

Further barriers are destination without direct connection (services) or less frequented intermodal services. There is not well-developed network of intermodal terminals in some regions related with transit duration and costs, longer road pre-carriage/on-carriage is needed. Total distance of transport becomes longer and costs become higher.

Lack of intermodal terminals is another barrier because most of the terminals in Slovakia are located in the southwest and southeast. Only the terminal in Žilina with limited extension possibilities (due to location in city centre) is available. Further terminal in Žilina – Teplička is not operated yet. The terminal in Ružomberok (close to cellulose plant) is almost unused. Some regions of Slovakia (north and northeast, south part of middle-Slovakia) are quite far from the intermodal terminals and motorways in these regions are not developed.

Further barriers is delay probability which is not lower than for road transport, but the effect is higher. Sometime there is a sufficient cargo volume but the costs for intermodal transport are simply too high.

Local legislation and restrictions e.g.:

- France: limited corridors for intermodal transport,
- allowed weight of road vehicles (and as a result - weight of cargo in ILU) is different in particular countries, not allowing optimization of the load unit. This factor limits possibility of practical use of higher payload of cranable semitrailers, because maximum allowed weight of road combination has to be followed both in the country of origin and the country of destination;

Other barriers:

- transport document do not always accompany the cargo or container when handled at a terminal;
- missing flexibility – a shipper is mostly bound to one operator or its partners;
- risk of failure to comply with the delivery time in case of just-in-time deliveries;
- the solution is heavily dependent on backflow;

- some destinations are not covered with intermodal services (France, Spain, Scandinavia, Balkan, middle and south Italy, localities in northeast part of Germany) – not covered at all or not in sufficient frequency which considerably determines flexibility and door-to-door transport time;
- a significant part of cargo flows to/from Slovakia bounds to/from neighbouring countries and using of intermodal transport is not advantageous for shippers (due to costs and transit time);
- if no suitable (direct) intermodal service is available, transport costs may be higher. The same situation is in case of short-distance transports;
- for products carried at elevated temperature – longer transit times cause a need to heating up before discharging;
- some of terminals haven´t conditions for chemical goods – storage of dangerous goods is restricted;
- System of single wagon or group of wagons isn´t applicable – complicated arrangement, long transport duration, higher risk of cargo damaging due to shunting shocks...[12];
- high financial amount (investments) is needed to acquisition of special equipment needed for intermodal transport operation without any guarantees of sufficient amount of transported cargo.

6 Possible Improvements inside companies to promote modal shift

Industrial companies reported following proposals to promote modal shift/extent the use of intermodal transport:

- expand sales to overseas countries (it would result into growth of traffic mainly on existing intermodal lines to maritime ports);
- modal shift from road to intermodal transport in regular relations of supply/distribution, which are simply accessible from intermodal terminals and do not need just-in-time deliveries;
- exchange of experience with start of terminal operation in company and transhipment systems;
- intermodal transports other than intermodal (loading of wagons in a production site on a railway siding – rail transport to warehouse in destination country – storage – road transport to final consignees) are complicated in logistics – operation of warehouses, transhipment of cargo in localities close to consignees etc. There are following possibilities to use intermodal transport:
  - using of intermodal terminals close to a site of consignor/consignee,
  - bigger production sites can also load/unload containers at the own area on railway sidings using own handling equipment. As an example can be mentioned the container side lifter used in one chemical company. Such type of handling equipment is suitable for operation in industrial areas due to its flexibility, lower space needed for handling in comparison with cranes or reach stackers operating in terminals.
- The companies in which the share of intermodal transport is already big, identify no further significant possibility for further increase of the share, because they already use intermodal transport where it is economically efficient. They have already attempted many different options to increase intermodality – the only limitation are costs. There needs to be a saving which is not always the case for transport routes which they investigate for. New depots are also required to be able to (off)load containers.

Logistics services providers reported following proposals for the improvement of modal shift:

- help with organisation of transhipment/operation of wagons/tank containers,
- ongoing (year on year out) investments into vehicle and wagon fleet suitable for intermodal transport,
- improvement of additional services offer – services in customs clearance, building, extending and modernization of intermodal terminals, modernization of terminal equipment, new locomotives for shorter transit durations, providing VGM declaration services.

7 Possible Improvements of external framework conditions to promote modal shift

Industrial companies propose to exchange experiences for start of transhipment – e. g. more chemicals companies became interested for system of side loaders; to improve loading capabilities of containers or other modalities at primary loading sites, and onsite intermodal terminal eliminating a shuttle transport between primary loading site and a railway/shipping terminal; lower railway infrastructure fee for intermodal transport; to establish state support of building company transhipment sites in areas of industrial enterprises (primary loading sites), acquisition of handling equipment etc. with grants; more support engaging of individual containers into block trains → motivate national rail operators and important private operators; to improve accessibility of intermodal terminals for road transport including remote marking of access routes; to complete missing intermodal terminals in some regions; to support construction of terminals and acquisition of handling equipment by state grants which is currently not enough (Higher support is e.g. in Czech republic (up to 50 % of costs) which brings better efficiency.).

From logistics service providers point of view there is a necessity of state investments into modernisation and improvement of infrastructure – enough capacity and speed has significant influence on transport duration.

8 Conclusion

A situation with intermodal transport and its use for transport of chemical goods is very similar in middle-European countries. Production sites of chemical industry (=sources of chemical goods transports) are usually relatively large and many of them are equipped with railway sidings. Intermodal transport is used mainly for transports from/to seaports within overseas transports in presence. Such transports are, in the field of chemical goods, already performed mostly intermodal and a potential of future growth is only in extension of amount of overseas transport in general. A better potential for intermodal transports growth is in continental transports for longer distances. Such potential is limited by shortcomings in offer. Existing services are focused mainly on seaports and partially on some centres in Rhine and Ruhr region, north Italy, Turkey etc. There is a lack of intermodal services to other regions (France, Spain, Great Britain, some regions in Germany, Balkan countries etc.) – where the connections from Slovakia are complicated or completely missing. There is also a lack of intermodal loading units (containers, swap bodies, intermodal trailers) for continental transports and an intermodal system of cranable semitrailers is not developed in Slovakia yet because some terminals in Slovakia are not equipped with handling devices and there is a lack of suitable wagons. Better situation is in continental transports of tank containers, where the respondent reports higher availability of capacities and better flexibility of intermodal transport in comparison with road transport.

The main factors causing the enterprises are willing to use intermodal transport are lower costs and/or favourable carriage-operational factors. Most of the enterprises expect longer transport duration of intermodal transport in comparison with road transport. But there are some specific cases in which the intermodal transport may reach shorter transport duration (e.g. in case of long transport distance, in relation with restrictions in road transport during weekends, public holidays, which do not apply on intermodal transport). Other aspects such as safety or environmental aspects are in general considered as less important for making decision about transport mode.

Following proposals resulting from previous analyses can be presented on behalf of ZCHFP SR (The Association of Chemical and Pharmaceutical Industry of the Slovak republic).

It is necessary to improve the offer of intermodal transport units for chemical goods – portable tanks, containers. It is necessary to create a support for a system of transport of cranable semitrailer by rail – this intermodal system is not developed in Slovakia yet. Because mostly large enterprises, equipped with railway sidings, operate in the field of chemical industry, it is suitable to improve and develop possibilities of transhipment rail – road in companies. This also solves a problem of dangerous goods storage limitations/prohibitions in some intermodal terminals.

This paper is supported by the research project "From horse-drawn railway to intermodal transport" within Visegrad Fund.

References

1 J. Zamecnik, J. Jagelcak, S.S. Peterssen, Communications: scientific letters of the University of Žilina **18**, 2, 76-82 (2016)

2 J. Vrabel, J. Jagelcak, J. Zamecnik, Caban, Procedia Engineering **187** (2017)

3 T- Skrúcaný, et al., *DYN-WIND'2017* (MATEC web of conferences **107**, 00072, 2017)

4 J. Kubáňová, B. Poliaková, *Transport Means 2016 – 20th International Scientific Conference on Transport Means*, 22-26 (Juodkrante, Lithuania, 2016)

5 O.Stopka, M. Chovancová, J. Ližbetin, V. Klapita, Naše more **63**, 3, 195-199 (2016)

6 J. Ližbetin, J. Ponický, V. Zitrický, Naše more **63**, 3, 161-169 (2016)

7 J. Ližbetin, O. Stopka, V. Cempírek, Naše more **63**, 3, 156-160 (2016)

8 J. Kubanova, C. Schmidt, Communications: Scientific letters of the University of Žilina **18**, 2, 104-108 (2016)

9 I. Kubasakova, J. Jagelcak. Communications: scientific letters of the University of Žilina **18**, 2, 109-112 (2016)

10 J. Jagelčák, I. Kubasáková, Naše more **61**, 5-6, 106-116 (2014)

11 R. Rievaj et al., Communications: scientific letters of the University of Žilina **18**, 2, 57-61 (2016)

12 J. Zamecnik, J. Jagelcak, Communications: scientific letters of the University of Žilina **17**, 4, 21-27 (2015)

Proposal of an Enhanced Safety System on Board of the Inland Vessel

Martin Jurkovič[1,*], *Tomáš* Kalina[1], *Robert* Turčan[1], and *Bruno* Gardlo[2]

[1]University of Žilina, Faculty of Operation and Economics of Transport and Communications, Department of Water Transport, Univerzitná 1, Žilina 010 26, Slovak Republic
[2]AIT Austrian Institute of Technology GmbH, Center for Technology Experience, Giefinggasse 2, 1210 Wien, Austria

Abstract. The share of human factor in the occurrence of navigation accidents represents a high proportion of the total accident rate. Elimination of the human factor responsible for the occurrence of navigational accidents would contribute to the overall safety of navigation on inland waterways. The main causes of human-induced navigation accidents include inattention, fatigue or poor concentration on vessel control. The design of the system for increasing navigation safety consists of a system for monitoring the life functions of a crew member as well as a position of the vessel towards the fairway / shore. The first level focuses on tracking the life function of the vessel's crew / boat master. The second level addresses the position of the vessel and the possibility of increasing safety by cooperating with the data provided by Automatic Identification System - AIS. The proposal of the system represents interconnection of both levels and brings high degree of control of the vessel in extraordinary situations.

1 Introduction

The system of monitoring and recording important life functions or parameters of the driver's state of consciousness is not new in the automotive industry. Volvo has been working on such a system since 2005, but it worked on the principle of tracking secondary data, such as directional control, braking frequency, involuntary lane departure, etc. The control unit with a predetermined computational algorithm then evaluated the nonstandard behavior based on the data obtained. At present, these systems are even more refined [1, 2].

A new generation of systems should bring direct driver monitoring while driving, based on the sensors built into the dashboard. The sensor monitors the direction in which the driver looks ahead, how far he has open his eyes or in which position is his head. From the scanned data, it is subsequently possible to evaluate whether the driver is tired, asleep or disintegrating. In the case of a negative result, the software will automatically ensure that the vehicle does not break out of the lane or collide with another vehicle [3].

Due to the impact of the human factor in navigational accidents, similar systems should be installed on inland waterway vessels as well. Currently inland waterway vessels are not

* Corresponding author: martin.jurkovic@fpedas.uniza.sk

equipped with an early warning system or collision tracking system. Implementing at least one of the system could reduce the number of human-induced accidents.

2 Methodology of system operation

The system of monitoring and evaluating important life functions is a complex and detailed processing of the obtained data, therefore it is suitable for use on inland waterway vessels. It would consist of several sensors divided into two levels of monitoring. For each level, the following parameters would be monitored:

First level:

- eye movement – percentage of eye lid closure and tracking of the eye,
- Brain activity (diffuse optical tomography - DOT method),
- breathing and heart activity (Emotional Intelligence EQ – Radio Method).

Second level:

- the relative position of the vessel against the fairway (AIS system required),
- slowdown and acceleration of the vessel (AIS required).

Both algorithms would be an integral part of a complex system where the software would collect, monitor, and evaluate information coming from subsystems at a lower rating level. Based on the outcome of the assessment, the system would take all necessary measures to reduce or eliminate the potential hazard. The co-operation of each level is shown in Fig. 1 [4, 5].

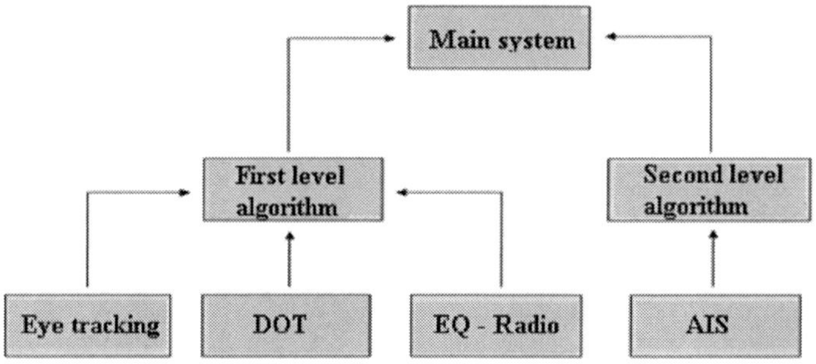

Fig. 1. Scheme of the monitoring system. Source: authors

2.1 Creating a network of sensors

The measurement of selected parameters would be ensured by the network of specific sensors. After collecting enough data, the data would be sent through CAN or "Controller Area Network" for evaluation to the control unit. CAN is a standard communication system used in automobiles to interconnect components such as an engine, transmission, chassis or brakes. CAN is a bus that is used for internal communication between sensors and functional units. It is used mainly in the automotive diagnostic system or in the industrial sphere for machinery and lines [6]. The network would consist of one complex or more specific CAN systems for each parameter.

2.2 Evaluation of obtained data

The evaluation would consist of two separate assessment algorithms, each for one level of tracking. In the system, both algorithms would be interconnected and co-operated. Their interaction is illustrated in Fig. 2.

 DOI: 10.1051/matecconf/201713400022

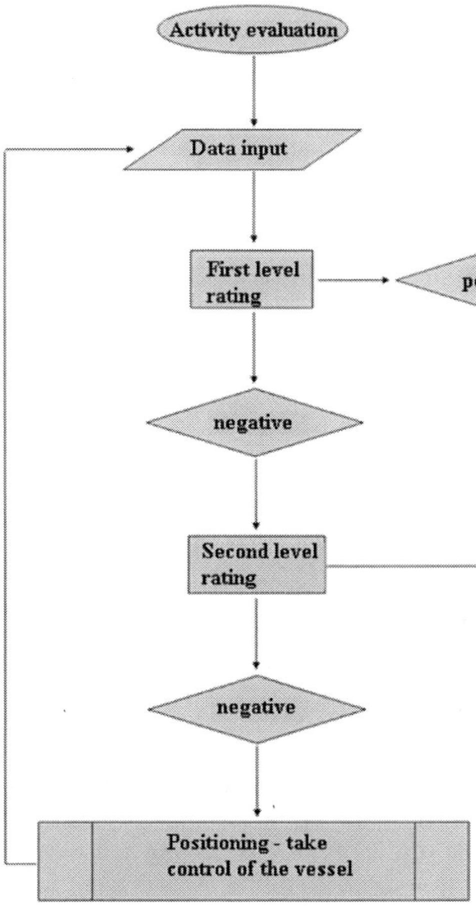

Fig. 2. Interaction of algorithms. Source: authors

3 Components of the monitoring system

The monitoring system would collect, monitor and evaluate information from individual subsystems. Based on its importance, the most appropriate measure will be adopted in terms of navigation and crew safety.

3.1 Eye movement – percentage of eye lid closure and tracking of the eye

Eye movement, more accurate motion and percentage closure of the eyelid, would capture the camera system installed on the bridge. Motion capture would be used by cameras with a certain scanning frequency. Due to the different velocity of eyelid movement during glans and fatigue, high-frequency scanning is required. For this reason, installation of a high-frequency camera is appropriate [7]. The algorithm of the system would work on the principle of recognizing the facial skin color as follows:

- recognition of skin and non-skinned pixels (facial and non-face recognition),
- rendering of the face part,
- pixel layout into multiple areas (definition of face and non-face areas).

Through a binary pixel set, the point that belongs to the face area is assigned a value of 1. The point that does not belong to the face area is assigned a value of 0. Based on the face's conversion to a black and white image, the outline of the shape, eyebrow, eyes, and part of the nose is drawn. The algorithm then calculates the coordinates of the eye according to the part of the eye (the eyebrow and the lower lid of the eye). Once the position of both eyes is detected, the system can record the intensity of the change from the closed to the open eyelid [7].

3.2 Monitoring of the brain activity

Monitoring of brain activity and measurement of individual essential functions would be performed using a relatively new, non-invasive brain scan method - DOT method. The method uses brain scan using LEDs. The results are comparable to magnetic resonance [8].

The method is referred to diffusion optical tomography and shows processes that are running at various levels of the brain. It works on the principle of transmitting light to the cerebral cortex. The light is then captured backwards. This process is followed by dynamic changes in the color of the tissue in the brain. Coloration occurs in ascending or declining cerebral activity, when neuronal activity increases in the brain.

To ensure the functionality of the system on the vessel (bridge), the following components are required:

- covering the entire bridge with LEDs,
- software for recording and evaluating brain activity.

3.3 Breathing and hearth activity

The analysis of breathing and cardiac activities without audio-visual or physiological data coming from the sensors on the human body is possible using EQ - Radio technology. The technology was developed and completed in 2016 at the Massachusetts Institute of Technology. It works on the principle of transmitting, reflecting and receiving a wireless signal to the human organism. The accuracy of heart rate measurement is the same as for the ECG. The error of measurement is only 0.3 percent [9].

The software processes the received data based on the selection of breathing rhythm and cardiac activity. When breathing, the system records a change in breathing rate (number of breaths and breaths over the reference period). Heart activity is evaluated based on a change in the number of heart strokes over the time. This can lead to the discovery of emotion as fear, excitement, disgust. The intensity of heart activity is directly proportional to fatigue. When falling asleep, heart activity decreases.

Analyzing and evaluating of the cardiac activity and respiration would consist of:

- positive / negative data evaluation;
- in the case of a positive evaluation, the whole cycle would be repeated, without warning,
- in case of a negative evaluation, the software will send the data to the control unit - the system will take all the necessary measures to eliminate the hazard.

3.4 The relative position of the vessel to fairway

The most optimal solution for tracking the vessel's relative position towards the fairway is the AIS (Automatic Identification System). AIS is used not only for inland navigation. AIS stores and displays position, speed, rate and vessel information (name, number, calling sign, vessel type). The system works on the principle of information exchange between coastal stations or directly between vessels [10, 11].

Vessels equipped with standard AIS stations receive and transmit data automatically at specified intervals. The advantage is cooperation (refill) with radar due to different characteristics of both systems. For integration into the second stage of the monitoring system, following information is needed:

- the precise position of the vessel,
- course and speed of the vessel.

The communication is via a digitized radio station, which can be complemented by a data transfer modulation device. Simple design is an advantage in terms of interconnection, since there is no need to install any additional elements into the vessel. The system would only be complemented by software for recording, monitoring and evaluating the optimal position of the vessel on the voyage. If the software evaluates that the position of the vessel to the voyage is not optimal, it would send the information to the monitoring system. The second evaluation algorithm would then alert the boat master [12].

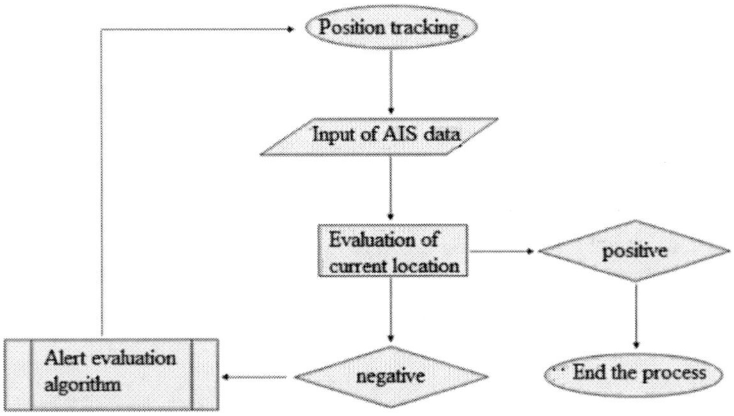

Fig. 3. Methodology of vessel tracking process in the voyage. Source: authors

An essential element in position tracking in combination with the evaluation algorithm is to determine the permissible deviation, since in the case of dynamic changes on the waterway it is not always possible to maintain the optimal system rate. For downstream and upstream navigation, it is necessary to specify a specific deviation for each type of voyage.

3.5 Acceleration and deceleration of the vessel

AIS can be used again to monitor vessel speed. The system provides information about the vessel speed based on satellite tracking. The system would have integrated software with its own evaluation algorithm. The software would process and evaluate critical vessel speed data. Depending on the negative or positive result, it would take appropriate measures as was mentioned above. In the case of a positive evaluation, the entire process starts from the beginning. On the other hand, in the case of a negative evaluation, the system would highlight a potential hazard.

A principal element of the speed tracking system is to determine the optimal speed of the vessel for individual sections of the waterway. The cooperation with the electronic map system is necessary, although speed monitoring is not necessary in all river sections. Therefore, only sections with increased risk of accidents would be monitored. Mainly:

- places with fords, shallows or straits,
- places with increased vessel density,
- anchorages and turning positions,
- port entrances, lock chambers,
- entrances to navigable bridges and tunnels,

- places where navigation is restricted.

4 Conclusion

At present, there is no system of early warning or monitoring of crew safety on inland vessels. The introduction of such system could significantly reduce accidents and related negative impacts. The design of the monitoring system consists of two levels. The first is to monitor the life function of the vessel's crew on the bridge. The latter tracks the optimal route, position and speed of the vessel in the voyage. Their interconnection and cooperation would ensure a high degree of control of the vessel in extraordinary situations. Modern, non-invasive systems have been selected to monitor life functions.

DOT Brain Measurement System or EQ - Radio System for Measurement of Heart and Breast Activity. Optimal tracking and location of the vessel would be ensured by AIS enhanced with software dedicated to evaluating the optimal and current position and speed of the vessel. It is a question of implementing such a system on vessels. New breakthrough technologies usually bring with them a huge cost of research and development that may be an obstacle to the implementation of such a project.

This paper is supported by the research project "From horse-drawn railway to intermodal transport" within Visegrad Fund.

References

1. R. Kampf, O. Stopka, L. Bartuska, K. Zeman, *Transport Means – Proceedings of the 19th International Scientific Conference on Transport Means,* 143-146 (Kaunas, Lithuania: Kaunas University of Technology, 2015)
2. M. Mikusova, J. Gnap, *CIT 2016* (Valencia, Spain, 2016)
3. T. Andrejcak, Pravda **7** (2014)
4. I. Kubasakova, R. Kampf, O. Stopka, Communications: scientific letters of the University of Žilina **16**, 2, 9-13 (2014)
5. J. Gasparik, V. Luptak, P. Mesko, *ICTTE – International conference on Traffic and transport engineering,* 388-392 (Belgrade, Serbia, 2016)
6. *CAN system*, A Serial Bus System, Available online: http://www.esd-electronics-usa.com/Controller-Area-Network-CAN-Introduction.html (2016)
7. J. Rakus, (STU Faculty of Electrical Engineering and Informatics, Bratislava, Slovak Republic, 2011)
8. A. Eggebrecht, Available online: https://www.ncbi.nlm.nih.gov/pubmed/25595183 (2016)
9. A. Conner, Simons, R. Gordon, *MIT*, 6-8 (Cambridge, USA, 2016)
10. *RIS, General Information,* Available online: http://www.ris.eu/general/what_is_ris_/ais (2009)
11. T. Skrucany, J. Gnap, Applied Mechanics and Materials **617**, 296-301 (2014)
12. J. Zamecnik, J. Jagelcak, J. Sosedova, *LOGI 2015*, 97-105 (České Budějovice, Czech Republic, 2015)

Information Support for Sustainable Tourism

Oldřich Kodym[1,*], *Jana* Kodymová[2], and *Simona* Matušková[3]

[1]College of Logistics, Přerov, Department of Master Studies, Palackého 1381/25, Přerov, Czech Republic
[2]VŠB – Technical University Ostrava, Faculty of Mining and Geology, Institute of Environmental Engineering, 17. listopadu 15/2172, Ostrava-Poruba, Czech Republic
[3]VŠB – Technical University Ostrava, Faculty of Mining and Geology, Institute of Economics and Control Systems, 17. listopadu 15/2172, Ostrava-Poruba, Czech Republic

Abstract. This paper focuses on possibilities of information provision for tourists so that they can get familiar more easily with new cultures and the "unknown" environment by means of commonly known and user friendly interfaces. Each object of interest can be equipped with a wireless identification tag that can serve as: 1) Pointer to already downloaded information and thus associate it with the real object in his proper environment. 2) Link for downloading of information in real time – if a wireless communication is available. 3) Direct provider of such information – if the identification tag has sufficient capacity. 4) Provider for feedback inclusive management of tourist movement so that the communication routes are not overloaded. Information support for such kinds of tourist areas is presented. Logistic model based on Petri nets is used for simulation and life cycle assessment following Ekoindikator 99 is evaluated. Advantages of described solutions are discussed.

1 Introduction

Tourism is now a major sector of the world economy, especially as it refers to the international trade in service. Tourism´s contribution to the world gross domestic product (GDP) is estimated at some 9 %, 1 in 11 of total jobs and to more than 50 % of international tourist arrivals in Europe.

In fact, tourism output is not a simple product but, rather, a wide range of goods and services interacting to fulfil a tourist experience that comprises both tangible parts (e.g. hotel, restaurant, airline) and intangible aspects (e.g. sunset, scenery, mood) [1]. The actual purchase and consumption/production of tourist services (e.g. airline ticket, meal, admission ticket) may often be incidental to "non-market" activities, such as independent sightseeing, hiking, or sunbathing [2].

Given the predictions of an increased role of tourism industries in the world economy, the environmental aspects of, and impacts generated by tourist activities should be accurately considered according to a Life Cycle Thinking (LCT) and logistics perspective [3].

* Corresponding author: oldrich.kodym@vslg.cz

DOI: 10.1051/matecconf/201713400023

To get a clearer, we can introduce the individual stages related to tourism using the following imagine.

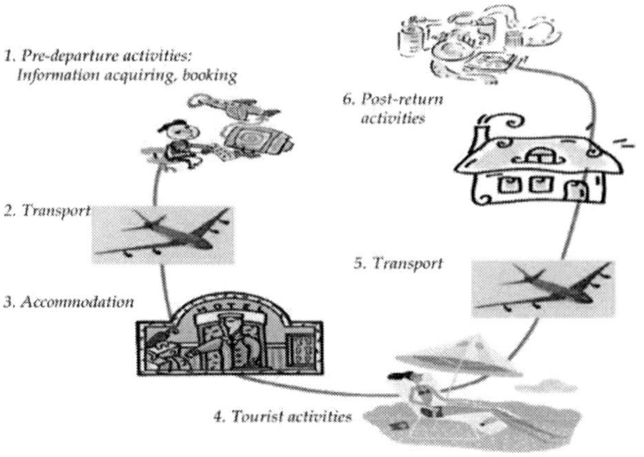

Fig. 1. Logistic point of view on life cycle of tourist. Source: [2]

Applying logistics principles the management of tourism affects the conditions of destinations and host communities, and more broadly, the futures of ecosystems, regions and nations (see Fig. 1). Logistic processes are often involved. Informed decisions at all scales are needed so that tourism can be a positive contributor to sustainable development in keeping with its role as a significant source of both benefits and potential stresses [4]. Well designed and managed tourism can make a significant contribution to the three dimensions of sustainable development, has close linkages to other sectors, and can create decent jobs and generate trade opportunities [5].

The tourism market is highly fragmented, geographically scattered and involves many and various actors along the value/supply chain: These range from a few very large tour operators (TOs) to the numerous SMEs which represent more than 95% of the organizations involved. The emergence of e-economy is a formidable opportunity for the European tourism industry to improve its competitiveness and to reinforce its sustainability. Information and communication technologies (ICT) are key enablers of a flexible response to evolving patterns of tourism and offering more attractive travel experiences, while increasing the quality and the efficiency of the value chain operations. Consumers are seeking more personalized tourism products and services, and are expecting intelligent and proactive access to relevant high quality information and services at anytime and anywhere in a mobile context [6]. ITs also transform the strategic position of organizations by altering their efficiency, differentiation, operational cost and response time. In particular, ITs have stimulated radical changes in the operation and distribution of the tourism industry [7]. This article will be focused on the use of information technology in "tourist activity" stage (holyday leisure time) and the possibility of involvement of ICT/logistics.

In the Czech Republic, there exist a lot of conceptual documents that concern issues of the sustainable development of tourism, for example the Environment in the Czech Republic, Programme for Country Revival, Regional Development Strategies, etc. This paper has rather focused on specific documents that relate directly to documents of the European Union, and also regard issues implied in the paper. Among these documents are [8-10] and others.

The sustainable development strategic framework is primarily oriented by sustainable development targets that are structured by 5 priority levels: 1. Community, people and health, 2. Economy and Innovation, 3. Regional Development, 4. Landscape, ecosystem, and biodiversity, 5. Stable Community. As regards our specific interest, the level, 3, (Priority, 3.2, Improving life quality of the populace), especially its goal, 4, 'Strengthen the role of sustainable tourism in the structure of regional economy', because there are lot of historical, natural, and cultural preconditions for development of tourism and spa industries. There exists an extensive and well-marked network of hiking trails in the country, which implies high potential for recreation and free-time activities. Nonetheless, regarding tourism in the Czech Republic, interests of visitors concentrate mainly on the country capital, Prague. For that reason, the concept highlights the necessity of strengthening technological, innovation, and knowledge accessibility of regional potentials, which would represent prerequisites of tourism sustainable development for the whole country.

In the framework of this concept [8], SWOT analysis of current tourism trends in the Czech Republic was performed. The analysis focused primarily on the issues that are directly related to those specified in this paper. We focus on the use of logistics information technology in "tourist activity" stage (holyday leisure time) and the possibility of involvement of ICT. From our point of view following opportunities are important:

- Development of IT across the whole industry of tourism
- Further development of country and winery tourism
- Modified logistics processes and supply chain management

In parallel to the opportunities of potential development of IT and country tourism, the concept, namely item, 1.3, Improving quality of tourism services, defines individual activities that should implement the objective, specifically tourism infrastructure development (transport, guiding, IS of tourist destinations). The concept highlights the importance of tourism information provision. Currently the access to such information is rather limited in the Czech Republic, which situation should be improved in future, especially as regards regional development and regional competitiveness improvement.

Nature trails and instructive paths in the Czech Republic: The practice of hiking trail marking dates back to the beginning of the past century and was started in Germany. The principal Alpine and Thuringia trails were initially simply marked, later better distinguished by mark shape and colour. The first network of marked trails in the Beskid Mountains was developed by the Czech Hiking Clubs in 1884. Nonetheless, the systematic marking started only in 1890. Originally, the trails were marked just for simple hiking, only later, in the second half of the 20th century, nature trails, instructive paths were added. Since 1990, there have been also cycling or ski trails [11].

2 Information system support

The concept of the Internet of Things (IoT), as utilized in the field of the logistic information systems, represents new substantial innovation trend. It is possible to use its services for interconnecting equipment, various devices, sensors, action elements, control elements, servers, and people that are located in physically and topologically different places into a unified, conceptually logical whole (see Fig. 3). This concept offers completely new possibilities to the developers of information systems. Logistics of tourism is to be developed this way.

The increasingly better provisions and speed of communication infrastructure services are no longer considered as SW or HW means but they are rather taken as network services per se. This approach is usually illustrated as a cloud, from which also the term, cloud computing. The providers can offer their services easily and on a large scale including items of protection against health and safety hazards. Many of them are tailored for support of information processes in logistics of tourism.

Possible scenarios for the tourist:

1. He / she pays entry at the desk, gets the bill with identification. This identification enables his access to information system to download all information into his mobile device (smartphone or tablet) or can borrow the device with information at the desk. He

can choose proper language version. He finds information point (board, desk, or table) in the area, identifies RDID or NFC tag and gets identification of the place (QR code or simple tag stores only the location / identification). Information is displayed on the screen. See Fig. 2a.

2. He / she pays entry at the desk, gets the bill with identification. This identification enables his access to information system. He can borrow the device with information at the desk. Tourist can choose proper language version. In the area he find information point (board, desk, table), identifies RDID or NFC tag and gets full information about the place (QR code has capacity only for basic text information, tag can store more complex information with images etc.). Information is displayed on the screen. See Fig. 2b.

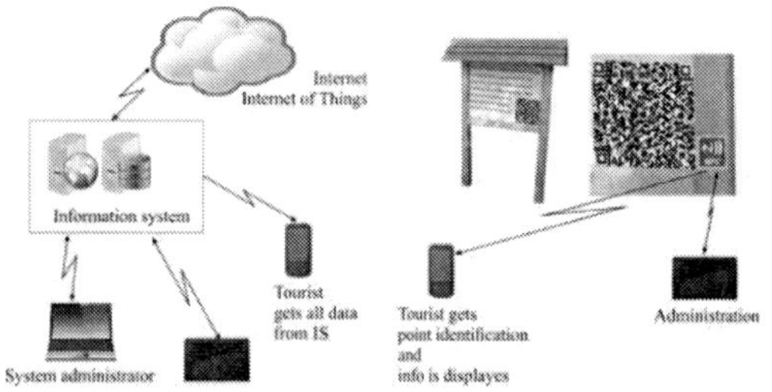

Fig. 2a. Communication infrastructure w/o internet access, without intelligent tag (unidirectional communication with tag). Source: authors

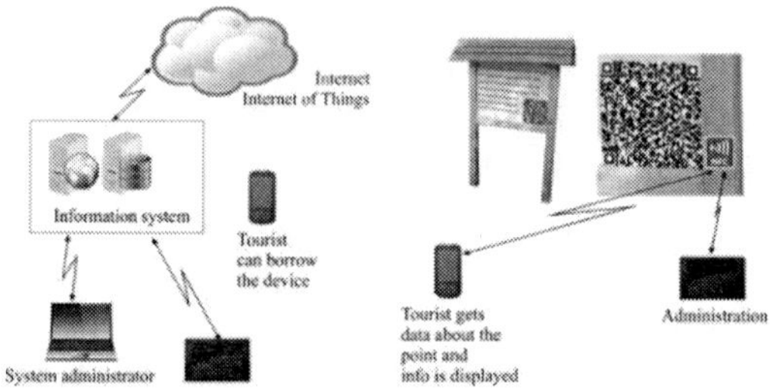

Fig. 2b. Communication infrastructure without internet access, with intelligent tag (bidirectional communication with tag). Source: authors

3. He / she pays entry at the desk, gets the bill with identification. This identification enables his access to information system via internet. Tourist can choose proper language version. In the area he finds information point (board, desk, and table), identifies RDID or NFC tag and gets identification of the place (QR code or simple tag stores only the location / identification). He gets complex information to his device. Information is displayed on the screen. See Fig. 3.
4. In case when information panel has its network connection (mostly applicable inside of buildings) important feedback can be gained in real time. This way information about number of tourists in particular points with time distribution can be obtained. It can be used for control of movement of tourists to avoid to concentrate in some places.

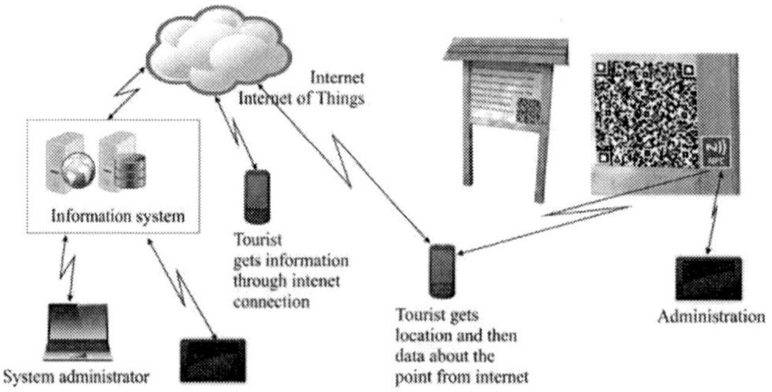

Fig. 3. Communication infrastructure with internet access (with tag and internet). Source: authors

3 Model and simulation of the load of tourist area

Introductory Reflection: People choose to go hiking in the mountains to recover from exhaustion and regain health and strength. A compact smartphone supersedes and integrates a lot of different devices which, in the past, could make a claim to the whole space of hikers' backpacks. There are several topic we have to involve into solution:

- Geolocation - Yes or Not?
- Identifiers as Model Data Source
- Motivation to Activate Smartphone at Information Board

The presence of the information board visitors and their movement on hiking trails can be modelled [12-15] by the programme, HPSIM, for modelling of Petri nets. Here we try to use it for creation of simplified model of tourist area and use it for simulation of visitor movement. The hiker movements are simulated by movements of tokens within the Petri nets [13].

The reason for such model implementation (see Fig. 4) was in easy accessibility of data and their low time processing demand as regards subsequent graphical presentation. Another advantage is in the possibility to represent the system functions synoptically and simulate various collision situations and drawbacks before its actual application.

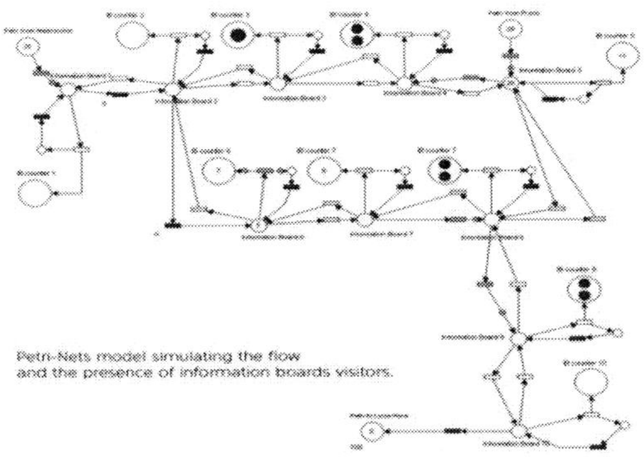

Fig. 4. Part of logistics model of the mountain tourist area used for simulation tourist area used for simulation. Source: authors

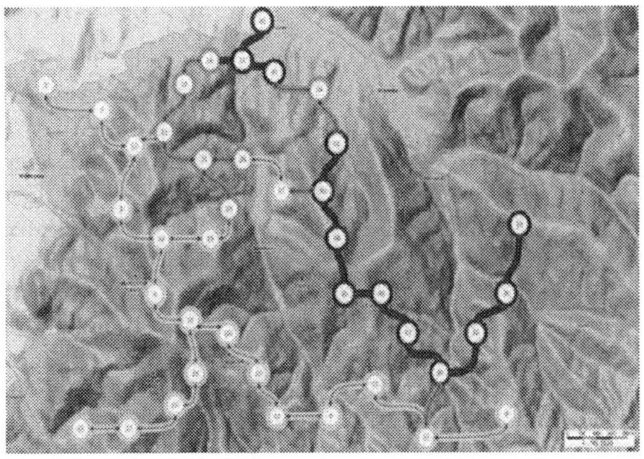

Fig. 5. Results of logistics model simulation (visitor movement in the mountain area). Source: authors

Aiming at better behaviour accuracy of the model (see Fig. 5), we decided to employ transitions (marked yellow and green in the chart) directed by exponential distribution of probability. The model gives a preferential treatment to those movement directions that are most popular as regards hiking trails of the sample landscape. Each information board has its own counter whose increments are affected by successive readings of the smartphone IDs present at the board.

The model data can be visualized by a synoptically arranged map, where attractiveness of individual information boards is colour highlighted. Yellow signifies less attractive tokens and gradually intensifying colour towards red means information board increasing popularity. In parallel to weather forecasting, intuitive perception of colour temperature is employed for the purpose.

4 Life cycle assessment

LCA is a primary analytical tool of industrial/logistics ecology, which constitutes the systematic view of local, regional, and global uses and flows associated with products, processes, and industrial and economic sectors.

The main scope of this study is a comparative Life Cycle Assessment of four possibilities of providing of information on the nature trails. Testing possibilities are following:

- providing information by a paper brochures

- providing information by an information panels
- providing information by a borrowed electronic equipment
- providing information by their own smart phone

This LCA focuses on the European markets. Therefore, the geographic boundary for the use, distribution, and disposition stages of the two systems is limited to the Europe. However, geographic boundaries for raw material extraction, material processing, and product manufacture are assumed worldwide, particularly for IT equipment.

When determining the functional units we used [14]. The fundamental unit was:

- for the paper brochure – standard A4 paper ½ printed by text and ½ printed by picture
- for the information panel – 1 information panel (equivalent amount of text and pictures as in the paper brochure)
- for the borrowed electronic equipment – 2,75 kB for printed text and 300 kB for pictures total 302,75 kB
- for the own smart phone – the same energy demands as for the borrower electronic equipment (302,75 kB) and the average energy necessary to download information

For the final life cycle inventory was used SimaPro 7.2 SW and Ecoinvent v2.1 DB.

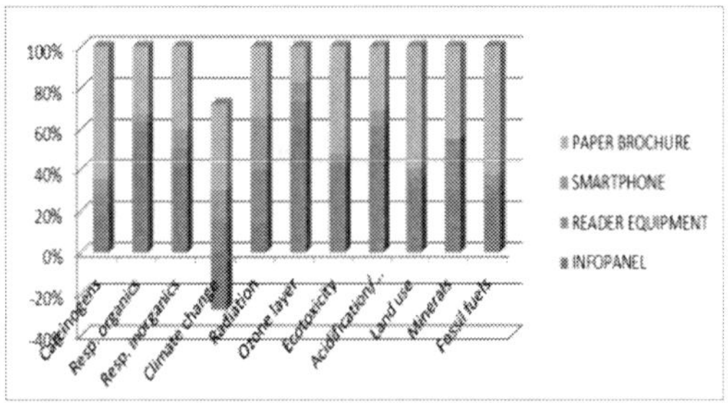

Fig. 6. Characterization of four assessed systems (midpoint category). Source: authors

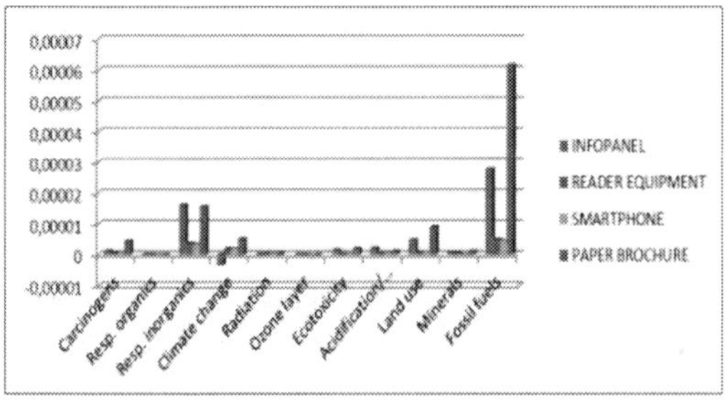

Fig. 7. Normalization of four assessed systems (midpoint category). Source: authors

Assessment of the potential impacts was made by Ekoindikator 99 method. The following charts presented the final comparison of individual different method of distributing information. Although the Ekoindikator 99 methodology is also endpoint methodology, there are presented only midpoint impact category.

The analyses performed (see Figs. 6 and 7) lead to the conclusion that printed information booklets, and also deployment of information boards are of the most detrimental effect on the distribution of information, which is both caused by high material costs of these measure realization (energy and resources needed for production of paper, steel, concrete, etc.). As regards the electronic devices of our assessment, electrical energy is of major impact (99.9 % for smartphone, 90 % for hired reading devices) as related to transfer, keeping, and visualization of data.

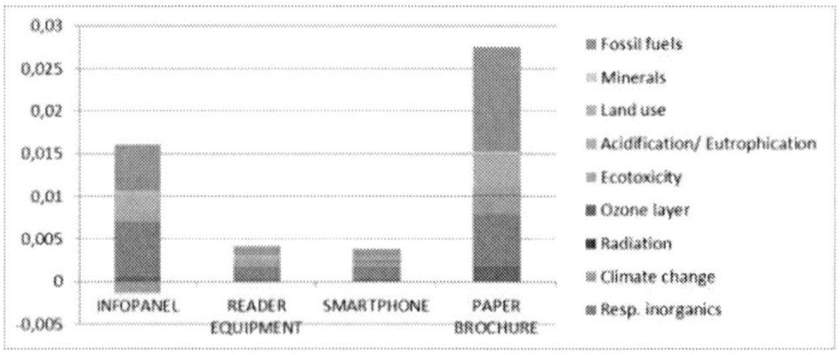

Fig. 8. Weighing of four assessed systems (midpoint category). Source: authors

For all the systems of our assessment, the consistence and uncertainty analyses were performed by the Monte Carlo method that detected calculation error does not exceed maximum limit specified (0,2 %).

The study implies especially these limitations and uncertainties: As regards paper booklets, the result is mainly influenced by the paper production method (inclusive input raw materials, as well as the way in which the booklet are dealt with after their life cycle). If electronic devices are concerned, the major negative factor is represented by taking no account of the material demands of the data storage devices (this information was not known, used estimated values), which will increase the influence of electronic devices. See Fig. 8. Although, only tentative data have been utilized, the study can maintain that, as regards electronic devices, electricity consumption and costs of data storage will constitute major impact factors.

5 Discussion

Identifiers as Model Data Source: The NFC smartphone penetration is accelerating and cannot be stopped. An option might be to equip the information board with a NFC chip, which could read a hiker's smartphone, when present at the board. The other way round, it could be also possible that a reading device within the board would read smartphone data of the hikers in the vicinity. No matter which of these possibilities is utilized, the number of visitors is counted and the length of their presence at the information board measured.

We can talk about the acceleration of the NFC smartphone penetration, but we can maintain that one hundred percent of them feature a camera. It might be more easily affordable to equip the information boards with identifiers that could be read optically. For example, it could be common bar codes. Almost any smartphone could read it and react accordingly, yielding IS input data of the same quality as would be the case of the NFC chips.

Motivation to Activate Smartphone at Information Board: Why should hikers use their smartphones in such manner that NFC or optical IDs are read. The information boards provide only static data about their surroundings. The board visitors might be motivated to activate their smartphones for reading an ID if they received some extra information better than that provided by the board itself. For example it might be an interesting video or some detailed images of the landscape. Good advice might be displayed about view preferences and individual landscape features or possible moments.

Following the provided logistics model and simulation example we can add some new "activities" to the area. For example to the effect of cross-country running, a contest oriented incentive for hikers to activate their smartphones, when present at an information board, might in a prize offered to those, who read all boards with their smartphones on the hiking trail (funicular pass discount, another bonus information, funny animations for children, etc.).

6 Conclusion

Nowadays, information systems for logistics can provide added value for tourists and for management of visited area too. The goal is to make the best condition for tourists to enjoy their stay and to get information and learn new things about the place, culture etc. Number of tourists increases every year and limits of tourist attractive place are reached or exceeded. Management of the area / place can increase logistics parameters as capacity of the place and allows more tourists to visit. Important spin-off are more working places for local people, less influence to the area. There is no necessity to develop special unique equipment. Hardware is ready, only software – user application is to be developed and spread.

This paper is supported by the research project "From horse-drawn railway to intermodal transport" within Visegrad Fund.

References

1. K.G. Debbage, P. Daniels, *The Tourist Industry and Economic Geography: Misse Opportunities, in Ioannides, The Economic Geography of the Tourist Industry: A Supply-side Analysis*, 17-30, (Routledge, New York, NY, USA, 1998)
2. A.A. Lew, C.M. Hall, A.M. Williams, A Companion to Tourism (Blackwell Publishing, Oxford, UK, 2004)
3. C. De Camillis, P. Peeters, L. Petti and A. Raggi, *Tourism Life Cycle Assessment (LCA): Proposal of a New Methodological Framework for Sustainable Consumption and Production, Visions for Global Tourism Industry – Creating and Sustaining Competitive Strategies* (2012)
4. UNWTO, *Indicators of Sustainable Development for Tourism Destinations. A Guidebook (English version)* (2004)
5. UN, *The Future We Want* (Rio de Janeiro, Brazil, 2012)
6. IST & EC, *Intelligent System for Tourism: Report on the R&D activities carried out in the specific programme Information Society Technologies (IST) of the Fifth Framework Programme*, Final report (2003)
7. D. Buhalis, Tourism Management **19**, 5, 409-421 (1998)
8. Ministry of Regional Development CR, *Concept of State Tourism Policy in the Czech Republic for the Period 2014 – 2020* (Prague, Czech Republic, 2013)
9. Ministry of Environment CZ, *State Environmental Policy CR 2012 – 2020* (Prague, Czech Republic, 2012)
10. Ministry of Environment CZ, *Sustainable Development Strategy* (Prague, Czech Republic, 2010)
11. Club of Czech Tourists, *The Development of Tourist Marking in our country and marking of tourist routes in most European countries*, Available online: http://www.kct.cz/cms/sites/default/files/users/user1/dokumenty/znackari/UT-M.pdf (2012)
12. O. Kodym, L. Kavka, M. Sedlacek, *16th International Multidisciplinary Scientific Geoconference*, 375-382 (Albena, Bulgaria, 2016), DOI: 10.5593/SGEM2016/B21/S07.048
13. F. Benes, *Use of RFID Technology to Identify the Location of Objects in a Defined Space* (Ph.D. thesis, VŠB – Technical University of Ostrava, Czech Republic, 2012)
14. G. Kozak, *Printed Scholarly Books and E-book Reading Devices: A Comparative Life Cycle Assessment of Two Book Options*, 252 p. (Ann Arbor, University of Michigan, USA, 2003)
15. G. Fedorko, Z. Čujan, *Proceedings of International Conf. on Industrial Logistics* (Zagreb, Croatia, 2014)

MATEC Web of Conferences **134**, 00024 (2017) DOI: 10.1051/matecconf/201713400024
LOGI 2017

Research on the Impact of Quality on Demand for Bus Transport

Vladimír Konečný[1,*], *Róbert* Berežný[1] , and *Miroslava* Bartoníková[1]

[1]University of Žilina, Faculty of Operation and Economics of Transport and Communications, Department of Road and Urban Transport, Univerzitná 1, 010 26 Žilina, Slovak Republic

Abstract. The demand for transport services is determined by several factors at the same time. The impact of these factors is varied. One of the most important factors is the quality of transport services. The aim of the paper is to identify and characterize potential methods for measuring and evaluating the impact of quality of service on passenger demand for suburban bus transport. These methods are applied to a specific transport system for assessing the impact of quality on demand. The results may contribute to the stabilization of the demand for bus services or they can serve as a basis for the improvement of transport services in the bus transport sector.

1 Introduction

In recent years the number of passengers transported by public passenger transport (PPT) has decreased not only in the Slovak Republic but also abroad. The aim is to slow down or to stop this negative development. In order to do this it is necessary to know the impact of the individual demand factors. One of the most important factors is the quality of transport services. They prove it current studies and projects in this area. E.g. Anderson et al. (2013) when assessing the sensitivity (elasticity) of passenger demand to change a particular factor found that the quality of the transport service is a significantly more influential factor than the cost of travel and also a more influential factor than the income of the population. Knowing the impact of the quality of transport services on demand can be valuable information for stakeholders (carriers, transporter, etc.) to improve the quality of transport services [1].

Before assessing the impact of the quality of services on demand in bus transport it is first necessary to quantify the quality level in an objective manner. This can be achieved by measuring and evaluating quality using quantitative methods. Measurement, evaluation and numerical expression of criteria that characterize the quality of public passenger transport. Also a numerical expression of the demand is needed. Consequently, it is possible to examine the impact of quality on demand through selected methods.

The aim of the paper is to identify and characterize potential methods by which it is possible to examine and assess the impact of quality on demand in bus transport.

* Corresponding author: vladimir.konecny@fpedas.uniza.sk

DOI: 10.1051/matecconf/201713400024

Subsequently, these methods are applied to a specific transport system for assessing the impact of quality on demand.

In Fig. 1 shows the stages of the methodology examining of the relationship between the quality of the transport services and the demand for them [2, 3]

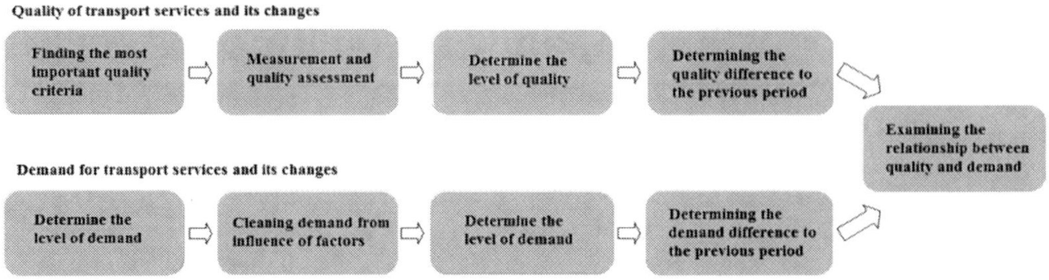

Fig. 1. Stages of the methodology for the examination of the relationship between the quality of transport services and demand. Source: authors

2 Analysis of possible procedures and methods of examining the impact of quality on demand

2.1 Methods of relationship analysis

2.1.1 Correlation analysis

The correlation analysis examines the tightness of statistical dependence between quantitative variables. Correlation analysis unlike regression does not indicate a causal-consequence relationship. The variable Y does not depend on the variable X but the two random variables X and Y change together. The regression analysis assumes that the variable Y is random and the variable X is fixed.

The correlation analysis tool is so-called coefficient of correlation (r_{xy}). Provides a measure of the tightness (degrees) dependent. Its definition is based on the consideration of the sum of the deviations of the individual values of the two correlated characters and their diameters. The relationship for calculating the correlation coefficient can be found in the literature (HINDLS 2013) [4].

The correlation coefficient reaches values from the interval <-1, +1>. In the case of independence between X and Y the correlation coefficient is 0. The positive values of the correlation coefficient refer to direct dependence and negative value for indirect dependence.

At the correlation coefficient r_{xy}, it is possible approximate schematic diagram of the statistical dependence intensity can be used. Values ranging from -0.1 to +0.1 say about linear independence values from - 0.4 to -0.1 or from +0.1 to +0.4 by weaker dependence values from -0.7 to -0.4 or from +0.4 to +0.7 with the mean dependence and values from -1 to -0.7 or from +0.7 to +1 by strong (or tight) dependency [4, 5].

2.1.2 Theory of demand elasticity

The notion of elasticity of demand is mainly used in the area of the economy but we can also meet this concept in the transport sector. Elasticity of demand is considered to be a simple and comprehensible quantitative measure of dependencies (responses) of one variable to the other.

a) Point elasticity

The point elasticity concept is used when we want to know the relative price elasticity of demand at a given point on the demand curve for decision-making and price changes. Through point elasticity it is possible to know the impact on revenue which is the total multiplication of the price with the quantity of demand (PxQ). Dominick Salvatore defines point elasticity of demand as: "Price elasticity of demand at a particular point on the demand curve." The elasticity point is determined as the distance from the Q axis to the point distance from the P axis. [4]

$$e_{x_i}^{point} = \frac{lim}{\Delta x_i \to 0} = \left(\frac{\frac{\Delta y}{y}}{\frac{\Delta x_i}{x_i}}\right) = \left(\frac{\frac{\partial y}{y}}{\frac{\partial x}{x}}\right) = \frac{\partial y\, x}{\partial x\, y} = \frac{d\,(\ln y)}{d\,(\ln x)} \quad (1)$$

b) Curve (arc) elasticity

We find the elasticity between the two points of the curve taking into account the average between the starting and final quantities and the average between the starting and the final price. It is calculated using the following relationship:

$$e_{x_i}^{arc} = \left(\frac{\ln y_2 - \ln y_1}{\ln x_2 - \ln x_1}\right) = \frac{\Delta\,(\ln y)}{\Delta\,(\ln x)} \quad (2)$$

Where y_1 and y_2 represent a change in demand before and after each change in the demand factor x_1 and x_2 (eg, travel prices, quality level, etc.). This estimate provides average elasticity at intervals $<X_1, X_2>$. Estimation requires two observations. One before and one after the change that occurs. An important function of such estimation is the convex relationship with the convex functional shape [4].

c) Linear elasticity

Linear elasticity is a very often used method for calculating elasticity. It is appropriate to use it if the monitored pointers are not expressed by function but by arranged pairs of points. It can be calculated based on the following relationship:

$$e_{x_i}^{line} = \left(\frac{\frac{y_2 - y_1}{1/2\,(y_2 + y_1)}}{\frac{x_{i2} - x_{i1}}{1/2\,(x_{i2} + x_{i2})}}\right) = \frac{(y_2 - y_i)(x_{i2} + x_{i1})}{(y_2 + y_1)(x_{i2} - x_{i1})} \quad (3)$$

2.2 Methods of causal analysis

2.2.1 Regression analysis

The meaning of regression analysis is to examine the two X and Y sign. The task is to model the dependence of the values of the Y variable on the values of the X variable. This is a pair regression task.

In regression modelling the task is to formulate which variable is to be considered independent (X) and which is the dependent variable (Y). The variable X should be the cause and the variable Y as a consequence of the phenomenon. In some cases it may be problematic to specify the cause and the consequence at the discretion of the investigator which of the variables will be the cause of cause X and consequence [4, 5].

a) Regression line

In the case of a line, the general model changes to the form: $y_i = b_0 + b_1x_i + e_i$ and the particular regression line model has the form $y_i = b_0 + b_1x_i$ where the estimate of y_i (i = 1, 2,...n).

Another task is to estimate values of parameters b_0 and b_1 based on the obtained values [xi, yi], i = 1, 2, ..., n (by measurement, observation, etc.). There are several parameter estimation methods. The most common method is the least squares method. Parameters b_0 and b1 are estimated so that the sum of the deviation squares is minimal (as small as possible) with respect to the parameters [4, 5].

b) Multicriterial regression analysis

Multicriterial regression analysis allows to express the influence of multiple independent variables (X, P, Q, R ...) on the observed indicator (Y) simultaneously. Multicriterial regression function should take into account all the important variables that result from the dependent variable Y. Each of the independent variables considered must be in a causal relation to the dependent variable. This dependence can be confirmed for example by calculating the correlation coefficients. The influence of independent variables on the dependent variables is determined by the parameter values of the universal multicriterial regression model.

The individual regression coefficients (b_1, b_2, ...,b_n) represent the average change of the dependent variable Y which is caused by the change of the respective independent variable. It is also important to reevaluate the inclusion of the b_0 coefficient in the proposed multi-criteria function i.e. from a logical point of view the model gives meaning if the independent variables X_1 to X_n reach 0 if the variable Y has a zero value or its value is nonzero a non-zero coefficient b_0.

3 Case study - the impact of quality of transport services on demand for suburban bus transport in the Žilina node

In this part of the article is processed a case study focuses on the application of selected methods, for assessing the impact of quality on demand in the suburban bus transport (SBT) in the Žilina node. The case study respects the phases of the methodology shown in Fig. 1 [3, 6].

3.1 Measuring the level of service quality

To assess the impact of quality on demand it is necessary to know the measured level of quality of services provided by the carrier in the surveyed period, the importance of the individual criteria of quality and passenger satisfaction with these quality criteria.

To identify the importance of individual quality criteria for passengers is appropriate to use a questionnaire survey. This method reflects the subjective perception of quality by the passenger. It is about finding passenger's opinions using questions asked by respondents. Table 1 shows the results of the survey of the importance of individual quality criteria for passengers carried out in 2013, 2014 and 2016 in the SBT in the Žilina node. As we can see over time there was a change in the importance of the requirements of passengers.

Measuring the quality of the services provided means the way of obtaining information about the performance or non-fulfilment of the quality criteria included in the set of measured criteria. Measurement provided services in the field can be performed by inspectors or by technical means. Table 1 shows the results of the quality measurement in 2013, 2014 and 2016 expressed as a percentage.

A questionnaire survey conducted in the Žilina node in 2013, 2014 and 2016 was also used to identify the satisfaction of passengers with the quality criteria. The statistical unit consisted of suburban bus passengers in the Žilina node who have real experience with services. The results are shown in Table 1.

Table 1. Standardized parameters of the most important quality criteria in the SBT in the Žilina node. Source: authors

Quality criteria	Importance of quality requirements (%)			Measured quality (%)			Passenger satisfaction (%)		
	2013	2014	2016	2013	2014	2016	2013	2014	2016
Punctuality	72	84	90	72	65	73	64	73	79
Information	69	79	84	80	77	99	56	73	93
Cleanliness	74	76	79	67	82	82	60	76	85
Driver behavior	65	77	84	64	88	94	67	68	85
$\sum$	**70**	**79**	**84**	**71**	**78**	**87**	**62**	**73**	**86**

3.2 Determining the level of demand for transport services

Determining the level of demand for transport services means to quantify the level of demand. The most common way of expressing the level of demand for public passenger transport is through the number of passengers transported or through the number of tickets sold. The number of passengers transported by type of fare and total is shown in Table 2.

Table 2. The number of passengers transported in SBT in Žilina node according to the type of fare in the years 2010-2016. Source: authors, based on [7]

Relativized number of transported passengers (demand in 2010 = 1.00)				
Year	**Full fare**	**Children/student fare**	**Other fare**	**Total**
2010	0.528	0.370	0.102	1.00
2011	0.490	0.343	0.105	0.938
2012	0.466	0.317	0.111	0.894
2013	0.433	0.299	0.104	0.836
2014	0.406	0.271	0.115	0.792
2015	0.397	0.264	0.105	0.766
2016	0.381	0.247	0.106	0.734

Note: Real demand is relativized. Demand in 2010 is 1.00, or 100%.

In order to assess the impact of the quality level on demand it is possible to use data on the evolution of real demand. However. past demand is affected not only by the level of quality but also by other factors of demand such as: fares, population incomes, employment rates, etc. Therefore, it is necessary to eliminate the impact of other factors on demand by cleaning demand values.

In Table 3 is real total relativized demand for the period 2013 to 2016 which is cleaned from the impact of income, the population and the impact of the introduction of free rail transport. The average monthly nominal wage in the Žilina node in individual years was used for cleaning the demand from the impact of income. The number of inhabitants in the districts served by carrier in the Žilina node was used for cleaning the demand from the impact of population. Estimation of the impact of free rail transport on the demand for bus transport was based on the information provided by the Bus Transport Association of Slovakia, the available statistical data on the demand for free transport and the research carried out in the Žilina node. In the first year of the introduction of free rail transport

(2014) there was a reduction in the number of passengers transported in SBT on average of 2.4%. In the following year, it was 4.8% and in 2016 a decrease of 7.5% [8].

Table 3. Cleaned relativized demand for SBT from the impact of selected demand factors. Source: authors, based on [7]

Year	Relativized real demand total	Cleaned demand from the impact of income	Cleaned demand from the impact of population	Cleaned demand from the impact of free rail transport	**Final cleaned demand**
2013	0.836	0.836	0.836	0.836	**0.836**
2014	0.792	0.810	0.811	0.830	**0.830**
2015	0.766	0.798	0.799	0.837	**0.837**
2016	0.734	0.780	0.781	0.839	**0.839**

3.3 Application of correlation coefficient to assess the impact of quality on demand

To calculate the coefficient of correlation between the level of quality and levels of demand in 2013, 2014 and 2016 was used relationship referred to in the literature (HINDLS 2013).

For measured quality (see Table 1), the correlation coefficient was 0.0209 and in the case of passenger satisfaction the correlation coefficient was 0.0409. In both cases the positive correlation coefficient was reached and therefore the relationship between demand and quality is direct, namely, the very low power of the relationship.

This method is appropriate to use if there is a sufficiently large set of ordered pairs of demand and quality values a minimum of 30 is recommended. In view of the limited availability of quality data for past years this method is only marginally applicable in this case.

3.4 Application of the method of linear elasticity to assess the impact of quality on demand

Elasticity of demand and quality can be examined in year-on-year changes and also in two and more yearly changes. To calculate the linear elasticity of demand is used relationship (3). The variable y represents the number of passengers carried in any given year and the variable x is the measured level of quality or passenger satisfaction with the quality of transport services in given year.

Table 4. Linear elasticity values for quality impact on demand. Source: authors, based on [7]

	Year-on-year changes		Two and more yearly changes			
	Real demand	Cleaned demand	Real demand		Cleaned demand	
	2014/2013	2014/2013	2016/2014	2016/2013	2016/2014	2016/2013
Measured quality	-0.5773	-0.0797	-0.6949	-0.6415	0.0795	0.0143
Passenger satisfaction	-0.3328	-0.0459	-0.4991	-0.4152	0.0683	0.0161

In examining the year-on-year changes and two and more yearly changes in real demand were calculated illogical negative values of elasticity. This would mean that with increasing quality would occur to a slight decrease in demand. While quality is considered to a factor increasing demand. The reason for this illogicality may be the fact that the year-on-year increase in the level of quality of the transport service may be due to a delayed increase in demand (with a delay of more than 1 year). It can also be caused by the existence of other factors of demand which impact on demand could not be cleaned.

By examining two or more yearly changes in cleaned demand were calculated logical positive elasticity. This means that the increase in the quality of bus transport services leads to an increase in the number of passengers. The elasticity of demand for measured quality is +0.0795 and +0.0143. The average value is approximately +0.047. This means that if the quality of the transport service improves by 1% in the short term there will also be an increase in the number of passengers by 0.047%. In the same way we could interpret the elasticity of demand in measuring passenger's satisfaction.

In both cases elasticity of demand values is less than 1. This is an inelastic demand.

3.5 Application of the method of regression analysis to assess the impact of quality on demand

With this method it is first necessary to estimate the parameters of the regression line y depending on x using the least squares method. Using relations for the regression line we then calculate coefficients b_0 and b_1. The independent variable *x* is the resulting cleaned demand (see Table 3) and the dependent variable *y* is the measured quality in the years 2013, 2014 and 2016 (see Table 1). Table 5 shows the auxiliary calculations necessary to determine the coefficients b_0 and b_1.

Table 5. Auxiliary calculations necessary to determine the coefficients b_0 and b_1. Source: authors

Year	Purified demand (x_i)	Measuring quality (y_i)	$x_i \cdot y_i$	x_i^2	y_i^2	$(x_i-\bar{x})^2$	$(y_i-\bar{y})^2$
2013	0.836	71	59.356	0.699	5 041	0.000001	58.783
2014	0.830	78	64.740	0.689	6 084	0.000025	0.445
2016	0.839	87	72.993	0.704	7 569	0.000016	69.439
In total	**2.505**	**236**	**197.089**	**2.092**	**18 694**	**0.000042**	**128.667**

Based on the above calculations, it is possible to calculate the coefficients b_0 and b_1. The coefficient b_1 is -0.463 and the coefficient b_0 is 79.053. On the basis of calculated coefficients, it is also possible to define the regression line equation, which has the form:

$$y = b_0 + b_1.x_1 = 79.053 + (-0.463)\ x$$

Using the regression line equation it is possible to estimate the future development of the demand when changing the value by measuring the level of quality of the transport services in the SBT.

4 Conclusion

This paper deals with the analysis of the various methods by which it is possible to assess the impact of quality on demand for bus transport. Consequently, these methods were applied to a specific case. In view of the limited scope of data on the level of quality of bus services in the past some methods for assessing the quality impact on demand could not be applied. The method of correlation coefficient, linear elasticity and regression analysis was used.

The linear elasticity method after cleaning the demand for suburban bus transport from the impact of other the most important factors (besides quality) confirmed the positive impact of the quality of transport services on passenger demand.

This positive effect may be the reason for further increasing the quality and stabilizing the constantly declining demand for bus transport. It can also serve to estimate the future demand due to a change in quality level planning.

It is necessary to continue the application of the methodology for measuring and evaluating quality in order to obtain objective data for longer time series.

This paper is supported by the research project "From horse-drawn railway to intermodal transport" within Visegrad Fund.

References

1. R. Anderson, et al., *Measuring and Valuing Convenience and Service Quality. A review of global practices and challenges from the public transport sector* (OECS/ITF, 2013)
2. J.T. Bekken, N. Fearnley, *Long – term demand effects in public transport* (Institute of Transport Economics, Norway, 2005)
3. M. Bartoníková, *Návrh metodiky pre hodnotenie vplyvu kvality služieb v autobusovej doprave na dopyt cestujúcich* (Diploma thesis, University of Žilina, Slovak Republic, 2017)
4. R. Hindls, S. Hronová, J. Seger, *Statistika pro economy* (Prague, Czech Republic 2003)
5. D. Markechová, A. Tirpáková, B. Stehlíková, *Základy štatistiky pre pedagógov* (UKF in Nitra, Slovak Republic, 2011)
6. R. Berežný, V. Konečný, *12th international scientific conference of young scientists on sustainable, modern and safe transport*, 40-45 (Procedia Engineering **192**, 2017), DOI: 10.1016/j.proeng.2017.06.007
7. *Internal data provided by the Žilina Self-governing Region*
8. *Statistical Office of the Slovak Republic*, Available online: http://www.statistic.sk/ (2017)
9. V. Konečný, *Kvalita služieb v cestnej doprave a zasielateľstve*, 186 p. (University of Žilina, Slovak Republic, 2015)
10. V. Konečný, M. Poliak, A. Poliaková, *Ekonomická analýza podniku cestnej dopravy*, 283 p. (University of Žilina, Slovak Republic, 2010)
11. V. Konečný, R. Berežný, Railway transport and logistics **13**, 1, 10-16 (2017)
12. V. Konečný, M. Kostolná, Perner´s Contacts **10**, 4, 63-71 (2015)

Modelling of Processes of Logistics in Cyberspace Security

Jiří Konečný[1,*], *Martina* Janková , and *Jiří* Dvořák[1]

[1]Tomas Bata University in Zlin, Faculty of Logistics and Crisis Management, Department of Crisis Management, Uherské Hradiště, Czech Republic

Abstract. The goal of this contribution is especially to familiarize experts in various fields with the need for a new approach to the system-defined model and modelling of processes in the engineering practice and the expression of some state variables' possibilities for the modelling of real-world systems with regard to the highly dynamic development of structures and to the behaviour of systems of logistics. Thus, in this contribution, the necessity of making full use of cybernetics as a field for the management and communication of information is expressed, and also the environment of cybernetics as a much needed cybernetic realm (cyberspace), determining the steady state between cyber-attacks and cyber-defence as a modern knowledge-based potential in general and specifically of logistics in cyber security. Connected with this process is the very important area of lifelong training of experts in the dynamic world of science and technology (that is, also in a social system) which is also expressed here briefly, and also the cyber and information security, all of which falls under the cyberspace of new perspective electronic learning (e-learning) with the use of modern laboratories with new effects also for future possibilities of process modelling of artificial intelligence (AI with a perspective of mass use of UAVs in logistics).

1 Introduction

Current global economy is faced with a number of challenges in the environment characterized by process engineering in dynamically developing areas and theories of virtual systems, information and communication technologies (ICT), cybernetics (theoretical, technical, and applied, such as in an environment for control and communication in the living and non-living-technical, technological and man-made systems), the theory of models and modelling of systems associated with new designing possibilities (Computer-aided design - CAD used 2D or 3D programs) and also modern construction of safe, optimal, reliable and efficient real systems (based on models of operations research) theoretical approaches to modern logistics in all programmes and also in information and communication technologies (ICT) and last but not least, in the developing areas of modern models for artificial intelligence [1] and learning and smart cyber-security systems (unmanned real resources in logistics).

* Corresponding author: konecny@utb.cz

Nowadays, world economy [3] is also significantly determined by implementation of new scientific and new research knowledge (especially new, efficient technology and logistics) introduced now into practice and also by developing modern means (teaching and laboratory) and tools for lifelong learning in general, particularly in the field of ICT.

The whole of the above mentioned process (as an area of process engineering is associated also with the developing progressive practices and their modelling on powerful IT (self-learning intelligent systems), new system knowledge and assorted possibilities of information security and new cyber-security (CS) from the perspective of system integration of attack and defence modelling, especially from the new perspective of current cyberspace for an information and knowledge based society.

The purpose of this contribution is also to formulate modern systemic uses of models and modelling in the new cyberspace (cyberspace) and formulate fully developed use of the resources of technical cybernetics in practice [5], [2] cyber-security and for further practice durable and upcoming smart means which are already present at the Faculty of Logistics and Crisis Management (FLK) in built laboratories: Cyber Security (LKB) for Bachelor's degree programmes and Applied Cyber-Security (LAKB) for the master's follow-up programmes.

2 The Current State of Logistic Model Solutions in Cyberspace Security

The current state of tasks underway (in the field of modelling in cyberspace security) stems mostly from the understanding of process engineering [4] of the emergence of knowledge society referred to in a number of publications and scientific tasks solved and some other publications [6]. Everything is determined by a wide range of processes across economic and other areas. Knowledge management in logistics with security systems and other programs, for example risk management, therefore becomes a modern and necessary key component of strategic management and also of dynamic development of modern and well-designed intelligent systems of cyber-security in modern society making use of logistics. In preparation of the social system [3] for the stated area of solution, we mainly focus on representation of the professional profile of researchers so that they are able to solve some of the problems associated with the formation of the model and the corresponding processes of modelling real-world systems and their significant surroundings and we sense the essential requirements on the design (especially through their system approach when they design systems) and with the application of creating modern models and with the possibility to model subsystems on the current modelling environment of a computer network, especially on digital computers for secure user environment in the cyberspace of management and of communication of information.

The stated approach to tasks will allow us to solve the present, necessary questions related to logistics, crisis management and risk control in the required electronic automated systems of management and also with the corresponding required basic cyber-safety.

The current conditions for the solution of the stated task are mainly determined by the professionals who have the knowledge of theoretical bases in mathematics, physics, computer science, ICT systems and models and communication (of tasks and specific areas of logistics, including logistics in ICT), system identification, risk analysis and assessment, management and regulation (in the areas of management, crisis management and security systems, information theory, economics (in the areas of microeconomics, macroeconomics), Cybernetics (in particular in the field of cyber-safety).

The current direction of the solution of this task is further subject to the knowledge of new world economy, in particular, of the new concept of process engineering in the world and the use of secure digital information and communication technologies in the new cyberspace and planning their activities, including crisis management of systems from the perspective of logistics environment.

Competitive environment is therefore the main driving force of contemporary society, and also the environment for new concept of cyber-security as part of the cyberspace of modern society. Cybernetics (as the "*area of management and communication in living and non-living organisms* "- Norbert Wiener, 1946) and in the modern system-defined environment, it is primarily new information technologies, enabling in a virtual environment to find information resources of the world and the necessary knowledge in the newly introduced cyberspace (cyberspace) defined in Act no. 181/2014 Coll. on cyber-security as amended by related laws (Act on cybersecurity). The starting point of the solution is also the application of this Act in the stated modelling and it also lies in increasing a systemically conceived interdisciplinary security in cyberspace and the protection of adequate infrastructure which is important for the functioning of the State and whose violation would led to damage or danger to the interests of the Czech Republic.

3 Metods

Based on information sources [3] of the world, we list some used selected contemporary modern methods for the production of models, including: Systemic approach to the stated chain of cyber-understanding of the new model which was used in accordance with the principles of general systems theory as a concept for thinking, analyzing real world situations, decision making and acting in terms of a complex reality. The systemic approach provided the methodological basis not only for problem-solving, but also enabled them to be dealt with more comprehensively with respect to other aspects that affect it. A systemic view of management brought about more extensive possibilities in particular in understanding the laws of phenomena and processes, and in researching these principles in its entire breadth. It further enables purposeful abstraction from certain factors and later the description of reality by appropriate models.

Systemic approachs OR/MS (Operatinons research/Management science) the modelling process as scientific solutions to complex problems in decision making. The aim of the aproach used was to improve the existing system and the design of a new system which would better fulfil some of the requirements placed on it. Problem formulation was the most important part of the process in the framework of the mentioned system engineering.

Solving the assigned tasks, it was especially strictly enforced that:

- the formulation of the problem is more important than its solution (theoretical basis of the problem solved, the definition of the system) with a practical experience,
- the definition of the system was the next step after formulation of the problem,
- The creation of a model was formed a basis for modelling approach OR/MS.

Generally, the creation of models is dealt with in theory of identification, in which we will use methods known from theory of artificial intelligence. (AI). They are represented by applications stemming from biology (neural networks and genetic algorithms), physics, mathematics and logic (such as technologies modelling and identifying chaos and technologies using fuzzy sets). These technologies form groups based on computer models of task solving with the supply of expert information (expert systems), on inductive learning and so on.

DOI: 10.1051/matecconf/201713400025

4 Results and discussion

Gradually obtained results of the research are regularly published and one of the most interesting results of modelling is according to [3] for example, this graph obtained as identified behaviour of the system:

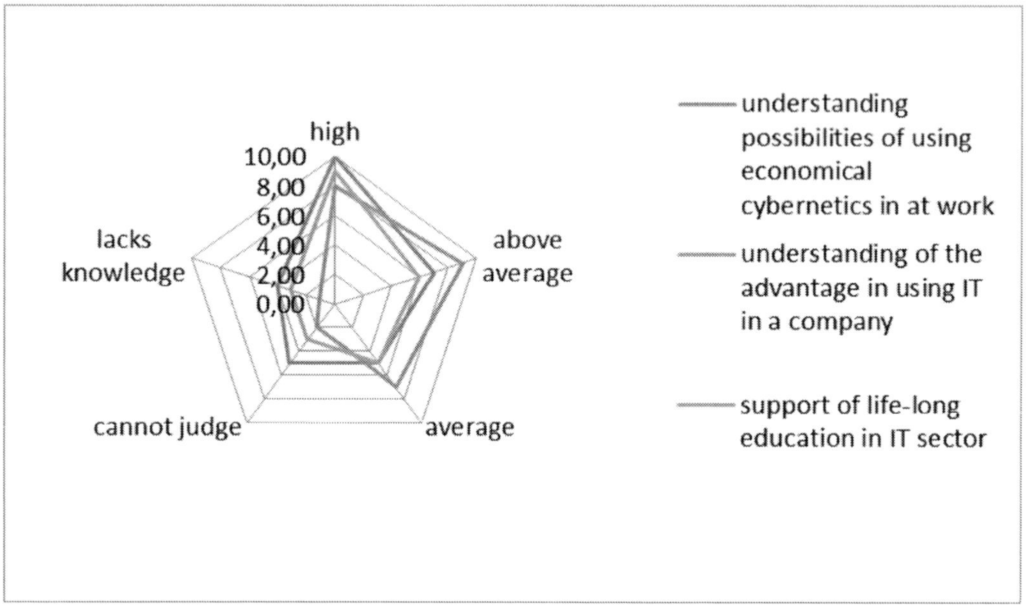

Fig. 1. Assessment of selected models of behaviour in cyberspace of companies. Source: [3]

Based on the detailed analysis of the results obtained by modelling, a comprehensive training program for professionals was created, including the following basic sections:

- System definition of security and protection of information and of communication assets (hardware, software, databases, communications networks, etc.).
- Defining cyberspace for possible recognition of vulnerabilities and identifying security threats.
- Modelling of a cybernetic system for a possible analysis of threat sources, of attacker and of the security attributes of information assets from the perspective of process engineering.
- Possibilities of early systemic detection of cyber attacks and defence against attacks (e.g. cyber-crimes, cyber-bullying, cyber-threats, cyber-terrorism, cyber-espionage, cyber-disinformation, etc.).
- Information sources and work with them in the implementation of legal, economic, technical and logistical standards for theapplicability of cyber security (Cybernetic Law, Information Security Management according to ISO 27000).
- Protecting social and technical system from highly efficient means of modern cyber warfare and from the possibility of new information weapons of mass destruction.
- Preparing the population for efficient use of resources and protection against cyber-attacks and restoration of safe assets in cyberspace and intelligent robotic systems.
- Technical, legal and economic environment in the dynamics of cybespace of the new information and knowledge-based society in the civilized world.

- Development and use of new means of cybernetics and their safety (UAV, robotic lines, 3D printers, modern cryptographic and optoelectronic devices and the use of bionic assets).

4.1 Systemic Process Modelling and Cyber Security

"*The Laboratory of Applied Cybernetics*"(LAKB) is a completely new perspective on solving very promising areas of application of cybernetics in safe environments with recognized and predicted (generated) and possible attacks and a balanced and therefore adequate adaptation of defence to these generated attacks and the resulting need for systemic definition of cyber-security for building (that is especially designing and constructing new and justified by a certain steady state between attack and defence for cyber security of the system). This will be in the near future already part of intelligent systems (with a proportional level of information) and importantly, with systemically conceived cyber-security on the social and artificial, therefore primarily technical level of applied cybernetics. Designing new systems without cyber security (as a background for all new systems) based on current scientific publications will no longer be possible (economically, technically, legally, ...). And so we must already now prepare experts for substantial knowledge and skill in the specific specialization (discipline) and for knowing how to systematically create a model system, model various variants of the reality prepared with its significant environment, how to systematically construct the desired hierarchical system level (appropriate means of hardware and software equipment modelling the environment, for example using appropriate CAD means, and other mathematical projects of promising models for real environments) and also formulate how to build a safe new real systems based on deep theoretical knowledge and knowledge acquired on (virtual) models in new laboratories already for the new knowledge-based society (in addition to information and other foster literacy among professionals, a new perspective on the optimal structure of the system and acquire the knowledge required for a knowledge-based literacy.)

The said proposal of LAKB has not been submitted anywhere in this way and thus has not been introduced into teaching at universities in the field of cyber-security and therefore cannot build on the experience of other departments in the modern area. The dynamics of the changing environment of industry and new promising technologies (not only in production but also in science and other areas related to the management of modern society based on knowledge literacy and new concept of robotics, such as mechatronics, nanotechnology and bionic systems). Thus an over-arching theme of cyber-security will make its projection into the very information and communication technologies (ICT) and lead us to this new view of qualified design of system approach of artificial intelligence for the modelling of cyber-security.

5 Conclusion

The contribution briefly introduces some of the possibilities of the developing principles of cyber-security and of the selected environment for the proposed workplace using artificial intelligence methods already in the design of LAKB.

The aim of the paper is primarily to point experts to the possible systemic definition of the real environment and also to the opportunity to express the environment of artificial intelligence in the newly conceived cyberspace of the modern society; and in the LAKB, such environment may be used as basis for new final theses of students and projects of scientific tasks solved by pedagogical researching staff of the university.

The modern approach is precisely in LABK intertwined cyber-security and interoperability of technical drawing in CAD and in other programs generating the attack and such conditions for the model and the modelling that would be inappropriate, uneconomical, dangerous to use in socio-technical environment, whereas the defending side needs to explore these inappropriate conditions to succeed through modelling in proving that cyberspace has to be described so as it is safe for use. This environment must above all be mastered by educators and provoke in students rational use of the environment (CAD and other programs) for the stated modelling and inducements of solutions of other variants. Modelling will create such a variety of situations that the qualified teachers must prepare instructions for the use of resources in such laboratories and devise so far only school tasks and later tutorials on "*practical*" but not secret tasks, which have solutions and lead to the validation of theoretical theoretical assumptions of modern design in a safe cyber-environment, and fulfilling the systemic the systemic conception of artificial intelligence for modelling of cyber-security.

The contribution represents an output of a specific research project, System definition of appropriate cyber-safety models for the needs of scientific and research activities and pedagogical at the Faculty of Logistics and Crisis Management of Tomas Bata University in Zlín.

This paper is supported by the research project "From horse-drawn railway to intermodal transport" within Visegrad Fund.

References

1. M. Janková, J. Dvořák, *EBES Conference*, 43-51 (Ekaterinburg: EBES, Russia, 2014)
2. M. Janková, J. Dvořák, *Mathematics, Information Technologies and Applied Sciences 2014*, 59-65 (Brno: MITAV, Czech Republic, 2014)
3. M. Janková, *Internetové nástroje pro celoživotní vzdělávání v sektoru IT (Internet Tools for Lifelong Learning on the Field of IT)*, 234 p. (Brno: Technical University in Brno, Faculty of Business and Management, Czech Republic, DDP, 2016).
4. V. Šefčík, J. Konečný, *Procesní inženýrství (Process Engineering), Bezpečné a spolehlivé vedení procesů (Safe and Reliable Process Management)* (Course Reader, Zlín: Tomas Bata University in Zlín, Faculty of Logistics and Crisis Management, Czech Republic, 2013).
5. J. Křupka, *Základy technickej kybernetiky, Liptovský Mikuláš (Basics of technical cybernetics, Liptovsky Mikulas)* (Armed Forces Academy of General M.R. Štefánik, Slovak Republic, 2008)
6. R. Petříková, *Moderní management znalostí (Modern Knowledge Management)* (Prague: Professional Publishing, Czech Republic, 2010)

Calculation of Selected Emissions from Transport Services in Road Public Transport

Vladimír Konečný[1,*], and *František* Petro[1]

[1]University of Žilina, Faculty of Operation and Economics of Transport and Communications, Department of Road and Urban Transport, Univerzitná 1, 010 26 Žilina, Slovak Republic

Abstract. The article deals with road public transport and its impact on the environment. According to the methodology given in EN 16258, CO_2 emission value has been calculated. The input data for the calculation and the results are shown in the tables. The declaration is created according to STN CEN / TR 14310, which contains recommendations for compiling environmental reports. Finally, the comparison of the environmental impact of a bus and a passenger car, when converted to one passenger, bus has a lower CO_2 emission than a passenger car in that section.

1 Introduction

In 2012 the European standard EN 16258 Methodology for calculation and declaration of energy consumption and GHG emissions of transport services (freight and passengers) was approved by the European Committee for Standardisation (CEN).

The purpose of this standard is to make it widely applicable across the transport sector and accessible to a very diverse user group. The use of standards provides a common approach to the calculation and declaration of energy consumption and emissions for transport services, irrespective of the difficulty of transportation technology and transportation process. The standard ensures that the declarations have greater consistency and transparency, and that the energy consumed and emissions produced correspond to the load, respectively occupancy of vehicles.

2 Way of expressing quantity of GHG emissions

For the calculation of greenhouse gas emissions standard considers, the unit of CO_2e (carbon dioxide equivalent) as carbon dioxide represents the largest share of the production of greenhouse gases. CO_2e value indicates the influence of each greenhouse gas on global warming using the conversion amount or concentration of CO_2, which would have similar effects. Compared with the current approach, to reporting greenhouse gas emissions standard provides:

- unification of the calculation methodology used uniform emission factors (unified emission factors are given in table 1) [1]

* Corresponding author : vladimir.konecny@fpedas.uniza.sk

- calculation of the production of all greenhouse gas emissions, calculated per unit of **CO_2e**
- taking into account both direct and indirect emissions and energy consumption of vehicle operation, allowing an objective comparison of their energy consumption and environmental impact

3 The principles of calculations of energy consumption and greenhouse gas emissions

For the calculation, it is necessary to know the operational characteristics of vehicles and transport services, such as fuel consumption of vehicles, transport distance, number of km of non-loaded vehicle, number of passengers, vehicles capacity.

3.1 Calculations for the vehicle operation system

If the transport service consists of several sections, it is necessary to identify the operation system of the vehicle (Vehicle Operation System - VOS) for individual sections, namely a number of categories, including working hours of the vehicle.

The calculation is based on the identification of a vehicle's consumption of a particular vehicle operation system (VOS). Conversion from total fuel consumption for the VOS into quantities of energy consumption and GHG emissions shall be made using following formulas:

- for well-to-wheels energy consumption of the VOS:

$$E_w\,(VOS) = F(VOS) \times e_w \tag{1}$$

- for well-to-wheels GHG emissions of the VOS:

$$G_w(VOS) = F(VOS) \times g_w \tag{2}$$

- for tank-to-wheels energy consumption of the VOS:

$$E_t\,(VOS) = F(VOS) \times e_t \tag{3}$$

- for tank-to-wheels GHG emissions of the VOS:

$$G_t\,(VOS) = F(VOS) \times g_t \tag{4}$$

where:

F(VOS) .... is the total fuel consumption used for the VOS (examples: F(VOS) equals five thousand litres of diesel; or F(VOS) equals thirty thousand kilowatt hours);

e_w .... is the well-to-wheels energy factor for the fuel used (example: for diesel, $e_w = 42.7$ MJ/l);

g_w .... is the well-to-wheels GHG emission factor for the fuel used (example: for diesel, $g_w = 3.24$ kgCO_2e/l);

e_t .... is the tank-to-wheels energy factor for the fuel used (example: for diesel, $e_t = 35.9$ MJ/l);

g_t .... is the tank-to-wheels GHG emission factor for the fuel used (example: for diesel, $g_t = 2.67$ kgCO2e/l).

Values for energy and GHG emission factors shall be selected in accordance with Annex A., see tables 1 and 2 in EN 16258:2012 *Methodology for calculation and declaration of energy consumption and GHG emissions of transport services (freight and passengers)* [1].

3.1 Calculations for the vehicle operation system

If the realized transport service consists of several sections (different customers, different numbers of passengers carried, the distance traveled with loaded and an empty vehicle and so on.), It is necessary to implement the conversion of energy consumption and emissions to stretch. The calculation is performed as follows:

- Identifies the VOS used to implement transport service on the field in question
- To quantify the total consumption of VOS
- Calculate the total energy consumption and emission levels for VOS according to formula (1), (2), (3) and (4)
- Calculate the share of energy consumption and emissions attributable to address segment of transport services (a dimensionless number) as the ratio of power attributable to the segment of transport services and power system operation of the vehicle:

$$S(leg) = T(leg) \div T(VOS) \quad (5)$$

Subsequently, this proportion will be used to calculate energy consumption and greenhouse gas emissions for solving section of transportation services:

$$E_w(leg) = E_w(VOS) \times S(leg) \quad (6)$$
$$G_w(leg) = G_w(VOS) \times S(leg) \quad (7)$$
$$E_t(leg) = E_t(VOS) \times S(leg) \quad (8)$$
$$G_t(leg) = G_t(VOS) \times S(leg) \quad (9)$$

where:

S(leg) .... is the factor used to calculate the share of the VOS's energy and emissions which is allocated to a transport service for the leg. This share is based on relative proportions of transport activity for the leg and for the associated VOS;
T(leg) .... is the transport service's transport activity for the leg;
T(VOS) .... is the transport activity of the VOS which is related to the leg;

Performance of vehicle operation system (T (VOS)) and the power of stretch of transport services (T (leg)) must be in the same unit terms. The standard recommended for freight and passenger traffic to use transport capacity. That means multiplying the number of passengers and the actual transport distance in terms of passenger kilometers in passenger transport, freight transport multiplying the quantity of transported goods and the actual transport distance in terms of tonne-kilometers.

4 The structure and content of the declaration of energy consumption and GHG emissions

The standard does not prescribe the form of the declaration, it only defines the requirements for their content. For declaring the recipient declaration standard user can use any medium that provides unambiguous results and related documents for the calculation such as web site. Declarations on energy consumption and GHG emissions of a transport service shall include:

a) *four results (E_W, G_W, E_t, G_t)* calculated according to previous clauses (when the energy in units of J, or MJ or GJ in CO_2e emissions in g or kg or t)
b) *supporting information*, which, among other things contain:

- transparent description of the method used, in case other factors of energy consumption and emission factors as listed in Annex A standards, they need to give reasons to justify the need to also use predefined values
- basic description of the services (source and destination of the route, the number of persons transported)
- used vehicle operation system (VOS) for each section
- the size of the vehicle fleet, vehicle categories

In formal terms, it is possible to create a declaration using the standard **STN CEN / TR 14310 freight services. Designation and reporting of environmental performance in freight transport network.** The aim of the standard is to provide guidance for preparing (creating) environmental declarations and reports. It contains recommendations on the content and structure of the documentation and evaluation of the impact of freight on the environment. It refers to road, rail and waterway does not apply to air transport [2].

5 External costs from transport services

External costs of transport are the costs associated with the negative effects of transport activities on the environment and human life, they are costs associated with deleting their effects. In particular, the pollution of air, water and soil disturbance of ecosystems, noise pollution, traffic accidents (of the damage not covered by insurance), takes up space and overloading capacity. Categories of external costs are accident costs, air pollution costs (emissions), noise costs, congestion costs [3].

The internalisation of external costs is an objectification costs and subsequently attributed to the those entities that cause them and are responsible for. Internalization is a tool to convert external effects of transport in monetary terms to those that actually cause these effects [3].

6 Case studies

The calculation for energy consumption and emissions was designed and tested in a model situation of passenger transport. Route: 21 km long and 34 stops. Vehicle: M3 category. Parameters of a vehicle: SOR BN 10.5, engine Iveco Tector (2. series), 185 kW, Euro IV, year 2007, diesel fuel, 3 doors for entry and exit, maximum occupancy 96 passengers.

The resulting calculation of greenhouse gas emissions and energy use for the direct transport are presented in tabular form in the form of output reports (table 1).

Table 1. Outputs of calculation of emission and energy consumption for route in road public transport (total). Source: authors

Indicator	Characteristic
Number of stops	34
Distance (km)	21
Vehicle, engine power, emission limit, year of production	SOR BN 10.5, Iveco Tector (2. series), 185 kW, Euro IV, 2007
Number of doors	3
Number of seats	30 + 1 (driver)
Number of places to stand	66
Tank volume (l)	180

DOI: 10.1051/matecconf/201713400026

Indicator	Result of calculations
Gross vehicle weight (t)	15
Average number of passengers	29.5
Fuel	Diesel
Fuel consumption per route (l)	5.67
Energy consumption in the transport (e_t(MJ))	203.553
Energy consumption in the transport (e_w(MJ))	242.109
CO_2 emissions during transportation (g_t($kgCO_2e$))	15.139
CO_2 emissions during transportation (g_w($kgCO_2e$))	18.371

In the next table, we can see model declaration with information about vehicle, route characteristics and final results of energy consumption and emission pollution expressed per one passenger in average.

Table 2. Outputs of calculation of emission and energy consumption for route in road public transport (per one passenger). Source: authors

Indicator	Characteristic
Distance (km)	21
Vehicle, engine power, emission limit, year of production	SOR BN 10.5, Iveco Tector (2. series), 185 kW, Euro IV, 2007
Tank volume (l)	180
Indicator	**Result of calculations**
Average number of passengers	29.5
Fuel	Diesel
Fuel consumption per route (l)	5.67
Energy consumption in the transport (e_t(MJ))	6.90
Energy consumption in the transport (e_w(MJ))	8.21
CO_2 emissions during transportation (g_t($kgCO_2e$))	0.51
CO_2 emissions during transportation (g_w($kgCO_2e$))	0.62

6.1 Leg of a transport service

Also we made a comparison concerning a certain leg of transport service between bus vehicle and passenger car. We followed energy consumption and emission production. Leg of transport service is represented by route from one bus stop to another. Total length is 8

DOI: 10.1051/matecconf/201713400026

km. Bus vehicle technical characteristics remains the same. Chosen passenger vehicle has these parameters: Škoda Octavia II, 1.9 TDI, 77 kW, Euro IV limit, year of manufacturing 2006. Final results of this comparison are shown in tables 3 and 4. The road elevation profile of the section of the transport service is shown in figure 1.

Fig. 1. Road elevation profile. Source: GoogleMaps

Table 3. Outputs of calculation of emission and energy consumption for route in road public transport (peak hour per one passenger). Source: authors

Indicator	**Characteristic**
Distance (km)	8
Vehicle, engine power, emission limit, year of production	SOR BN 10.5, Iveco Tector (2. series), 185 kW, Euro IV, 2007
Tank volume (l)	180
Indicator	**Result of calculations**
Average number of passengers at peak	36
Fuel	Diesel
Fuel consumption per route (l)	5.67
Energy consumption in the transport (e_t(MJ))	5.65
Energy consumption in the transport (e_w(MJ))	6.73
CO_2 emissions during transportation (g_t($kgCO_2e$))	0.42
CO_2 emissions during transportation (g_w($kgCO_2e$))	0.51
Total CO_2 external costs	**0.0045 €**

The Handbook of external costs was used to determine the amount of money generated for total emissions. It states that 1 tonne of CO_2 emissions are at 4.81 €. This value is multiplied by the total emissions produced, resulting in the external costs of one bus passenger amounting to 0.0045 € (see table 3). The same procedure was used to determine the amount of external costs per one passenger in car. Value is 0.046 €, see table 4.

DOI: 10.1051/matecconf/201713400026

Table 4. Outputs of calculation of emission and energy consumption for route by passenger car (in road public transport [car + driver]). Source: authors

Indicator	Characteristic
Number of stops	0
Distance (km)	8
Vehicle, engine power, emission limit, year of production	Škoda Octavia II, 1.9 TDI, 77 kW, Euro IV, 2006
Tank volume (l)	55
Indicator	**Result of calculations**
Average number of passengers at peak	1 (driver)
Fuel	Diesel
Fuel consumption per route (l)	1.62
Energy consumption in the transport (e_t(MJ))	58.16
Energy consumption in the transport (e_w(MJ))	69.17
CO_2 emissions during transportation (g_t($kgCO_2e$))	4.33
CO_2 emissions during transportation (g_w($kgCO_2e$))	5.25
Total CO_2 external costs	**0.046 €**

7 Conclusion

Road public transport in Slovakia is less preferred and has little passenger support. The advantage of a passenger car lies in greater passenger comfort and higher transport speeds to the desired destination. The results of the model show that road public transport has one very important asset and is more environmentally friendly than a passenger car. Comparison of the bus and passenger vehicle with external costs is the difference in prices per kilogram of emissions produced up to 1050% higher than the passenger car. The results are shown in tables 3 and 4. This comparison is for the CO_2 emission component. We also know other negative components like No_x, CO, PM, etc., which also have their monetary expression as external costs (see [3]). Better promotion and transfer of passengers to public transport would have less unacceptable emissions to the environment. It would also reduce the number of passenger cars in city centers. Passenger cars, which are capacitively used on average by 2 passengers, are among the largest environmental pollutants in cities.

This paper is supported by the research project "From horse-drawn railway to intermodal transport" within Visegrad Fund.

References

1. EN 16258:2012, *Methodology for calculation and declaration of energy consumption and GHG emissions of transport services (freight and passengers)* (CEN, Brussels, Belgium, 2012)
2. STN CEN/TR 14310 Freight transportation services, *Declaration and reporting of environmental performance in freight transport chains* (SÚTN, Bratislava, Slovak Republic, 2003)
3. F. Petro, V. Konečný, *12th international scientific conference of young scientists on sustainable, modern and safe transport*, 677-682 (Procedia Engineering **192**, 2017), DOI: 10.1016/j.proeng.2017.06.117
4. V. Konečný, F. Petro, Logi – scientific journal on transport and logistics **7**, 1, 51-65 (2016)
5. V. Konečný, F. Petro, R. Berežný, Perner´s Contacts **11**, 4 (2016)
6. *Update of the Handbook on External Costs of Transport*
7. S. Eggleston, *Emissions: energy, road transport*, Available online: http://www.ipcc-nggip.iges.or.jp/public/gp/bgp/2_3_Road_Transport.pdf (1998)
8. *Guide on Calculating GHG emissions for freight forwarding and logistics services* (CLECAT, 2012)
9. J. Gnap, V. Konečný, *Výpočet produkcie emisií z nákladnej dopravy* (University of Žilina, Slovak Republic, 2010)

MATEC Web of Conferences **134**, 00027 (2017) DOI: 10.1051/matecconf/201713400027
LOGI 2017

The Impact of 3D Printing Technology on Supply Chain

Lukáš Kubáč*[1], and *Oldřich* Kodym[2]

[1]Technical University of Ostrava, 17. listopadu 2172/15, 708 00 Ostrava, Czech Republic
[2]College of Logistics, Přerov, Palackého 1381/25, 750 02 Přerov, Czech Republic

Abstract. 3D printing technology has emerged as one of the most disruptive innovations to impact the logistics industry and the global supply chain. Some claim that the technology merely enhances some aspects of production process, while others argue that technology will revolutionize and replace existing manufacturing technologies. Whether revolutionary or evolutionary, 3D printing technology is recognized as an important trend that will significantly impact supply chains. The objective of this article is to explore basic issues related to 3D printing technology and possibilities for altering manufacturing and supply chain.

1 Introduction

Since the Industrial Revolution, manufacturing has been synonymous with factories, machine tools, production lines and economies of scale. So it is startling to think about manufacturing without tooling, assembly lines or supply chains. However, that is what is emerging as the future of 3D printing services takes hold. Today you can make parts, appliances and tools in a wide variety of materials right from your home or workplace. Using a computer, simply create, modify or download a digital 3D model of an object. 3D printers are being used to economically create custom, improved, as well as design of parts that could not be manufactured before, and right there where they will be used. A single printer can produce a vast range of products, sometimes already assembled.

1.1 What is 3D printing

3D printing, also known as additive manufacturing – AM (the terms 3D printing and additive manufacturing have become interchangeable), is an additive technology used for making three dimensional solid objects up in layers from a digital file without the need for a mould or cutting tool. 3D printing uses a computer aided design (CAD) to translate the design into a three-dimensional object. The design is then sliced into several two-dimensional plans, which instruct the 3D printer where to deposit the layers of material. Additive process, of depositing successive thin layers of material upon each other, producing a final three dimensional product [1]. A wide variety of materials can be utilized

* Corresponding author: lukas.kubac@vsb.cz

- plastics, glass, metal, polymers, wax, sand and glue mixes, nylon, ceramic, edible material, and even human tissue.

Three basic ingredients are essential for creation of a 3D object - a digital model, feed material and a 3D printer (fig. 1).

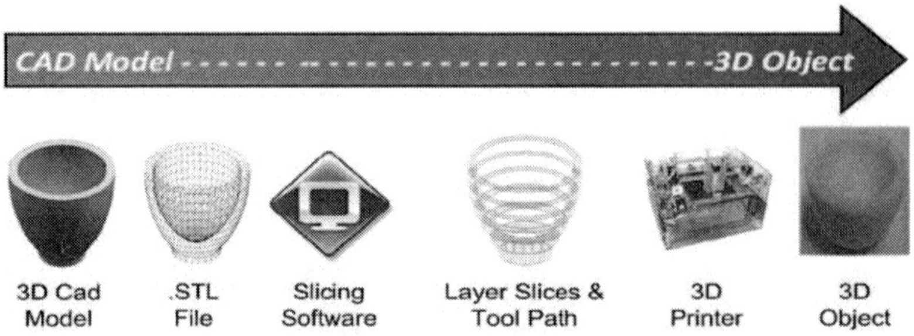

Fig. 1. Generalized additive manufacturing process. Source: [2]

1.2 Basic advantages and disadvantages of 3D printing

Advantages

- Less setup costs and less tooling costs
- Lower production costs
- Lower development time and costs (increase the pace of design to product)
- Eco friendly: less influence on $CO2$
- Lower cost price for small volumes and complex shaped parts
- Shorter delivery time
- Complex products relatively easy to print
- Flexibility (printing on-demand, customisation)

Disadvantages

- Lower production speed
- Higher investment costs
- Unclear responsibilities regarding intellectual property
- Liability - warranties, certification
- Process variability/ quality
- Not efficient for large series
- Limited choice of material (sometimes)
- Limited dimensions of product
- Complexity of the Supply Chain (3D printing is added to the Supply Chain)

Logistics service providers will only be allowed to store and print out the data models if the manufacturers trust them and are prepared to give them the 3D data models. What's certain is that in future, the "value" of a product will reside in a digital file. Manufacturers will try to protect their intellectual property by inserting copy protections and assigning licensing rights [3].

There are still some limitations to be compensated, such as durability and other material related characteristics of the printed components, printing time, and the price of raw materials and commercially available printing equipment. However, it seems that technology evolution is beating the limitations one by one.

3D printing is not yet conducive to mass production due its typically higher cost per unit and slower production time per unit. If the new technology is to completely transform

DOI: 10.1051/matecconf/201713400027

global industry, 3D Printing must be able to mass produce goods in the same volumes as traditional manufacturing techniques.

2 Challenges

Like any disruptive new opportunity, on-demand AM manufacturing networks have challenges that must be addressed [4]. Challenges exist in three major areas:

Technology and Engineering

- Different design principles and stress data for each different AM technology
- Intensive investment in engineering training and process maturity is needed
- Slight changes in material properties can make one-to-one replacements difficult (sometimes)
- Extensive post-processing (heat-treat, machining…) are presently at lower levels of automation

Data Management

- Future geographical and/or organizational separation of the design and production organizations requires a secure transfer of approved design data (i.e. build file, material specifications)
- Design may only be printed on a qualified printer and must be authorized to print the part
- Design and build data must be traceable against source requirements

Business Impacts

- Local printing in comparison to centralized printing offers significant availability/lead time advantages for spare parts supply
- Supporting platforms over decades of use
- Intellectual property protection for design and printing data
- Opens the door to new business models and modes of operation
- Designs may be licensed for print by authorized end users
- Subscription models for parts catalogues

3 Impact of 3D printing on the Supply Chain

The impact of AM technologies on the global setup of supply chains can be very disruptive. The technology has the potential to eliminate the need for both high volume production facilities and low-level assembly workers, thereby drastically reducing supply chain cost [5].

In terms of impact on inventory and logistics, we can print on demand. Meaning we don't have to have the finished product stacked on shelves or stacked in warehouses anymore. Whenever we need a product, we just make it. And that collapses the supply chain down to its simplest parts, adding new efficiencies to the system. Those efficiencies run the entire supply chain, from the cost of distribution to assembly and carry, all the way to the component itself, all the while reducing scrap, maximizing customization and improving assembly cycle times.

3.1 Traditional supply chain vs AM model

The supply chain traditional model is founded on traditional constraints of the industry, efficiencies of mass production, the need for low cost, high volume assembly workers, and so on. But 3D printing bypasses those constraints. 3D printing finds its value in the printing of low volume, customer specific items, items that are capable of much greater complexity than is possible through traditional means. This at once eliminates the need for both high volume production facilities and low level assembly workers, thereby cutting out at least half of the supply chain in a single blow. From that point of view, it is no longer financially efficient to send products across the globe when manufacturing can be done almost anywhere at the same cost or lower. The raw materials today are digital files and the machines that make them are wired and connected, faster and more efficient than ever. And that demands a new model of supply chain [6].

With support local sourcing, the 3D printing technology has the potential to tear established global supply chain structures apart and reassembles it as a new, local system. Furthermore, the technology creates a close relationship between design, manufacturing and marketing. The technology could transform the global supply chain to a globally connected, but totally local supply chain [7].

Let's take a look at the differences in today's more traditional supply chain versus what it could look like in the future with more 3D printing and the supply chain:

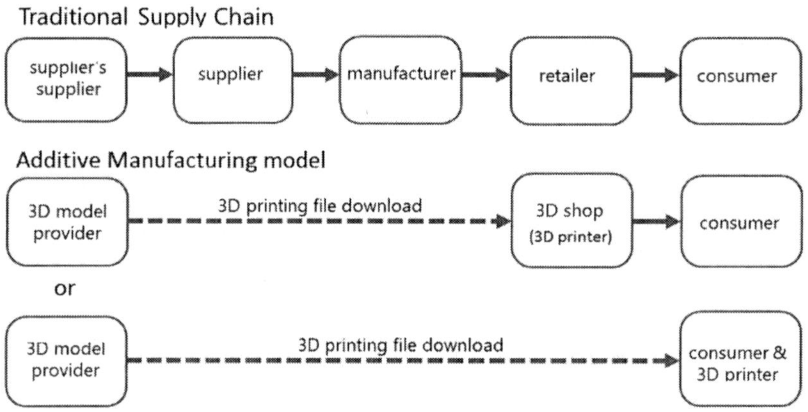

Fig. 2. Supply chain – Traditional and AM model. Source: [8]

Traditional Supply Chain

- Products are mass produced (e,g. in China)
- Manufactured goods are 'pushed out' and distributed through warehouse network to customers
- Long lead time
- High transport costs
- Large carbon footprint

Additive Manufacturing model

- Customised production
- 'Pulled' by end customer demand
- Locally printed and distributed
- Short lead time
- Low transport costs
- Low carbon footprint

Generalized processes in traditional supply chain and Additive Manufacturing model see Fig. 2.

3.2 Advantages of AM model

Cost Savings

- Eliminate the need for large bulk inventory
- Eliminate the need for high volume production facilities
- Reduce transportation cost
- Eliminate penalty for redesign
- Reduce the size of an economical lot
- More economical and effective packaging solutions
- Vetting out designs: offer customized designs at lower cost
- Reduce labor inputs: eliminate low level assembly workers
- Reduce required tooling and machining centers
- Economical mass customization

Speed Responsiveness

- Eliminate the time lag between design and product
- Shorter lead time
- Enabling on-demand manufacturing
- Improving process flexibility
- Supply chain disintermediation

Quality Improvement

- Reduce production waste
- Improve quality
- Incorporate customer feedback
- More optimum products across many industries
- Eliminate excess parts that cause drag and add weight
- Management of demand uncertainty

Environmental Impact

- Less negative impact on environment
- Reduce carbon footprint
- Reduce the waste that accrue in traditional manufacturing

With 3D printing, manufacturing will become more agile and better able to react to customer demands. This means there will be less work-in-progress and finished product in transport and in stock and less obsolescence of existing stock. Although the cost per unit may be higher, with reduced storage and less outdated product, the overall supply chain system costs may be lower than that of traditional manufacturing supply chains. Products can be made closer to the consumer [9, 10]. 3D printing supports a build-to-order, mass customization environment, and in some cases it will allow for product differentiation to be postponed in the supply chain process. There will be a dramatic increase in information transmission. The feedstock, often very versatile and useful for several products, becomes the inventory. This process is ideal for small batches and uncertain demand [7].

Opportunities for 3D printing are not entirely captured by the reduction of manufacturing costs, even though it is likely for low-volume parts that the manufacturing costs can be reduced because of lower setup and tooling costs. AM offers opportunities for new marketing strategies and distribution channels. Instead of manufacturers producing products based on forecasted demands, 3D printing allows for manufacturing based on real-time demand. This results in shorter lead times from order to delivery. The supply chain will be much more flexible in order to accommodate surges in product demand [11].

Global networks of 3D printing installations will give enterprises the ability to respond rapidly to shifts in market demand and to introduce new products quickly and inexpensively. More versatile design processes could unleash a new wave of product innovation. In the longer term, the technology has the potential to redefine traditional manufacturing methods. The concept of making products in large, complex plants could become outmoded as companies adopt the more flexible additive manufacturing model.

3.3 Implications for the logistics industry

The implications of this new manufacturing technology for the logistics industry could be massive:

- Potentially a proportion of goods which were previously produced in China or other Asia markets could be 'near-sourced' to North America and Europe. This would reduce shipping and air cargo volumes.
- Customisation of products would mean that inventory levels fall, as goods are made to order. This would have the effect of reducing warehousing requirements.
- Downstream logistics would also be affected. Build-to-order production strategies could fundamentally impact the manufacturer-wholesaler-retailer relationship. In the future the shopping experience could also be vastly different. In some sectors, retailers will either cease to exist or become 'shop windows' for manufacturers, keeping no stock of their own. Orders are fulfilled directly by the manufacturer, and delivered to the home of the consumer [12].
- A major new sector of the logistics industry would emerge dealing with the storage and movement of the raw materials which 'feed' the 3D Printers. As 3D Printers become more affordable to the general public, the home delivery market of these materials would increase.
- The Service Parts Logistics sector would be one of the first to be affected. At present billions are spent on holding stock. If you can download a design from the online library, print 3D and then enter it in a very short time window, then global and national warehouses as well as front warehouses will not be needed [9].

3.4 Applications in Industries Supply Chain

The breadth and impact of AM continues to expand as the technology gains acceptance and functionality, making it a feasible means of production in a variety of industries. In the past few years, many companies have adopted AM and are beginning to rip the real benefits from the technology. Healthcare, automotive and aerospace industries are among the sectors with the greatest transformative potential. AM technology has been pushed beyond the realm of prototyping and has become an effective means of advancing the way parts and tools are produced in the Aerospace industry. AM makes it possible to have objects printed in remote locations, as delivery of goods is no longer a restriction. This benefit of AM makes it possible for the use for technology in Space. NASA has been testing AM in zero gravity in hopes of establishing on-demand manufacturing for astronauts. This would allow component parts for maintenance and repair of the international space station to be manufactured in Space. This would decrease the need for shuttle to make trips to the international space station to deliver parts, thus greatly reducing the lead-time on replacement parts. A reduction in leadtime would imply a reduction in inventory and a reduction in costs.

Fig. 3. Even basic 3D printers can be reprogrammed to create new designs at near zero marginal cost. Source: [13]

4 Conclusion

3D printing started, mainly as a means to create prototypes. The recent technological advancements and applications of 3D printing suggest that the technology has potential to revolutionize many facets of everyday life. The impact of AM on supply chains takes many forms, including simplified production processes, reduced material waste for leaner manufacturing, increased flexibility, reduced costs, faster reactions to demand, and the ability to decentralize production.

3D printing opens up new opportunities and is slowly emerging as a valuable way to improve supply chain efficiencies. AM has the potential to reduce costs and increase profits. As a result, supply chain can quickly react to changes in the marketplace. It opens up new opportunities and provides many possibilities for companies looking to improve manufacturing efficiency. AM significantly streamlines traditional methods and has the potential to become the norm over decades to come. The technology provides freedom of design using standard CAD software, and it is not limited to manufacturing technology. It also enables cost effective product customization.

The technology has improved processes in many industries, including aerospace, automotive, industrial products, consumer products, defence, architecture, or healthcare. Experts believe that this technology will keep growing at a fast pace and play a major role in the future of supply chains.

However before that, 3D print technology has to overcome one big obstacle. To provide more information among people about all its abilities and innovations, there are still views the 3D printing from the time of its beginnings more than 5 years ago. Today the situation is different, and the claim that 3D printing increases the consumption of raw materials is misleading. The opposite is true. 3D printing will greatly impact waste logistics. But positively. A well-designed production process using 3D printing results in significant savings in consumed material and the amount of waste generated. As production will be closer to the consumer, there will be less need for packaging and shipping materials than ever before. And less waste in production is then given by the very essence of the additive process. Another erroneous view of 3D printing is that its main potential is for products made of plastics and usually components to final products. That's not true. The main potential of the technology is the use of a wide range of materials, not just plastics. For instance Metal 3D Printing has the ability and big potential to produce intricate, streamlined components with physical properties that can sometimes exceed those of parts manufactured by traditional means. Consequently, the technology has the potential to completely shift the way that we fabricate critical components. With it, we can create lightweight objects with unique geometries capable of decreasing material waste and energy consumption.

In other words, there's a "knowledge gap" between the technology available and the people must close before 3D printing can be a common method of manufacturing product.

This paper is supported by the research project "From horse-drawn railway to intermodal transport" within Visegrad Fund.

References

1. What Is 3D Printing, *3dprinting.com*, Available online: https://3dprinting.com/what-is-3d-printing/ (2017)
2. T. Campbell, C. Williams, O. Ivanova, B. Garrett, *Could 3D Printing Change the World? Technologies, Potential, and Implications of Additive Manufacturing* (Washington, DC: Atlantic Council of United States, 2011)
3. H. Lipson, M. Kurman, *Fabricated: The New World of 3D Printing* (John Wiley & Sons, Hoboken, USA, 2013)
4. B. Berman, Business Horizons **55**, 155-162 (2012)
5. G.R. Janssen, I.J. Blankers, E.A. Moolenburgh, A.L. Posthumus, *The Impact of 3-D Printing on Supply Chain Management* (TNO: Innovation for Life, April 2014)
6. M. Gravier, *3D Printing: Customers Taking Charge of the Supply Chain* (Industry Week, April 12. 2016)
7. H.J. Nyman, P. Sarlin, *47th Hawaii International Conference on System Science*, 4190-4199 (Waikoloa, 2014)
8. A. Wieland, *scrmresearch.org*, Available online: https://scmresearch.files.wordpress.com/2014/12/3d_printing_supply_chain.png (2014)
9. C. Alexander, *Five Ways 3D Printing Is Transforming the Supply Chain* (Supply & Demand Chain Executive Blog, 8 April 2015)
10. H. Paris, et al., CIRP Annals – Manufacturing Technology **65**, 29-32 (2016)
11. R. Bogue, Assembly Automation **33**, 307-311 (2013)
12. R. Crandall, *Where Will Additive Manufacturing Take Us?* (APICS, 26 June 2016)
13. A. Gramulewicz, *Raw Intelligence: How 3d Printing Is Transforming Supply Chains. Circulate*, Available online: http://circulatenews.org/2017/01/how-3d-printing-is-transforming-supply-chains/ (2017)

Logistics Cost Calculation of Implementation Warehouse Management System: A Case Study

Tomáš Kučera[1,*]

[1]University of Pardubice, Faculty of Transport Engineering, 53210, Studentská 95, Pardubice, Czech Republic

Abstract. Warehouse management system can take full advantage of the resources and provide efficient warehousing services. The paper aims to show advantages and disadvantages of the warehouse management system in a chosen enterprise, which is focused on logistics services and transportation. The paper can bring new innovative approach for warehousing and presents how logistics enterprise can reduce logistics costs. This approach includes cost reduction of the establishment, operation and savings in the overall assessment of the implementation of the warehouse management system. The innovative warehouse management system will be demonstrated as the case study, which is classified as a qualitative scientific method, in the chosen logistics enterprise. The paper is based on the research of the world literature, analyses of the internal logistics processes, data and finally enterprise documents. The paper discovers costs related to personnel costs, handling equipment costs and costs for material identification. Implementation of the warehouse management system will reduce overall logistics costs of warehousing and extend the warehouse management system to other parts of the logistics chain.

1 Introduction

At present time companies try to minimize the costs of logistics. The concept of warehousing is very closely linked to logistics and distribution. Warehousing is within the logistics chain, one of the activities cannot be omitted. Warehousing addresses many crucial issues, inventory levels, ordering cycles, warehouse equipment and their spatial distribution, distribution of warehouse and inventory management. At high inventory levels incurred by companies fixed cost for each additional unit of inventory. Possession of inventory on hand may be necessary for the company due to preserve their source of supply. Warehousing is one of the most important parts of the entire logistics system. Warehousing is a link between manufacturers and customers, it provides storage for many products (e.g. raw materials, parts, semi-finished products, components and finished products). Warehouses allow to bridge space and time. Manufacturing inventories ensure optimal continuity of the production. Inventories of goods for ensuring smooth supply of the final customer.

* Corresponding author: tomas.kucera@student.upce.cz

2 Theoretical background and methodology

With the constant technological advancement, the companies started to invest in new systems to support competitiveness in the market in which they operate. The trend of reducing the cost of warehousing is the maximum use of warehouse space, the minimum number of transfers, effective solution to the warehousing points or reducing the number of species of stored inventory. These costs are significant, if the free warehouse space on someone's rented, or if with a higher warehousing amount of enterprise must pay for additional rented premises, or energy, or heat. Warehouse managers are usually interested in providing high quality services to their customers at minimum cost. From a tactical, strategic and operational point of view, the main issues concern both the warehouse and the inventory management [1, 2].

Warehousing systems and material handling systems play a pivotal and critical role in the supply chain, and requirements for warehousing and handling operations have significantly increased in recent years [3].

Warehousing refers to activities associated with inventory management. These activities take place within the warehouse area, it is mainly the receipt of goods, purchase goods handling, picking, gathering and activities related to sorting [4, 5].

Warehouse management is not just a purchase, warehousing and transport within the warehouse; this is a more complex system that exceeds the physical boundaries of warehouses. Warehouse Management System (hereinafter WMS) is used to increase performance and warehousing for better management decisions for holding only strictly necessary supplies needed for the manufacturing process [6, 7]. WMS can take full advantage of the resources and provide efficient warehousing services provided by the plan, organize, control and coordination processes [8]. WMS is a necessary approach for every warehouse. An automated warehousing system provides less effort, more efficient, and reliable results compared to manual handled system. WMS is designed to help reduce costs through effective warehouse processes [9]. The tool known as WMS is a system that facilitates the handling operations and storage materials through defined parameters, but the deployment of this system need to go through steps that impact positively and negatively the company's various sectors [10].

The need for automating the warehouse arises from the fact that manual handling may cause human errors which may affect the warehouse utilization [11].

The case study, as one of the qualitative research methods, was chosen because it counts among the most frequently used methods within the research focused on the implementation of different management approaches into practice of organizations [12]. Case study is defined as an empirical survey of the current phenomenon in its natural environment, using multiple sources of evidence. Study cases are selected based on pre-defined conditions of the case [13, 14].

The aim of this paper is to show advantages and disadvantages of the warehouse management system in a chosen logistics service provider (Ewals Cargo Care), which is focused on logistics services and transportation. The paper will bring new innovative approach for warehousing and presents how the logistics enterprise can reduce logistics costs, mainly personnel costs. This is especially thanks to the implementation of warehouse management system. Data used in the case study is based on real logistics processes in logistics service provider. The paper discovers costs related to personnel costs, handling equipment costs and costs for material identification.

3 Case study: Warehouse management system in logistics service provider

The case study is focused on new innovative approach for warehousing and presents how logistics service provider (Ewals Cargo Care) can reduce logistics costs, thanks to the implementation of warehouse management system. Ewals Cargo Care (hereinafter ECC) is a transport and logistics company operating in the market for more than 110 years. For its customers, primarily in the automotive, electronics, paper and packaging industry, manufacturers and distributors of consumer electronics provides services both in the field of transport, as well as services related to warehousing and handling of goods [15].

The company's goal is always to find the optimal solution for shipping process on the way from supplier to customer, with the possibility of combining shipment, transhipment or use consolidation warehouses. ECC fully uses its years of experience, skills and knowledge to find and implemented for customers the most effective solution with the least impact on the environment [15].

The case study compares two models. The first one is the existing warehouse system and the second one is the new innovative WMS. For the design of WMS implementation was used analysis of current warehouse management based on fixed warehouse positions and common software that keeps track of inventory about individual stock items. The current system of fixed positions does not allow variable use of empty warehouse cells for current needs income and material warehouse. Proposal for the implementation of WMS in turn allows almost 100% utilization of warehouse capacity and reducing the administrative handling time needed to process goods receipt, put-away, warehousing and distribution of goods to the production. For the current calculation of measured values required for the proposal to introduce managed warehouse was used ECC customer production plan in 2017. Of these production plans, can derive data needed to analyse the personnel and technical complexity in the case of implementation of WMS.

Tab. 1 shows the personnel costs in the existing warehouse system. Personnel costs are governed by wage regulation in the ECC. Based on the number of employees (32 people) are determined by the total personnel costs, which amounts to 1,183,000 CZK per month.

Table 1. Personnel costs – current state – existing warehouse system. Source: author, based on [15]

	1.Shift	2.Shift	3.Shift	Total (Person)	CZK/Person	Total
Forklifts driver	**6**	**6**	**6**	**18**	36 000 CZK	**648 000 CZK**
Operator of material income	**2**	**1**	**1**	**4**	32 000 CZK	**128 000 CZK**
Administrator	**2**	**2**	**2**	**6**	38 000 CZK	**228 000 CZK**
Shift Supervisor	**1**	**1**	**1**	**3**	42 000 CZK	**126 000 CZK**
Manager of Warehouse	**1**	**0**	**0**	**1**	53 000 CZK	**53 000 CZK**
Total	**12**	**10**	**10**	**32**		**1 183 000 CZK**

Required operating personnel warehouse in 2017 could be changed about the implementation of WMS. In tab. 2 lists the personnel costs in the warehouse after implementation of WMS. Out of 32 employees after the implementation of software and hardware WMS has been optimized six workers. Personnel costs would be reduced from the amount 1,183,000 CZK to 967,000 CZK per month.

Table 2. Personnel costs – state after implementation of WMS. Source: author, based on [15]

	1.Shift	2.Shift	3.Shift	Total (Person)	CZK/Person	Total
Forklifts driver	5	5	5	15	36 000 CZK	540 000 CZK
Operator of material income	1	1	1	3	32 000 CZK	96 000 CZK
Administrator	2	1	1	4	38 000 CZK	152 000 CZK
Shift Supervisor	1	1	1	3	42 000 CZK	126 000 CZK
Manager of Warehouse	1	0	0	1	53 000 CZK	53 000 CZK
Total	10	8	8	26		967 000 CZK

In tab. 3 is an overview of the necessary handling equipment in the warehouse before and after the implementation of WMS. The changes that have occurred since the implementation of WMS are saving one forklift (yellow highlighted box in the tab. 3). Number of forklifts can be reduced because WMS reduces the number of forklift drivers. Monthly savings is 24,242 CZK.

Table 3. Handling equipment costs. Source: author, based on [15]

					Existing System	WMS
Type of Forklift	From	To	Days	Daily Rate	Price/Month	Price/Month
EFG 110 FN407280	01.01.17	31.01.17	31	620 CZK	19 220 CZK	19 220 CZK
EFG 110 FN350228	01.01.17	31.01.17	31	620 CZK	19 220 CZK	19 220 CZK
EFG 110 FN407311	01.01.17	31.01.17	31	620 CZK	19 220 CZK	19 220 CZK
EFG 216 FN419164	01.01.17	31.01.17	31	782 CZK	24 242 CZK	0 CZK
EFG 216 FN419166	01.01.17	31.01.17	31	782 CZK	24 242 CZK	24 242 CZK
EFG 216 FN419167	01.01.17	31.01.17	31	805 CZK	24 955 CZK	24 955 CZK
EFG-216 MP FN372153	01.01.17	31.01.17	31	805 CZK	24 955 CZK	24 955 CZK
EFG 220 FN406164	01.01.17	31.01.17	31	805 CZK	24 955 CZK	24 955 CZK
EFG 220 FN324425	01.01.17	31.01.17	31	805 CZK	24 955 CZK	24 955 CZK
ERE 225 98007140	01.01.17	31.01.17	31	326 CZK	10 106 CZK	10 106 CZK
ERE 225 98007152	01.01.17	31.01.17	31	326 CZK	10 106 CZK	10 106 CZK
H50D					34 310 CZK	34 310 CZK
H20D					23 480 CZK	23 480 CZK
Total Handling Equipment Costs					283 966 CZK	259 724 CZK

Tab. 4 shows the hardware and software costs of implementation WMS. These costs include software and hardware equipment (radio frequency terminal equipment, radio frequency network including installation and the necessary printers for printing identification cards). The monthly depreciation of software is 36,944 CZK and maintenance is 7,875 CZK. Monthly depreciation on hardware is 19,168 CZK. Total monthly costs for software and hardware of WMS are 63,987 CZK.

Table 4. Hardware and software costs of implementing WMS. Source: author, based on [15]

Software (SF)			Depreciation period of 36 months	
Product			Price	
Delivery Chain Integrator			700 000 CZK	
Implementation Services			450 000 CZK	
Expanding Implementation Services			150 000 CZK	
Travel Costs			30 000 CZK	Monthly Depreciation
Total Software Costs			**1 330 000 CZK**	**36 944 CZK**
Price for maintenance				
Product			Price for a year	Monthly Payment
Delivery Chain Integrator			**94 500 CZK**	**7 875 CZK**
Hardware (HW)			Depreciation period of 36 months	
Product	Number	Price/unit	Price	
Radio frequency (RF) terminal	9	30 000 CZK	270 000 CZK	
Accessories	9	5 237 CZK	47 133 CZK	
RF network including installation	1	300 000 CZK	300 000 CZK	
Printer	2	36 455 CZK	72 910 CZK	Monthly Depreciation
Total Hardware Costs			**690 043 CZK**	**19 168 CZK**
Total Monthly Costs for Hardware and Software				**63 987 CZK**

Tab. 5 expresses the logistics costs for material identification in the ECC warehouse. From the available data to determine the total number of printed identification cards (hereinafter IDC) is 660,294 pieces. Print of one IDC worth 0.61 CZK, so they are identifiable costs to the total number of IDC calculated the amount to 402,779 CZK. Costs of hardware and software (column HW and SW) are fixed monthly 29,637 CZK. This amount includes the use of personal computer and printers including consumables. Total yearly costs for material identification are 758,423 CZK.

Table 5. Logistics costs for material identification, HW and SW equipment – existing warehouse system. Source: author, based on [15]

Month	Number of IDC	IDC Costs	HW and SW	**Total**
1	59 432	36 253 CZK	29 637 CZK	**65 890 CZK**
2	57 436	35 036 CZK	29 637 CZK	**64 673 CZK**
3	67 187	40 984 CZK	29 637 CZK	**70 621 CZK**
4	55 804	34 040 CZK	29 637 CZK	**63 677 CZK**
5	61 649	37 606 CZK	29 637 CZK	**67 243 CZK**
6	60 996	37 207 CZK	29 637 CZK	**66 844 CZK**
7	40 073	24 445 CZK	29 637 CZK	**54 082 CZK**
8	41 201	25 133 CZK	29 637 CZK	**54 770 CZK**
9	58 316	35 573 CZK	29 637 CZK	**65 210 CZK**
10	55 231	33 691 CZK	29 637 CZK	**63 328 CZK**
11	57 078	34 817 CZK	29 637 CZK	**64 454 CZK**
12	45 892	27 994 CZK	29 637 CZK	**57 631 CZK**
Total	660 294	402 779 CZK	355 644 CZK	**758 423 CZK**

Tab. 6 shows the costs for material identification in case of implementation WMS. IDC costs will change, because new IDCs are cheaper. One identification card costs only 0.24 CZK. Changes that occur with this implementation shall remain in hardware and software. Monthly operating costs of hardware and software of WMS are from tab. 4 totally 63,987 CZK per month. As is apparent from a comparison of the tables, the implementation of WMS will increase the total cost of the material identification from 758,423 CZK to 926,318 CZK per year.

Table 6. Logistics costs for material identification, HW and SW equipment – WMS. Source: author, based on [15]

Month	Number of IDC	IDC Costs	HW and SW	**Total**
1	59 432	14 264 CZK	63 987 CZK	**78 251 CZK**
2	57 436	13 785 CZK	63 987 CZK	**77 772 CZK**
3	67 187	16 125 CZK	63 987 CZK	**80 112 CZK**
4	55 804	13 393 CZK	63 987 CZK	**77 380 CZK**
5	61 649	14 796 CZK	63 987 CZK	**78 783 CZK**
6	60 996	14 639 CZK	63 987 CZK	**78 626 CZK**
7	40 073	9 618 CZK	63 987 CZK	**73 605 CZK**
8	41 201	9 888 CZK	63 987 CZK	**73 876 CZK**
9	58 316	13 996 CZK	63 987 CZK	**77 983 CZK**
10	55 231	13 256 CZK	63 987 CZK	**77 243 CZK**
11	57 078	13 699 CZK	63 987 CZK	**77 686 CZK**
12	45 892	11 014 CZK	63 987 CZK	**75 001 CZK**
Total	660 294	158 471 CZK	767 848 CZK	**926 318 CZK**

Tab. 7 expresses the total cost evaluation, the use of an existing type of warehousing at fixed positions. The right part of the table shows implementation of the WMS. The total savings would be 6 people, so it would reduce personnel costs. WMS could reduce the costs of the warehouse equipment of the forklift. Costs for material identification would increase slightly, but the total yearly cost savings would amount to 2,715,009 CZK.

Table 7. Yearly evaluation of total costs. Source: author, based on [15]

	Existing Warehouse System	Warehouse Management System
Number of personnel	32	26
Costs of personnel	14 196 000 CZK	11 604 000 CZK
Handling equipment costs	3 407 592 CZK	3 116 688 CZK
Costs for material identification	758 423 CZK	926 318 CZK
Yearly savings	**2 715 009 CZK**	

4 Conclusion

WMS brings many positive impacts. WMS means in practice that all movement of the material is controlled by software, material is identified by a bar code transmitted with a radio frequency terminal. Workers are minimized errors and increasing the productivity. Processes performed by the warehouse staff can monitor and evaluate at any moment. The system minimizes the complaint allows the merging of orders and guarantees compliance with FIFO (First In First Out).

The aim of this paper was reduced the logistics costs of warehousing in the ECC and propose the implementation of WMS for reasons of logistics cost savings. With the implementation of WMS will be savings on the cost side. Overall it will be reduced the necessary operating staff consists of 32 to 26 workers, also will be reduced the required handling (warehouse equipment) and slightly will be increased the logistics costs for material identification. Total yearly savings are 2,715,009 CZK.

Acknowledgements: The work was created in connection with the scientific research project of the University of Pardubice no. SGS_2017_009. The author is grateful for their support.

This paper is supported by the research project "From horse-drawn railway to intermodal transport" within Visegrad Fund.

References

1. F. Guerriero, O. Pisacane, F. Rende, App. Math. Mod. **39**, 23-24, 7375-7389 (2015)
2. R. Kampf, S. Lorincová, M. Hitka, Z. Caha, Nase More **63**, 3, 120-125 (2016)
3. R. Manzini, Y. Bozer, S. Heragu, Inter. J. of Prod. Econ. **170**, 711-716 (2015)
4. S. Apak, H. Tozan, O. Vayvay, Tech. Vjesnik-Technical Gazette **23**, 5, 1439-1446 (2016)
5. F. Nemec, S. Lorincová, M. Hitka, L. Turínska, Nase More **62**, Spec. Issue, 131-138 (2015)
6. T.C. Poon, K.L. Choy, H.K.H. Chow, H.C.W. Lau, F.T.S. Chan, K.C. Ho, Exp. Sys. with Appl. **36**, 4, 8277-8301 (2009)
7. J.Y. Shiau, M. C. Lee, Com. & Ind. Engin. **58**, 3, 382-392 (2010)
8. Y. Feng, Inter. Con. on Ind. Con. and El. Engin. 1258-1260 (2012)
9. A.M. Atieh, H. Kaylani, Y. Al-abdallat, A. Qaderi, L. Ghoul, L. Jaradat, I. Hdairis, *Proceedings of the 48th CIRP Con. Man.* **41**, 568-572 (2016)
10. D.R. Novaes, D.C. Souza, M.A. de Veras Filial, C.V. Giordano, L.C. Goncalves, Rev. En. Pes. **5**, 2, 223-239 (2016)
11. S. Seifermann, J. Boellhoff, J. Metternich, A. Bellaghnach, *Var. Man. in Manuf.: Proc. of the 47th CIRP Con. in Man. Sys.* **17**, 588-593 (2014)
12. Z. Hoque, Brit. Accoun. Rev. **46**, 1, 33-59 (2014)
13. S. Kaae, B. Sondergaard, L. S. Haugbolle, J. M. Traulsen, Phar. W. & Sc. **32**, 1, 36-42 (2010)
14. K. R. Yin, *Case Study Research: Design and Methods* (Sage Publications, 2014)
15. Ewals Cargo Care. *Internal enterprise data* (Prague, Czech Republic, 2017)

Warehouse Location Problem in Context of Delivery Time Shortening

Nina Kudláčková[1,*], and *Jan* Chocholáč[1]

[1]University of Pardubice, Faculty of Transport Engineering, Studentská 95, 532 10 Pardubice, Czech Republic

Abstract. The current trend in logistics and the distribution chain is the pressure to reduce the delivery time, which was triggered by the e-commerce trend. To shorten the delivery time, it is necessary to choose the location of the warehouse or distribution center to maximize efficiency. Therefor is this article focused on warehouse location problem and chosen methods to solve this problem. It is also focused on use of the multi-criteria method to evaluate the alternatives for the warehouse location because it will be necessary to go even further in this area and find a combination of existing models or find new one which will be based not on costs but on delivery time. That will help companies solve the allocation problem to meet the requirements for shorter delivery times.

1 Introduction

Most companies are currently either directly or at least indirectly involved in a trading system that we call global. The global concept of business is quite challenging, and businesses are challenging how to manage the corresponding processes with products from the perspective of production, distribution and logistics.

The current trend in logistics and the distribution chain is the pressure to reduce the delivery time, which was triggered by the e-commerce trend.

This is a change in the delivery requirements of customers, where standard delivery is performed within 1-2 days or in the shortest possible time, free or almost free of charge.

Such a market trend needs to be adapted to the behavior of the whole logistics chain, because supply speed can be an important factor in gaining a competitive advantage.

To shorten the delivery time, it is necessary to choose the location of the warehouse or distribution center to maximize efficiency. This is an important strategic decision that has a long-term impact on the further development of the supply-demand chain.

Very important attribute that affects the supply-chain efficiency aspects is the number, size and location of the distribution centers, as well as the decision from which distribution center to serve a given range of consumers or customers.

Based on the importance of this logistical problem, a great deal of research has been devoted to the development of mathematical models to find a suitable location for the distribution center. However, most of them are based on cost minimization.

* Corresponding author: Nina.Kudlackova@upce.cz

Current trends, however, place particular emphasis on shortening the delivery time, even at the expense of higher costs. It is therefore necessary to shift current models of decision making on the location of the warehouse network from minimizing costs to minimizing delivery times.

This raises the need to find in the current spectrum of theories of geography experts such models that are based both on distance assessment and on minimizing costs.

The decision making process for location of the warehouse network could be based on the models below.

2 Globalization and global logistics

While Rodrigue defines globalization simply as a term commonly used to express the mutually evolving economic, social and cultural relationships between individual world regions, The International Monetary Fund defines globalization in several levels. In the most general it characterizes it as a force that is increasingly affecting the lives of people of all countries in the world. More specifically, it defines globalization as the historical process, the result of human innovation and the technological process. In the opinion of the International Monetary Fund, it refers to the growing integration of economies around the world, particularly through the movement of goods, services and capital across borders. The IMF also recognizes the globalization of another plane, namely cultural, political and environmental [1, 2].

Globalization has caused many problems in logistics. For example, there have been prolonged delivery times, especially since transition times have become uncertain. There has been a significant increase in break-bulk items and also an increase in consignment consolidation requirements. In some companies, there has been a direct conflict between the process of globalization and the introduction of just-in-time (JIT) procedures under the new global strategy.

Global companies tend to prolong delivery times and inventory levels due to the large distances of the markets and the complexity of logistics operations. In companies doing business on the JIT philosophy is the situation opposite aiming at shortening delivery times and stock levels as much as possible.

The solution is the level of compromise between the cost of order fulfillment and stockholding and the costs resulting from the uncertainty of product deliveries. Thus, there is a general shift in competition. They no longer compete with companies as such, but with their distribution chains. The solution is therefore in solving the warehouse location problem and optimizing the distribution strategy [1, 3].

3 Chosen methods of warehouse location problem

The creation of a warehouse or distribution center model can be understood as a classical locational task in the field of operational theory of graph theory. As Cura states, it is possible to divide the localization tasks into static and dynamic. Static localization tasks focus more on space, dynamic to time factor [4, 11].

3.1 Weber model

The Weber model is focused on minimizing the transport costs of the company. This model defined so-called localization factors, such as raw materials or labor costs.

The so-called Weber's problem is a typical example of solution with one service center. The purposeful function for optimal solution of these tasks is expressed by equation (1):

$$min\ Z = \Sigma_{(i \epsilon F)} = w_i \cdot d\ (X, P_i) \quad (1)$$

The function minimizes the total cost Z.

A set of all demanding customers is marked as F, scale weights transferred on the customer costs i are marked as w_i. X a P_i, and represent searched position of the service center and the customer's position i, so $d\ (X,\ P_i)$ expresses the distance between the two locations.

The task with one operator center can be converted to task with more centers. The new form describes the following equation (2).

The set of all demanding customers is marked D, the rated weight of the connection between the customer i and the service center j is, then w_{ij}, $d\ (X_j,P_i)$ is the distance of the searched service center j and given customer i.

$$min\ Z = \Sigma_{(i \epsilon F)} \Sigma_{(j \epsilon D)} w_{ij} \cdot d\ (X_j, P_i) \quad (2)$$

By extending the task of evaluating the weight of the connection between the individual service centers we get the equation of the purpose function (3), where v_{ik} indicates the weight between the service center i and k.

$$min\ Z = \Sigma_{(i,k \epsilon D, i<k)} v_{ik} \cdot d\ (X_i, X_k)\ \Sigma_{(i \epsilon F)} \Sigma_{(j \epsilon D)} w_{ij} \cdot d\ (X_j, P_i) \quad (3)$$

In addition to the previous tasks, the locational allocation task, in addition to the previous tasks, optimally assigns customers to the optimally deployed service centers, provided that their demand is met.

The analytical model of the optimal solution can be formulated using the following purpose function (4) and the set of limiting conditions (5) – (9).

$$min\ Z = \Sigma_{(i \epsilon F)} \Sigma_{(j \epsilon D)} c(i,j) \cdot y_{ij} \quad (4)$$

$$\Sigma_{(j \epsilon F)} x_j = p \quad (5)$$

$$\Sigma_{(j \epsilon F)} y_{ij} = 1\ \forall i \in D \quad (6)$$

$$x_j \geq y_{ij}\ \forall i \in D,\ \forall j \in F \quad (7)$$

$$x_j = \{1,\ if\ we\ place\ a\ center\ at\ one\ of\ the\ possible\ locations\ j,\ \forall j \in F; 0,\ other\ situations\} \quad (8)$$

$$y_{ij} = \{1,\ if\ one\ of\ the\ available\ places\ j\ serve\ the\ customer\ i,\ \forall i \in D,\ \forall j \in F; 0,\ other\ situations\} \quad (9)$$

F represents the set of possible locations of the serving point (which is called depo), D represents the set set of demanding customers; $c\ (i,j)$ costs associated with assigning the customer i to the service location j. The condition (5) expresses that the p service points are open, the condition (6) ensures the assignment of each customer, and the limiting condition (7) allows customers to be assigned only to those service locations which are in operation. The restrictive conditions (8) and (9) define binary conditions for variables [5-7]

3.2 Uncapacitated warehouse location problem

As Michel & Van Hentenryck states, the uncapacitated warehouse location problem is simply defined as a network of W + S knots and W.S link nodes, where S is number of trades and W is number of warehouse candidates. [8]

Each warehouse has certain fixed costs $f_i \geq 0$, and the costs of transport from the warehouse to the store are marked as $c_{ji} \geq 0$.

The task is to find a subset of warehouses and store assignments to stores so that total fixed costs and transport costs are minimized. This task can be simplified to simply location of subset of warehouses, while each trade should of course be assigned to a particular warehouse and that to the one that has minimum shipping costs for the store. In other words, this model determines a set of warehouses as opened, and each store will be assigned by single store in that set where c_{ji} will be minimal.

This means that the decisive variable $x_i \in \{0,1\}$ $i = 1, \ldots \ldots, W$ (12) means, that the store i is opened, when $x_i = 1$ (11), and otherwise it is closed. The mathematical model for solving this problem is, according to experts Michel and Van Hentenryck, formulated as the following (10):

$$min\ \Sigma_{(i=F;W)} f_i \cdot x_i + \Sigma_{(j=1;S)} min\ [(1-x_i) \cdot M + C_{ji}](X,P_i) \quad (10)$$

$$\Sigma_{(i=1;W)} x_i = 1 \quad (11)$$

$$\Sigma_{(i=1;W)} x_i \in \{0,1\}\ i = 1, \ldots \ldots, W \quad (12)$$

M stands for a large enough number, such as $C_{max} + 1$ is, for example [8, 9].

3.3 P-median model

The selection of the location of the warehouse is basically a selection and subsequent assignment of distribution center in a certain economic area, which is characterized by a certain degree of certain demand.

Network distribution analysis is one of the first steps in choosing a location for the center. The role of assignment of placement by the P-median model is, according to Huifenga and Aigonga, explained as follows: the number of supply stations P is chosen to satisfy n requests from the total number of m supply stations in order to minimize distance, time or cost.

The assumption is the demand i where ω_i is the quantity (weighted value) of this demand i and d_ij is the distance between the supply station j to the demand i. The formula is as follows (13):

$$min\ (\Sigma_{(i=1;n)} \Sigma_{(j=1;m)} \omega_i \cdot \alpha_{ij} \cdot d_{ij}) \quad (13)$$

And at the same time

$$\Sigma_{(j=1;m)} \alpha_{ij} = 1,\ i = 1, 2, \ldots, n \quad (14)$$

$$\Sigma_{(j=1;m)} (\Pi_{(i=1;m)} \alpha_{ij}) = P,\ P \leq m \leq n \quad (15)$$

α_{ij} is distribution coefficient (14) and (15), which says that if the demand i is satisfied by the supply station j, the value of this coefficient is 1, otherwise it is 0.

The formula is as follows (16):

$$\alpha_{ij} = \{1, i\ \text{is suppliled through}\ j\ ;\ 0, \text{other situations}\} \quad (16)$$

This basic storage location model always includes only individual factors however there are many factors to select the current location of the warehouse, so the use of this model would be inadequate [9, 10].

4 Storage location model minimizing delivery costs and warehouse maintenance

In the Warehouse location problem, according to Thomas Nagy, the company is generally considering opening a warehouse in certain potential (candidate) locations to supply their existing stores.

Each possible warehouse has the same maintenance costs and the capacity is the maximum number of stores the store can supply.

Each store must be supplied with just one open stock.

In practice, the fact is that the cost of delivering goods to a store depends on the location of the warehouse.

The aim of this model is to determine which warehouses to open and which of these stores should supply different stores so that the sum of warehouse maintenance and delivery costs is minimized.

Let i be the number of stores, j express the number of stores, Cij is the table containing the costs associated with the supply, Sij matrix of indicators in case the store i is supplied in stock j, cf represent fixed costs, and Oj is a vector of truth value indicating which warehouses are open.

The model is as follows (17):

$$min\ \Sigma_{(i,\,j)} S_{ij} \cdot c_{ij} + d_f \cdot \Sigma\, 0_j \tag{17}$$

Condition that the trade must always have one store (18):

$$\forall_i\, \Sigma_{(j)} S_{ij} = 1 \tag{18}$$

Condition that the warehouse has a certain capacity (19):

$$\forall_i\, \Sigma_{(j)} S_{ij} \leq Capacity\, j \tag{19}$$

A condition that the trade may only be stocked on stock, when the store is opened (20):

$$\forall_{i,j}\, \Sigma\, S_{ij} \leq 0_j \tag{20}$$

As can be seen from the above selected models, the traditional approach to addressing the warehouse location problem is to minimize costs [7, 11].

For example, according to Michel and Van Hentenryck, these models are able to simultaneously find a selection of suitable devices and meet the requirements of a given number of customers to minimize total cost of placement, operation and transport [8].

However, most cost-minimization approaches are addressed at the level of supply companies and do not include customer benefits. More advanced models addressing the warehouse location problem are based on maximizing profits. Again, no attention is paid to the customer's requirements, but this does not necessarily have to be met [8, 9].

In several other models, customization features such as product availability, delivery frequency, and certain elements of delivery time requirements can be found but it is still not enough. It seems we will have to link existing models of find a new one, which will reflect customization and will lead to delivery time shorting, because above mentioned trend is further developing

The decision to place the warehouse is still one of the most important decisions when creating a new or expanding existing distribution network, however, companies increasingly take the customer's wishes into account and transfer the issue to distribution and logistics centers rather than to their own supply and demand chain.

5 Conclusion

As can be seen from the above, none of these models are currently tailored to meet the conditions of globalization and customer requirements.

No comprehensive systematic research framework can be found in the current literature to address the location of the distribution equipment so that it meets the customer's requirements to shorten the delivery time while meeting the requirements of the company. Many models created for the warehouse placement do not represent the consequences of criterion dependency. Most multi-criteria decision-making methods are based on the concept of adding a criterion to the assumption of independence, but in reality, not always all criteria are independent.

Multi-criteria decision models based on the use of the Choquet integrator with the evaluation of criteria by the degree of veracity and the subsequent evaluation of alternative locations for the location of the warehouse in order to determine the main and partial criteria and set the hierarchy of the warehouse location problem and then use the multi-criteria method to evaluate the alternatives for the warehouse location are a step in the right direction, however, it will be necessary to go even further and find a combination of existing models and find those one which will be based not on costs but on delivery time. That will help companies solve the allocation problem to meet the requirements for shorter delivery times.

This paper is supported by the research project "From horse-drawn railway to intermodal transport" within Visegrad Fund.

References

1. J.P. Rodrique, *The geography of transport system* (New York, Hofstra University, USA, 2013)
2. International Monetary Fund, *Globalization and Inequality* (2007), Available online: http://www.imf.org/external/pubs/ft/weo/2007/02/pdf/c4.pdf
3. R. Farahani, R. Zanjirani, S. Rezapour, L. Kardar, *Logistics Operations and Management – Concept and Models*, 469 p. (Waltham, USA: Elsevier, 2011)
4. T. Cura, Computers & Industrial Engineering **37**, 5, 3943-3952 (2010)
5. A. Máca, L. Kalecký, *Location – alocation role and centrality*, 271-276 (Think Together 2013, Prague: Česká zemědělská univerzita, Czech Republic, 2013)
6. T. Demirel, N.C. Demirel, C. Kahraman, Expert Systems with Applications, 3943-3952 (2010)
7. S. Nickel, *Location Theory – Discrete Location Problems, Facility Location and Strategic Supply Chain Management,* (Saarland University, Germany, 2013)
8. L. Michel, P. Van Hentenryck, European Journal of Operational Research **157**, 3, 576-591 (2004)
9. T. Özcan, N. Celebi, E. Sakir, Expert Systems with Applications **38**, 8, 9773-9779 (2011)
10. J. Huifeng, X. Aigong, The International Archives of the Photogrammetry, Remote Sensing and Spatial Information Sciences, XXXVII.(B2), 545-548 (2008)
11. T. Nagy, *Warehouse location problem – CSP model*, Available online: http://www.bitlib.net/show.php?id=3369316 (2004)

MATEC Web of Conferences **134**, 00030 (2017) DOI: 10.1051/matecconf/201713400030
LOGI 2017

Human Capital Management in Transport Enterprise

Alžbeta Kucharčíková[1,*], and *Martin* Mičiak[2]

[1]University of Žilina, Faculty of Management Science and Informatics, Univerzitná 1, 010 26 Žilina, Slovakia

Abstract. Human capital management represents a strategic and planned approach to the management of the most valuable resource that organizations have, namely employees and their human capital. Human capital management system supports the various processes of the organisation and helps to achieve organisational goals using effective use of human capital and effective investment in human capital as investment in education and training, health and relaxation, safety at works and ergonomics. The article aim is to explain the nature of human capital management, characterize the relationship between the human capital management and human resources management, and identify some approaches to measurement for human capital management.

1 Introduction

Present is represented by a constant pressure on the competitiveness of enterprises. Increasing of overall productivity of the company is one of the factors enabling the creation of added value and achieving long-term growth of the company [1]. Competitive fight is forcing businesses to bring new ideas, products or services to the market [2]. The quality of human resources in relation to competitiveness and economic growth is currently a highly topical theme [3]. The workforce of an enterprise consists of employees of various ages with different personality types. A manager should identify generational differences. Subsequently, the management style, leadership and employee motivation should be adapted forasmuch as well-motivated employees are able to influence the efficiency of business processes in a right way [4].

For this reason, the managers draw their attention also to the detection, assessment and comparison of the efficiency of enterprise inputs. Human capital is one of them. The concept of human capital management (HCM) emphasises and draws attention to the current knowledge of man, their meanings and use for the enterprise and their systematic development and strengthening. It also represents the ability to exploit the potential of people, both for daily work and for improving the quality and innovation. It includes the procedures and methods of work with people and care about their development. Human capital management is carried out by all managers and has an impact on all of human

* Corresponding author: Alzbeta.Kucharcikova@fri.uniza.sk

capital, and thus to all employees in the enterprises. This also applies to transport companies.

The HCM concept can also be linked to corporate social responsibility (CSR) activities. Moravčíková and Križanová [5] emphasize that CSR has become an important means of improving business relations, addressing social problems, consolidating customer relationships and, ultimately, increasing the company's competitiveness in the long run and spreading good reputation

2 Objective and methodology

The objective of the paper is to explain and analyse the views on the human capital management, to characterise the relations between human capital management and human resources management (HRM), to identify some approaches to measurement of the HCM aimed at increasing the efficiency of human capital and enterprise performance. There was the procedure applied from the general to the specific in order to achieve the objective. In view of the fact that it is a later concept of work with people, their knowledge, skills and abilities, there is often a confusion of the concept of human capital management and human resources management, not only in practice, but in literature as well. For this reason, we have identified and gathered different views on the explanation of the relationship of these concepts of work with people in enterprises, we have finally pointed out the possibility of applying some of the metrics in the context of the implementation of human capital management in the transport enterprises practice.

3 Human capital management and human resource management

The successful operation of any enterprise is largely determined by the abilities of people who work in it. Providing the enterprises to ensure their competitiveness in domestic and global market, it is necessary to realise the value and importance of people in the process.

At the beginning, for a better understanding of the issues, it is necessary to define the difference between human capital and human resources. While *human resources* are considered to be people in the process, the *human capital* is a summary of abilities, skills and knowledge the people have. The fundamental difference lies in the fact that the human capital is available to everyone, regardless of whether s/he is employed, while human resources are the bearer of human capital available for enterprise. It is for the enterprise itself, whether it develops and uses its human resources and human as much as possible in order to achieve a competitive advantage in the market.

At the present time, employee motivation is an atmosphere that meets the interests and needs of employees in the company, stimulates and influences their work performance in order to meet the needs of the organisation as well as their personal needs [6].

Investment in human capital are oriented to development of knowledge and skills of employees. Enterprises invest in the human capital when they expect future benefits. Investment in human capital are realised as investment in education and training, investment in health and relaxation, investment in safety at works and ergonomics. Education is the process of purposeful mediation, creating an active and acquisition of knowledge, practical experience, creating interest and attitudes. Investment in improving the health of employees as human capital include: organizing the relax stays, regular medical visits, swimming pools tours and building a fitness centres, regular fluid supply, financial participation on food, etc. Investment in safety and ergonomic are related to the fact that ergonomics helps harmonize things that interact with people in terms of people's needs, abilities and limitations. Regular monitoring of anthropometric data on employees contributes to the design and evaluation of tasks, jobs, products, environments and systems in order to make them compatible with the needs, abilities and limitations of people. By investing in improving working conditions and safety at works, by improving the quality of work equipment, the enterprises prevent accidents, workers' health damage. This is very important in transport enterprises.

The economic subjects invest in the human capital, when the benefits exceed the costs of education activities. Making decisions about investing in human capital is part of the HCM process. In order to be able to define the relationship of the human capital management to human resources management, it is necessary to define both concepts.

According to Storey [7] there are a number of definitions that define *human resources management* (HRM), whereas each book and each organisation has its own definition. Price (2011), in his book, defines HRM as the philosophy of people management, which is based on the belief that human resources are an important prerequisite for the continued success of organisation. If the organisation can effectively use and develop them, they can represent one of the largest competitive advantages.

Šikýř [8] defines the human resource management as concept refers to practices that enable any organization to attract, employ and develop enough capable and motivated employees, including such practices as job design and analysis, human resource planning, employee recruitment and selection, performance management and appraisal, employee compensation, employee training and development or employee care.

In defining the essence of human capital management we were based on different approaches to its definition. According to Accounting for People Task Force [9], human capital management is a strategic approach to the management of people, which focuses on the critical questions on the success of the organisation. This approach considers management of people to be a high level of strategic decision-making and consists of a systematic analysis, measurement and evaluation of how people are progressing in creating values.

Nalbantian [10] in the framework of the management of human capital emphasised the measurement as an important aspect. Whereas he defines human capital as a summary of knowledge, experience, creativity and other properties, in his opinion, human capital management provides tools for measuring the value of human characteristics for the needs of an effective enterprise management.

Donkin [11] also supports the consideration of measurements within the implementation of HCM. He also argues that the organisational strength of human capital management is situated in 3 areas:

1. development and application of relevant quantitative and qualitative measurements,
2. assembly and interpretation of results,
3. using the information obtained as a strategic advantage.

Enterprises concentrating efforts of management on the following three areas will be in the best position on the strategic arrangement of the employment policy. Good management of human capital is therefore only about the learning, understanding, use and adaptation.

Baron and Armstrong [12] define **three specific objectives** of human capital management. At the beginning the objective of human capital management is to obtain, analyse, and on the basis of the facts acquired, report on the direction of the strategy and processes chosen within human resources. Then, compare the results of the measurements taken with a pre-determined, and assess whether the strategy and management of human resources processes used are pointing to the desired output. The final objective resides in the strengthening of the human resources management strategy chosen and the processes used create the value for the enterprise based on the people.

According to Kearns [13] these are the people in the management of human capital who add the value to the enterprise, while the concept of human resource management considers people the costs of the enterprise. According to his opinion, human resource management is carried out by the personnel department providing support services to other functional departments within the enterprise. While human capital management is a strategic and focused on the enterprise as a whole.

Mayo [14] defines the basic difference between HCM and HRM lies in the fact that the management of human capital sees a person as an asset and the management of the human resources perceives a person rather a cost item.

In contrast to previous claims, according to the CIPD (professional organisation for human resource management and human resource development professionals), strategic management of human capital is identical with the management of human resources. Both concepts have the same basis. It is about both the concepts see people rather benefits than the costs. For this reason, it is necessary to adopt an integrated and strategic approach to the management of people. Given the fact that people are the foundation of each enterprise, they should also concern all parts of the organisation, not only personnel department. The group argues that HCM strengthens the value arising from strategic management of human resource using:

- strengthening the need for developing strategies for the management of human resources and also the processes through which the value of people is being created,
- pay attention to a sense of management through measurement and subsequently determining the emphasis on clear relationships between the intervention in the field of human capital and the success of the organisation [15].

According to Armstrong [16], the relationship between human capital management and human resources management lies in the fact that the HCM concept supports and complements the HRM, because HCM provides the measurement in order to ensure the best possible connection of activities and measures in the area of human resources with the objectives of the organisation. It emphasises the belief that people have to be understood rather as the wealth than a cost item. The concept puts an emphasis on strategy and processes in the area of human resources management, which ultimately lead to the achievement of the overall objectives of the organisation.

It should be noted that the HCM concept does not apply only to enterprises. For example Cahyaningsih and Sensus in several works [17-18] devoted to the implementation of HCM in terms of government and government institutions - ministries.

According to the above views, the concept of human capital management can be seen as a shift from the traditional perspective of enterprises per employee as one of the cost items. Employees are taken as assets to which it is necessary to invest (educate, develop knowledge, abilities, skills of workers) for the purpose of increasing the value, performance, competitiveness of the enterprise. However, the problem is that there is no exact method for the valuation of the assets in the enterprise. It is a strategic and planned approach. Human capital management system supports different processes leading to the use of human capital towards the effective achievement of the corporate objectives. For this reason it is necessary to manage and develop this capital.

4 Measurements for human capital management

The role of human capital management is to combine the information that the management has available on the human capital in its enterprise for a number of purposes. It takes care to identify the relationship between the information and financial results. They tell us about what is going on in the enterprise. On the other hand, the data on the human capital explain why this has happened. Consequently, it is necessary for the management to get only relevant information, which will be the key to the management and improvement of the enterprise then. Only employees who are able to use their capital more efficiently can increase their share in achieving the objectives of the enterprise [19].

The measurement of human capital is an important approach to its management. The successful companies are trying to connect measures as the causal models, which identify the causal relationships. They interconnect what causes the enterprise's success with its results.

4.1 HCM value matrix

HCM value matrix is a tool that was first used only for the measurement of HCM. Then, using this matrix there started to structure the HCM strategies toward creating value. The measures focused on added value of management are largely oriented to the achievement of corporate objectives. It is usually the outputs relating to the ability of the people, their motivation, and engagement. The matrix provides a very useful framework for the measurement which may shed light on the enterprises what is measured. A major benefit is that it is focused on high value and strategic differences. The objective of the tool is to use such units, which are the most appropriate for a specific enterprise a measurement is in progress. People management as an activity is generally rather measured in terms of quality, time, or yield (fig. 1) [20].

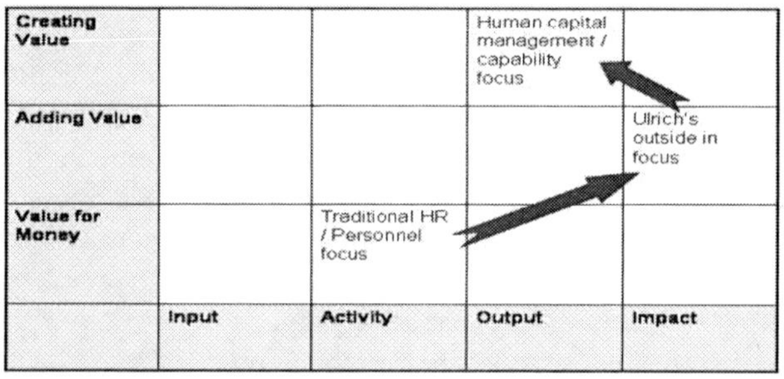

Fig. 1. HCM value matrix. Source: [20]

4.2 Human capital development framework

Most of the indicators relating to the human capital give an account of volumes about the human assets of the organisation. They do not say, however, about whether the processes are also aligned with its business activity. Just for this purpose, there are developed indicators of human capital management and the effectiveness of human capital management.

One of the methods is the methodology of human capital development framework (HCDF). Only the enterprise, which is trying to be successful, must link investments it puts into its workers with the result of its enterprise activities. Accenture is the author of the above method, which deals with the analysis of human capital in enterprises. HCDF is built on the basis of *4 levels* reflecting the relationship between human capital and enterprise performance.

The first level consists of measuring results using traditional measurements of the activity of the enterprise, such as EVA, market share, etc. *The second level* represents the factors that to a large extent, affect the performance of the enterprise, such as the quality of production, innovation, and customer satisfaction. *The third level* analyses the competencies of human capital. There are the qualities and skills that are relevant to the achievement of the performance of the enterprise. They are found using factors such as the evaluation of education, and knowledge management. The last *fourth level* is different in each organisation. There are the individual variables analysed on the basis of the best approaches in the industry and adjusted according to the results of the evaluation of the enterprise's workers. When all the data collected, there comes the evaluation of the organization's ability to effectively use the human capital of the enterprise to achieve enterprise objectives. In the case of low value of indicator, there is the apparent lack of human capital management; on the other hand, high value represents the high level of qualification and the sophistication of the management system [21].

4.3 HCM and BSC method

If managers at all levels of management of the enterprise want to decide in right way they have to continuously monitor and evaluate the enterprise's performance. Sometimes, however, it may be difficult to show a clear link between the efficiency of managing human capital and the success of the business [22]. The Balanced Scorecard (BSC) method can be used to better implement the human capital management concept. BSC works with financial and non-financial value criteria mutually interconnected and it emphasises customer satisfaction, ability to retain high-quality employees and achieve effective enterprise processes, what has a major impact on the long-term performance of the enterprise. This method is focused on four main areas: financial perspective, customer perspective, perspective of the internal processes and employment perspective (also known as the perspective of learning and growth).

Learning and growth perspective is oriented for the measurement of worker potential, their motivation and the system of objectives in relation to the workers of the enterprise. Through this perspective, enterprises are creating objectives and measures aimed at promoting learning and growth of employees.

Matrix BSC emphasises the need for the need to invest in the future. If the enterprise wants to achieve the long-term objectives, it has to invest in its infrastructure consisting of people, systems, and processes. Considering the enterprise to be able to meet the objectives set in the area of learning and growth, it is necessary, in addition to the classical recruitment parameters, to focus on the following areas: education, motivation, flexibility, ability to work in a team, target orientation. Learning and growth perspective consists of 4 main parts (fig. 2): acquisition, promoting, developing and maintaining. All parts are dedicated to the field of human resources, while the development is aimed at the development of HC in enterprises [19].

Authors Chodasová et al. point out that only employees with a rising value of human capital bring benefits to the enterprise in the form of higher productivity [23]. This, together with new technologies and innovations in the field of production, leads not only to the growth of efficiency but also to the competitiveness of enterprises in the market [24].

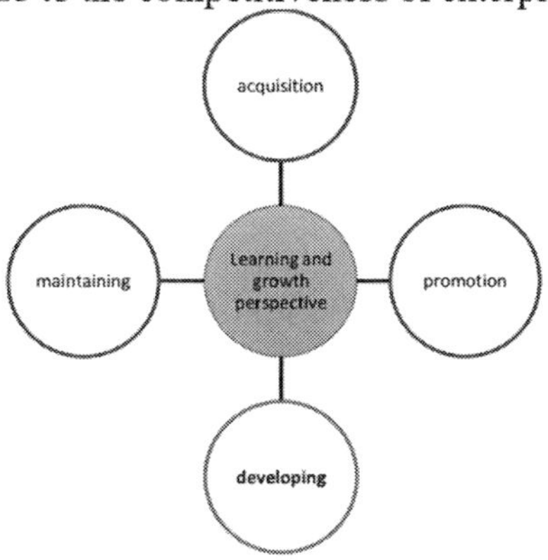

Fig. 2. Learning and growth perspective. Source: authors, based on [19]

The main objective of the enterprise is to find ways to achieve some of the advantages of human capital and find where and what kind of the investment of people are the most profitable. This is intended to increase the performance and competitiveness of the business on the market. Therefore, the enterprise itself needs to respond to basic practical questions related to human capital, and based on them develop a detailed strategy.

5 Conclusion

In the present, the information and communication technologies allow to enterprises to use increasing amount of information. Such businesses include transport companies. The problem seems to be the company's ability to evaluate this information and get the maximum benefit from it [25]. Here can help the human capital management concept. Human capital management is an approach, when the people in the enterprise are seen as intangible asset, which forms part of the value of the enterprise and not as a cost item. The ambition of the authors was to explain the essence of human capital management, to characterise the relations between human capital management and human resources management, and identify some approaches and the human capital management metrics towards increasing the efficiency of the use of human capital aimed at improving the performance and competitiveness of transport companies.

As part of the implementation of human capital management, this means seeking to link business performance with effective use of specific knowledge, skills and capabilities of its employees. However, in order to implement the HCM concept in the practice of a transport company, performance indicators need to be adapted to the particular conditions of transport business, in the context of enterprise strategy.

This article was created as part of application of project: VEGA No 1/0652/16 Impact of spatial location and sectorial focus on the performance of businesses and their competitiveness in the global market and APVV-16-0297 Updating of anthropometric database of Slovak population.

This paper is supported by the research project "From horse-drawn railway to intermodal transport" within Visegrad Fund.

References

1. Z. Tekulová, M. Králik, Z. Chodasová, *ESPM 2015 – Proceedings of the International Conference on Engineering Science and Production Management*, 545-549 (Tatranská Štrba, High Tatras Mountains, Slovak Republic, 2015), DOI: 10.1201/b19259-97
2. S. Lorincová, M. Potkany, *ESPM 2015 – Proceedings of the International Conference on Engineering Science and Production Management*, 157-161 (Tatranská Štrba, High Tatras Mountains, Slovak Republic, 2015), DOI: 10.1201/b19259-30
3. Z. Kozubíková, ISCKM Conf. 222-230 (2016)
4. R. Kampf, S. Lorincová, M. Hitka, O. Stopka, Sustainability **9**, 1561, (2017), DOI: 10.3390/su9091561
5. K. Moravčíková, A. Križanová, *ICMIBI-ASSB 2016 – International Conference on Applied Social Science and Business* **65**, 20-25 (Windsor, United Kingdom, 2016)
6. M. Hitka, Z. Zavadská, D. Jelacic, Z. Balážová, Drvna Industrija **66**, 3, 235-239 (2015)
7. J. Storey, *Human Resource Management: a Critical Text* (Thomson Learning, 2007)
8. M. Šikýř, *MSED Conf.*, 1405-1414 (2015)

9. *Accounting for People Task Force*
10. H.R. Nalbantian, *10th ASHRM conference* (Bahrain, 2010)
11. R. Donkin, *Human capital management*, Available online: http://www.richarddonkin.com/Archive/x_human_capital_management.htm
12. A. Baron, M. Armstrong, *Human Capital Management: Achieving Added Value Through People*, (Kogan Page Publish., 2007)
13. P. Kearns, *Evaluating the ROI from Learning* (CIPD Publish., 2005)
14. A. Mayo, *Human Recources or Human Capital?* (Gower Publish., 2009)
15. J. Martin, *Key Concepts in Human Resource Management* (London: Saga Publ., UK, 2010)
16. M. Armstrong, *Řízení Lidských Zdrojů* (Prague: Grada Publ., Czech Republic, 2007)
17. E. Cahyaningsih, D.I. Sensuse, A.M. Arymurth, W.C. Wibowo, *SCORED 2016 – IEEE Student Conference on Research and Development*, (Kuala Lumpur, Malaysia, 2016), DOI: 10.1109/SCORED.2016.7810038
18. E. Cahyaningsih, D.I. Sensuse, W.P. Sari, *ICITSI – International Conference on Information Technology Systems and Innovation* (Bandung, Indonesia, 2015), DOI: 10.1109/ICITSI.2015.7437718
19. J. Fitz-Enz, *The ROI of Human Capital: Measuring the Economic Value of Employee Performance*, (New York: Amacom, 2009)
20. J. Ingham, *Strategic Human Capital Management: Creating Value Through People,* (New York: Elsevier, 2007)
21. C.J. Sesil, *Applying Advanced Analytics to HR Management Decision* (New Jersey: FT Press, 2014)
22. M. Šikýř, F. Bušina, *MSED Conf.* 1477-1486 (2014)
23. Z. Chodasová, Z. Tekulová, L. Hľušková, S. Jamrichová, Procedia – SBS **174,** 3170-3177 (2015)
24. A.E. Gorokhova, J.M. Šafránková, V.D. Sekerin, *MSED Conf.* 477-486 (2015)
25. M. Ďurišová, E. Tokarčíková, E. Malichová, T. Potkanová, *Transport Means – 20th International Scientific Conference on Transport Means*, 169-174 (Juodkrante, Lithuania, 2016)

Road Tax as an Economic Tool of the Support for the Development of Multimodal Transport in Selected EU States

Lenka Ližbetinová[1,*], *Pavel* Fábera[2], *Tsolmon* Jambal[1], and *Zdeněk* Caha[1]

[1]Institute of Technology and Business in České Budějovice, Faculty of Corporate Strategy, Okružní 517/10, 370 01 České Budějovice, Czech Republic
[2]Technical University of Košice, Faculty of Mining, Ecology, Process Control and Geotechnologies, Letná 9, 042 00 Košice, Slovakia

Abstract Multimodal transport is an opportunity for a more ecological approach to the environment in EU states. Member states are interested in supporting its development and therefore in shifting the performance percentage from the road transport to other types of transport. States use various different tools to support the multimodal transport, partial or total exemption from road tax on motor vehicles used in multimodal transport is one of them. The goal of our contribution is to compare the conditions of the road tax reduction in the Czech Republic, Slovakia, Austria, Germany, Poland and Croatia.

1 Introduction

High attention is paid to multimodal transport in developed countries and it has a strong support there. Ecologization of transport belongs to crucial and supported programmes within the European Union as well. The importance of multimodal transport has only grown in recent years. Multimodal transport allows us to regulate the disproportionate growth of transport and the negative impact of the excessive road traffic on the environment [1, 2]. The distribution of transport work has so far been developing to the disadvantage of environment friendly types of transport [3, 4]. Fig. 1 shows the development of multimodal transport in 2011 – 2015. We can say that the total number of consignments transported by UIRR member operators increased by 0.75%. The output (in tonne-kilometres) grew by 5.23% v [5].

* Corresponding author: lizbetinova@mail.vstecb.cz

Fig. 1. Development of continental multimodal transport in the EU. Source: [5]

Multimodal transport has not been fully appreciated in the Czech Republic yet and is still considered complementary to road, rail and water transport. On the other hand, it has more than a 40-year tradition in the territory of the Czech Republic [6, 7]. The volume of multimodal transport has been increasing since 1993, in 2013 it formed approximately 2.49 % of the whole volume in tons by all types of transport. Rail and water transports are the most environment friendly types of transport [8,9,10]. The government tries to reduce such kinds transport that have the least favourable impact on the environment. Competitiveness of multimodal transport requires improvement of technologies to reduce the critical transport distance [11]. Multimodal transport, as a system with all the aspects, like terminals, fleets, technologies, public logistic centres, has to become one of the priorities of the transport policy [12, 13].

Governments apply multiple different tools to support multimodal transport. The programme "Support to modernization of the construction of multimodal transport terminals" is one of the most important ones in the CR at present. This programme is included in the Operation Programme Transport for 2014 – 2020. Among other tools governments also support the use of multimodal transport by means of the road tax. Our article focuses on an analysis of the reduction of this tax in selected EU states.

2 Comparative analysis of road tax in selected EU countries

The goal of our article is to compare the conditions of road tax reduction in the Czech Republic, Slovakia, Austria, Germany, Poland and Croatia. Comparison of the support for multimodal transport from the road tax point of view is elaborated in the article on the base of the legislations of these selected states.
Directive of 17 March 1999 on the charging of heavy goods vehicles for the use of certain infrastructures (amended particularly by Directive 2011/76/EU).

Directive 1999/62/EC of the European Parliament and of the Council dealing with the charging was established for prevention of the disruption of the competition among forwarders in member states, for the unification of the charges for utility vehicles exceeding a certain total weight and for determination of their minimum rates and finally as a support for using vehicles that are considerate to both roads and the environment. On the base of the

Directive the EU states are obliged to set the rates in such manners so that the tax rate for each vehicle category and subcategory (as defined in Annex I to the Directive) is not lower than a minimum rate set in the above-mentioned Annex. The article also presents the regulations that reduce the tax for vehicles used in multimodal transport in the selected countries [14, 15].

2.1 Tax legislation in the selected states

Czech Republic - according to Act No. 16/1993 Col. on road tax as amended (2013) multimodal transport is understood as transport of goods in a single transport unit. A container, a changeable trailer body and a rolling container, or a truck, a trailer with or without a tractor, where rail or inland water transport is used, can be such a transport unit. The railway or water section has however to exceed the distance of 100 km in a straight line and the first or the last section has to be on road:

• Between the loading or unloading point and the nearest railway station suitable for reloading or a multimodal transport terminal.

• Between the loading or unloading point and an inland port unless the distance exceeds 150 km in a straight line (Act No. 16/1883 Col. on road tax as amended, 2013). A road tax reduction of up to 100 % can be applied to a motor vehicle that is solely used in the start or end section of multimodal transport in the Czech Republic. To a motor vehicle that makes in multimodal transport during a tax period:

• from 31 to 60 journeys the reduction reaches 25 % of the tax,
• from 61 to 90 journeys the reduction reaches 50 % of the tax,
• from 91 to 120 journeys the reduction reaches 75 % of the tax,
• more than 120 journeys the reduction reaches 90 % of the tax.

If the distance travelled on rail within the territory of the Czech Republic exceeds 250 km the journey is considered double for the taxation purposes. To gain the tax reduction a taxpayer has to prove the realized journeys by transport documents with the data confirmed in the multimodal transport terminals, the station of loading or unloading or the inland port suitable for reloading [16].

Slovakia – the road tax is governed by Act No. 361/2014 Z. z. on road tax and modification and amendment of certain acts in Slovakia. According to this Act a 50 % tax reduction can be applied to a vehicle that was used at least 60 times in the start or the end section of multimodal transport in a tax period in the conditions of the Slovak Republic. Like in the CR, if a distance of more than 250 km is made on rail within the territory of Slovakia, such modal transport is counted twice for the purpose of the tax rate [17].

Austria - The road tax is based on the road tax act of 1992 in Austria (BGBI. No. 449/1992, as amended on 28/09/2017). The conditions and the extent of the advantages for forwarders that use motor vehicles in multimodal transport are defined in sec. 2. Full tax exemption can be achieved for vehicles within unaccompanied multimodal transport (road/railway) that **solely** perform collection and distribution and meet the following technical conditions:

• The total weight of the vehicle exceeds 3.5 tons.
• The motor vehicle or trailer only performs collection and distribution within multimodal transport in a calendar month.
• Only containers longer than 20 feet, equipped for rail transport can be used as the transport units.

To a vehicle (with a trailer or a semi-trailer) heavier than 3.5 tons used within multimodal transport a 15 % tax reduction will be applied per each railway journey, the reduction will however not exceed the amount paid as the tax for the particular calendar year [18].

Germany – the road tax in Germany is governed by the road tax act amended by a decree of 26 September 2002 and the latest amendment of 24 March 2007. Sec. 4 of the road tax act on the tax refund defines 4 categories of tax reduction in multimodal transport. A written application for the tax refund has to be filed and the vehicle (loaded or unloaded) has to be transported more than 124 times by rail. A part of the road tax is returned if the above number of journeys is not achieved, namely as follows:

- from 32 to 62 journeys 25 % of the annual tax will be returned,
- from 63 to 93 journeys 50 % of the annual tax will be returned,
- from 94 to 123 journeys 75 % of the annual tax will be returned.

If the distance done by rail transport exceeds 400 km, the journey is counted as double. If it exceeds 800 km, it is counted three times as much.

To be able to use the tax reduction a forwarder has to produce all the necessary confirmations of the rail transport for each vehicle separately to prove the meeting of the conditions [19].

Poland – the road tax in Poland is governed by Act (Art. 8 – 12b) on local tax and charges of 12 January 1991 together with the Collection of laws of 2014, pos. 849 as amended. Four categories related to the number of journeys were determined for the road tax reduction in Poland. A written application for the tax refund has to be filed. Forwarders that use means of transport in multimodal transport are entitled to the refund of the whole amount or a part of the road tax paid. The amount of the returned tax is:

- from 20 to 49 journeys 25 % of the tax rate will be returned for the tax period,
- from 50 to 69 journeys 50 % of the tax rate will be returned for the tax period,
- from 70 to 99 journeys 75 % of the tax rate will be returned for the tax period,
- 100 or mode journeys 100 % of the tax rate will be returned for the tax period.

To be able to use the tax reduction a forwarder has to produce all the necessary confirmations of the rail transport for each vehicle separately to prove the meeting of the conditions [20].

Croatia – the road tax is governed by the Act on financing local and regional governments, by the Act on public roads NN 84/11, 22/13, 54/13, 148/13, 92/14, which is complemented by Article 7 of the Annual public road charge tariff, NN 124/2009 (Republic of Croatia, 2009). Owners of road vehicles registered in Croatia are entitled to a tax exemption if their vehicle has made at least 80 journeys to or from a railway terminal for multimodal transport within 12 months since the latest technical check-up. Vehicles used solely for collection and distribution within multimodal transport are fully exempt from road tax if they are taxed individually [21, 22, 23].

2.2 The results of the comparison of tax reduction applicable to multimodal transport

From the viewpoint of the tax support for accompanied multimodal transport the forwarders are best motivated in Austria (Fig. 2), where the total proportion of the performance within multimodal transport belongs to the highest. Where a road vehicle is used for multimodal transport a 15 % discount from the monthly road tax is provided per each railway journey, the reduction however does not exceed the amount paid as road tax per the particular calendar year. Full tax exemption can be achieved for motor vehicles within unaccompanied multimodal transport (road/railway) that solely perform collection and distribution in this country during a particular month.

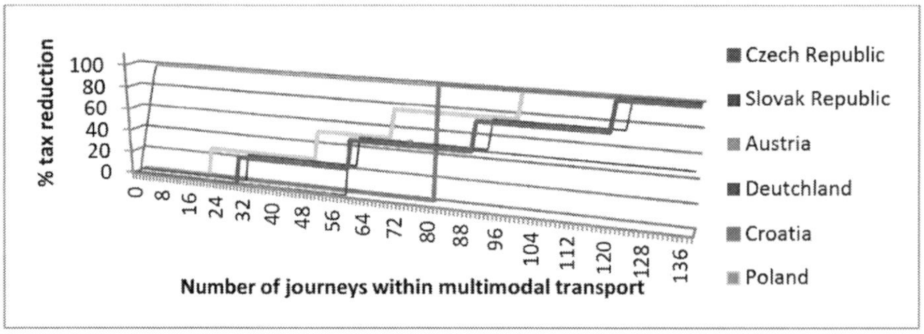

Fig. 2. Tax advantage on the road tax for the use of a vehicle within multimodal transport in selected EU countries. Source: authors

A favourable tax advantage also exists in Croatia. Full exemption from this tax is already applicable from 80 journeys to or from a railway terminal within multimodal transport during 12 months. Although Croatia is the youngest of the examined countries to join the European Union it has introduced necessary steps to the support of multimodal transport. Introduction of new legislation that meets the conditions of the European Union was among the important tasks for Croatia.

Poland, Germany and the Czech Republic are further states that were analysed. These states are mentioned together as the tax advantage is similar there. There are always four categories within 25 % and 100 % set in these countries. Each category depends on the extent of the use of a vehicle within multimodal transport. The tax system in Poland can be classified as the first. The full exemption applies to 100 journeys and the lowest category, 25 %, reduction applies to 20 – 49 journeys.

The tax system applied in Germany already enables full exemption from road tax on 124 journeys. 20 – 49 journeys are the lowest limit with 25 % tax reduction. The Czech Republic closes the "4-category" group. The full exemption is only applied if a motor vehicle is only used for transport in the start and end sections of a multimodal transport journey. Unless the above-mentioned condition is met, the highest tax reduction, namely 90 %, applies to 120 journeys and the lowest reduction of 25 % applies to 31 – 60 journeys.

Slovakia has the lowest support in terms of the tax advantage. There is only a single road tax reduction, namely 50 %, for a vehicle that was used at least sixty times in multimodal transport during a taxation period.

We must however not forget that unlike in all the other countries, where the taxation period is one year, there is only a one-month taxation period in Austria. Austria is also the only state where full tax exemption is applied to unaccompanied transport. Full tax exemption also exists in the Czech Republic if a vehicle is solely used for multimodal transport. A tax subject has to file certified documents from multimodal transport terminals to be entitled to the tax reduction. Full exemption from the road tax can be applied in the Czech Republic, Austria and Croatia when vehicles are used solely in multimodal transport during a taxation period. All the countries enable achievement of 100 % tax reduction if particular conditions are met, except for Slovakia, where the maximum reduction limit is only a half (50 %).

3 Conclusion

The European Union pressurize the member countries to reduce externalities and to use alternative methods of transport. This idea leads to the possibilities to reduce the road tax when environment friendly vehicles are used, laid down by the legislations of the individual countries. Forwarders using their vehicles for multimodal transport are also given advantages. This legislation differs in the individual states and the conditions are often differentiated for sole use of vehicles in multimodal transport or scaled according to the extent of application. Motivation for forwarders only represent a part of the possibilities that a government can utilize for the development of multimodal transport itself. Nevertheless, a suitably designed complex of all the tools is a strong factor of the motivation for forwarders in selection of the transport method, which is why a suitable adjustment of all the elements that influence their decision making is so important.

This paper is supported by the research project "From horse-drawn railway to intermodal transport" within Visegrad Fund.

References

1. R. Kampf, P. Prusa, C. Savage, Transport **26**, 4, 425-432 (2011)
2. J. Gnap, V. Konecny, Procedia Economics and Finance **34**, 344-351 (2015)
3. J. Novák, V. Cempírek, I. Novák, J. Široký, *Kombinovaná přeprava* (Univerzity of Pardubice, Czech Republic, 2015)
4. J. Ližbetin, Z. Caha, Procedia Engineering **161**, 1197-1203 (2016)
5. UIRR, *UIRR Report European Road-Rail Combined Transport 2015-2016* (2017)
6. R. Kampf, S. Lorincová, M. Hitka, Z. Caha, Naše more **63**, 3, 120-125 (2016)
7. J. Cejka, L. Bartuska, L. Turinska. Open Engineering **7**, 1, 55-59 (2017)
8. V. Cempírek, J. Široký, P. Nachtigall, WorldCargo News **26**, 1355-0551 (2006)
9. J. Ližbetin, Z. Caha, Naše more **62**, Special Issue, 97-100 (2015)
10. K. Klarić, K. Greger, M. Klarić, T. Andrić, M. Hitka, J. Kropivšek, Drvna industrija **67**, 3, 241-248 (2016)
11. T. Skrucany, J. Ponicky, M. Kendra, J. Gnap, *ICTTE – 3rd International Conference on Traffic and Transport Engineering*, 744-749 (2016)
12. M. Vojáčková, *Intermodální přeprava v ČR*, Available online: https://www.czechtrade.cz/o-czechtrade/reference/czechtrade-media/intermodalni-prepravy-v-cr-26-6-2015/ (2015)
13. H. Chalupová, Vohánková, J. Široký, Applied Mechanics & Materials, **803** (2015)
14. European Commission, *Directive 2011/76/EU* (2011)
15. European Commission, *Directive 1999/62/EC* (1999)
16. Czech Republic, *Act No. 16/1883 Col. on road tax as amended* (2013)
17. Slovakia, *Act No. 361/2014 Z. z. on road tax* (2014)
18. Austria, *BGBI. No. 449/1992, as amended on 28/09/2017* (2017)
19. Germany, *Motor Vehicle Tax Act* (2015)
20. Poland, *Ustawa o podatkach i oplatach lokálních* (Art.8-12b.) (2014)
21. Republic of Croatia, Croatian Chamber of Economy, *Roads Act NN 84/11, 22/13, 54/13, 148/13, 92/14* (2014)
22. Republic of Croatia, Ministry of Maritime Affairs, Transport and Infrastructure, *Rules on the amount of annual fees for use of public roads which is payable on registration of motor vehicles and trailers, number 35/2011, and 53/2011* (2015)
23. Republic of Croatia, Croatian Parliament, *Act of Combined Traffic* (2014)

Motivation of Employees of Transport and Logistics Companies in the Czech Republic and in a Selected Region of the PRC

Lenka Ližbetinová[1,*], *Miloš* Hitka[2], *Chenguang* Li[3], and *Zdeněk* Caha[1]

[1]Institute of Technology and Economics in České Budějovice, Faculty of Business Strategy, Okružní 517/10, 370 01 České Budějovice, Czech Republic
[2]Technical University in Zvolen, Faculty of Wood Sciences and Technology, Masaryka 24, 960 53 Zvolen, Slovak Republic
[3]North China University of Technology, School of Economics and Management, No.5 Jinyuanzhuang Road, Shijingshan District, Beijing 100144, P. R. China

Abstract. Improvement of employee performance is an inseparable part of an enterprise. It ensures the fulfilment of goals and the company profitability. The main goal of the paper is to present the results of a research focused on comparison of the motivation factors of employees of transport and logistics companies in selected regions. The research carried out in 2017 examined the level of the motivation in the conditions of transport and logistics companies in the Czech Republic and the Peking Municipality in the PRC. These territories have different historic and cultural backgrounds, which leads to a high assumption of the existence of different views of the importance of motivation factors not only in the sphere of transport. The contribution also brings comparison of the preferences from the gender point of view.

1 Introduction

Intellectual capital is usually defined as the sum of the resources of a company. It includes knowledge, information, intellectual property, experience and human resources that might contribute to the production of the wealth of a company [1]. Intellectual capital consists of three basic mutually linked components, namely the human capital, the structural capital and the relational capital [2]. Human capital represents the human factor in a company [3, 4]. It is a combination of intelligence, skills and knowledge that it can provide to the company. Researchers claim that human capital is the most important value of a company, as it is the resource of creativity, brings new innovations, changes and improvements [5]. Companies often concentrate on the human capital as a source of new possibilities and innovation [6].

Human capital also brings the emotional relation to a company. A psychological contract is an emotional relation between an employer and an employee and in measures the organization commitment [7]. The psychological contract origins from the works by

[*] Corresponding author: lizbetinova@mail.vstecb.cz

Argyris [8], who used this term to describe the relation between a group of employees and their superior. Rousseau [9] considers the psychological research to the mental models of employees, mutuality and effectiveness the basic elements of psychological theories. A psychological contract consists of two factors. Employee satisfaction is the first one and employee commitment is the other. Both these factors capture the emotional relation of individual employees to their organization.

Managers have an unenviable and difficult role. The factors that motivate a manager do not necessarily motivate his/her employees. It is necessary to view each employee individually [10, 11]. Each employee is different and unique in terms of the age, education, needs and expectations. All that has to be taken into account in the right motivation of employees [12]. Motivation supports the success of employees in their fields and their coping with the everyday tasks [13, 14]. With the aim to improve the general organizational abilities and meeting the company goals managers have to manage employees adequately and smoothly, to achieve the required work success [15]. The monitoring of and the support for the individual achievements is an important precondition of the monitoring and strengthening of the overall success of the business system. [6, 16].

These general rules also apply to the environment of transport [17] and logistic companies [18]. Their specifics (the specifics of transport services) however create a precondition determining modifications within employee motivation [19]. Transporting and logistic companies employed 6.9 % of the economically active people in the Czech Republic in 2015, of which 74.1 % were men [20], and 71 % in Slovakia, while the proportion of men was 76.5 % there [21]. Numerous companies fight with the lack of quality human resources and look for the possibilities how to cover this deficit [22]. Another problem they also face is, how to hold the quality labour force that has proven to be useful and is trained. The identification of what these employees prefer in the work relations and what motivates them to a higher performance and what, on the other hand, demotivates them, is the answer [23, 24].

Our research focuses on the motivation factors of employees in the environment of transport companies active in the CR, compared with the employee preferences ascertained in the Peking Municipality in China. The distribution sphere in China, a continuously growing world producer, does not lag behind the production. This requires a strong logistic background. This sphere is also characterized by a higher economic level, which has influence on the traditional Chinese values and attitudes. This homogeneous Asian society is particularly characterized by a strong power distance, the seniority principle, strong interpersonal relations, collectivism and higher corruption rate [25]. The original traditional social patterns are dodging a modern business environment. It is therefore more likely that there are employees with similar preferences in motivation in this territory than in less developed territories of the PRC. On the other hand, the Czech Republic is a country that passed the transformation from the original communist regime to the market economy conditions. It is characterized by Central-European culture, which also influences the employee preferences in an employment to a high extent [26].

2 Material and methods

The aim of our contribution is to compare the levels of the preferences of motivation factors between employees of transporting and logistic companies in the Czech Republic and in the Peking Municipality. The gender of the respondents is taken into account in the comparison. With regard to the different social roles and attitudes we examined which motivation factors have stronger influence on women and which have stronger influence on men. The paper focuses on looking for the differences in the motivation factors between the genders and the different nationalities of the employees, which affect their performance.

The contribution is based on the data obtained through questionnaires distributed personally during 2017. The questionnaires [27] are divided into two basic parts, where the first part deals with the respondents' personal details (age, gender, years with the company, achieved education and the company size). Table 1 presents the structure of both the sample sets, from the Czech Republic and from the Peking Municipality in the PRC. The second part is dedicated to the evaluation of the importance and the fulfilment of the motivation factors from the respondent's point of view. 30 motivation factors in total were evaluated my means of the Likert scale from 1 to 5 (where 1 meant unimportant/dissatisfied and 5 meant very important/very much satisfied).

The respondents were chosen by a random choice method from among the employees of transporting and logistic companies from both the countries. The data comparison was carried out by means of the weighted arithmetic means. Within the data evaluation we also used mathematic-statistic comparison by means of a two-sample T-test. Two-sample T-test is applied to the evaluation of various experiments where we do not know the mean value of the basic set and we only know two sets of selected data. Student's T-test for dependent samples tests the hypothesis of the average distribution of the pair values. The test is usually used to verify whether the average difference of pairs ascertained from samples can be only random (the variables are independent) or it is statistically significant (the variables are dependent). An interval estimation of the mean pair difference provides more information than the test itself, as it determines the limits within which the real mean difference occurs with the required confidence (probability, most frequently 95 %) [28].

Table 1. Characteristics of respondents. Source: authors

		Czech Republic	**Peking Municipality**
		Number of respondents	Number of respondents
Gender	Male	632	430
	Female	538	469
Age	less than 30 years	373	483
	31-40 years	294	231
	41-50 years	313	139
	over 50 years	190	46
	Total	**1170**	**899**

3 Results and discussion

In Table 3, we can see which groups of motivation factors influence the categories of women and men with regard to the nationality. In table 2, we can see which group of motivation factors most influences employees of transporting and logistic companies and their work performance in the Czech Republic and in the region in Peking. The highest stress is put on the relational motivation factors in both the nationalities. We can therefore state that both the nationalities improve the work performance if they feel good at the workplace, have good relation with their colleagues and their superiors are a support to them. Financial motivation factors are the second most important group. If employees are sufficiently and fairly appraised, they increase their work performance. Nevertheless, focusing solely on the aspect of an increase of finance might lead to a worse attitude to work among employees, as they might only be driven by financial incomes (Baldoni, 2005). Among Czech employees of transporting and logistic companies the work factors are the third most important group, while among the respondents from the Peking Municipality there are the career aspects. Employees can be motivated by the opportunity to apply their own abilities, they want to educate themselves, self-actualize and they can be motivated by recognition by their superiors. Employees are least influenced by the company vision, the development of the region, communication of the company performance and the company's relation to the environment. The difference between men and women in the comparison of the basic motivation groups is minimum in both nationalities.

Table 2. Comparison of motivational factor types by gender and nationality (in weighted arithmetic mean). Source: authors

Factors	Career		Work		Social		Relational		Financial	
Region	Czech Republic	Peking Municipality	Czech Republic	Peking Municipality	Czech Republic	Peking Municipality	Czech Republic	Peking Municipality	Czech Republic	Peking Municipality
Total	3.93	4.01	4.03	3.97	3.79	3.93	4.37	4.13	4.26	4.09
Women	3.92	4.02	4.07	4.01	3.78	3.99	4.45	4.17	4.34	4.12
Men	3.94	4.00	4.01	3.93	3.79	3.87	4.31	4.09	4.19	4.05

Table 3 present the analysis of employees of transporting and logistic companies from the territories of the Czech Republic and the Peking Municipality from the gender point of view. The part 1 of Table 3 compares the influence of motivation factors on men and women working in the transportation sphere in the Czech Republic. As we can see, both genders are motivated to the best work performance by the atmosphere at the workplace. From the above-mentioned facts, we can conclude that the friendlier the atmosphere at the workplace is the more effective the employees are for the company. The relational motivation factors can most efficiently influence the increase of the employee performance according to the results. The difference between men and women is particularly visible in the opportunity to apply one's own abilities and the elimination of stress at the workplace.

Men are motivated to better work performance by the opportunity to apply their own abilities and women care more about the elimination of stress at the workplace. Recognition is another factor in which the genders differ. Women need recognition. It motivates them to better performance. They try to work as efficiently as possible to achieve recognition. Men are not motivated by recognition to such an extent as women. They care more about independent decision making and self-actualization. These results lead us to an idea, namely that women need a support from their colleagues, recognition from their superiors and the environment to increase their work performance. Unlike men, who want to decide independently and are motivated by self-actualization and their own ideas, they want to be of benefit to the company. The motivation factors that do not have a special influence on the employee performance are more or less the same among men and women. They are mainly the social factors, like the company's relation to the environment, the region development, the company vision, the prestige of the company, but also the physical effort at work and the prestige of the position.

Table 3. Motivational factor preference in the Czech Republic and the Peking Municipality by gender. Source: authors

Czech Republic				Peking Municipality			
Men		**Women**		**Men**		**Women**	
Atmosphere in the workplace	4.45	Atmosphere in the workplace	4.53	Good work team	4.15	Good work team	4.23
Supervisor's approach	4.37	Supervisor's approach	4.52	Atmosphere in the workplace	4.12	Atmosphere in the workplace	4.22
Good work team	4.36	Good work team	4.45	Further financial reward	4.1	Basic salary	4.19
Basic salary	4.34	Job security	4.39	Recognition	4.1	Further financial reward	4.18
Job security	4.3	Fair appraisal system	4.39	Work performance	4.09	Recognition	4.17
Further financial reward	4.16	Basic salary	4.38	Communication in the workplace	4.07	Supervisor's approach	4.12
Work performance	4.15	Communication in the workplace	4.29	Moving up corporate ladder	4.04	Stress /elimitation of stress in the workplace/	4.12
Work object and type of work	4.11	Further financial reward	4.25	Work object and type of work	4.02	Moving up corporate ladder	4.11
Communication in the workplace	4.09	Stress /elimitation of stress in the workplace/	4.21	Fair appraisal system	4.02	Free time	4.1
Opportunity to apply one's own ability	4.08	Work object and type of work	4.19	Competences	4.02	Region's development	4.1

The following analysis, which is summarized in part 2 of Table 3, deals with employees in the sphere of transportation and logistics in the Peking Municipality. The relational motivation factors, like for example good work team, the atmosphere and communication at the workplace took the first positions again, both among men and women. Employees care about the atmosphere and the relations at the workplace. Pleasing environment and contentment at the workplace can motivate them to higher performance. Employees better concentrate on their work, are more productive and more efficient. As we can see, there is an obvious difference between men and women. Men care about both, the relational motivation factors and the work motivation factors. They are motivated to better performance if they have certain authorities and are informed on the work procedures and the object of the performed work. Women, on the other hand, want to avoid work stress; they want to eliminate it to the minimum. They are motivated by the approach of their superiors. Better results come with their support. Free time and the region development is a substantial motivation factor for women. With sufficient regeneration they are able to show their abilities and skills in the performance of their work task to the full extent.

3.1 Analysis by means of the Student's T-test

We applied the Student's T-test, which is applicable to the calculation of the differences between averages to the testing of the differences between the employees of the two countries. We analysed the motivation preferences of employees of transporting and logistic companies in the Czech Republic and in the Peking Municipality (PRC) in our T-test. We tested how the individual motivation factors influence the employees in the two different countries and we tried to find out whether these factors are statistically significant or whether at least their difference is statistically significant.

Table 4. Analysis of Czech employees and employees from the Peking Municipality by means of the Student's T-test. Source: authors

	μ CR	μ PRC	standard deviation CR	standard deviation PRC	F test	P level
Atmosphere	4.48	4.08	0.744	0.947	1.620	0.000
Team	4.40	4.04	0.817	0.955	1.366	0.000
Further reward	4.20	3.91	0.875	1.008	1.327	0.000
Physical efforts	3.69	4.07	0.994	0.969	1.053	0.328
Job security	4.35	3.94	0.803	1.033	1.655	0.000
Communication	4.19	3.97	0.936	1.024	1.196	0.000
Name of the company	3.83	3.85	1.064	1.030	1.067	0.218
Opportunity to apply one's own abilities	4.04	4.01	0.914	1.019	1.242	0.000
Work object and type of work	4.14	4.05	0.821	1.036	1.594	0.000
Communication of the result	3.92	3.97	0.958	1.047	1.194	0.000
Working hours	4.03	3.92	0.963	1.003	1.085	0.106
Work environment	4.01	4.01	0.906	0.999	1.215	0.000
Work performance	4.10	3.99	0.850	1.033	1.476	0.000
Career ladder	3.88	3.89	0.915	1.028	1.262	0.000
Authorities	3.79	4.03	0.970	0.972	1.004	0.934
Prestige	3.75	4.14	0.996	0.956	1.085	0.119
Management approach	4.44	4.11	0.775	0.996	1.651	0.000
Decision making	4.00	3.94	0.914	1.033	1.278	0.000
Self-actualization	4.00	3.97	0.931	1.024	1.210	0.000
Social benefits	3.84	3.85	0.992	1.030	1.078	0.138
Appraisal	4.21	4.01	0.975	1.019	1.093	0.080
Safety	4.13	4.05	0.964	1.036	1.157	0.004
Psychical stress	3.90	3.97	1.002	1.047	1.092	0.081
Vision of the company	3.80	3.92	1.037	1.003	1.070	0.199
Region development	3.67	4.01	1.101	0.999	1.216	0.000
Education	4.05	3.99	0.964	1.033	1.149	0.006
The environment	3.65	3.89	1.113	1.028	1.172	0.003
Free time	3.93	4.03	1.012	0.972	1.084	0.124
Recognition (work)	3.90	4.14	1.024	0.956	1.146	0.009
Basic salary	4.36	4.11	0.870	0.996	1.310	0.000

Note: Statistically significantly differences are highlighted in green.

The analysis by means of the Student's T-test (Table 4) revealed the fact that a lot of the examined motivation factors are statistically significantly different. We can particularly see a statistically significant difference in the motivation factors like the atmosphere at the workplace, the work team, further financial reward, the physical efforts at work, the job security, communication at the workplace, the work object and type of work, working hours, work performance, authorities, prestige, management approach, fair appraisal, safety at work, company vision, region development, the environment, free time, recognition and the basic salary. The influence of the above-mentioned motivation factors on employees differ in the two countries.

The analysis also revealed factors that do not differ statistically significantly and their influence on employees is nearly identical in both the countries. These factors are: the name of the company, the opportunity to apply one's own abilities, communication of the achieved results, the work environment, the career ladder, decision making, self-actualization, social benefits, psychical stress, education and personal growth.

4 Conclusion

Each human and so each employee as well, is unique. He/she tries to satisfy his/her needs and desires. We can expect people from different countries to have different needs. We tried to give an insight into the difference between employees from the Czech Republic and employees from the Peking region that work in the sphere of transportation and logistics.

The results show that the relational motivation factors most influence the performance of employees in the Czech Republic, but those in the Peking Municipality as well. The differences between the employees are obvious. Comparison between men from the Czech Republic and men from the Peking Municipality led us to a conclusion that men in the Czech Republic are motivated to high work performance by the basic salary and by the job security. Men in the Peking Municipality on the other hand put stress on recognition, authorities and the career ladder. The basic salary is of course a strong motivation factor for them as well. Upon comparison of women in the Czech Republic and those in the Peking Municipality we can definitely conclude that women in both the countries care about recognition, superior's approach and the effort to eliminate stress at the workplace. The differences that we can see are, like among men, in the job security. Free time and the region development are other obviously different factors. Women from the Peking Municipality care more about free time, recreation and regeneration. The efforts for the development of the region they live in can motivate them to increased work performance.

The presented research results can provide an insight into the issue of work motivation in the sphere of transportation and motivation and thus create a portfolio for a proposal of motivation programmes. A lot of companies expand to or cooperate with China at present. It is therefore more than important to learn about the basic cultural differences, but also the differences of preferences among employees from the two cultures.

This research was supported by VEGA No. 1/0024/17, Computational Model of Motivation.

This paper is supported by the research project "From horse-drawn railway to intermodal transport" within Visegrad Fund.

References

1. N. Bontis, Technology Management **18,** 621-642 (1999)
2. K.E. Sveiby, *The New Organizational Wealth: Managing & Measuring Knowledge-based Assets Barrett-Kohler* (San Francisco, CA, USA, 1997)
3. D. Weberová, P. Štarchoň, L. Ližbetinová, Vision 2020: *Innovation Management, Development Sustainability, and Competitive Economic Growth – Proceedings of the 28th International Business Information Management Association Conference* (2016)
4. M. Vetrakova, L. Smerek, E & M Ekonomie a Management **19**, 1, 62-73 (2016)
5. R. Zámečník, Marketing and Management of Innovations **4**, 325-341 (2016)
6. E. Carson, et al., Journal of Intellectual Capital **5**, 3, 443-461 (2004)
7. P.R. Massingham, W.Y.L. Tam, Journal of Intellectual Capital **16,** 2, 390-418 (2015)
8. C. Argyris, Dorsey Press. Michigan, 60-10149, 89-126 (1960)
9. D.M. Rousseau, Journal of Occupational and Organizational Psychology **74**, 4, 511-541 (2001)
10. S.P. Robbins, T.A. Judge, *Organizational behavior. Global and Southern African Perspectives,* 234 p. (South Africa: Cape Town, 2008)

11. K. Stachová, Z. Stacho, G. Pajtinková Bartáková, Business: Theory and Practice **16**, 264-270 (2015)
12. J. Drábek, S. Lorincová, J. Javorčíková, Issues of human resource management, 113-136 (2017)
13. S. Lorincová, J. Schmidtová, J. Javorčíková, Drvna Industrija **67**, 4, 351-362 (2016)
14. H. Urbancová, J. Urbanec, Acta Universitatis Agriculturae et Silviculturae Mendelianae Brunensis **60**, 4, 387-396 (2012),
15. A. Olšovská, L. Mura, M. Švec, Polish Journal of Management Studies **13**, 110-120, 2016,
16. A. Kucharcikova, L. Konusikova, E. Tokarcikova, Communications: Scientific Letters of the University of Zilina **18**, 49-54 (2016)
17. I. Kubasakova, R. Kampf, O. Stopka, Communications: Scientific Letters of the University of Zilina **16**, 2, 9-13 (2014)
18. J. Ližbetin, L. Černá, M. Ľoch, Naše more **62**, 147-152 (2015), DOI 10.17818/NM/2015/SI11
19. E. Nedeliakova, M. Panak, *International Conference on Engineering Science and Management*, 195-198 (Zhengzhou, China, 2016)
20. Czech Statistical Office, Available online: https://www.czso.cz/ssu (2016)
21. Slovak Statistical Office, Available online: https://www.czso.cz/ssu (2016)
22. R. Mysková, Scientific Papers of the University of Pardubice **6**, 75-79 (2001)
23. A. Grenčíková, J. Španková, *Proceedings of the 1st International Conference Contemporary Issues in Theory and Practice of Management*, 103-108 (Czestochowa, Poland, 2016)
24. L. Mura, V. Gašparíková, Acta Universitatis Agriculturae et Silviculturae Mendelianae Brunensis **58**, 3, 157-164 (2010)
25. G. Hofstede, *Cultural consequences: international differences in work-related values* (1980)
26. M. Potkány, L. Krajčírová, Procedia Engineering **192**, 731-736 (2017)
27. M. Hitka, Organisational Culture and Motivation as Instigators for Employee Engagement, (ES TU Zvolen, Slovak Republic, 2009)
28. J. Gravetter, Frederick, L.B. Wallnau, *Statistics for the behavioral sciences, Cengage Learning* (2016)

MATEC Web of Conferences **134**, 00033 (2017) DOI: 10.1051/matecconf/201713400033
LOGI 2017

Proposal for Evaluating a Connection Quality within Transport Networks

Vladimír Ľupták[1,*], *Jozef* Gašparík[2], and *Mária* Chovancová[1]

[1]Institute of Technology and Business in České Budějovice, Faculty of Technology, Okružní 517/10, 370 01 České Budějovice, Czech Republic
[2]University of Žilina, Faculty of Operation and Economics of Transport and Communications Univerzitná 1, 010 26 Žilina, Slovakia

Abstract. The paper deals with a proposal of the connection quality evaluation on a particular transport network within passenger transport. Currently, there is no relevant methodology for a comprehensive evaluation of timetables in general and no selected indicators for evaluating the connection quality on a transport network. The aim of the paper is to outline the way to evaluate the timetable in passenger transport in terms of connectivity and connection. From a passenger's perspective, it is necessary to evaluate the availability of opportunities to travel between selected locations on a railway network. The transport offer from the place A to the place B basically affects the transportation time, the number of transfers and the number of travel opportunities. This represents the starting point for designing a set of criteria in order to evaluate the overall connection quality.

1 Introduction

The offer of connection, assigned by the transport enterprise, depends on the demand for the carriage of passengers who need to move to another location between the two points on a transport network for objective or subjective reasons. Figure 1 shows the areas of quality evaluation as well as their schemes. In terms of connections quality evaluation on a transport network, operational processes and transport processes have an important role, and the paper is focused on them. Supplementary services, awareness and comfort are supportive areas complementing the quality of the transportation process [1-3].

* Corresponding author: luptak@mail.vstecb.cz

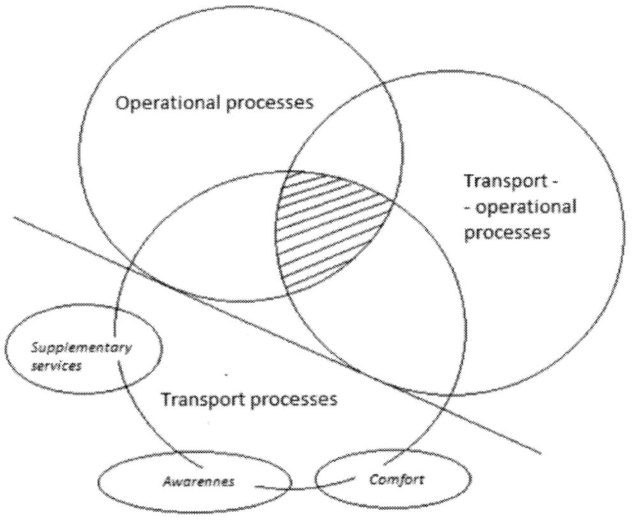

Fig. 1. Coincidence of transport processes with impact on the connection quality on a transport network. Source: [1]

The transport system, its structure and processes depend on the direct and indirect effects of surrounding. The quality of the provided transport services is given by a set of aspects, referred to as quality indicators: regularity, reliability, safety, speed, economy, reasonable cost, ecology, comfort, performance and availability. Quality is a dynamic phenomenon related to a particular object or process over time. Current quality status is a function of real conditions, possibilities and intent in real time, with individual aspects developing in desirable or undesirable ways. This is caused due to the objective necessity of continuous quality management [1, 2, 4-6].

2 Proposal of the connection evaluation methodology on a transport network

The resulting suggestion of the paper is to integrate the synergies and drawings of the qualitative indicators characterizing the connection on a transport network as well as the elements of the connection quality between the tariff points on the transport network. The methodology of connection quality evaluation aims to cover the possibilities of achieving any pair of tariff points on the selected transport network comprehensively [4-7].

The proposed methodology is based on the evaluation of defined criteria for the connection between selected tariff points on a network. The evaluation of a particular connection is the fundamental part of such a methodology. It is necessary to determine whether connections are evaluated during a working day or weekend, or process the evaluation separately for the selected working days, Saturday or Sunday. Consequently, summary indicators are evaluated for connections within a transport relation and within an entire network [7, 8].

The new methodology, which includes partial quality evaluations of the connection on a transport network, should offer a comprehensive evaluation defined by successive steps. The interconnection of the individual indicators must allow for an equalized evaluation of the connection in terms of the transportation quality. It means to take into account all the connections on a selected network. The following steps represent a part of the evaluation process:

1. selection of a transport network,
2. selection of the set of connections,
3. selection of relevant tariff points on a transport network,
4. selection of evaluation indicators,
5. evaluation of individual relations,
6. evaluation of all the connections within the relation,
7. network connection evaluation.

2.1 Selection of a transport network

In this step, it is necessary to define a transport network (railway, road, air, water, public transport, etc.). In the next step, a selected transport network will be subject to the quality evaluation itself. This selection can be specified according to the mode of transport, operator and transport network, or other criteria. These may also include the integrated transport system lines. The network is defined by the set of tariff points (vertices of graph) and the set of road transport relations connecting the tariff points (edges of graph) [7, 9].

2.2 Selection of a set of connections

In the second step, it is necessary to select the connections. This selection may be divided according to the carrier providing transport on the selected network, depending on the mode of transport, or to choose a complete search across all available connections between all the carriers [8, 9].

2.3 Selection of relevant tariff points on a transport network

In order to comprehensively assess the connection between the tariff points (stations, stops, interchanges) in the timetable, it is necessary to determine suitable quality indicators characterizing the performance indicator of this connection [1, 10, 11].

2.4 Selection of evaluation indicators

The following factors are identified when evaluating the connectivity and connection quality of transport relations [12-18]:
- number of connections N_s,
- average waiting time of the passenger,
- distance of the L_i,
- type of train and number of trains,
- transportation time T_{pi},
- number of transfers N_{pi},
- transfer time T_{wi},
- time to reach the destination T_{Di},
- speed of transportation V_{Pi},
- speed of reaching the destination V_{Di}.

2.5 Evaluation of individual relations

After processing a connection evaluation within one relation, it is necessary to evaluate the given relations between the tariff points in a network. For simplification, the relevant selection of tariff points is required. To evaluate a specific connection in a selected relation, a table of all connections must be compiled, as shown in the Table 1 [1-3, 7]

Table 1. General table for connections evaluation on a transport relation. Source: [1-3, 7]

Serial number n-th connection	Station X leaving [hh:min]	Station Xj arrival [hh:min]	Average waiting time W_i [h]	Distance L_i [km]	Train category	Transportation time T_p [h]	Number of transfers N_p	Transfer time in total T_w [min]	Time to reach the destination T_D [h]	Speed of transportation V_P [km.h^{-1}]	Speed of reaching the destination V_D [km.h^{-1}]
1	T_{odXi}	T_{prXj}	W_i	L_i		T_{pi}	N_{pi}	T_{wi}	T_{Di}	V_{Pi}	V_{Di}
2											
3											
.	.	.	.	.	.	.	.	.	.	.	.
.	.	.	.	.	.	.	.	.	.	.	.
.	.	.	.	.	.	.	.	.	.	.	.
n											

2.6 Evaluation of all the connections within the relation

When evaluating the connection within a relation, the selected indicators are aggregated to the average values, which serves to resulted evaluation of the entire relation This also represents the basis for a comprehensive connection quality evaluation. Table 2 shows the proposed indicators monitored during 24 hours to evaluate the connection quality on a network [1-3, 7].

Table 2. General table for the entire relation evaluation. Source: [1-3, 7]

Serial number n-th connection	Station X leaving [hh:min]	Station Xj arrival [hh:min]	Average waiting time W_i [h]	Distance L_i [km]	Train category	Transportation time T_p [h]	Number of transfers N_p	Transfer time in total T_w [min]	Time to reach the destination T_D [h]	Speed of transportation V_P [km.h^{-1}]	Speed of reaching the destination V_D [km.h^{-1}]
1											
2											
3											
.	.	.	.	.	.	.	.	.	.	.	.
.	.	.	.	.	.	.	.	.	.	.	.
.	.	.	.	.	.	.	.	.	.	.	.
n											
	Average values per connection in the relation:						ø N_p	ø T_w	ø T_d	ø V_P	ø V_D

2.7 Network connection evaluation

In the last step, it is necessary to summarize the partial average values of monitored indicators on relations to the matrix table, as shown in Table 3. All the evaluated relations on a network can be evaluated in accordance with the monitored indicators or the weighted arithmetic mean of amount of passengers, thus the quality of the connection or the compiled timetable on a selected transport network is assessed [1-3, 14].

Table 3. General table for the connection quality evaluation on a transport network. Source: [1-3, 14]

Monitored quality indicator of the connection on a transport network	Tariff point A	Tariff point B	Tariff point C	. . .	Tariff point X	Tariff point Y	Tariff point Z
Tariff point A				. . .			
Tariff point B				. . .			
Tariff point C				. . .			
...	...	...	...	. . .	...	...	...
Tariff point X				. . .			
Tariff point Y				. . .			
Tariff point Z				. . .			

After identifying all the average monitored indicators, the average number of transfers over an entire transport network can be calculated. When evaluating the transport quality on a transport network, many aspects can be taken into consideration, such as [16, 19-22]:

a) based on the arithmetic average

- the average number of transfers,
- average time to reach,
- average speed of transportation,

b) average speed of reaching the destination using a weighted arithmetic mean of the number of passengers.

3 Conclusion

For the field of evaluating the connection quality on a transport network, a number of standards, regulations and directives are implemented in national law and legislation. However, each passenger or potential passenger perceives the quality of connection and accessibility in a subjective way. Currently, emphasis is placed on ensuring the quality of the transport process focusing on means of transport, stations and stops. In order to objectify the quality of the connection in any transport system, it is necessary to comply with these standards, however for the connection quality evaluation from the transport point of view, it is necessary to develop a uniform methodology. Until now, such an uniform methodology has not been designed in the Slovak Republic or EU. For example, different timetables are drawn up every year in railway transport conditions and have led to controversy about the availability quality of different tariff points on the network.

The above-mentioned ideas and the fundamentals of the connection quality evaluation support the promotion of the EU transport policy objectives in the field of transport, i.e. promoting the mobility of the population by environmentally friendly modes of transport. These objectives are expressed particularly in the White Paper - Single European Transport Area, where at the national level, the Slovak Republic expresses its opinion through the transport policy of the Slovak Republic, the Transport Development Strategy of the Slovak Republic up to 2020, the Strategic Plan for Development of Transport Infrastructure of the Slovak Republic by 2020, Strategies for the development of public passenger and non-motor transport by the SR by 2020 and several other documents approved by the Slovak Government.

The methodology outlined in the paper fully covers the possibility of achieving any pair of tariff points on a selected transport network. The methodology offers not only a connections evaluation within the selected relation, but also an objective evaluation of the connections availability between two selected tariff points based on qualitative indicators such as average number of transfers, average waiting time, average transfer rate and average speed of reaching the destination. It evaluates the quality of the provided passenger travel opportunities in the selected territory by selected indicators.

This paper is supported by the research project "From horse-drawn railway to intermodal transport" within Visegrad Fund.

References

1. J. Gasparik, O Stopka, L. Peceny, Nase more **62**, 3, 114-118 (2015), DOI: 10.17818/NM/2015/SI5
2. L. Bartuška, L. Černá, J. Daniš, Naše more **63**, 3, 93-97 (2016), DOI: 10.17818/NM/2016/SI3
3. O. Stopka, I. Šimková, V. Konečný, Naše more **62**, Special Issue, 126-130 (2015), DOI: 10.17818/NM/2015/SI7
4. L. Černá, V. Zitrický, J. Ponický, *MME – 34th international conference - Mathematical methods in economics*, 126-131 (Liberec, Czech Republic, 2016)
5. Y.G. Zhang, Q. Y. Liu, D.Y. Lei, X.P. Xu, J.Z. Zhao, *Joint 2016 – ICEME 2016 – International Conference on Economics and Management Engineering and EBM 2016 – International Conference on Economics and Business Management*, 481-486 (China, 2016)
6. T.J. Mlinaric, T. Plesa, I. Balagic, *CETRA – 2nd International Conference on Road and Rail Infrastructure,* 875-880 (Dubrovnik, Croatia, 2012)
7. B. Gorlewski, Logi – Scientific Journal on Transport and Logistics **2**, 1, 16-26 (2011)
8. N. Aydin, Transport Policy **55**, 87-98 (2017), DOI: 10.1016/j.tranpol.2017.02.001
9. B. Abramovic, V. Zitricky, P. Mesko, Logi – Scientific Journal on Transport and Logistics **8**, 1, 1-10 (2017), DOI: 10.1515/logi-2017-0001
10. Z. Bryniarska, L. Zakowska, *CSUM 2016 – 3rd Conference on Sustainable Urban Mobility,* 25-32 (Transportation Research Procedia **24**, 2017), DOI: 10.1016/j.trpro.2017.05.063
11. E. Semchugova, V. Zyryanov, N. Negrov, A. Nikitina, *SPBOTSIC-2016 – 12th international Conference – Organization and Traffic safety Management in Large*

Cities, 584-590 (Transportation Research Procedia **20**, 2017), DOI: 10.1016/j.trpro.2017.01.094

12. C. Oprea, E. Rosca, M. Popa, A. Ilie, O. Dinu, M. Rosca, *IMANEE 2016 – 20th Innovative Manufacturing Engineering and Energy Conference*, UNSP 012098 (IOP Conference Series-Materials Science and Engineering **161**, 2016), DOI: 10.1088/1757-899X/161/1/012098
13. V.I. Sopov, V.V. Biryukov, *9th Russian/Korean International Symposium on Science and Technology* - Book Group Authors, 396-399 (Novosibirsk, Russia, 2005), DOI: 10.1109/KORUS.2005.1507741
14. C. Hokstok, Logi – Scientific Journal on Transport and Logistics **4**, 1, 63-80 (2013)
15. A.O. Baba, O. Bamaarouf, A. Rachadi, H. Ez-Zahraouy, International Journal of Modern Physics C **28**, 5, 1750064 (2017), DOI: 10.1142/S0129183117500644
16. J. Ližbetin, L. Černá, M. Ľoch, Nase More **62**, 3, 147-152 (2015), DOI: 10.17818/NM/2015/SI11.
17. D.U. Kim, J.D. Ryoo, J.H. Lee, B.C. Kim, J.Y. Lee, Etri Journal **38**, 1, 18-29 (2016), DOI: 10.4218/etrij.16.0115.0024
18. S. Kudlac, J. Majercak, C. Mankowski, *12th International Scientific Conference of Young Scientists on Sustainable, Modern and safe Transport*, 510-515 (Procedia Engineering **192**, 2017), DOI: 10.1016/j.proeng.2017.06.088
19. R. Kampf, O. Stopka, I. Kubasakova, V. Zitricky, *WMCAUS 2016 – World Multidisciplinary Civil Engineering-Architecture-Urban Planning Symposium*, 1538-1544 (Procedia Engineering **161**, 2016), DOI: 10.1016/j.proeng.2016.08.623
20. J.G. Dias, G. Trindade, Social Indicators Research **129**, 3, 1383-1399 (2016), DOI: 10.1007/s11205-015-1168-2
21. M. Weiszer, G. Fedorko, V. Molnar, *CLC 2013 – Carpathian Logistics Congress,* 330-335 (Krakow, Poland, 2014)
22. L. Guihery, Research in Transportation Economics **48**, 298-304 (2014), DOI: 10.1016/j.retrec.2014.09.056

DOI: 10.1051/matecconf/201713400034

Draft Deployment of Traction Units with Active Tilting System for Regional and Long-distance Transport on Non-modernized Railway Tracks

Vladimír Ľupták[1,*], *Ondrej* Stopka[1], and *Karel* Jeřábek[1]

[1]Institute of Technology and Business in České Budějovice, Faculty of Technology, Okružní 517/10, 370 01 České Budějovice, Czech Republic

Abstract. The paper presents the perspectives of suburban traction units with tilting system particularly in regional transport to decrease the travel times and increase the competitiveness of railway transport. It aims to outline the importance of its use in this mode of transport, especially in the case when the condition of infrastructure is unable to provide higher efficiency possibilities for railway enterprises and investments for tracks optimization are not available or insufficient, i.e. enterprises cannot be more efficient in utilization of classic traction units of conventional design. The paper is focused especially on using traction units with tilting system within a selected regional transport section when analyzing the operational and technical characteristics and comparison of time savings after these units deployment.

1 Introduction

A travel time of passengers from a boarding station to a destination station is considered the main barrier for a high competitiveness of passengers' railway transportation. This is not due to carrier's ability to provide fast, comfortable and safe transportation only, but also due to infrastructure conditions in which a railway carrier operates. One of these conditions is represented by the maximum track speed which cannot be increased by its reconstruction in many cases due to the current deficiency of economic profitability or natural conditions under which the track is built.

Thus, the deployment of modern fleet (rolling stock) of conventional construction to the operation may not bring a sufficient and desired effect which ultimately cannot contribute to the achievement of intended competitiveness. An entry of tilting units into service in this case represents one option in order to reduce travel times on the railway track section. Due to occurrence of tilting device, train can pass the curves at higher speed, and thus reduce travel times, and at the same time, the comfort level can be increased when traveling [1, 2].

* Corresponding author: luptak@mail.vstecb.cz

2 General principles of the railway tracks selection

For the best utilization of tilting units regarding passenger transport, the current tracks should be selected in addition to modernized main corridors which are held in harsh and poor natural conditions and the direction is characterized by many curves. That is why it is possible to increase technical speed of passenger trains in such conditions easily and this step leads to increase of its efficiency and attractiveness [1, 3].

Railway tracks are generally divided into five speed zones (see Table 1) expressing the upper track speed limit and its related demands on infrastructure maintenance. As the tilting system can be used only for speed of conventional construction vehicles over 70 km.h^{-1} (included), further, there is no need to consider the tracks included in the first speed zone SZ1 [1].

Table 1. The speed zones division. Source: [1]

Zone	Speed
SZ1	up to 60 km.h^{-1}
SZ2	from 60 km.h^{-1} to 90 km.h^{-1}
SZ3	from 90 km.h^{-1} to 120 km.h^{-1}
SZ4	from 120 km.h^{-1} to 160 km.h^{-1}

Track speed of each railway track is limited by many factors (criteria) such as [4-8]:

- small curve radius,
- insufficient inclination of outer rail in a curve (superelevation),
- improper outlook conditions for horizon ahead,
- inappropriate length or shape of curve transition (for radius and inclination),
- technical condition and low safety level of switches,
- state of the railway superstructure and substructure, etc.

Thus, the suitability evaluation for deployment of tilting units is required, needs to be done and undergo a thorough analysis of the railway track according to the current standards for the geometric position of the normal gauge tracks.

2.1 The suitability evaluation for deployment of specific traction units

Units with tilting system serve especially to shorten travel times without the high costs for upgrading an existing infrastructure. Active tilting devices allow 20 to 30 % travel time savings with higher speed driving along curves at tilting up to 8 ° against the level of wheel and tracks contact, or to 6.5 ° without the comfort of passengers being decreased [1].

According to the current valid STN EN 73 6360 [9] for geometric position of normal gauge tracks while assessing the state of infrastructure in terms of deploying the units with tilting system, it is required to perform a thorough analysis of the route based on all the relations arising from strictly injunctive or recommended values. These relations and limits are subject to this standard in which certain principles are selected [8-10]:

- tilting system operates by the speed limit at least v_t = 70 km.h^{-1} in the curves with inclination. In the sections with tracks speed limit less than mentioned above for conventional construction vehicles, it cannot be taken into consideration the increase of speed limit when using the tilting units,

- to take advantage of tilting system, the value of track inclination in curves needs to be at least p = 30 mm. In folded curve with an inclination, intermediate transitions should have an inclination difference at least $\Delta p = 30$ mm,
- the value of insufficient inclination (the difference between theoretical and actual inclination) for tracks without solid parts is reduced to $I_k = 250$ mm,
- the value of insufficient inclination for tracks without solid parts compared to the value of inclination should be at least $I_k = 3.0p$ (in the most cramped proportions up to $I_k = 3.5p$)
- for the curves without inclination, using the tilting units in higher speed limit than the basic limit cannot be considered,
- in order to take advantage of tilting system, these conditions are required:
  - curves with inclination must have the outer transitions,
  - folded curves with an intermediate inclination transition must have an intermediate radius transition,
  - inclination and radius transitions must have the same proper length,
  - the shape and course of curvature must increase the same way for both inclination and radius transitions,
- outer transitions with a linear increase of curvature and inclination shall comply with requirements of the STN EN 7363 60, the same applies for intermediate transition curves.

Trains with the tilting system allow for a reduction of travel times on such tracks which meet the requirements of standards for increase of the tracks speed, however, mostly thanks to its tilting on the tracks with limiting factor of reaching higher speeds right in a curve. All the train wagons are equipped with an active hydraulic system, partially offsetting the effect of lateral acceleration on passengers when passing through the curves to significantly increase the average speed and convenience of traveling on curved railway tracks [7, 11].

2.2 Application of several principles into the selected railway section Bratislava (main station) – Banská Bystrica

In order verify the suitability of abovementioned units, and thus, achieve the desired results, the railway section Bratislava main station - Palarikovo - Surany - Zvolen passenger station - Banska Bystrica was used when considering intensive directional disparity in the vast length due to the river Hron [1, 12].

Figure 1 depicts an existing static speed (velocity) profile (marked as green), which varies in the range from 70 to 100 $km.h^{-1}$ in the section Palarikovo - Banska Bystrica and from 100 to 140 km.h-1 in the section Bratislava main station (hl. st.) - Palarikovo. Static speed profile for considered tilting units (marked as blue), in order to analyze track geometry, represents discontinuous and inefficient progress with frequent alternation of velocity extremes, therefore, it was necessary to create reduced profile (marked as red), which respects the fundamental dynamic options of adhesive rolling stock [1, 12, 13].

Subsequently, this course can be forwarded for further analysis by specifying the travel dynamic of the unit with tilting system and depicting the curve course (flowchart) of actual speed depending on traveled distance. In this case, the position km 0.000 is the position of station Bratislava main station and km 230.200 represents the position of station Banska Bystrica [1, 14].

For such a speed profile, it would be necessary to establish the curve inclinations at this railway track section for three curves with their transitions, and for one curve, in order to increase the projected value of inclination when considering the current length of inclination transitions of curves corresponding to the newly proposed value of inclination (superelevation) [1, 14, 15].

2.3 Comparison of selected indicators

As for the calculation of new qualitative indicators acquired after implementing the units with tilting system into operation, three basic types of passenger trains of category express train (Ex), long distance fast train (F) and commuter local passenger train (Ps) are proposed, and therefore, existing trains, which are operated under the current timetable regardless of their regularity, are supposed to be deployed [15-17].

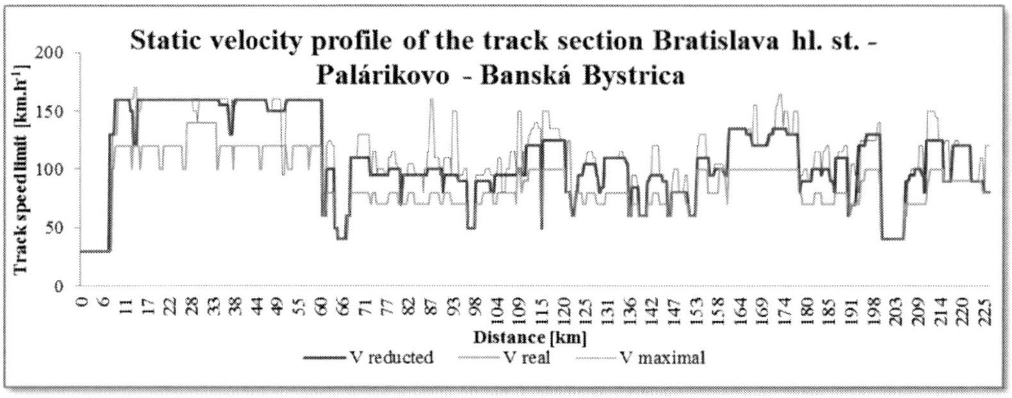

Fig. 1. Static velocity profile of selected railway track section. Source: authors

New travel times were calculated (determined) indirectly using the approximate method of estimation from static speed profile which was selected for the initial orientation detection of time savings due to implementing the tilting units. Travel time under static speed profile of current speeds was calculated and compared with actual travel times of passenger trains under the current timetable due to the dynamic difference coefficient to the static speed profile. The resulting approximate travel time saving for this 223 km long railway track section is about 20 minutes [8, 12, 18, 19].

Comparison of current and proposed technical speeds for each passenger train category is graphically summarized in following Figure 2.

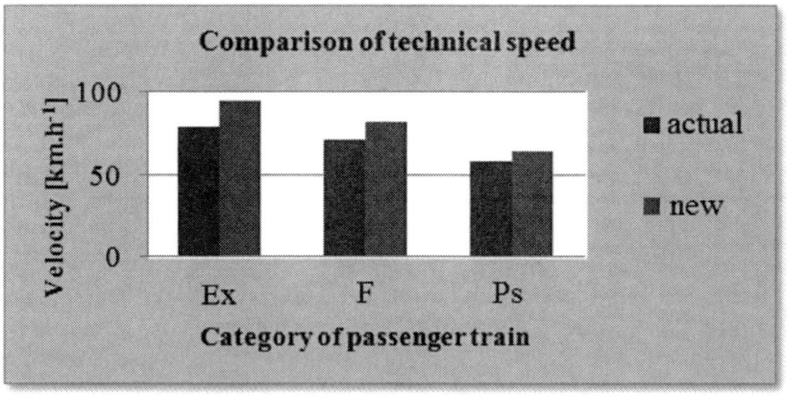

Fig. 2. Comparison of technical speeds. Source: authors

As can be seen in case of the express train (Ex), using tilting units, technical speed increases from current 78.93 km.h^{-1} to speed of 94.93 km.h^{-1}. Technical speed of the fast train (F) also increases from original 70.47 km.h^{-1} to speed of 81.87 km.h^{-1}, and as for the local passenger train (Ps), technical speed increases the from original 57.79 km.h^{-1} to speed of 64.03 km.h^{-1}.

In regard to sectional travel speeds (technical speed with respect to delay times) of passenger trains, it is the similar scenario in terms of deploying the tilting units (see Figure 3) [8, 12].

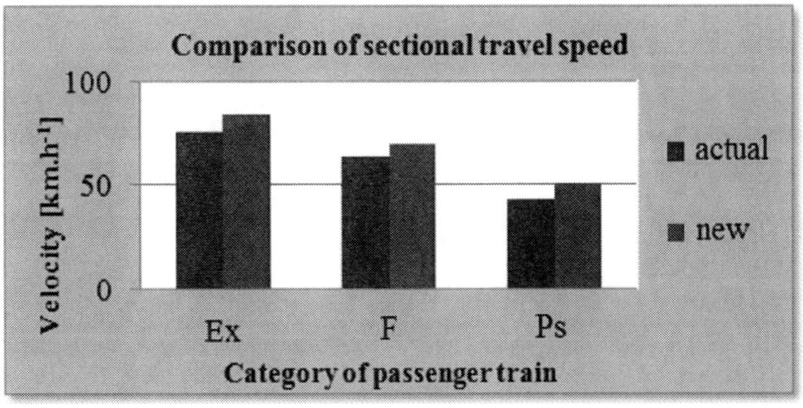

Fig. 3. Comparison of sectional travel speed. Source: authors

As can be seen in case of the express train (Ex), deployment of tilting units improves its sectional travel speed from current 75.48 $km.h^{-1}$ to speed of 84.01 $km.h^{-1}$. Sectional travel speed of the fast train (F) also increases from original 63.36 $km.h^{-1}$ to speed of 69.69 $km.h^{-1}$, and regarding the local passenger train (Ps) the sectional travel speed increases from original 43.43 $km.h^{-1}$ to speed of 50.78 $km.h^{-1}$.

3 Conclusion

Performing these analyzes concerning railway track section Bratislava main station - Banská Bystrica, it was found that it is likely to deploy regional tilting units to the operation on given track in the vast majority of its length. For their deployment, it would be necessary to make some low-cost steps in order to adjust the railway infrastructure. With the approximate method of estimating the travel time savings, it was found an indicative shortening the express train travel time (Ex) approx. 20 minutes which represents a fundamental value for increasing competitiveness compared to other transport modes within passenger transport [8, 20, 21].

After deployment of traction units with active tilting system, it leads to greater flexibility of passenger carriers which might significantly compete to road transport, not only by shortening driving (travel) times, but also by increasing the comfort and competitive prices and services [12, 22].

Implementation of such a measure would have a beneficial effect not only from a society-wide point of view, but also in terms of emerging railway transport standards.

This paper is supported by the research project "From horse-drawn railway to intermodal transport" within Visegrad Fund.

References

1. V. Zitricky, J. Gašparík, L. Pečený, Transport Problems **10**, Special Edition 2015, 59-72 (2015)
2. L. Bartuška, L. Černá, J. Daniš, Naše more **63**, 3, 93-97 (2016)

3. E. Brumercikova, B. Bukova, F. Brumercik, *Transport Means – 20th International Scientific Conference on Transport Means*, 745-750 (Juodkrante, Lithuania, 2016)
4. B. Abramovic, Promet-Traffic & Transportation **25**, 2, 191-192 (2013)
5. K. Giannakos, Engineering Structures **122**, 296-309 (2016), DOI: 10.1016/j.engstruct.2016.04.058
6. K.F. Li, W.N. Liu, Z.W. Han, Z.Z. Wu, *Proceedings of the Institution of Mechanical Engineers Part F*, Journal of rail and rapid Transit **229**, 7, 798-814 (2015), DOI: 10.1177/0954409714521801
7. S.G. Lee, *International Conference on Control, Automation and Systems* **1-4**, 1229-1232 (Seoul, South Korea, 2008)
8. J. Lalinska, J. Gasparik, D. Sipus, Logi – Scientific Journal on Transport and Logistics **8**, 1, 74-81 (2017), DOI:10.1515/logi-2017-0009
9. *STN 73 6360. Geometric position and track configuration of the normal railway tracks gauge (Geometrická poloha a usporiadanie koľaje železničných dráh normálneho rozchodu) (1999)*
10. V. Zitricky, L. Cerna, B. Abramovic, Procedia Engineering **192**, 994-999 (2017), DOI: 10.1016/j.proeng.2017.06.171
11. H.Y. Kim, J.H. Lee, S.H. Han, N.J. Lee, B.T. Kim, C.G. Kang, *Proceedings of the Institution of Mechanical Engineers Part F*, Journal of rail and rapid Transit **231**, 3, 370-378, (2017), DOI: 10.1177/0954409715627841
12. L. Peceny, J. Gasparik, V. Gaborova, *ICTTE – 3rd International Conference on Traffic and Transport Engineering*, 462-466 (Belgrade, Serbia, 2016)
13. M. Weiszer, G. Fedorko, V. Molnar, *CLC 2013 – Carpathian Logistics Congress*, 330-335 (Krakow, Poland, 2014)
14. O. Havlena, M. Jacura, T. Javořík, M. Svetlík, L. Týfa, Transport Problems **9**, 4, 97-104 (2014)
15. M. Halas, J. Gasparik, L. Peceny, Logi – Scientific Journal on Transport and Logistics **4**, 2, 28-38 (2013)
16. Z. Cujan, G. Fedorko, Open Engineering **6**, 426-431 (2016), DOI:10.1515/eng-2016-0057
17. T. Kalina, M. Jurkovic, H. Binova, B. Gardlo, Communications: Scientific Letters of the University of Zilina **18**, 2,26-29 (2016)
18. M. Chovancova, V. Klapita, *Transport Means – 20th International Scientific Conference on Transport Means*, 112-116 (Juodkrante, Lithuania, 2016)
19. J. Vrabel, J. Jagelcak, J. Zamecnik, J. Caban, *TRANSBALTICA 2017 – Transportation Science and Technology: 10th International Scientific Conference*, 89-99 (Procedia Engineering **187**, Vilnius Lithuania, 2017), DOI: 10.1016/j.proeng.2017.04.354
20. J. Siroky, Communications: Scientific Letters of the University of Zilina **10**, 3, 19-25 (2008)
21. M. Halas, V. Zitricky, P. Blaho, Logi – Scientific Journal on Transport and Logistics **4**, 1, 53-62 (2013)
22. R. Kampf, P. Průša, C. Savage, Transport **26**, 4, 425-432 (2011), DOI: 10.3846/16484142.2011.635424

MATEC Web of Conferences 134, 00035 (2017) DOI: 10.1051/matecconf/201713400035
LOGI 2017

Logistics Assessment of Functional Performance of Material Flows in Railway Transport

Jozef Majerčák[1,*], and *Peter* Majerčák[2]

[1]University of Žilina, Faculty of Operation and Economics of Transport and Communications, Department of Railway transport, Univerzitná 1, Žilina 010 26, Slovakia
[2]University of Žilina, Faculty of Operation and Economics of Transport and Communications, Department of Economics, Univerzitná 1, Žilina 010 26, Slovakia

Abstract. The article includes the proposal of new methodology assessment of functional performance of material flows as a tool for management decision-making in the selection of suitable suppliers of logistical services in railway transport. This selection is based on evaluation of key indicators and determination of their weights. The evaluation of the indicators and determination of their weights is realised by mathematic methods.

1 Introduction

Theories about the logistics evaluation of material flow of rail transport are overlooked by all stakeholders. In Slovakia, while we try to apply foreign knowledge and search the best solutions for our region (for example methods KPI, KPD, LPI, VDA 17, MS 9000 respectively MMOG) [1-9].

2 The proposal assessment methodology

The methodology is conceived to be usable for any company that will need to assess the functional performance of material flow in the implementation by different suppliers based on objective indicators. The procedure we propose is depicted in the following breakdown (see Fig.1).

* Corresponding author: jozef.majercak@fpedas.uniza.sk

 DOI: 10.1051/matecconf/201713400035

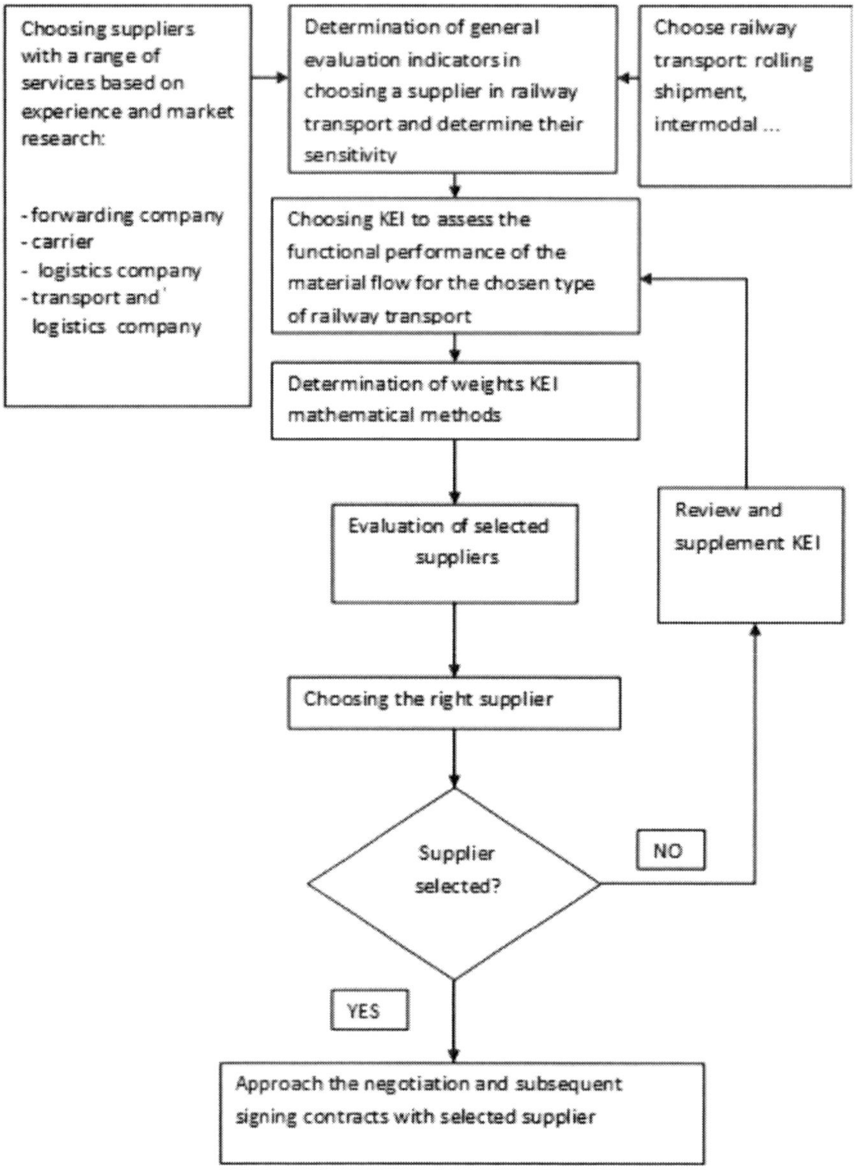

Fig. 1. The procedure for selecting the suitable supplier of logistics services in railway transport. Source: authors

3 Determination of general evaluation indicators in choosing a supplier in railway transport

Table 1 shows the general evaluation indicators used in daily operations. Collection of these of indicators is open and can be added at any time for new indicators. The order of indicator in the table is a random and does not prioritize the individual indicators.

DOI: 10.1051/matecconf/201713400035

Table 1. The general evaluation indicators for choosing a supplier of railway transport. Source: authors

Order (random)	**General evaluation indicators**
1.	The reliability compliance with the time of delivery (transit)
2.	Total transit time (from border to border)
3.	Total price from sender to recipient
4.	Quality of forwarding services
5.	The possibility of find shipments
6.	Providing regular monitoring of movement shipments
7.	Insurance of transported shipments
8.	Guarantee the protection of goods against loss, damage or breach of
9.	The range and quality of services provided at the state border in railway transport
10.	Possibility to provide customs operations necessary for the further handling of goods
11.	Staff qualifications and competence
12.	Language skills of staff
13.	Economic and financial stability of supplier
14.	The quality of the sale of services, ease to negotiate competitive prices
15.	Possession of the necessary equipment. Ability to ensure the equipment for handling goods
16.	Ability to provide the type or kind of vehicle for further transportation - such as the transshipment
17.	Readiness supplier negotiations on the revision services
18.	The complexity of the order process required services
19.	Readiness supplier negotiations on the revision price
20.	Availability of storage
21.	Secure store against theft
22.	Insurance of goods in the store
23.	Warehouse staff training required for handling the types of goods
24.	The owner of the certificate of quality ISO
25.	A member of FIATA
26.	A member of Association of Logistics and Freight Forwarding of the Slovak Republic
27.	Additional services required for the treatment of goods before delivery to the customer (for example, assembling packages)
28.	Possibility to provide palletizing goods during transshipment
29.	Possibility to provide strapping goods during transshipment
30.	Possibility to provide foiling goods during transhipment
31.	Possibility to provide selection of suitable wagons for loading
32.	Possibility to provide sorting of goods according to specified criteria
33.	Affinity shipment
34.	Supplementary criteria for supplier selection

4 Determination of sensitivity for the assessment of individual indicators

By determining the sensitivity of indicators identifies a threshold when the indicator crosses the border between the two evaluations.

Determination of the sensitivity of indicators:

a - The supplier can arrange the selected indicator accurately according to requirements without the need to use the agreed tolerances the supplier can propose new quality or cost-effective solutions.

b - The supplier can arrange the selected indicator, as required by agreed tolerance without intervention of customer.

c - New supplier with good references.

d - The supplier can arrange the selected indicator, does not respect the agreed limits of tolerance, the intervention of the customer.

e - The supplier can arrange the selected indicator, exceeds the limits of the agreed tolerance and waits for the customer intervention, is required constant monitoring of suppliers to meet contractual obligations.

f - New customer without reference.

g - The supplier cannot provide the selected indicator, despite contractual obligations to the client.

5 The selection of key indicators of evaluation and determination of their weights mathematical methods – model example

In determining the estimated weights KEI, we used more mathematical methods:

• The order method,
• The scoring method,
• Method pairwise comparisons - Fuller method,
• Method of quantitative pairwise comparisons of indicators (Saaty´s method).

We select KEI to test the procedure for assessing the functional performance of the material flows implemented by multiple vendors. From general evaluation indicators (Table 1) we choose the key evaluation indicators.

For example (KEI):

I1. - The reliability compliance with the time of delivery (transit).
I4. - Quality of forwarding services.
I5. - The possibility of find shipments.
I7. - Insurance of transported shipments.
I20. - Availability of storage.

We will consider the following order of importance, KEI: 1,4=5,7,20 which means that I1 indicator will be most important, for it will I4 and I5 indicators as important, then followed I7 and at the end of I20. The indicator that is least important.

For all the methods used to determine the weights we will be based on the same initial assumptions.

We choose weights always so that the total weight of all KEI selected for a specific model example equal to one. If the weight for the i-th indicator I tag vi for i = 1, 2, ...u, where "u" is the number of indicators, then for the relationship will apply:

$$\sum_{i=1}^{u} vi = 1 \, , \; vi \geq 0 \qquad (1)$$

When determining the weights of indicators allocate more weight to the criterion which is important. We weight assigned to the selected indicators under the terms of the sum of weights equal to one (as stated above) or by using mathematical methods that were used for the model example.

6 Comparison of the estimated weights for the key evaluation indicators by different methods

In Table 2, we make a direct comparison between different methods of estimating weights. The Saaty´s method the most sensitive assigns appropriate assessment of individual indicators [10,11], so we decided to use the results of this method in further calculations.

Table 2. Comparison of estimated weights for KEI the various mathematical methods. Source: authors

Indicators	**Weights indicators**			
	The order method	**The scoring method**	**Fuller method**	**Saaty´s method**
I1. The reliability compliance with the time of delivery (transit)	0.33	0.31	0.40	0.38
I4. Quality of forwarding services	0.23	0.25	0.25	0.22
I5. The possibility of find shipments	0.23	0.25	0.25	0.22
I7. Insurance of transported shipments	0.14	0.13	0.10	0.12
I20. Availability of storage	0.07	0.06	0.00	0.06

7 Data collection to assess

- Own database (experience),
- The references,
- Inquiry (agency).

8 Evaluation of selected key indicators

I use for the evaluation of KEI scale ranging from 1 to 5. The higher will be evaluation, more supplier will meet the needs of the individual KEI.

Evaluation:
5 - Meets my needs, I don´t have further comments.
4 - Still meets my needs but with some minor reservations.
3 - Still meets my needs but with substantial reservations.
2 - Meets my needs minimum, but I don´t have to choose, now.
1 - Totally satisfying for my needs and is unacceptable.

Assignment to the sensitivity of evaluation:
5 - a.
4 - b, c.
3 - d.
2 - e, f.
1 - g.

We consider about six suppliers of services in the model example [12]. We evaluated KEI for each supplier selected on experience while respecting of sensitivity assigned the evaluation.

Overall evaluation of the individual KEI for the particular supplier:

$$C_{i,d} = v_i \,.\, H_{i,d},\ for\ i = 1\ to\ u \tag{2}$$

where :

$C_{i,d}$ - An overall assessment of the i-th key indicator (I) d-th supplier where $d = 1.2 \ldots.z$ and z - means of the number of suppliers.
u - Numbers of indicators.
v_i - The weight of the i - th KEI.
$H_{i,d}$ - Evaluation of i - th KEI , d - th d supplier.
$\sum C_{i,d}$ - The sum total evaluation KEI - final evaluation d - th supplier.

The supplier with the highest sum of evaluations KEI will head order suppliers and I give him 1st place [13]. A detailed overview of our assessment of suppliers Table 3 provides.

Table 3. Evaluation of suppliers Source: authors

	Indicators					The order suppliers
	I1.	**I4.**	**I5.**	**I7.**	**I20.**	**∑ Ci,d – final evaluation suppliers**
Weight vi	0.38	0.22	0.22	0.12	0.06	
Hi, evaluation 1st supplier	5	4	4	2	5	1.
Ci,d, overall evaluation I	1.90	0.88	0.88	0.24	0.30	4.20
Hi, evaluation 2nd supplier	4	5	1	2	5	5.
Ci,d, overall evaluation I	1.52	1.10	0.22	0.24	0.30	3.38
Hi, evaluation 3rd supplier	4	3	5	4	1	2.
Ci,d, overall evaluation I	1.52	0.66	1.10	0.48	0.06	3.82
Hi, evaluation 4th supplier	3	4	4	5	5	3.
Ci,d, overall evaluation I	1.14	0.88	0.88	0.60	0.30	3.80
Hi, evaluation 5th supplier	2	4	4	3	5	6.
Ci,d, overall evaluation I	0.76	0.88	0.88	0.36	0.30	3.18
Hi, evaluation 6th supplier	5	2	3	5	1	4.
Ci,d, overall evaluation I	1.90	0.44	0.66	0.60	0.06	3.66

9 Conclusions

In further work we will elaborate in more detail using entropic methods [3, 10] to determine the weights of individual indicators for cases where we do not clearly identified priorities between individual KEI. In further work we will elaborate in more detail using entropic methods [3, 10] for determining weights of individual KEI in cases where we do not clearly identified priorities between indicators. In this case, individual KEI assigns different probabilities. Preliminary calculations show that the process is very sensitive to the objectivity of input information, which then affects a significant degree of uncertainty (entropy) and thus the weight of the selected KEI. Higher sensitivity could this method ever closer to objective evaluation weights the key evaluation indicators (KEI).

This paper is supported by the research project "From horse-drawn railway to intermodal transport" within Visegrad Fund.

References

1. D. Holcová, R. Pokorný, *Multicriteria evaluation of variants*, Available online: www.fzp.ujep.cz/~Pokornyr/01_Materialy/KREK_vkv.ppt (2010)
2. J. Královenský, J. Majerčák, *Transport status in logistics* (EDIS, Žilina, Slovak Republic, 2001)
3. J. Majerčák, *Entropy of transport process in multimodal transport conditions,* 90-96 (Eurokombi – Intermodal, 2009)
4. J. Majerčák, J. Plevko, *Logistics chain evaluation using the LPI*, 220-225 (Company diagnostics, controlling and logistics, 2010)
5. J. Majerčák, *Supply chain in logistics chains*, Available online: www.scss.sk/smpmcd/files/semestralne_projekty2/logistickastrategia podniku/zdroje/supplychain_v_lr.pdf (2011)
6. J. Majerčák, *The function efectiveness of transport*, Available online: www.ilogistics.cz/files/docs/efektivitaDopravy.doc (2010)
7. L. Ojala, T. Solakivi, *CSCMP's Annual Global Conference* (Turku School of Economics, Finland, 2009)
8. J. Piatka, *Integrated managerial system and Entropy method* (2008)
9. *World bank,* Available online:http://info.worldbank.org/etools/tradesurvey/mode1a.asp (2009)
10. Š. Kudláč, V. Štefancová, J. Majerčák, Procedia Engineering **187**, 749-755 (2017), DOI: 10.1016/j.proeng.2017.04.433
11. J. Majerčák, Š. Kudláč, M. Panák, *Transport Means 2016 – 20th International Scientific Conference,* 65-70 (Kaunas University of Technology, Lithuania, 2016)
12. L. Černá, V. Zitrický, J. Daniš, Open Engineering **7**, 1, 6-13 (2017), DOI: 10.1515/eng-2017-0002
13. Z. Záhumenská, J. Gašparík, *12th international scientific conference of young scientists on sustainable, modern and safe transport*, 976-981 (Procedia Engineering **192**, 2017)

Remarks to the Reliability Assessment and to Human Actions – Especially Car Driver

Zbigniew Matuszak[1,*], *Marek* Jaśkiewicz[2], *Dariusz* Więckowski[3], and *Józef* Stokłosa[4]

[1]Maritime University of Szczecin, Faculty of Marine Engineering, ul. Wały Chrobrego 1-2, 70-500 Szczecin, Poland
[2]Kielce University of Technology, Faculty of Mechatronics and Mechanical Engineering, al. Tysiąclecia Państwa Polskiego 7, 25-314 Kielce, Poland
[3]Automotive Industry Institute, ul. Jagiellońska 55, 03-301 Warszawa, Poland
[4]University of Economics and Innovation in Lublin, ul. Projektowa 4, 20-209 Lublin, Poland

Abstract. The description attempt of the system person-vehicle-environment as an element of the system person-technical object-environment has been made. In the system analysis the reliability of the machine exploiter – a car driver has been taken into account. It has been pointed out that technical system operators (in this case – the drivers) have to be considered important in the structure of reliability of these systems. The operational and biological reliability characteristic of a human being – technical system operator has been carried out. The necessity for conducting different psychological tests for assessment of the quality of the human being – operator of each technical system – has been indicated.

1 Introduction

It is a human being who has the main impact on causing road accidents. When looking at the impact of the human factor exclusively on causing road accidents it equals 63%, and counting it together with the impact of the environment (road) and the vehicle the value may reach 90÷95%. The possibility that the vehicle was the mere cause of the accident is relatively small as it equals only few percent [1]. There has been a study showing the impact of the drivers on the road accidents reaching 36÷41% in a non-built-up area, 39÷53% in a built-up area, with the impact of the vehicle reaching 15÷25%, and the impact of the environment being at the level of 38÷45% in a non-built-up area and 31÷34% in the city [2].

The majority of the statistical data shows a minor impact of both the vehicle and the environment on the road accident [3].

In many statistics, the impact of the environment can be omitted since a wet surface or a road covered with snow does not pose a direct threat these days. So the impact of the factors like bad surface condition or snow on the road on the accidents occurrence is quite low. Under normal road conditions an accident happens usually as a consequence of few

* Corresponding author: z.matuszak@am.szczecin.pl

causes happening at the same time and which influence its beginning and further course of action. But definitely the driver can contribute to the accident to a greater extent [4].

All these factors play together so the straight-forward determination of the accident causes is not always possible. Complicated relationships between these factors, features to a particular accident and individual features of the driver contribute to the fact that speaking in general about particular causes is still difficult [4]. In order to present the figure of the driver different personal features, like for instance: concentration skills, intellectual abilities, resistance to stress, or reaction time are taken into consideration. They are often measured by means of special tests.

The reaction of a driver on the road situations can be different depending on the circumstances: it can be logical, illogical or irrational [1, 5-8]. During the logical behaviour his reactions can be overseen. But in other cases the driver can make turns not adjusting them to a given situation. Starting such turns or responding in a wrong way can escalate the danger [4]. Such spontaneous behaviour of the driver is often ascribed to less experienced drivers.

Such states of mind as vagueness, absence of mind, tiredness may provoke actions inconsistent with rational and logical thinking of the drivers. These, in turn, may spark the events directly leading to an accident. The risk of the danger caused by the driver occurring on the road is determined with many factors including his psychological and physiological features. These features can differ dynamically during the drive [4].

Psychological features may heighten or lower the level of risk that the accident will happen. They include, above all, ability to oversee things, ability to drive in a hazardous manner (for instance fast), maintaining common sense, kindness, hostility, neurotisation. Under physiological features we distinguish: condition of eyesight and hearing, level of tiredness, manual skills.

The features mentioned above are not ascribed to a particular driver for good. They may change dynamically depending on the conditions, for instance, time of the day, season, traffic conditions, etc. In the theory of human's reliability there are quantitative and qualitative definitions. The operator's reliability (in this case reliability of a driver) in the quantitative sense means the ability described by the probability to perform given tasks. In the qualitative sense though, according, it is the ability to perform tasks with the minimal risk of making a mistake and in a given time. Reliability may be understood as a function of adequacy between internal factors of the operator (temperamental features, personality, motivation, qualifications), and external factors (tasks, technical quality and machine exploitation, methods of work, physical environmental conditions, social climate, work atmosphere).

A person who is an element of the system *person-machine-surrounding* may be in general considered, among others, as [9]: machine operator in the utilising processes (pilot, driver, etc.); one operating the machine in the operating process (condition control, current and proactive repairs); element of the system realizing certain functions (explorer, observer); person using the machine functions (e.g. passenger of a particular means of transport). Seeing a person as an exploiter of a machine – technical device – it is of utmost importance to perform the operating actions both on time and without any faults. This affects the efficiency of the system and, in many cases, safety of the operator and other people as well as technical elements of the system or of the natural environment. Safety should characterize the whole system: a person affects system safety as much as deadline unfaithfulness, lack of precision, and failure of his actions generate emergency situations [10].

Fallibility of a person (people) during work in the marine environment is one of the main reasons of accidents and breakdowns. This fallibility usually stems from bad adaptation of objects themselves and their equipment to the needs and abilities of a human being.

In assessment of the operator's work it is the result of his work and the way it has been achieved that count most, if it is significant from the perspective of processes taking place in the operating objects as well as the economy of their exploitation process. Sometimes also the time of realizing operating action is important, for example in the situations when the object condition changes quickly, like during turning of the operating installation and in emergency situations, in which the time of reaction to the external or internal events is crucial. It leads to the dichotomy of the reliability models of human beings: reliability which does not depend on time, and depends on time.

The model of reliability which does not depend on time looks as follows:

$$R=P(Z=1,\ W=1) \tag{1}$$

where:

Z – discrete random variable so that:

$Z = 1$ when the operator's task has been realized, and 0 when it has not;

W - discrete random variable so that:

$W = 1$ when the task was performer properly, and 0 – when improperly.

So the time-dependent reliability looks as follows:

$$R(t) = P(Z=1,\ T<t) \tag{2}$$

where:

T – continuous random variable presenting the time of task realization by the operator;

t – positive real number so that $t \in R^+$.

Operator's reliability can be determined for the specific tasks. We state that a task consists of minimal collection of actions corresponding to exercises completion of which realizes a particular goal. Some actions must be preceded with other to be realized. It is also possible to have selective actions which are implemented to the task realization procedure only when it is necessary for realizing a task.

Model of reliability which does not depend on time corresponds to the routine actions performed in the process of ship installations, such as starting-up of some devices, setting them aside, switching devices and installations, changing of work parameters and settings of automation systems. Time-dependent reliability model is used for assessment of work of a ship mechanic (operator) in emergency situations or situations in which not completing a task within a given time results in posing a danger for installations and objects being around. One can distinguish cases in which mechanic's reaction to any interruptions or breakdowns includes only routine actions and cases in which he can rely only on his own knowledge. In most cases the time for task realization underlies log-normal distribution [11]. Operator's reliability according to [11] may be determined on the basis of the studies realized on simulators, since in natural conditions one is not able to achieve necessary data to estimate operator's reliability characteristics. It is so due to the fact that the tasks he performs in difficult situations may cause waste.

2 The attempt of determining human reliability

A human being is a system of high level of complexity with excess reliability structure [9]. From the cybernetic point of view this system can be seen as autonomous as it is able to perform self-control and to counteract against losing the ability to control. Besides this, it is capable to keep functional balance regardless of the changes in the surrounding and it aims at maintaining its existence as well as acts on behalf of its own interests.

In the human organism all the functions of autonomous system are performed. These are [9]: affecting the surrounding (*effectors*), receiving information from the outside (*receptors*), consumption of energy from the outside (*alimentator*), storing and processing information (*correlator*), processing and storing energy (*accumulator*), maintaining functional balance (*homeostat*).

A person, from the reliability point of view can be characterized with: the mistakes of receiving information from the outside (it can result from imperfection of *receptors* and interruptions of the information taken from the outside); mistakes of processing information what can result from the excessive amount of information which exceeds the capabilities of correlator or not preparing the *correlator* to process certain information; mistakes or insufficient interfering with the surrounding what may stem from the deterioration of the functional balance or exceeding the abilities of the *effectors*; losing or limiting the ability to act, as a result of damage of the organs or losing or limiting the ability to adapt to the changes going on in the surrounding.

From the point of view of the mentioned features one may assess such human reliability traits like faultlessness and effectiveness. Both of them determine human operational reliability and faultlessness determines biological reliability.

Human biological reliability is the ability to act in a given period of time and in given conditions of influence of the surrounding. It is measured by the probability of maintaining the ability to act in a given period of time and in given conditions. When we use the technical term "damage" to describe a human being, we may use the notion of reversible damage (for instance, a temporary loss of the ability to act due to stressful situations or an illness) or non-reversible damage (for example, improper functioning of some of the internal organs of a human body).

Reliability of human work can be measured with the probability of success in the work or task realization on a given level of system functioning, in a given time period, described with the requirement about the duration of work.

A proper functioning means to fulfil the task in a given time. Improper functioning, on the other hand, may be the reason for the machine breakdown or may change the course of fulfilling the task. A person is characterized with a certain time of receiving information T_1 (which depends on the abilities of *receptors*), time for information analysis T_2 (which depends on the abilities of *correlator*), decision making time T_3 (which depends on the cooperation between *correlator and effector*), decision performing time T_4 (which depends on the abilities of the *effector*). A task will be performed on time if a predefined time dedicated to perform the task will meet the inequality

$$T_1 + T_2 + T_3 + T_4 \leq T_0 \tag{3}$$

where:

T_0 – predefined time dedicated to perform the task.

During realization of the following basic tasks (information receiving, information processing, decision making, decision performing) a person (exploiter, operator, ...) may make mistakes. The mistakes an exploiter of a technical system may encounter include: design mistakes depending on too low design quality; mistakes made by the crew not performing the expected tasks properly or performing unexpected procedures; executive mistakes appearing on the stage of manufacturing, resulting from bad quality of work, using incorrect material, producing the good without complying with the requirements; mistakes of technical operation in the exploitation process due to bad quality of repair or assembly; control mistakes, for instance, treating faulty product as a good one; storing mistakes relating to wrong stock keeping or transport.

Some reasons of incorrect acting of human being (exploiter) include: improper qualifications of a crew; performing wrong operating tasks; improper working conditions; insufficient or improper working tools; improper stimulation of a faultless work.

During the work of exploiter the probability of his faultless work is shown by the relation

$$R_0(t) = \exp\left[-\int_0^t \lambda_0(\tau)d\tau\right] \tag{4}$$

where:

$\lambda_0(t)$ - intensity of making mistakes by a person in time t.

The ability of a person to correct their mistakes is the probability that a mistake acceptable at performing a task will be corrected after time t at certain load and surrounding conditions corresponding to the realized task. The probability of correcting a mistake is presented by the relation

$$R_{kb}(t) = \exp\left[-\int_0^t \mu(\tau)d\tau\right] \tag{5}$$

where:

$\mu(t)$ – frequency of correcting a mistake.

Operating reliability of an operator-exploiter can be defined as a probability of performing a task comprising N actions on time and faultlessly and this might be described by the relation

$$R_0 = \left(\prod_{i=1}^{N} P_i\right) P\left(\sum_{i=1}^{N} T_i \leq T_0\right) \tag{6}$$

where:

P_i – probability of performing i-action properly,
T_i – realization time of i–action,
T_0 – time for performing a task.

In case a crew acts under stress, the variables T_i (i = 1,2 ....., N) are stochastically dependent, so the probabilities P_i are functions of stress, so the practical application of the relation (6) becomes very difficult. For determining operational reliability the methods of computer simulation are used.

3 Assessment of human activities

In human activities modelling for the needs of integrated risk assessment of the system it is necessary to take these factors into account which affect human behaviour the most. They are called Performance Shaping Factors – PSF [12] and they can be divided into external and internal. External ones include whole working environment, especially construction of devices and machines, written and oral instructions. Internal PSF are connected to the individual features of people, their skills, motivation and expectations. Psychological and physiological stresses are the factor of working environment, where the requirements imposed on the person by the system are not compatible with his skills and limitations.

On the other hand, in human behaviour modelling four levels of stress factors can be distinguished [12] very low (insufficient to maintain a proper level of vigilance during performing the tasks); optimal (which makes performing the task easier); moderately high (hinders, but in a moderate way, to perform the tasks optimally); very high.

First three levels relate to the task stress connected to low, optimal, or high workload. The highest level is represented by the stress resulting from hazards, it is caused by the emotional reactions to the situation in which a task is being performed.

Being aware of the consequences of a mistake may decrease or increase the probability of making one. Often, the consequences of human mistakes may be compensated by other elements of the system or other human activities. It is a recovery factor – RF that controls it in the system. A human reservation is a typical example of this factor.

The specificity of driver's work requires to spend some time on the chosen functions and human organism functioning.

A human organism realizes some activities, like for example ones connected to the brain activity by secretion of the glands, functions of the skeleton and muscle movements. As a result of a decision made by the nervous system the effect of movement may occur. Planning of a movement is a complex issue which includes: choosing a right muscle or their group and adjusting it or them to perform a certain movement or to ensure a proper position (tension); choosing the level of force; informing about the relation between the position of muscles and initial conditions of the movement; being precise in achieving a right track of movement; achieving a right speed of movement.

It is up to nerve centre, so the collection of neurons, whether the movement will occur, what the attention period (latent stimulation) for the movement there is and what force the effector will be stimulated with. During movement higher brain functions are realized: learning of movements, thinking which has a goal of movement as its criterion, movement memory whose criterion is based on the optimization of the movement; information circulation; comparison mechanism and its feedback; reactions tracking. During every work one can divide actions into movements: position movements, which include moving body parts from one position to another; repetitive – repeating of the same activity all of the time; continuous – taking place constantly in a given time; serial – many different, singular, quite independent movement units performed in certain order; static – performed by some, fixed muscle groups which do not contain movement elements (keeping them in a right position).

Reaction time is dependent, among others, on different factors: movement characteristics, signal characteristics (space and time features); type of a functioning factor and type of a sense used; relative intensity of the trigger evoking reaction both at the source of emission and on the side of a receiver (differentiating of a trigger with respect to the background); disturbance in the transfer channel; resistance power of an operated device; duration time of the trigger, meaning the way it accumulates and disappears; pause time in between the following signals; explicitness and amount of information carried by the signal; state of readiness of a person reacting; psychical and manual predisposition of the receiver; climate conditions; individual features of a receiver (his health state; eating habits, etc.).

4 Conclusion

In the reliability analysis of a human factor – the name which refers to the person in a technical system – three types of dominance behaviour can be distinguished:
- Proficiency in performing more or less aware or reflexive activities based on the remembered models of behaviour;
- Rules resulting from remembered or written principles compiled in case of occurring forecasted symptoms and exploitative situations;
- Knowledge including identification of the situations and states of the device on the basis of the received and processed information about the work parameters of devices and systems.

Estimating the reliability of a person as an element of the technical system one has to take a wide scope of factors into account:
- Probability of mistake which can occur at performing every operation included in the activity process;
- Possibility to forecast the most significant mistakes that can be made in the process of utilizing and operating of the devices;
- Frequency of breakdowns of devices and technical systems due to a human factor;
- Taking into consideration human mistakes which are reversible;
- Probability of the correct functioning of a device (system) in case of making a mistake.

DOI: 10.1051/matecconf/201713400036

In the analysis of technical system reliability the damages caused by the operator are usually neglected if they do not involve high costs.

The circadian rhythm is the most important in adaptation of the living organisms to the conditions of the external world. Social factor for people and alteration of lightening for animals are its basic synchronizers. Deil rhythm of body temperature [2] was described as a first biological rhythm. It has a sinusoidal character. It shows maximum values in the afternoon hours and minimum values during the night with relatively high individual deviations.

Adaptation of living organisms to the environment takes place with the participation of biophysical factors of cyclical character. These factors are called synchronizers or *time determinants*, since they have major influence of the course of biorhythms. Diel rhythm is synchronized not only with light but also with other environmental influences. Thanks to the latter property and function the synchronizers are *time donors*, and their role is not equal. Chronobiologists rejected the idea of an external character of a diel rhythm. According to the studies by [11, 13] eliminating the alteration of phases: day (light) – night (darkness) does not result in the interruption of counting time by the human body, yet it interfered with it slightly. The length of day and night in such conditions differs by 1 hour from a 24-hour astronomical day and night. In this way a term circadian rhythms was coined. The majority of these rhythms is of endogenous character. It means they are not a passive reflection of the external world conditions but are deeply coded in the particular cells of the organism. These cells use their own *biological clock* which adjusts itself to the surrounding conditions. Different factors affect internal rhythm of physiological activities and they adjust the *biological clock* to the current surrounding conditions. For a human being it is not only photoecological factors or time of eating that play the role of synchronizers of biological rhythms. It is mainly social, civilization, and psychological factors called social synchronizers which are crucial here.

Among biological rhythms one may distinguish short, circadian weekly, monthly, seasonal and many-year rhythms.

According to [12] the methods of study of *personality psychology* may be divided into basic ones and auxiliary ones. Basic methods established and developed during the development of the psychology itself, and their methodological assumptions and technique were worked out by the psychologists.

Basic studying methods of *personality psychology* include, among others:

- Psychological observation allowing for the objective assessment of occurring certain activities and determining the course of these activities. The activities occurring spontaneously in the examined person are the subject of these observations;
- Psychological experiment allowing for evoking the activities examined, observing and measuring them;
- Analysis of the effects of the action which occurred during learning process. Such an analysis allows for collecting findings about the course of activity and personal features shown by the examined person;
- Psychological interview allowing for collecting information about the environment, life history, state of health, education and family situation. When the interview is properly directed one can see the reactions about the interests, emotions, attitudes and personal features;
- Introspection (self-observation) in which a person examined is asked questions about their mental experiences which is then a base for describing examined mental phenomena;
- Survey;
- Questionnaires and personal inventories, meaning improved and more comprehensive forms of surveys;

- Tests allowing for measuring many different activities, mental functions and personal features; measuring the level of intelligence, skills, inclinations to mental disorders…

Putting all these studies together allows for the analysis of social contacts within a group and mechanisms underlying its functioning. Out of these three elements of the system person-vehicle-surrounding it is the vehicle which is the best known and best mathematically described. The influence of a road surrounding is described better and better. A clear disproportion is visible here. Yet, the activity of a diver-person is still not entirely analysed and much less modelled.

This paper is supported by the research project "From horse-drawn railway to intermodal transport" within Visegrad Fund.

References

1. J. Wicher, *Bezpieczeństwo samochodów i ruchu drogowego* (Warszawa: WKiŁ, Poland, 2012)
2. A. Reński, *Bezpieczeństwo czynne samochodu. Zawieszenia oraz układy hamulcowe i kierownicze* (Warszawa: Oficyna Wydawnicza Politechniki Warszawskiej. Poland, 2011)
3. L. Prochowski, et al., *Podstawy rekonstrukcji wypadków drogowych* (Warszawa: WKiŁ, Poland, 2014)
4. R.S. Jurecki, *Badania i modelowanie reakcji kierowców w sytuacjach zagrożenia wypadkowego*. Monografie Studia Rozprawy Nr M84 (Kielce: Wydawnictwo Politechniki Świętokrzyskiej, Poland, 2016)
5. T. Toledo, H.N. Koutsopoulos, M. Ben-Akiva, Transportation Research, Part C: Emerging Technologies **15**, 2, 96-112 (2007)
6. G.N. Bifulco, L. Pariota, M. Brackstione, M. Mcdonald, Transportation Research, Part C: Emerging Technologies **36**, 352-366 (2013)
7. R.G. Hoogendoorn, B. Arema, K.A. Brookhuis, Transportation Research, Part C: Emerging Technologies **36**, 581-603
8. S.H. Hamdar, H.S. Mahmassani, M. Treiber, Transportation Research, Part B: Methodological, **78**, 32-53 (2015)
9. L. Hempel, *Człowiek i maszyna. Model techniczny współdziałania* (Warszawa: WKiŁ, Poland, 1984)
10. J. Jaźwiński, J. Borgoń, *Niezawodność eksploatacyjna i bezpieczeństwo lotów* (Warszawa: WKiŁ, Poland, 1989)
11. E.M. Dougherty, J.R. Fragola, *Human Reliability Analysis* (New York: John Wiley & Sons, USA, 1988)
12. P.H. Lindsay, D.A. Norman, *Procesy przetwarzania informacji u człowieka. Wprowadzenie do psychologii* (Warszawa: PWN, Poland, 1984)
13. T. Dzierżykray–Rogalski, *Rytmy i antyrytmy biologiczne* (Warszawa: Wiedza Powszechna, Poland, 1986)

Modification of the Internal Transport in the Tyre Factory

Zbigniew Matuszak[1,*], *Damian* Sokołowski[2], *Marek* Jaśkiewicz[3], *Branislav* Šarkan[4], and *Józef* Stokłosa[5]

[1]Maritime University of Szczecin, Faculty of Marine Engineering, ul. Wały Chrobrego 1-2, 70-500 Szczecin, Poland
[2]Maritime University of Szczecin, Faculty of Economics and Transport Engineering, ul. H. Pobożnego 11, 70-507 Szczecin Poland
[3]Kielce University of Technology, Faculty of Mechatronics and Mechanical Engineering, al. Tysiąclecia Państwa Polskiego 7, 25-314 Kielce, Poland
[4]University of Žilina, Faculty of Operation and Economics of Transport and Communications, Department of Road and Urban Transport, Univerzitná 1, 010 26 Žilina, Slovakia
[5]University of Economics and Innovation in Lublin, ul. Projektowa 4, 20-209 Lublin, Poland

Abstract. The production plant of tyres and rubber goods has been characterized in the article. The production process structure in general as well as the production process of tyres has been presented. The analysis of the internal transport system used in the production process of large tyres (OTR - Off The Road). The modification of the present transport system used in the large tyres production has been proposed. Both systems: the present one and the modified one have been compared using SWOT method (Strengths, Weaknesses, Opportunities, Threats). Not only the advantages but also the critical remarks concerning the proposed system have been enumerated.

1 Introduction

Internal transport as an element of the production and storing system underlies production targets and tasks present in the factory. It requires specific rules and principles of functioning.

The organisation of the internal transport should be adjusted to the specificity of a given factory, and what is more it should ensure safe and reliable movement of loads at the minimum costs. Effective organisation of internal transport in a factory should guarantee transfer of certain amount of goods on shortest possible ways, at the maximum use and minimum wear of means of transport.

Information and decision flows are strongly connected to the technical systems realizing physical material flows in the field of transport [1-3].

Looking at the internal transport from the perspective of the system allows for cheaper and better management of transport structures. Using methods and techniques of system analysis we can solve the problems of the internal transport more rationally because they

* Corresponding author: z.matuszak@am.szczecin.pl

analyse the whole scope of transport activities from the single means of transport to the transport system when material flow is completed with information flow [4-6].

An efficient system of managing internal transport resources, human resources as well as material resources, is necessary for collision-free and efficient course of production processes.

The modification of the internal transport system of the factory was presented on the example of one of the leading tyre manufacturers in Poland.

The company is one of the key tyre and rubber goods manufacturers. They produce different tyre types, such as: large tyres (OTR – Off the Road), truck tyres (TBR), delivery car tyres (AGR), aircraft tyres, tyres for agricultural and industrial equipment. The company also manufactures rubber and rubber and metal semi-finished goods, like rubber compounds, capes, rubber fibres, mats, inner tubes. The company is also involved in the contract production of special materials for the Polish Armed Forces but the information about the production details cannot be disclosed.

On the company premises there is a Development Centre of the Tyre Industry established in 1957. It deals with the analysis of tyres and rubber goods as well as research works on the construction and manufacturing technology of rubber-fibre products carrying high loads and untypical tyre types. The Centre belongs to the company therefore all the products manufactured in the company are analysed and certified there without the need for using external centres for this purpose. The Centre also offers the possibility of assigning approval numbers to the products for other companies.

The whole company covers the area of 12 hectare, with its facilities occupying 66 850 m^2, including office spaces, storage infrastructure, production floors and research buildings.

2 Description of company's production

The factory manufactures the following product groups:

- tyres: large (OTR) for the industry (earthmoving equipment, road works machines and mining machinery); for trucks and delivery cars (TBR), for fork lift trucks and industrial trucks, for aircrafts and helicopters, for agricultural equipment;
- semi-finished goods: rubber compounds; capes, tyre cords; rubber fabrics;
- other products: protectors; inner tubes; shock absorbers; rubberised wheels; mats; rubber end stops;
- special products for the military.

The first group includes large OTR tyres (Off-The-Road), which are used in special equipment and vehicles outside the public roads. They are used in the industry for: wheel loaders, loading and output loaders, dump trucks, wheel tractor-scrapers, cranes, portal cranes, container gripping devices, diggers, and forest works devices. These are tyres over 0.28 m wide, and their height, depending on the type varies from 1 m to 2.5 m. The tyres can weigh from few hundred kilograms up to few tones.

3 Production infrastructure

The production infrastructure in a described factory is very complex. The production itself takes place on a few production lines and different stages of the process are strongly interconnected.

The premises of the factory occupy 66 850 m^2, which includes: production area - 23 400 m^2, raw materials warehouses - 2 500 m^2, finished goods area - 3 000 $m^{2,}$ additional warehouses - 6 600 m^2. On the company premises there are also car parks and multi-purpose yards.

Factory structure includes mixing plant and rolling mill, where semi-finished goods, including rubber are manufactured. In other buildings there is a production of finished goods, meaning: OTR tyres, aircraft tyres, special materials, TBR tyres and fork lift trucks/ agricultural tyres.

Production division results from the fact that different types of tyres require different production equipment. There are also separate buildings dedicated for the production for the request of Polish Armed Forces.

The company has got an extensive machine park, which includes, among others: injection moulding machines, hydraulic press, extruders, autoclaves and special equipment and machinery used for preparing the surface for rubber coating. A wide range of devices allows for realization of mass orders as well as special ones.

Every production of a final product begins from the delivery of raw materials or semi-finished goods required for production. They are then connected together (and packed) and given a final shape.

Rubber is a main semi-finished good in a factory; it consists mainly of india rubber which along with soot, sulphur, oils, boosters and other additives creates a paste which is blended in a mixer. Such a compound is seasoned and underlies quality control. Vulcanisation process is all about pouring a compound of right raw materials in high temperature (over 150°C) onto the rollers called calenders in the rolling mill. Then this rubber is rolled into special flaps which are then interlaid with a special paper. Such a rubber is a semi-finished good which is either used in further production or sold as such. So rubber production consists of two stages: weighing and composing a compound as well as vulcanisation (rolling mill).

Tyre production, on the other hand, includes four different stages: two initial ones look similar to rubber production stages, since a tyre consists mainly of rubber.

A second stage includes preparation of a cord made of nylon fibres (network of fibres inside a tyre) and construction of bead wire (element clamping a tyre against the rim) and other additional materials, like band or filler. A cord, also named as technical fabric is coated from both sides with a slim film of a proper rubber compound. Depending on the material used we distinguish different cords: nylon, viscose, polyester, glass, and steel. A cord is manufactured on the device called calender, and the process itself is called calendaring. Then a ready-made cord is cut into pieces. Both wide and narrow rubber pieces of a tyre undergo calendaring.

This is followed by tyre building, which is based on joining all the elements in order to achieve a raw tyre ready for vulcanisation. This is how a raw tyre, which then undergoes quality control, is made.

The last stage of tyre production is called vulcanisation. It is based on placing raw tyres in special moulds, in vulcanising presses where, under high pressure of membrane and high temperature, they are pressed against a mould. This is how a final shape and look of a tyre is achieved. A ready-made tyre undergoes another quality control test.

In the company they produce radial and diagonal tyres (the difference between those two lies in the arrangement of cord, tubed tyres (marked as TT) and tubeless tyres (marked as TL).

The other rubber and metal-rubber goods, although are produced in smaller amounts than tyres and do not belong to the main line of products, represent a significant part of company's sales. The production of shock absorbers, rubber end stops and membranes is based on the similar stages like tyre production. First stage includes supplying the semi-product, meaning rubber, then the production made by specialised machines: injection moulding machines, extruders, hydraulic presses and autoclaves starts. Rubber is an elastic material, prone to high temperatures what enables us to shape it anyhow.

The production in the factory takes place in three-shifts, the company currently produces 250 thousand tyres a year but still the production capacity is much higher, which is over 544 thousand tyres a year.

4 Internal transport characteristics in the company

Internal transport infrastructure is quite extensive since flexible connections between the following production stages of rubber products are required.

In the internal transport they use groups of means of transport used in production process as well as during warehousing and storage. Some of the means of transport are used in both phases.

Means of close transport at the production include: band conveyors, bucket conveyors, roller conveyors, tractors, trains, different trucks used during handling of semi-finished goods at the production stage.

Means of internal transport used during warehousing and storage on the other hand comprise of hand pallet trucks, fork lift trucks, electrical platform trucks, and pallet transport trucks. Due to the untypical character of production many devices of close transport have a non-standard design. This design of trucks was done on the request of the company itself.

A finished good – in this case OTR tyres – is transported by the fork lift truck, with the use of special methods, or by transport platform trucks.

The first method requires using a hook and transport belts; second one assumes putting a tyre on the forks of a fork lift truck; the third one (the most common) is based on using a special rotary gripper for tyre transport; and the last method (developed by the company) consists in using a gripper of a special profile which is placed inside a tyre.

5 Analysis of the internal system

Material flow in a production plant starts from the place where the material is delivered by the external transport. It goes through the warehousing station, already transported by the means of close transport until it is unloaded. Then the material undergoes machining and assembly steps until it is stored in the warehouse again or it is immediately sent as a finished-good.

In the company a mixed material flow is used. In the first stage of production it is a continuous flow, based on band and bucket conveyors, and in the further stages the means of closed transport are used.

In the analysis of transport process examination we calculate what follows [1]:

- labour consumption of the material and information flow process, taking means of transport, equipment and human labour into account;
- whatever is needed to realize the process of means of transport and people, taking schedule of cycles and processes into consideration.

The material undergoes transport movements, so called transport cycles which are performed by the means of close transport during production process. During rubber production in the factory there are 7 up to 8 transport cycles, assuming that the material has been stored in the warehouse. A rubber is a semi-finished good, used in further production therefore the transport cycle of the final product is longer, which is over 8 cycles. Counting it together with rubber production, the production of a tyre requires minimum 16 transport cycles.

Transport cycle is based on the serial production structure, where production machines are placed according to the technological process, and the whole production process is done on the beat. Unfortunately, this type of production has significant drawbacks: the longest operation imposes the time of its duration on the whole cycle, and the breakdown of one of the devices inhibits work of the whole system. Therefore, the number of devices and machines in the company is greater than the maximum demand for them.

Table 1 presents the data for the analysis of the means of internal transport for OTR tyres production.

Table 1. Using means of transport during OTR tyres production. Source: authors

No.	Means of transport	Usage [%]		Sum [%]
		Production	Warehousing	
1.	Band conveyors	30%	-	30%
2.	Bucket conveyors	30%	-	30%
3.	Platform truck with a drawbar for treaded tyre transport + hand lift truck	20%	10%	30%
4.	Special hand transport truck for cope and other material transport	15%	-	20%
5.	Special transport truck of I TYPE or II TYPE for bead wire transport	10%	-	15%
6.	Truck/ special platform for packed tyres transport	20%	-	20%
7.	Overhead one-track conveyor for transport of packed tyres to the press	10%	-	10%
8.	Tractor + platform (bellows)	20%	-	20%
9.	Platform transport truck for mould transport	5%	20%	25%
10.	Electrical platform truck or fork lift truck	20%	20%	40%

The table has been compiled on the basis of data about the means of transport used at the production of OTR tyres made available to the authors. The usage of the devices and machines expressed as a percentage means the ratio of the equipment used to the maximum possibilities of usage of this equipment. Usage of vulcanising presses remains on the level of 40%, and usage of human resources on the level of 70%.

Band and bucket conveyors are in fact used in 60% since the rubber they produce is used also on the other production lines, but it is also sold as a semi-finished good.

Internal transport devices are used whenever they are needed in a given time. Currently OTR tyres production constitutes 20-30% of the whole production in the company and it used only 30% of means of transport and 40% of machine capacity (so whenever a greater demand appears, it is possible to double the production).

The fragmented production structure and employing the internal transport employees hinders the cost analysis of the close transport and optimum usage of the costs. According to the data made available by the company the authors know that the factory employs 40 people on the positions connected to the internal transport. When dividing these people into departments (production lines), there are only five people for one department and five people responsible for warehousing of the finished goods.

In order to make sure that internal transport is handled properly, by the sufficient number of people, the company shifts some employees from the positions where there is temporarily no demand for work (for instance, when the production demand is low) to different departments.

Figure 1 presents the current structure of the OTR tyres production system. The graph shows that three press stations are provided with the packed tyres via the overhead conveyor and the other three vulcanisation presses are dependant from the hand or electrical trucks.

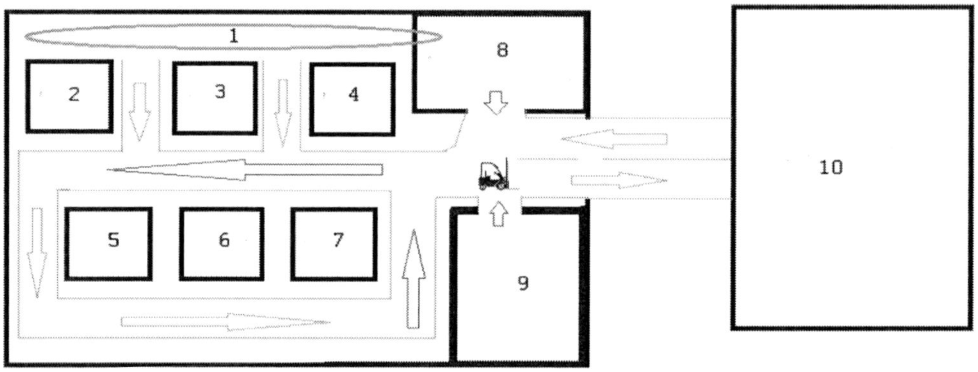

Fig. 1. OTR tyres production system structure: 1 - overhead conveyor; 2-7 – presses; 8- packing station; 9 - bead wire warehouse; 10 - treaded rubber and tyres warehouse. Source: authors

Analysing the above graph one can notice that there might be congested traffic on the way between the vulcanisation presses, since transport road is narrow and it is used by the trucks transporting finished goods from six presses and trucks transporting goods to the presses. In case of a breakdown of one of the trucks, the driver informs the department manager about the fact and the manager looks for a spare means of transport in other departments and reports the breakdown to the maintenance workshop. The situation may get worse of the truck breakdown occurs in more than one department at the same time.
When the production in some departments is not fluent, all means of transport are not equally busy. Using only 50% of the available means of transport results in higher maintenance and exploitation costs.

Taking the data about the usage of internal means of transport and the remarks about the work into consideration, a process of designing a new system of internal transport at OTR tyres production has been conducted. The project is based on a slight modernisation of transport processes, without a significant interference with the production process. It relies on the circular system, which is more efficient and fluent than the horizontal roll-on/ roll-off system. Production process has been changed with regards to bead wire warehousing which was shifted to the centre of production floor and constitutes an integral part of tyres packing stations, which were also shifted to the centre of the floor. Vulcanization presses remained where they were, only the presses 1, 2 and 3 were moved to the inside of the axis. The overhead conveyor was shifted to the line situated in the centre of the production floor and it now leads to the tyre packing station through the vulcanisation presses. Transport roads are now circular, and the roads with congested traffic have an extra two-way lane. Such a solution allows for passing by another truck which is currently being loaded or is broken.

The project also suggests which means of transport should be used for which activities, so the transport of bellows takes place via the overhead conveyor, and finished goods are transported by the trucks only. Thanks to this solution the transport system is kept in order and transport layout is not collisional.

New project structure of internal transport of OTR tyres production system is shown in Fig. 2.

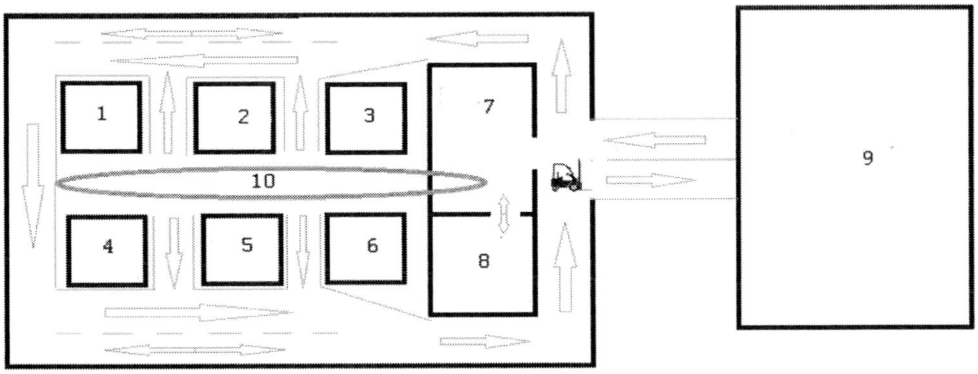

Fig. 2. New structure of OTR tyres production system: 1-6 – presses; 7 - packing station; 8 - bead wire warehouse; 9 - treaded rubber and tyres warehouse; 10 - overhead conveyor. Source: authors

A preparatory analysis of modernisation of the close transport process at the OTR tyres production may be conducted with SWOT analysis (Strengths, Weaknesses, Opportunities, Threats) – Table 2, which may point out both strengths and weaknesses of the proposed solution.

Table 2. SWOT analysis for the project of new system of internal transport in OTR tyres production. Source: authors

	Positive factors	**Negative factors**
	Strengths	Weaknesses
Internal factors	- shortening of production time, - using one means of transport for a given process, - maintaining order, - eliminating collisions from the transport ways, - decreasing a risk of congested traffic in the transport process,	- longer transport ways for transporting finished goods, - increasing exploitation costs, - one direction of movement for all means of transport, - dependence on one means of transport,
	Opportunities	Threats
External factors	- production increase, - decrease of production costs, - easy production process, - innovative system, - increasing production flexibility, - increasing competitiveness,	- the system may not prove right, - delays in the realization of orders, - the costs of system implementation may be too high,

6 Conclusion

Internal transport is an indispensable element of production and warehousing processes. Modernisation of transport systems is one of the elements, the improvement of which may be a cheap and easy way to improve a production process and to decrease its costs [7-10]. The modernisation of current internal transport system at OTR tyres production requires certain funds. Also the number of cost studies and analyses should be made, since they can answer the question whether the implementation of the project will pay off on the first place. But from the perspective of time analysis, the modernisation of the system is highly advisable, as it shortens the production time of the tyre (from one warehouse to another) down to 40 minutes in a worst case, the current time (in the optimistic variant) being 50 minutes. The big bulk of material does not allow us to include the detailed calculations).

The transport project described has many advantages, such as: no congested traffic on the roads; the roads do not cross with each other, what eliminates the risk of collision; bead wire warehouse is situated close to the packing station; it makes transport more effective; shortens transport time, and makes the whole system easier. Though, there are some weaknesses of the project: it makes a transport way longer, it forces one-way traffic of the trucks and, in special cases, and transport of a finished tyre from the press can take longer.

This paper is supported by the research project "From horse-drawn railway to intermodal transport" within Visegrad Fund.

References

1. Z. Korzeń, *Logistyczne systemy transportu bliskiego i magazynowania. Infrastruktura, technika, informacja* (Poznań: Instytut Logistyki i Magazynowania w Poznaniu, Poland, 1998)
2. M. Jacyna, K. Lewczuk, M. Kłodawski, Journal of KONES, Powertrain and Transport, BK **22**, 3, 49-58 (2015)
3. T. Govidaraj, E. Blanco, D. Bodner, L. Goetschalckx, L. McGinnis, G.P. Sharp, *Proceeding of the 2000 IEEE International Conference on Systems, Man and Cybernetics*, 1099-1104, 8-11 (Nashville, Tennessee, USA, 2000)
4. J. Fijałkowski, *Transport Wewnętrzny w Systemach Logistycznych - Wybrane Zagadnienia* (Warszawa: Oficyna Wydawnicza Politechniki, Poland, 2000)
5. W. Wawrzyński, M. Jacyna, Journal of KONES, Powertrain and Transport **22**, 4, 319-327 (2015)
6. J.L. Chen, M.C. Chen, Ch.W. Chen, Y.Ch. Chang, Computer Communications **30**, 2070-2086 (2007)
7. S. Krzyżaniak, et al., *Organizacja i monitorowanie procesów magazynowych* (Poznań: Instytut Magazynowania i Logistyki, Poland, 2014)
8. R. Raczyk, *Środki transportu bliskiego i magazynowania* (Poznań: Wydawnictwo Politechniki Poznańskiej, Poland, 2013)
9. Ł. Wojciechowski, A. Wojciechowski, T. Kosmatka, *Infrastruktura magazynowa i transportowa* (Poznań: Wyższa Szkoła Logistyki, Poland, 2009)
10. D. Pyza, Archives of Transport **21**, 3-4, 125-148

Robotization of Logistics Processes

Nikoleta Mikušová[1,*], *Zdeněk* Čujan[2], and *Eva* Tomková[1]

[1]Technical University of Košice, BERG Faculty, Logistics Department, Park Komenského 14, 040 01 Košice, Slovakia
[2]College of Logistics, Přerov, Palackého 1381/25, 750 02 Přerov, Czech Republic

Abstract. Recently, it is possible to register a relatively large increase of robotized workplaces also in the fields which not suppose the using of robots. Their implementation was limited due to complexity of work activities, including for example handling of various goods in large quantities in various combinations and often also in the limited space.

1 Introduction

Industrial robots were until recently immobile and non-intelligent machines, respectively their task was repeating of simple repetitive tasks with the high level of accuracy. These abilities are sufficient for simple production processes but they are not sufficient for most logistics activities. In logistics, the requirements for industrial robots are much higher. Robots in logistics have to have a distinctive ability in a variety of goods, must be able to distinguish various parts in an infinite number of combinations, must be able to perceive their surroundings, move and interact with the environment and work with people. Increasing flexibility and lower cost of robotized workplaces create prerequisites for their use in logistics activities such as order picking, packing and shipping, as well as deliveries. In the field of logistics 80% is still manually operated. Problems of unfavourable demographic developments and the associated declining number of employees can be replaced by robotic workplaces in some logistics activities.

2 Relationship among human, robotization and mechanization

For humans, it is significant that compared to the current simple mechanization and using control technology (IT technologies), it excels with its natural intelligence and functional mobility – from kinematics we express the mobility by the number of degrees of freedom. The human hand has among shoulder joint and at the end of one of the fingers 110 degrees of freedom. If we take into account all fingers, we will get 270 freedoms. It is obvious (Fig. 1), that the mutual comparison of the level of human intelligence, but especially its physical (power, speed) and functional (mobility) capabilities with the capabilities of industrial robots and manipulators, is significantly higher in these automation devices. A robot is an automated or computer-controlled integrated system, capable of autonomous, target-oriented interaction with the natural environment according to human instructions [1]. This

* Corresponding author: nikoleta.mikusova@tuke.sk

interaction is based on the perception and recognition of this environment and manipulation with objects, or in moving in this environment [2]. The basis for the construction of robots and manipulators are mechanisms, i.e. mechanical devices that serve to transform of movement and transfer of forces and provide displacement of the objects (tools, etc.) in a way, that results from the required function of the system [3]. This function is for example technology of the logistics process. Mechanisms of robots and manipulators are formed by the set of moveably connected members, one moves and one creates a frame. These mechanisms are derived from opened spatial kinematic chains which contain binary members. Members are bound together by kinematic couples.

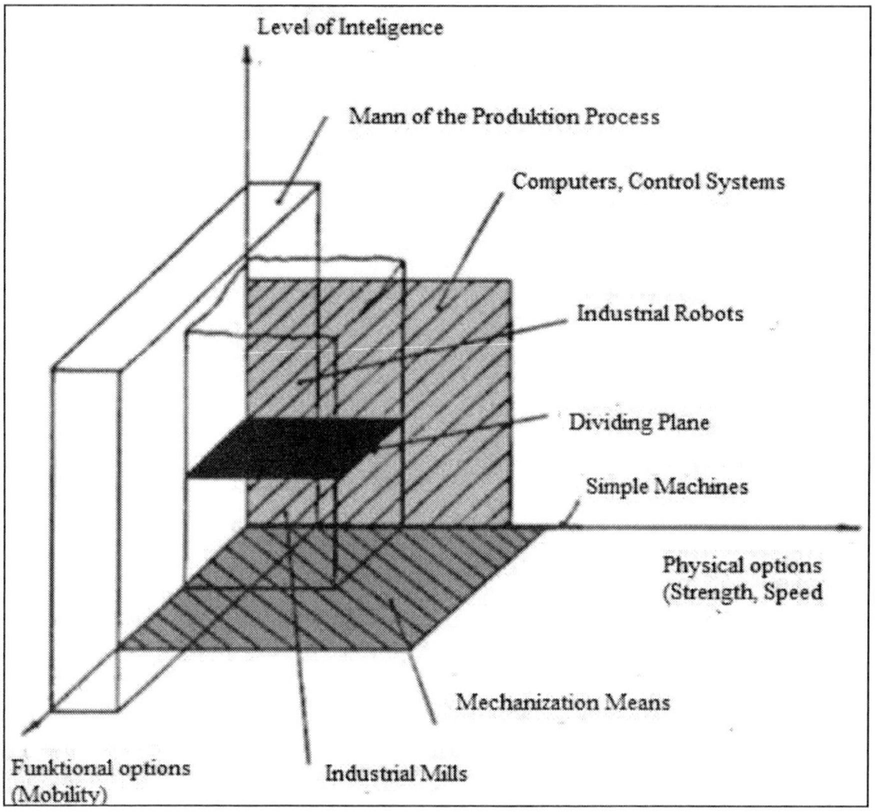

Fig. 1. Interaction among human and machine. Source: [2]

2 Basic kinematic construction of robots (positioning)

The mechanical construction of the industrial robot is created by movement system that is usually divided into two separate parts:

- the first part is the main movement system which provides positioning of the centre of gravity of the object in the space (positioning system) and
- the second part is presented by accessory movement system (orientation system) which determines displacement – orientation of the object in the space [4].

The orientation system often extends also the function of the main positioning system and it is structurally concentrated in a subgroup that is referred as the wrist. At the output of the whole movement system is located the working head adapted to the way of handler or robots for realization of handling, or other technological operations.

The positioning device with 10, 20 or 30 degrees of freedom is joined to the robotic positioning device. Technological head is joined to the orienteering device. The total number of degrees of freedom is different, from 40, 50, 60 and more. An angular kinematic structure with a torus (torus is created by rotation of the circle around the axis) by the spatial workplace is currently the most widespread in the construction of universal robots which have within the frame of the wrist two or three rotational movements.

3 Robotization in intra-logistics

Robots find application by loading and unloading of semi-trailers, or containers. At present, robots are able to analyse different sizes of shipments and subsequently optimize operation of loading or unloading of the goods. Development of new generation of intelligent robots also accelerates progress in the areas of scanning technology, faster computer technologies, analysis of data, accumulators, clouds utilization and thus greater flexibility. The use of new technologies extends the ability of apply and use of robots in their quantity and their individual species. Dozens of robotics specialists in the world develop solutions for more complex warehouse works. Robots facilitating this work have various functions and forms. From chassis automatically transport small shelves to warehouse staff, mobile carts that follow the warehouse employee, and also mobile robotic arms that can pick simple items from racks. Automated warehouses are usually designed for store of pallets and boxes or crates.

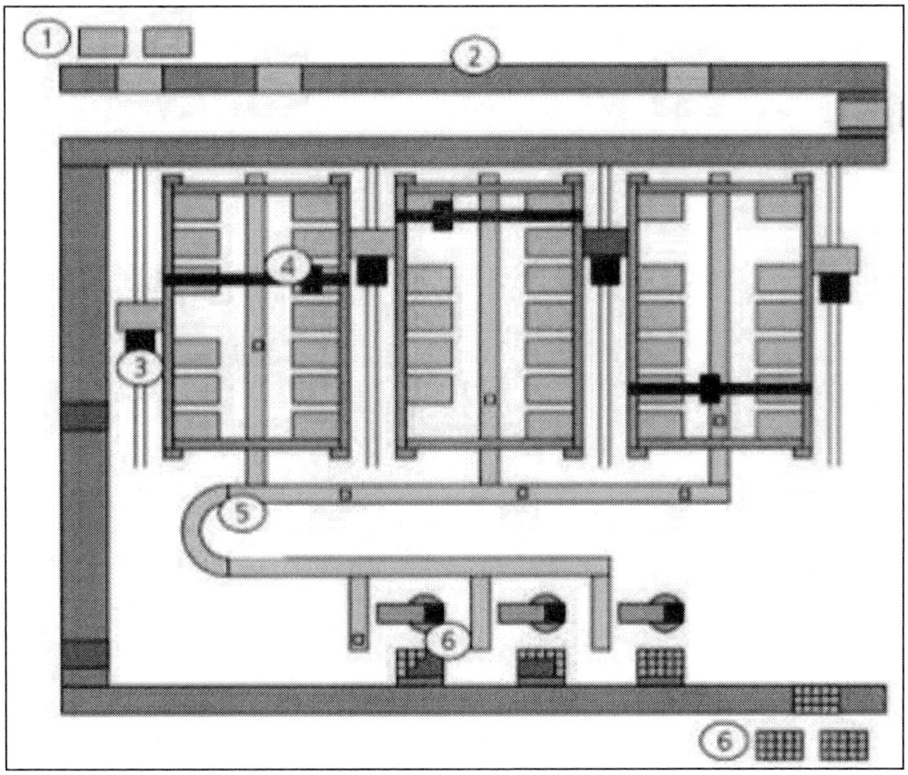

Fig. 2. Layout of automated storehouse. Source: [5]

Warehouses working in automatic mode are designed in a way that minimizes the need for human factor. Their conception, data management, overview and order in warehouse and storage processes are mainly designed to eliminate errors by storage that are often caused by a human factor. By realization of automatization and robotization of logistics activities in a warehouse (storage, packaging), all mechanically controlled processes, control operations are under the management of the control system . Therefore, the time for storage or removing of pallets is shortened. Shortening of storage operations is by realization of automatization and robotization is up to a third against classically controlled warehouse.

The pallets with goods (Fig. 2) are at the entry (1) placed on the conveyor belt (2), from which they are stored by automatic collator (3). The height collator can operate in fully automatic mode, in long-distance or local semi-automatic mode. By manipulation with pallets the portal handler (4) is used, and this stores the pallets to the belt conveyor (5). Conveyor transfers pallets to the robotized workplace (6), with the process of depalletization. Goods are transfer to completion according to the the customer's order [5].

The pallets with goods (see Fig. 3) are at the entry (1) located on the conveyor belt (2), from which the goods on the pallets are taken by the collator (3) and stored to the shelf system. Depalletization is performed on robotic workplaces (5), where the pallets with goods are moved by a belt conveyor (4). Classification of goods is realized manually at the workplace (7) after removal from the conveyor (6).

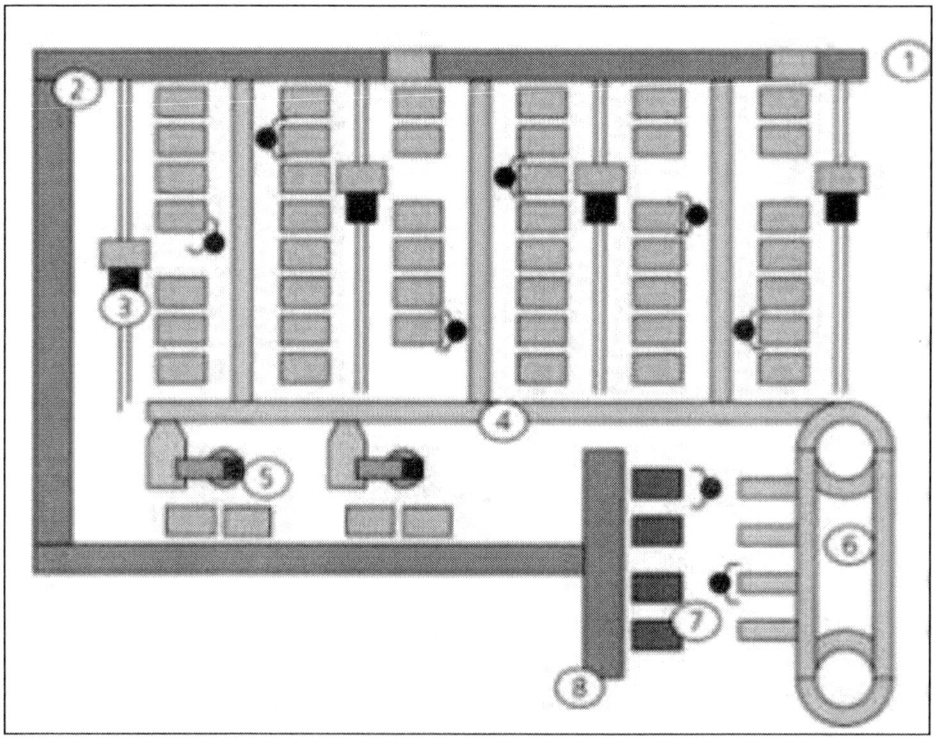

Fig. 3. Scheme of automatic and manual handling with goods. Source: [5]

For enterprises which have only globally determined aims of logistics development and provide bigger flows, it is strategically advantageous to orientate investments to automated technologies [6]. In practice, it usually means gradual building of a larger automated system [7] (Fig. 4).

Fig. 4. Robotized workplace in logistics centre. Source: [8]

4 Robotization of production and assembly line

Development of robotization of production lines (Fig. 5) is not dependent only on the possibilities of the relevant technical means, but it is also influenced by the development of its own technologies [9]. Development and realization of technologies can be expected in flexible systems created by multifunctional production machines and robots.

Fully automatic operation of production machines, equipment and system cannot be imagined without automatic manipulation, which in the most cases directly determined the automated operation of technological means.

Fig. 5. Robotized welding workplace of car body. Source: [10]

The structure of the elementary technological workplace is formed by production mean for the realization of one technology, input and output hopper of objects for processing, manipulator of object, hopper of tools and handler of tools. Various variants of the workplace layout are based on a different level of integration of constructional solution of these partial functional parts. By connection of these elements of elemental technological workplace by automatic transport or handling means, it is created an automatic production system (Fig. 6).

Fig. 6. Robotized welding line of car bodies. Source: [10]

Trend in the development of automated production and assembly workplace is affected by the necessary minimization of transport and handling operations. The minimization of these activities influences the direction of development of new production machines and equipment, including transport and handling equipment (Fig. 7).

Fig. 7. Robotized assembling of car bodies parts. Source: [10]

5 Robotization of transport, transfer and handling means

Material handling and its transport in production enterprises is one of the moments effecting the total costs for production of the product [11]. Reducing production costs in the areas that can be influenced by the use of new technologies and organization of work is one of the ways how to reach the chosen goal.

Reducing of the rate of live work in routine, and also difficult activities, that are periodically repeated is the easiest by the use of robot handler, unmanned vehicles and robots, too. This fact is presented in production enterprises by using of transport routes by material processing, from material receiving to store, transport of semi-finished products among operators, then storing of the final product in the warehouse and its allocation by sale.

AGV - Automated Guided Vehicles (Fig. 8) – are used for material handling and they serve in a variety of applications with autonomous operation of logistics systems without human intervention [12]. The advantage of this solution is above all that there is no need to set a road/way because the vehicle can ride directly on the floor, if it fulfils certain criteria, such as flatness, continuity, small inclination and other parameters, determined by producer as conditions for trouble-free operation [13]. This means that this system is flexible and its rebuilding, or extension is very easy [14]. The most common concept of this vehicles contains one guided, and at the same time driven wheel, or axle with two wheels and the other wheels are supportive [15].

Fig. 8. Intelligent hoisting set CEIT. Source: [16]

Vehicles can be guided by several ways which always have their pro-and-con. Selection of suitable navigation is good to consult with a producer, with experiences about application for different environments and requirements [17, 18].

6 Conclusion

Lack of labour power is currently the biggest motivation for implementation of automatized production lines and robotic workplaces. At present, lack of labour power is a very strong motivation for greater implementation and use of robots and by this way to achieve the needed increasing of productivity and meet customer requirements. By effective use of robots and robotized workplace, robots are four to six times more efficient than human service. But the latest robot solutions are fully automated and often they are combined with human service (for example material completion), it brings automatized systems for reducing of handling costs by 20 to 40 %. These values depend on many factors, such as complexity with order processing and technical preparation of production, order size, production and storage capacity, etc.

This work is a part of these projects VEGA 1/0063/16, KEGA 018TUKE-4/2016.

This paper is supported by the research project "From horse-drawn railway to intermodal transport" within Visegrad Fund.

References

1. H. Neradilova, G. Fedorko, *12th international scientific conference of young scientists on sustainable, modern and safe transport*, 638-643 (Procedia Engineering **192**, Elsevier science BV, Amsterdam, 2017)
2. J. Skařupa, *Průmyslové roboty a manipulátory* (TU Ostrava, Czech Republic, 2012)
3. V. Molnar, K. Pacutova, *ICTTE 2016 – Proceedings of the third international conference on traffic and transport engineering*, 570-575 (Belgrade, Serbia, 2016)
4. R. Konflíček, *Roboty a pružné výrobní systémy* (VUT Brno, Czech Republic, 2012
5. D. Fritsch, K. Wöltze, *Logistik. wt Werkstatttechnik on line*, **96**, 624 (2006)
6. J. Fabianova, P. Kacmary, V. Molnar, P. Michalik, Open Engineering **6**, 270-279 (2016)
7. L. Bartuska, O. Stopka, Chovancova, J. Lizbetin, *Transport Means – 19th International Conference on Transport Means*, 107-111 (Kaunas University of Technology, 2016)
8. *Průmyslové roboty a manipulátory*, Available online: http://www.kod.tul.cz/predmety/AOV/Vyuka/prednaska_robot.pdf (2017)
9. J. Strohmandl, UPB Sci. Bull. Ser. D Mech. Eng. **76**, 223-230 (2014)
10. *Robot*, Available online: http://www.mmspektrum.com/multimedia/image/87/8701.jpg (2017)
11. I. Kubasakova, O. Stopka, R. Kampf, Communications: scientific letters of the University of Žilina **2**, 9-13 (2014)
12. H. Neradilova, G. Fedorko, Open Engineering **6,** 700-710 (2016)
13. A. Kuptcova, P. Průša, G. Fedorko, V. Molnár, Transport Problems **11**, 21-31 (2016)
14. O. Kodym, *ICCC –13th International Carpathian Control Conference*, 296-299 (High Tatras, Podbanské, Grand Hotel Permon, Slovak Republic, 2012)
15. R. Danel, V. Kohut, O. Kodym, M. Repka, V. Kebo, M. Kijonka, *SGEM 2013 – 13th International Multidisciplinary Scientific Conference*, 151-158 (Albena, Bulgaria, 2013)
16. Z. Čujan, *3. vědecká konference pro doktorandy* (College of Logistics, Přerov, Czech Republic, 2017)
17. Z. Čujan, *Logistika výrobních technologií* (College of Logistics, Přerov, Czech Republic, 2013)
18. R. Kampf, L. Lizbetinova, K. Tislerova, Open Engineering **7**, 26-30 (2017)

Identifying the Barriers for Development of Inland Waterway Transport: a Case Study

Dejan Mircetic[1,*], *Svetlana* Nikolicic[1] , *Sanja* Bojic[1], and *Marinko* Maslaric[1]

[1]University of Novi Sad, Faculty of Technical Sciences, Trg D. Obradovica 6, 21000 Novi Sad, Serbia

Abstract. Inland waterway transport represent environmentally friendly transport mode accompanied with other potential benefits, such as increased transport safety and increased cost savings. In order to encourage further development of inland waterway transport in this paper focus is set on the identification of the barriers that limit and prevent development of inland waterway transport. More specifically, emphasis is placed on the barriers that are preventing the cargo shifting from land to waterway transport system. Research conducted in case of Republic of Serbia investigates several barriers which are classified in four main categories: goods, logistics, infrastructure, framework conditions. Framework conditions barriers are further subdivided and classified in four subcategories: political/legal, environment, economy, technology. Results indicate that the most significant negative influence on the further development of the inland waterway transport in Serbia comes from the political/legal barriers. This result implies that the government need to pay more attention to development of inland waterway transport, as well as to make more effort to enhance the cooperation with stakeholders.

1 Introduction

This paper focuses on the detection of barriers which limit development of inland waterway transport. The main objective of the study is to identify key barriers for shifting cargo from land to waterway transport system. Inland waterway transport (IWT) represents economical and environmentally friendly mean of transportation of goods. In order to support increasing orientation towards intermodal transportation, European Union is encouraging the development of IWT through variety of studies and project calls. Accordingly, presented research is part of one of such project which aims at supporting the Danube-Black Sea region to become an attractive gateway region for maritime and IWT between Central Europe and Black Sea, the Caspian region and the Far East by facilitating increased intensity and quality of cooperation between ports, regions and other key actors. The joint effort shall improve accessibility of both the ports and the region and strengthen interoperability between maritime and inland waterways as well as with their hinterland. Together with raising the awareness of the possibilities of intermodal transport, this will

* Corresponding author: dejanmircetic@gmail.com

lead to shifting existing and attracting new cargo flows to environmentally friendly transport systems [1].

At present, aging infrastructure and inefficient, non-transparent intermodal services limit the potential of the waterway transport system. This research is built upon the belief that the cooperation of public authorities, ports and their related associations is the key success factor in order to raise quality, reliability and efficiency of the IWT system. Cooperation in itself will not yet lead to the envisaged results. Therefore in this paper a questionnaire is conducted in order to identify the main barriers that limit the development of IWT in the case study country. Case study is conducted in the Republic of Serbia, hence conclusion and results of this research are mostly limited to this country. However, results of this paper could be used for purpose of comparison with situation in some other European (landlocked) countries. There are many barriers which could be classified into different groups. For example, in research [2] authors identified 182 different administrative and regulatory barriers which obstruct the proper functioning of the European IWT industry. In our research, barriers are classified in four main categories, and framework conditions barriers are classified in four subcategories, as shown in Figure 1.

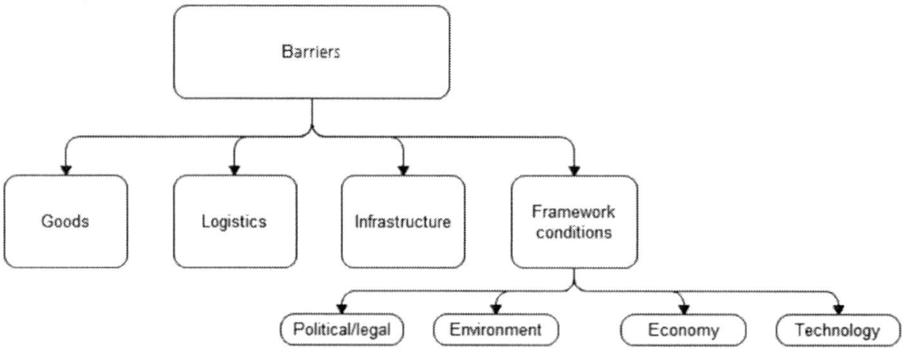

Fig. 1. Barriers for shifting the cargo from land to waterway transport. Source: authors

Beside dispersion of the barriers, in this research there is dispersion of the stakeholders which evaluate highlighted barriers. Stakeholders come from four different areas: public authority and sectoral agencies, infrastructure service providers and logistics service providers, industrial and trading companies and R&D/University. Dispersion of the stakeholders was necessary in order to have as much as possible comprehensive judgment of influence of different barriers from stakeholders which are practically daily encountered with them.

The remainder of this paper is organized as follows: next section gives short theoretical background and introduces the situation in the case study country regarding the IWT. In Section 3, methodology steps for questionnaire are provided. Section 4 deals with the results of the questionnaire and provides the insight about most influential barriers. Section 5 provides final remarks about influence of different barriers, and provide the concluding remarks.

2 Theory background

2.1 Inland waterway transport

Inland waterway transport could be the backbone of the future European intermodal transport chains, due to the fact that it can ship heavy as well as a large amount of commodities in combination with price advantages. Besides, IWT have still free shipment capacities. In Europe around 14.000 km of approximately 29.000 km of IWT are used for freight carrier. Also, Europe has around 1.500 inland ports and more than half of those are located in the main inland waterway TEN-T corridors. IWT represents the only means of land transport which does not suffer congestion problems like that of rail or road within Europe. In general, IWT are underused, but inland navigation is not considered as a truly competitive alternative to other means of land transport [3]. Estimates suggest that IWT would carry up to 425 million tons per year, including the accession countries, in the European IWT network, if the necessary action towards an integration of IWT into managed intermodal logistics chains were undertaken [4].

IWT offers a very energy efficient, reliable and safe mode of transport. It has a growth potential not yet fully explored and according to the European Commission's Transport White Paper, inland waterway by linking the European seas will have an increasing role in the freight sector [5]. However, there has been a problem of overcoming different kind of barriers for further development of IWT sector as well as problem of underinvestment in inland waterway infrastructure and suprastructure. Hence, in order to develop and implement an advanced European concept to manage intermodal transport chains with IWT as a core transport mode, we need to indentify and overcome barriers of all kinds.

2.2 IWT in case study country

IWT is especially important for landlocked countries, and which in the same time have navigable inland waterways. One of those countries is the Republic of Serbia, which has several navigable rivers, among which Danube is the most important one. Through the territory of the Republic of Serbia, there is a European Corridor-The Rhine-Danube Corridor, which connecting the central region around Strasbourg and Frankfurt via Southern Germany to Vienna, Bratislava, Budapest and the finally the Black Sea. Part of the Danube flowing through Serbia plays an important role in the functioning of this corridor [6]. More detailed characteristics of inland waterways in Serbia are provided in Table 1.

Table 1. The length of inland waterways in the Republic of Serbia, 2015. Source: [7]

	Length of waterways (km) per ship capacity				
	150t	400t	650t	1500t	3000t +
Danube	588	588	588	588	588
Sava	211	211	211	211	-
Tisa	164	164	164	164	-
Tamish	41	3	3	3	-
Bega	67	67	35	9	-
Danube-Tisa-Danube Canal	522	505	449	13	-

For the functioning of the IWT, except navigable waterway, important infrastructural complexes are inland ports as well. In the Republic of Serbia there are 8 inland ports on the Danube, 2 on the river Sava and one on the river Tisa. In the developed regions, inland ports have developed and transformed from the places of integration of river and road transportation modes, where traditionally the processes of transhipment and storage of goods were performed, into centres that combine all important logistics activities and transport, i.e. into logistics centres. Unfortunately, this does not apply to the inland ports in Serbia [6].

DOI: 10.1051/matecconf/201713400039

3 Methodology

In order to strengthen the cooperation and the commitment to work together towards the identification of the barriers, the methodology foresees the incorporation of all relevant stakeholders in the Republic of Serbia. Methodology of data/information collection was conducted through questionnaire which was carried during the regional workshop in June 2017. Questionnaire included feedback-loops and direct input to relevant topics from stakeholders. A questionnaire consisted from tree groups of question:

- general information,
- barriers for IWT use,
- opportunities for IWT use.

The first set of questions contains general information about the respondent, such as: name of organisation, type, key competence, category of stakeholders, etc. The second group of questions contains a list of barriers for development of IWT. Questions are organized according to the hierarchy from Figure 1. Minimum rating for each question was 0, while maximum was 3 (3=most significant factor for not using the inland waterway). In the third group of questions, it is required from stakeholders to objectively assess the possibilities for massive use of inland waterways. A total of 21 questionnaires was collected. Distribution of collected questionnaires by the type of stakeholders (sector) is shown in the Table 2.

Table 2. Participation of respondents by sector. Source: authors

	Sector	Number of respondents	%
1	Public authority and sectoral agencies	4	19,05
2	Infrastructure and logistics service providers	9	42,85
3	Industrial and trading companies	4	19,05
4	R&D / Universities	4	19,05
	Total	21	100,00

4 Results of analysis

Comparative analysis of the results obtained by each sector and each group of questions is shown in the Figure 2.

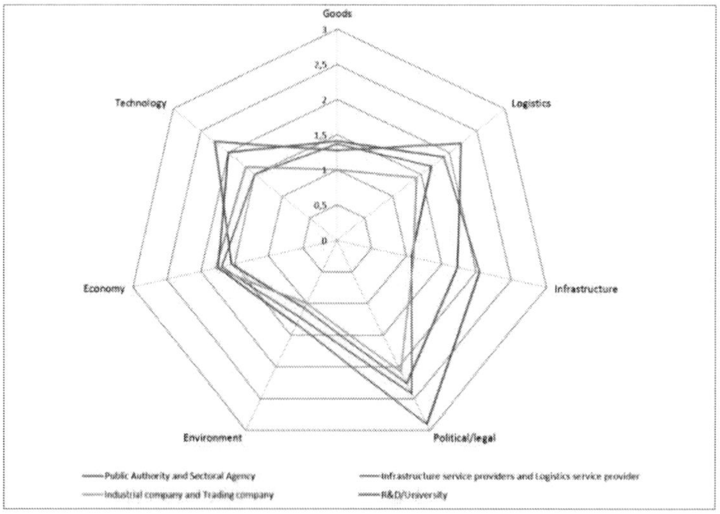

Fig. 2. The average rating for each group of questions. Source: authors

In general the perception of the most stakeholders was that the *political/legal* subgroups of barriers are the most critical. However, the overall picture is not quite clear regards the type of stakeholders. A survey has shown that there is almost an even split between on the one hand the group of stakeholders having no problems and on the other hand the group of stakeholders having problems with some kind of barriers. The main difference is between group of "*Public authority and sectoral agencies*" and other group of stakeholders. The basic conclusions which could be made from the figure above are:

• All participants think that *political/legal* is the most important group of barrier for not using the inland waterway; then follow *logistics* and *technology* groups of barriers;

• Stakeholders from public authority and sectoral agencies have evaluated the *technology* as the most important group of barriers and as the least significant the *environment*;

• Infrastructure service providers and logistics service providers have evaluated the *political/legal* as the most important barrier and as the least significant the *infrastructure*;

• Industrial and trading companies have evaluated the *political/legal* as the most important group of barriers and as the least significant the *goods* and *environment*;

• R&D/University have evaluated the *political/legal* as the most important group of barriers and as the least significant the *environment*.

Since the *political/legal* barriers is the most influence on further development of IWT on the Danube river in the Republic of Serbia, more details about the structure of stakeholder's answers for this group of question is provided in the Figure 3.

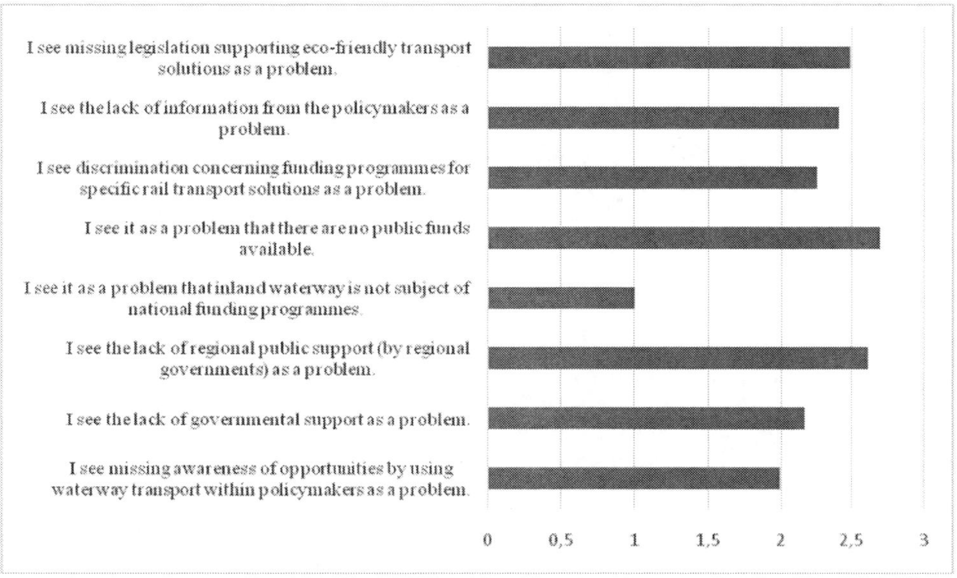

Fig. 3. Stakeholders answers regarding the influence of policy making (*political/legal* subgroup of questions) on the usage of IWT. Source: authors

From Figure 3, it is easy to notice that questions 4 and 6 ("*I see it as a problem that there are no public funds available*" and "*I see the lack of regional public support (by regional governments) as a problem*"), are the most influential on further development of IWT. Also questions 1 and 2, had high average rate from stakeholders, which implies that stakeholders think that government is not sufficient supporting eco-friendly solutions in transport, and that it is not providing information to the stakeholders.

Stakeholders also sugest following measures to the policy-makers, in order to improve the current situation in IWT:

• more incentives for safer, cheaper and environmentally friendly IWT transport,

• reduction of port taxes and adjustment of the regulations with the IWT users ship owners,

- enhancing the cooperation between ports and initiatives for offering the joint transport by more shippers and container operators, in order to attract new cargo flows on IWT,
- Increasing number of incentives which are aiming on improving the quality of services and promoting IWT,
- Introducing the price list and subsisting fees by government for smaller loads.

5 Conclusion

Because of their efficiency and reliability IWT is considered today as one of the main mitigation option available for humanity to reduce carbon footage. However, inland waterway infrastructure, modal interaction and accommodation of wide range of goods is still poorly developed. The lack of integration and fragmentation of its infrastructure, the operational, political, and technology issues are often considered as some of the obstacles to its more intensive growth. In this paper the main barriers that exist in the Serbian IWT industry were identified. This was done by directly approaching market parties, industry organisations and public authorities, as a main group of stakeholders. The stakeholder's answers revealed that government, as the main policy maker, has the most significant negative influence on the development of the IWT in the Republic of Serbia. In the future research it will be interesting to compare results from other project member countries, in order to see is there any correlation with the presented results. A number of actions/measures that could be taken to solve or at least diminish the impact of identified problems are possible and have to be proposed in future research.

This research was supported by Danube Transnational Programme and Autonomous Province of Vojvodina, Provincial Secretariat of Finance, through project: "*Regional and Transport Development in the Danube-Black Sea Region towards a Transnational Multiport Gateway Region (DBS Gateway Region)*", number: DTP1-050-3.1.

This paper is supported by the research project "From horse-drawn railway to intermodal transport" within Visegrad Fund.

References

1. *Interreg Danube Transnational Programme DBS Gateway Region*, Available online: http://www.interreg-danube.eu/approved-projects/dbs-gateway-region (2017)
2. C. Peeters, *Study on administrative and regulatory barriers in the field on inland waterway transport* (2007)
3. T. Backalic, M. Maslaric, Transport Problems **7**, 4, 13-24 (2012)
4. *European transport policy for 2010: time to decide (White paper)*, (European Commission, Luxemoburg, 2001)
5. P. Mortimer, J. Ribeiro, P. Kula, S. Balik, O.A. Mistode, Transport Problems **9**, Special Edition, 71-88 (2014)
6. S. Nikoličić, M. Maslarić, J. Strohmandl, D. Mirčetić, *Logistics performance and transport infrastructure, Paper presented at the 3rd Logistics International Conference* (Belgrade, Serbia, 2017)
7. *Statistical Office of the Republic of Serbia Bilten: Saobraćaj i telekomunikacije u Republici Srbiji, 2015* (Serbia, 2017)

MATEC Web of Conferences **134**, 00040 (2017) DOI: 10.1051/matecconf/201713400040
LOGI 2017

Analysis of Workplaces Supply in a Production Company

Vieroslav Molnár[1], *Petr* Průša[2,*], *Jan* Strohmandl[3], and *Michal* Legát[1]

[1]Technical University of Košice, BERG Faculty, Letná 9, 04200 Košice, Slovakia
[2]University of Pardubice, Department of Transport Management, Marketing and Logistics, Studentská 95, 532 10 Pardubice, Czech Republic
[3]Tomas Bata University in Zlín, nám. T. G. Masaryka 5555, 760 01 Zlín, Czech Republic

Abstract. The paper deals with the analysis of the supply of workplaces and the description of the related activities. Workplaces are supplied using specific supply means. The individual supplies used in the supply system, a description of the main activities and the principle of work are characterized. There are five alternative proposals for improving the supply of production workplaces. A proportional index method was used to select the optimal variant of the supply solution for the production workplaces. The optimal solution proposal is to add the transport lanes in the production hall.

1 Introduction

In today's modern times, the workplaces supplying is an important part of the especially production companies, that want to succeed in the global market [1]. At present, different supply systems are used in the world, and one of them is the system of supplying workplaces with supply means [2]. The successful functioning of the company in a competitive struggle cannot be realized without functional supply logistics [3, 4]. Supplying workplaces is a complicated process both in terms of technical difficulty and security. Supplying ensures the transfer of material from the warehouse to the place of production and moving the finished products from the place of production to the place of storage [5]. Transferring material or goods in this space has a certain distance that takes some time [6]. Every manufacturing company is trying to shorten this time by appropriate measures [7]. The company also strives to find optimal solutions that do not have a big economic impact on the business [8]. Such solutions can be determined by using various methods that evaluate different options on the basis of the given criteria [9].

By analysing the company's supply system, it is possible to identify deficiencies that cannot be detected without defining its system elements and elements that operate on the system [10, 11].

* Corresponding author: Petr.Prusa@upce.cz

2 The current state of workplaces supply

If a company wants to maintain its competitiveness, it must take prompt responses to the constantly changing market conditions. That is the reason the modern logistics solutions and external procurement are used [12, 13], which are mainly focused on company costs saving. In a surveyed enterprise, production planning is realized in a week ahead interval and for this reason, a supply scheme is created managed by a production program [14]. This is related to the utilization of individual supply stations and also to all supply trains T1÷T5 that transport material to production lines at times recorded by the enterprise information system.

2.1 Train T1

Train T1 consists of a tractor and three gantry wagons that are used to transport the full transport boxes from the production hall to the warehouse and transport empty boxes from the warehouse to the production hall. Wagons are characterized by their simple handling, high stability and the control of all wheels as required. An advantage is the loading of transport boxes from both sides of the wagon. In the wagons, a wheeled trolley of the exact dimensions is used, which fits on the gantry wagon included the loaded transport box (Fig. 1). Gantry wagons are also provided with transport position indicators.

Fig. 1. Train T1 with the gantry wagons left and train T2 with transport basket. Source: authors

2.2 Train T2

Train T2 supplies the production workplaces with a medium size material of higher weight (Fig. 1). The material is stored in transport baskets. Baskets with the material are picked by a stacker located in the aisle between shelf racks with stored baskets. The stacker places them on transport wheeled trolley adapted for the wagons of this train. The transport capacity of the wagons is limited to 6 transport baskets. Wagons are flexible. Their weak point is the unloading of the transport baskets only from the right side of the wagons.

2.3 Train T3

Train 3 supplies production workplaces with materials of smaller dimensions and lower weight in crates (Fig. 2). There are used four types of crates in the company, which have different dimensions and volumes for different small materials and are stackable with each other (Fig. 2). The train is equipped with a specific wagon intended only for the transport and supply of workplaces by transport crates and a smaller wagon adapted for a metal cage to collect empty transport crates that are collected at the production workplaces and transported back to the warehouse.

DOI: 10.1051/matecconf/201713400040

Fig. 2. Train T3 and the crates left and train T4 right. Source: authors

2.4 Train T4

Train 4 (Fig. 2) is used to supply manufacturing workplaces with small specific materials stored in the maximum of 15 crates. The material is picked by entering its position on the control panel using four modern cabinet warehouses according to the production requirements.

2.5 Train T5

The purpose of the Train T5 (Fig. 3) is to supply workplaces exclusively in the case of a priority production requirement for a particular type and quantity of material stored in transport baskets and crates (stored in metal cages). This train does not enter the production hall except for priority requirements.

Fig. 3. Train T5. Source: authors

2.6 Loading and unloading of trains

The design of the production hall and the points of loading and unloading by Train T1-T4 are shown in Fig. 4.

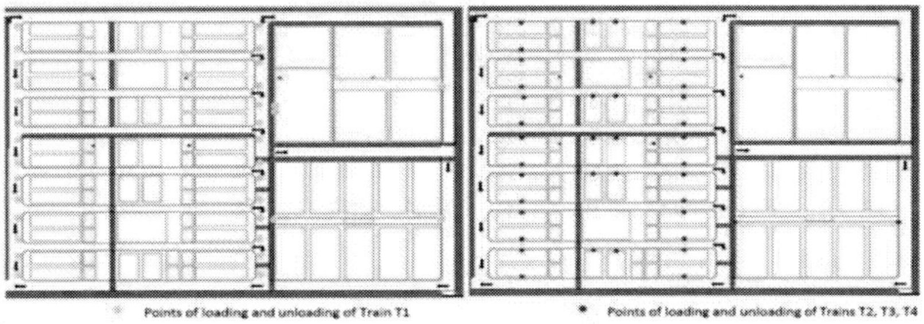

Fig. 4. The design of the production hall and the points of loading and unloading by Train T1-T4.1. Source: authors

3 Alternatives to improving workplaces supply

The most important factor in the proposed alternatives for improving the supply of workplaces in the selected manufacturing enterprise is their economic efficiency. The cheap and the most immediate improvement in supply efficiency must count on the fact that production takes place continuously during all business days. In the next section, five possible alternatives to improving the supply of workplaces will be presented.

3.1 Adding transport lanes in the production hall

Transport routes in the production hall contain a single lane allowing only one supply mean to be driven without the possibility of being overtaken by the other means due to safety at the workplace. When overtaking in one narrow lane, there is a high probability of possible collisions with other supplying means of workplaces, production lines and, last but not least, with employees. For these reasons, traffic jams arise in the production hall (Fig. 5).

Fig. 5. Traffic jam in the production hall. Source: authors

If such a case occurs in the production hall, all the trains standing in the row must wait until the first in the column moves to the next position where the same situation may arise if the train standing behind goes in the same direction. Therefore, by solving this relatively frequent supply efficiency situation, it is precisely the new transport routes creation or extension them of two lanes (Fig. 6) and, at some places in the production hall, the introduction of bi-directional lanes with the possibility of overtaking.

Fig. 6. A demonstration of a two-lane transport route. Source: authors

3.2 Reducing the number of unloading places at production lines

The essence of the proposal to reduce the number of unloading places at production lines is to minimize the movement of a worker with the supplying means. Any unnecessary movement of the worker additionally increases the supply time of workplaces. In this case, one production line would only be supplied in one place or the supply means would concentrate the material according to the specified production requirements at a precisely defined location in close proximity to the production line from where it would be available for the whole production line (Fig. 7).

Fig. 7. Reserved place for material unloading. Source: authors

3.3 Supply means differed by priority

The proposal of driving supplying means according to certain priority criteria is to drive visually labeled supplying means, which have a preference in movement along the transport routes in the production hall. The main purpose of this proposal is to prevent the situation from occurring when an urgent demand on the production workplaces is to be met and this prompt fulfillment is not possible because of an obstacle in the form of any supplying means that, for example, standing in the lane (Fig. 8). The solution is to prioritize the supply means to transport at intersections and at certain points of transport routes in the production hall.

Fig. 8. Blocking of priority production requirement. Source: authors

3.4 Hybrid supply means

Hybrid supply means are formed by the creation of four supply means, which are a combination of four trains T1 ÷ T4. They would be able to meet all production requirements. In their optimal deployment, they would be able to flexibly supply high-performance production lines.

3.5 Automated supply means

Automated supply means are reliable in supplying production lines and able to flexibly change their transport routes as required. They are guided by magnetic strips glued on the floor throughout the manufacturing plant. The equipment includes automatic unloading and loading of material as well as a security scanner, which indicates movement in its surroundings (Fig. 9).

Fig. 9. Automated supply means. Source: [15]

DOI: 10.1051/matecconf/201713400040

4 Results and discussion

Authors and several experts participated on criteria defining and evaluating and classification individual proposals. Table 1 shows which criteria have each suggested alternative. Its basic evaluating of alternatives.

Table 1. The strengths and weaknesses of proposed solutions. Source: authors

Criteria	Suggested alternatives				
	1	2	3	4	5
Flexibility	X	X	X	X	X
Quick installation		X	X		
Economic efficiency	X	X	X		
Capacity utilization	X	X	X	X	X
Security	X	X		X	X
Reliability	X		X	X	
Damage to material	X		X	X	
Accuracy	X		X	X	

The weight values of the selected criteria were determined by the direct method for classifying the weights into groups.

The criteria were assigned the weights: 1. Very important criterion - 9, 2. Medium significant criterion - 6, 3. Less important criterion - 3. The total sum of standard values equals generally 1.

Table 2. The weight allocation and their transfer to a standard value. Source: authors

Criteria	NONSTANDARD VALUE w_i	STANDARD VALUE α_i
Flexibility	6	6/60 = 0,1
Quick installation	6	6/60 = 0,1
Economic efficiency	9	9/60 = 0,15
Capacity utilization	9	9/60 = 0,15
Security	9	9/60 = 0,15
Reliability	9	9/60 = 0,15
Damage to material	6	6/60 = 0,1
Accuracy	6	6/60 = 0,1
Total	60	1

Partial usefulness is designed by direct access. 1 - does not satisfy the criterion; 2 - satisfies the criterion.

Table 3. Classification of individual proposals. Source: authors

Criteria	α_i	1		2		3		4		5	
Flexibility	0.1	2	0.2	2	0.2	2	0.1	2	0.2	2	0.2
Quick installation	0.1	1	0.1	2	0.2	2	0.2	1	0.1	1	0.1
Economic efficiency	0.15	2	0.3	2	0.3	2	0.3	1	0.15	1	0.15
Capacity utilization	0.15	2	0.3	2	0.3	2	0.3	2	0.3	2	0.3
Security	0.15	2	0.3	2	0.3	1	0.15	2	0.3	2	0.3
Reliability	0.15	2	0.3	1	0.15	2	0.3	2	0.3	1	0.15
Damage to material	0.1	2	0.2	1	0.1	2	0.2	2	0.2	1	0.1
Accuracy	0.1	2	0.2	1	0.1	2	0.2	2	0.2	1	0.1
Total	1		1.9		1.65		1.75		1.75		1.4
Ranking			1.		**4.**		**2-3.**		**2-3.**		**5.**

Table 3 shows that the optimal solution proposal is to add transport lanes in the production hall that was ranked as top-of-the-range. This proposal is effective in order to maintain the current supplying of production workplaces, which would enable overtaking the supply means in the production hall, thus achieving lower times of production requirements fulfilments. This proposal will not have a significant impact on the company's finances, but it brings a high added value in terms of accelerating the supply of production workplaces by supply means. Proposal no. 2 - reducing the number of material unloading - has ranked on the fourth place. The advantage is the flexibility that would make it possible to minimize the number of unloading in the production hall in order to improve delivery times by supply means. The disadvantage of this solution is less reliability. The least suitable solution is automated supply means, due to the high initial investment costs.

5 Conclusion

The contribution of the article is to point to one of the options to determine the optimal alternative to improve the process of supplying workplaces in the manufacturing company, taking into account the economic efficiency, which is for a company a fundamental condition for implementing the proposed solution. The paper presents the characteristics of the supply means and the proposed solutions for raising efficiency in the supply of production workplaces. The proportional index method evaluates the criteria and the optimal alternative to improve the supply of workplaces.

This work is a part of these projects VEGA 1/0063/16, 1/0403/18, KEGA 018TUKE-4/2016.

This paper is supported by the research project "From horse-drawn railway to intermodal transport" within Visegrad Fund.

References

1. H. Neradilova, G. Fedorko, Open Engineering **6**, 700-710 (2016)
2. M.C. Cooper, D.M. Lambert, J.D. Pagh, Int. J. Logist. Manag. **8**, 1-14 (1997)
3. D. Pan, W. Deng, F. Shao, *ICLEM 2010,* 4424-4430 (American Society of Civil Engineers, Reston, 2010)
4. U. Clausen, M. ten Hompel, M. Klumpp, *Efficiency and Logistics* (2013)
5. J.T. Mentzer, S. Min, L. Michelle Bobbitt, Int. J. Phys. Distrib. Logist. Manag. **34**, 606-627 (2004)
6. N. Brnjac, B. Abramovic, M. Maslaric, Promet-Traffic & Transp. **22**, 303-307 (2010)
7. P. Ficzere, Z. Ultmann, Á. Török, Transport **29**, 278-284 (2014)
8. E. Tomková, N. Husáková, J. Strohmandl, *Transport Means 2015 – 19th International Conference on Transport Means*, 694-697 (2015)
9. B. Chramcov, R. Bucki, S. Marusza, Adv. Intell. Syst. Comput. **210**, 423-434 (2013)
10. I. Kubasakova, R. Kampf, O. Stopka, Communications **16**, 9-13 (2014)
11. R. Kampf, L. Lizbetinova, K. Tislerova, Open Engineering **7**, 26-30 (2017)
12. D. Malindžák, *Teória Logistiky* (TU Košice, Slovak republic, 2007)
13. A. De Grahl, E. Hartmann, *Success Factors in Logistics Outsourcing* (2012)
14. H. Neradilova, G. Fedorko, Procedia Engineering **192,** 638-643 (2017)
15. CEIT, Available online: https://www.5dimensions.sk/wp-content/uploads/2015/10/agv-3d-visualization-ceit.jpg (2017)

MATEC Web of Conferences **134**, 00041 (2017) DOI: 10.1051/matecconf/201713400041
LOGI 2017

Shunting Locomotives Fuel and Power Resources Decrease

M.O. Mussabekov[1], *G.B.* Bakyt[2], *A.M.* Omirbek[1], *Eva* Brumerčíková[3,*], and *Bibiána* Buková[3]

[1]Kazakh Transport and communication Academy, Faculty of Automation and Telecommunications, Shevchenko Street 97, Kazakhstan
[2]Karagnada State Technical University, Mira Avenue 56, Karagandy 100000, Kazakhstan
[3]University of Žilina, Faculty of Operation and Economics of Transport and Communications Department of Railway Transport, Univerzitná 1, 010 26 Žilina, Slovakia

Abstract. The article discusses about the increase in operational fuel consumption of shunting locomotives in the cold season. Possible solutions to this problem through the use of diesel engines with dry crankcase are considered. The technique of evaluating the thermal state of the diesel engine and the calculation of fuel consumption of heating boiler are provided.

1 Introduce

Railway transport is carrying out more than 70% of transportation in Kazakhstan, and, therefore, it is one of the major consumers of fuel and power resources (F&PR) of the country. More than 50% of transportation process is carried out by locomotives.

An integral part of transportation process is shunting operation. It grows along with the development of transportation, increase in network spread and number of stations performing technical and commercial operations. Shunting operations are performed by diesel locomotives only. The main operator of transportation process in Kazakhstan railroads, providing with locomotive traction, is JSC "Locomotive", and its annual fuel consumption makes 700÷800 thousand tons. Taking into account the fact that 40÷50% of all railroads operating costs in Kazakhstan consist of fuel consumption and locomotives power supply, it is obvious that saving of fuel and power resources (F&PR) is an important reserve for energy efficiency increase.

2 The total fuel quantity, consumed by diesel locomotives

The total fuel quantity, consumed by diesel locomotives, is based on fuel consumption by operation type

*Correspondig author : eva.brumecikova@fpedas.uniza.sk

 DOI: 10.1051/matecconf/201713400041

$$E_T = \sum_{i=1}^{n} G_i \tag{1}$$

where: Gi – locomotive fuel consumption by operation type (cargo, passenger, shunting and administrative), kg.

Diesel locomotives effective use common criteria is specific fuel consumption per 10 000 tons kilometers (gross).

$$e_t = \frac{E_T}{A} \cdot 10^4, \quad \frac{kg}{10^4\ t\ km\ gross} \tag{2}$$

where: A – annual locomotive scope of work by operation type, million ton kilometers (gross) [1].

The scope of work of JSC "Locomotive" diesel locomotives in 2015 was 208264.6 million tons kilometers (see Fig. 1) with 727,117 tons of fuel consumption and average specific fuel consumption 34.9 $\frac{kg}{10^4\ t\ km\ gross}$ (see Fig. 2).

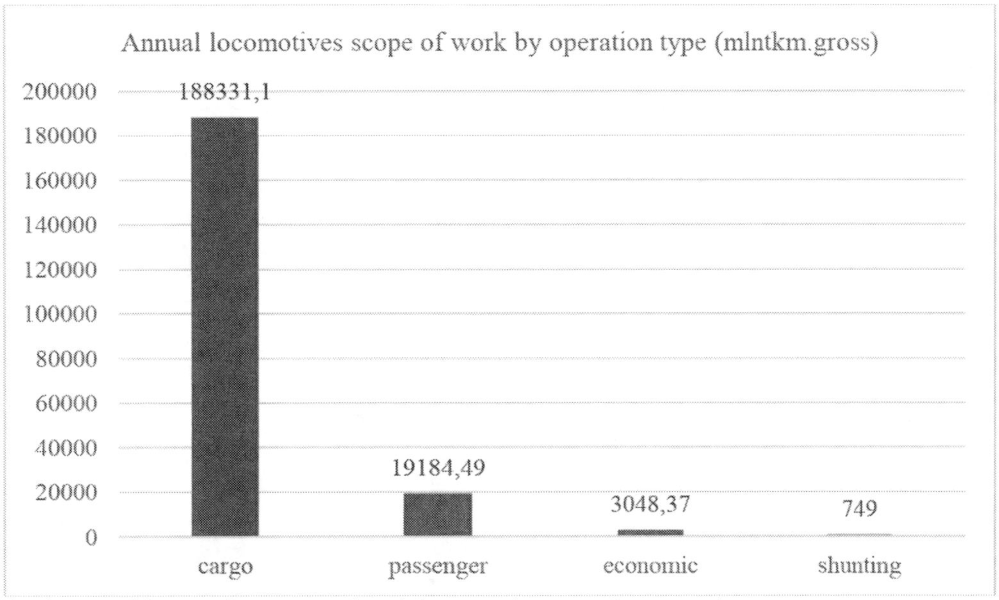

Fig. 1. Annual locomotives scope of work by operation type. Source: authors

DOI: 10.1051/matecconf/201713400041

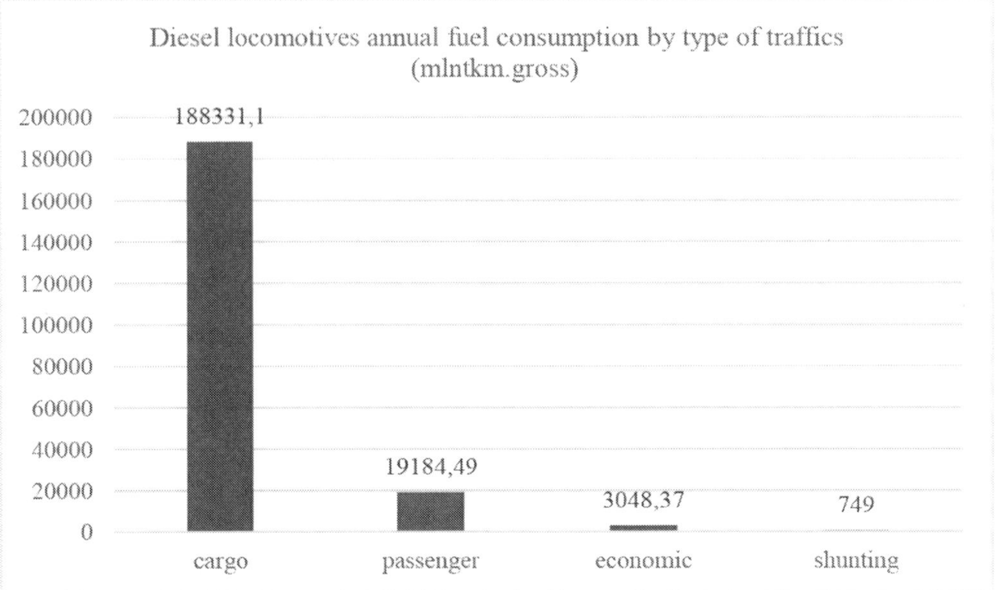

Fig. 2. Diesel locomotives annual fuel consumption by type of traffics. Source: authors

3 Efficiency of diesel locomotives

Efficiency of diesel locomotives utilization capacity is determined by DGU (diesel generator unit) - K load factor.

$$K = \frac{N_{int.p}}{N_{nom}} \quad (3)$$

where: $N_{int.p}$ - intermediate power, kW;
N_{nom} - nominal power, kW.

Locomotives GDU (diesel generator unit) loading at different operation types have the following average values:

- cargo K = 0.7 ÷ 0.9,
- passenger K = 0.5 ÷ 0.7,
- administrative K = 0.3 ÷ 0.4,
- shunting K = 0.2÷0.3.

This information shows the non-economic operation mode of locomotives DGU, especially at shunting and administrative operations.

The characteristic feature of shunting locomotives operation is the fact that idling and small loadings make 60÷75% of overall work duration, DGU nominal mode makes 1.5% of locomotive overall operating time [2].

Such shunting diesel locomotives engine operational modes distribution is noted also in some other operations [2-6]. It is determined by train and shunting operational conditions and temporary (seasonal) factors as well. Fuel consumption at the idling mode of the diesel engine is 30 ÷ 50% of the total fuel consumption in operation [2, 4]. As a result, shunting diesel locomotives specific fuel consumption exceeds specific expense of mainline diesel locomotives in cargo operations by 10.6 times (Fig. 3).

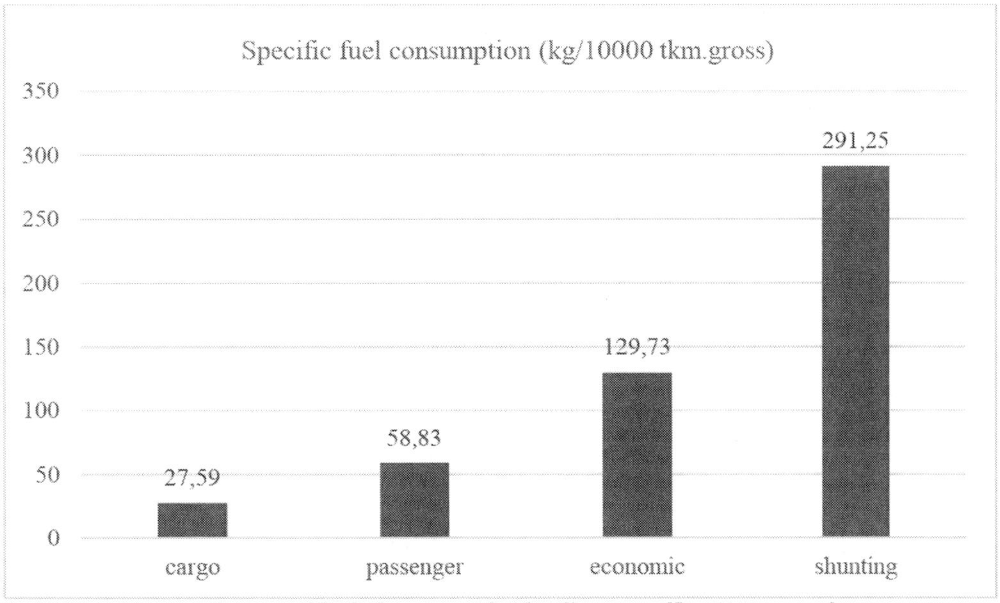

Fig. 3. Diesel locomotives specific fuel consumption by diverse traffics. Source: authors

Increased fuel consumption is not the only negative consequence of engine operations with small loadings and idling. It is combined with an intensive carbonizing in turbo compressor that leads to the rotor imbalance, productivity and air pressure decrease and, as a result, to the increased depreciation of supporting bearings in turbo compressor, engine fuel consumption increase in all capacity range, increased thermal stress of sleeve assembly, and increase of emissions release into the atmosphere. Currently, it is offered a large number of projects to reduce the fuel consumption by the shunting locomotive engine at idling during the hot idle, reduce carbon formation by reducing the overhaul periods of turbo compressor or injecting water into it, etc. [5]. Considerable decrease in operational efficiency and respectively shunting diesel locomotives huge fuel consumption occurs at low air temperature. Danger of water freezing in cooling system and high oil viscosity at low temperatures do not allow switching the engine off. The use of antifreeze in the existing locomotive diesels is economically and functionally impossible. There are very many various heating systems of switched off diesel engines. For example, American company Kim Hotstart's system is widely used; where the main heating sources are electric heating tubes installed in diesel cooling system and powered by additional DGU of small capacity [6].

The system of autonomous diesel heating of the locomotive called "Gulf Stream" which is developed by "Delway Management" [4] uses as a source of heating the heat exchange units (fuel combustion energy is converted directly to heat).

Fundamental faults of these systems are the following:

- low efficiency due to energy conversion,
- frequent maintenance of additional DGU,
- use of the additional device required for oil heating in a diesel engine.

The main consuming part of locomotives is the diesel engine.

The analysis of chronometer data shows that:

- first – idling corresponds to waiting time of locomotive, i.e. the locomotive stands idle,
- second – nominal power (for example 880 kW of diesel PD1M – TEM2 locomotive) practically is not used, i.e. shunting locomotives DGU is operated with very low loading coefficient.

The most optimal solution of fuel consumption decrease at shunting operations is shut down of diesel engine when diesel locomotive is waiting for operations. It also reduces the emissions of harmful substances by diesel locomotive.

4 Diesel engine with a dry crankcase

The use of diesel engine with a dry crankcase in shunting locomotives allows fixing diesel shutdown problem and its thermal state support. In railway transportation practice, there is an experience of diesel engines use with a dry crankcase at locomotives with hydro mechanical transmission (HMT).

The conceptual diagram of cooling system of diesel engine with dry crankcase based on the innovative patent [7] is provided in Figure 1.

The main condition for diesel shutdown in winter is water temperature support in the diesel cooling system, so water heating and circulation are provided as follows:

- locomotive operates in normal operations mode – water circulation is carried out by the main water pump,
- diesel engine is switched off – water circulation in the cooling system is carried out by additional pump from the power supply (segment a-4-4-b).

The switched off diesel and oil in a separate oil tank are warmed up by water heated in a special heating boiler, that operates using diesel fuel.

Water circulation is carried out along the outline – diesel – radiators – heating boiler - additional pump - oil tank - cooler of boosted air-diesel. Oil from the oil tank to the diesel engine and back into the tank is pumped over pipelines by oil pumps (figure 4).

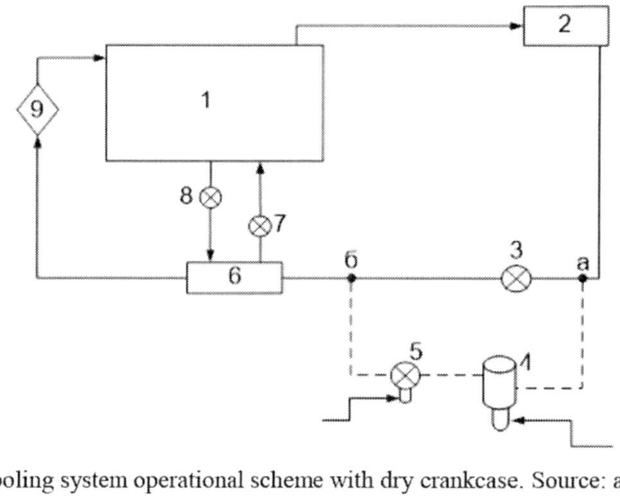

Fig. 4. Diesel cooling system operational scheme with dry crankcase. Source: authors

Legend to image: 1-diesel, 2 – cooling radiators, 3 – main water pump, 4 - water heating boiler, 5-additional water pump, 6 – oil tank, 7-oil boost pump, 8 – oil discharge pump, 9 – boost air cooler.

The electric motor of the recycling pump 5 is powered by the power supply – the energy store (ground electrical equipment or ultra-condenser). At the started diesel engine, the cooled in radiators water, that cools oil, is supplied to the oil storage tank.

The heating boiler efficiency evaluation for diesel locomotive thermal condition support is determined by comparison of fuel consumption at diesel idling and fuel consumption required for heating boiler operation [8].

Required heat for engine heating [9]:

$$Q = \frac{C_Д \cdot \Delta T}{\tau} + \alpha \cdot F \cdot \Delta T \quad (4)$$

where: $C_д$ – modal diesel thermal capacity, kJ/grad

$$(C_д = C_{met} \cdot M_{met} + C_o \cdot M_c + C_o \cdot M_c) \quad (5)$$

where: C_{met}, C_o, C_c – specific thermal capacity of metal, oil, coolant, kJ/kg•grad respectively; M_{met}, M_o, M_c - weight of metal, oil, coolant (water) respectively.

ΔT –required engine heating, $\Delta T = t_{кор.п.} - t_0$

$T_{base\ b.}$ –temperature of base bearings is accepted as equal to –20°C,

t_o –initial engine temperature is accepted as equal to –40°C

t – engine thermal pre-starting procedure duration,

α –heat transfer coefficient from engine surface to the environment (depending on wind speed and quality of engine warm keeping means within 15…80 kJ/ m²•hour • grad),

F – area of the engine cooled surface, m²:

$$F = i \cdot (\pi \cdot D \cdot S + \frac{2 \cdot \pi \cdot D^2}{4}) \cdot K_n \quad (6)$$

where: i – number of engine cylinders,

D, S - diameter and engine stroke, respectively

K_n – coefficient of actual engine surface is accepted as equal to 1.25 … 1.3.

Based on found boiler heating capacity, it is required to check the temperature of cylinders head block at the end of warming up [10].

$$t_{head} = \frac{Q \cdot t}{C_{head} \cdot M_Д \cdot 60}) \quad (7)$$

where: c_{head} modal specific heat capacity of cylinders head block,

$$C_{гол} = \frac{B}{S/D} \quad (8)$$

where: B - the constant coefficient accepted for the cast iron head –0.45, aluminum alloy casting – 0.5.

Switched off diesel thermal control is exercised by cylinder head temperature that has to be not less than 40°C.

Fuel consumption for engine heating at a certain ambient temperature in fixed time:

$$B_m = \frac{Q}{H_u \cdot \eta_n} \quad (9)$$

where: H_u –fuel efficiency (for diesel fuel 42.5•103 kJ/kg),

η_n – heating efficiency.

Switched off diesel thermal control is exercised by cylinder head temperature that has to be not less than 40°C.

Upon receipt of operation start command, the operator starts the diesel engine and puts locomotive in motion.

5 Conclusion

Shunting operations are characterized by long-term operation of diesel locomotives at idle mode and on low-load conditions, which amounts to 60 ÷ 75% of their daily operation and is accompanied by an increased specific fuel consumption. The timing analysis showed that the specific consumption of diesel locomotives in shunting operations exceeded the value of the specific consumption of diesel locomotives in train operation by 10.6 times, and the load factor of the engine exceeded k = 0.2 ÷ 0.3. At the same time, environmental indicators deteriorate along with the increased fuel consumption.

The possible solution to this problem is by using diesel engine with dry crankcase as power installation of the diesel locomotive according to an innovative patent. № 31244, it will allow to maintain the thermal state of the muffled main diesel engine during cold weather, thereby reducing fuel consumption and emission of harmful substances with exhaust gases at idle. The technique of an estimation of a diesel thermal condition and calculation of fuel consumption of a heating boiler is given.

The paper is supported by the VEGA Agency by the Project 1/0095/16 "Assessment of the quality of connections on the transport network as a tool to enhance the competitiveness of public passenger transport system", that is solved at Faculty of Operations and Economics of Transport and Communication, University of Zilina.

This paper is supported by the research project "From horse-drawn railway to intermodal transport" within Visegrad Fund.

References

1. L. Jedlinsky, J. Caban, L. Krzywonos, S. Wierzbicki, F. Brumercik, Journal of Vibroengineering **17**, 175-187 (2015)
2. V.A. Gapanovich, *Railway transportation energy saving* (The textbook is edited by PhD in Technical Sciences, 2012)
3. J. Gasparik, V. Zitricky, Transport **25**, 387-393 (2010)
4. A.I. Molchanov, I.L. Povarkov, L.A. Muginstein, K.M. Popov, All-Russian Scientific Research Institute of Railway Transport Bulletin **2** (2004).
5. A.I. Volodin, E.I. Skovorodnikov, A.S. Anisimov, *Case study of thermodynamic, economic and ecological characteristics of diesel locomotive engines in operation: Monograph* (2011)
6. Y.E. Prosvirov, et.al, *Innovative technologies in locomotive economy: Monograph* (2012)
7. M.O. Mussabekov, G.B. Bakhyt, A.M. Omirbek, *Innovative patent No. 31244 "Shunting diesel locomotive"* (16.02.2016)
8. A.I Volodin, M. Zheldorizdat, *Evaluation methods of technical condition, operational profitability and ecological safety of diesel locomotives: Monograph* (2007)
9. N. Nestorova, S. Goncharuk, V. Anisimov, A. Anisimov, V. Shvartcfel, Procedia Engineering **165**, 1547-1555 (2016)
10. D. Barta, M. Mruzek, M. Kendra, P. Kordos, L. Krzywonos, Advances in Science and Technology **10**, 240-247 (2016)

Cycling Traffic Development in Slovakia in the Terms of the City of Košice

Hana Neradilová[1,*], and *Tomáš* Ragulský[2]

[1]College of Logistics, Přerov, Palackého 1381/25, 750 02 Přerov, Czech Republic
[2]Technical University of Košice, BERG Faculty, Park Komenského 14, 040 01 Košice, Slovakia

Abstract. Cycling traffic has been booming in urban agglomerations recently. This is due to the fact that using bicycles helps solve existing transport problems. The advantage is that it supports the limitation of the use of cars in city centers and thus reduces the negative impact of transport on the environment. It can be realized in various forms - using private bicycles or bike-sharing. Košice includes among cities where introducing bike-sharing is considered. Košice is a city that has its specific transport problems, whose partial solution may be to increase the support of cycling when one of the options is the introduction of a city bike-sharing system. The article offers a closer perspective at this issue in adaptation to real urban conditions. It introduces the possibility of introducing support for cycling traffic in the city of Košice based on the selection of suitable locations for building the stations and the subsequent evaluation of selected localities based on the determined criteria using the proportional index method.

1 Introduction

Cycling traffic and cycle tourism have an irreplaceable place in the field of transport today. Using cycling is constantly raising, thus increasing the requirements for integrating cycle routes, off-street cycle paths and the availability of historic city centers on a bicycle. Bicycle gradually regains its status as a significant part of the transport infrastructure and is no longer perceived only as a form of sporting activity. In many cities, it becomes a faster and more efficient mode of transport than a car. The full inclusion of bicycles in the city transport system is a necessary way to support the fluent flow of traffic. The current trend in cycle tourism is to increase the number of cycling routes and cycle paths. One of the options for pedestrians, cyclists, and drivers to co-operate is the introduction of public bicycle sharing system - bike-sharing, which works as a public automated bicycle rental. It is a network of mutually interconnected bicycle stands that serve the general public.

In the context of increased efforts to increase the involvement of cyclists in public transport, it is necessary to solve important aspects such as the safety of bicyclists in mixed urban traffic, which is addressed by Apasnore et al. [1] in relation to the impact of the lateral spacing between bicycles and vehicles and other factors. Other authors [2] examined

* Corresponding author: hana.neradilova@vslg.cz

bicycle sharing issues in Polish conditions based on the attitude of the Poles towards cycle traffic and the idea of a self-service public bike rental system has been widely presented and the components of the architectural system have been discussed. The authors presented the principle of bike-sharing, as well as its transport accessibility in cities in Poland. Cairns et al. [3] presented interesting findings in the study of a change in people's behavior in relation to the use of electric bikes if people could use these vehicles for several weeks free of charge. In their study, they concluded that almost a third of the surveyed number of people in the future intends to use this mode of transport much more often, and nearly three-thirds would like to own the electric bicycle as an efficient mode of transport. Waard et al. [4] investigated the influence on the behavioral change of cyclists when the common map or smartphone navigation is used for orientation. Prateli et al. [5] have dealt with improving bicycle mobility in urban areas through ITS technologies. In connection with cycle traffic and its involvement in urban transport, Wang et al. [6] have investigated the effect of the distance of moving targets on the emergence of different types of conflicts between vehicles and bicycles at signaling crossroads. In solving the issue of launching public bicycle system the experiences from different world destinations such as Taipei [7], Fargo in North Dakota [8] and Beijing [9] are beneficial. When introducing a bike sharing system, it is also important to address the issue of proper station distribution, for example, using a mathematical model [10] or interpolation techniques [11].

There is often a distinction between mobile and static cycling infrastructure in literature. Measures to improve cycling are usually related to mobile infrastructures, such as cycling routes. This includes the construction of cycling paths (on-street, off-street) paths or multifunctional crossings. The stationary bicycle infrastructure includes all types of bicycle parking, ranging from simple bicycle racks to high-quality bicycle stations with a wide range of services. Therefore, if there is a purpose to support cycling in a city or community, more attention should be paid to improving conditions for stationary cycling infrastructure. Building bicycle parking facilities enormously improve the conditions for cyclists.

2 Cycling usage in urban agglomerations

As part of the development of cycling traffic in urban agglomerations, it is necessary to consider several factors - the possibility of parking the bicycles, the location, and choice of bicycle rack type, the purpose of building a certain station for bicycles, etc. Proper cycle racks are such racks that allow a bicycle to be locked on frame and wheel. Parking does not only include the right cycle racks but also the locations where the cycle stands are situated. In general, stands should be located at every type of civic amenities such as school, office, sports facility, theatre, cinema, shop, and others. Parking in terms of location may be in the associated traffic area on the pavement, but it can also be located in the main traffic area in the parking lane or in an unused location on the road, where, for example, it fulfils the role of the calming element [12]. Bicycle racks should be placed in front of any type of civic amenities available on a bicycle. Co-operation with the citizens themselves has proven very well in selecting new locations when citizens are asked about where do they the stations with bicycle racks need. This way was used for example in the city of Zilina, where the design of new sites for cyclists was based on suggestions from citizens.

The main principles for the appropriate placement of bicycle racks are to place them as close as possible to the entrance, place them in a visible place or where a large movement of persons is around, and place them within the reach of surveillance cameras. Another condition is a good access to cycling path or route, the possibility to protect the bicycle from damage and to enable the locking of the frame and the front or rear wheel with a short wheel lock. Cycle racks should not endanger passers, for example, to restrain pedestrians walking on pavement.

 DOI: 10.1051/matecconf/201713400042

From a safety point of view, cycle stands can be divided into three types. These are cycle racks that allow safe locking of the wheel and frame or locking into a lockable and guarded shelter or box. Cycle racks that allow only wheel or frame locking and those that do not allow locking the bicycle only parking into, for example, grooves. In this case, the bike needs to be guarded personally and supervised.

From a time perspective, short and long-term parking is distinguished. Short-term parking includes a parking time up to 3 hours, what may include dealing certain matters at the office, post office, shopping, etc. Long-term parking means a period longer than 3 hours and represents the type of parking when the cyclists put off the bicycle for a longer period of time, for example, time at work and school, during the night at the house.

2.1 The current state of the bicycle traffic system in Slovakia

Many Slovak cities and municipalities have not even a single kilometre of cycling paths. Even where they are built, unsatisfactory technical solutions often impedes their usability. This is related to the lack of a high-quality and detailed technical regulation that would be devoted to the design and building of cycling infrastructure and the minimum attention paid to cycling. The technical condition of road infrastructure in cities and towns in Slovakia is unfavourable. Absent of segregated communications network, cyclists are forced to use road communications where they are endangered by cars. This fact discourages many people from using cycling. Negative features include road inequality and frequent confining of cyclists by cars passing through cycling routes. These elements aggravate the comfort of cycling and also increase the risk of accidents and injuries.

In Slovakia, there is also the lack of cycling infrastructure documentation, which confirms the fact that the potential of cycling is still underestimated. For making out this documentation, 138 cities were addressed in the year 2012, of which only 67 cities have participated the survey. According to data from these 67 cities, the total length of cycling routes was nearly 150 km. Up to three-fifths consists of separate communications - either exclusively for cyclists (15%) or more often for pedestrians and cyclists with separate traffic (23%) and with mixed traffic (22%). A third of the total length is formed by off-road cycle paths built in the associated transport area, of which the movement of cyclists is separated on 16% and on more than 18% is movement of cyclists mixed with the movement of pedestrians. Only 6% of the cycling communications lengths in the cities participating in the survey are located in the main traffic area (cycling lane, cycling route, on-street sharrow), (Fig. 1).

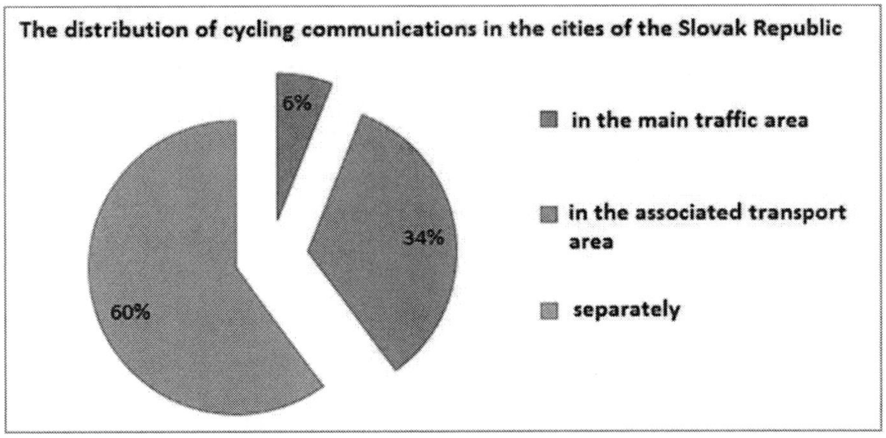

Fig. 1. The distribution of cycling communications in the cities of the Slovak Republic. Source: authors

Cities often do not deal with cycling issues in obligatory parts of regional planning documents. Only 29 of the city of the 67 cities involved in the survey have mentioned cycle transport in their urban plan. Only four cities - Trnava, Moldava nad Bodvou, Piešťany and Banská Bystrica - have developed a separate non-motorized transport conception document and 19 cities have a technical study of bicycle routes. Two cities, Moldava nad Bodvou and Banská Bystrica, have static cycling traffic studies. The role of the cycle-coordinator is currently set up by the city authorities in Banská Bystrica, Bratislava, and Trnava.

2.2 Cycling infrastructure in Slovakia

Cycling route is a marked trail for cyclists in the terrain along existing roads. Cycling route can be guided by intravilan, extravilan, in forest, mountain, or field. At the same time, after their legislation in the sense of Slovak legislation, cycling routes must be fitted with orientation cycling marking.

There are 530 marked cycling routes in Slovakia with a total length of more than 10,000 kilometres. The main axes consist of 2,926 km national long-distance cycling trails, international routes and Eurovelo routes, of which 82.6 km are now marked on the territory of the Slovak Republic. They are defined in regional planning documents and the cycling traffic conception document that respect broader relationships and major national corridors. Regional and local routes account for 70% of all cycling routes of nearly 7,000 km. The network of cycling routes in Slovakia is run by 74 subjects, mostly civic associations. The leading part of the cycle routes network is managed by the associations covered by the Slovak Cycle Club. Partially the cycling routes administration was taken over by the municipalities - the city of Bratislava, the city of High Tatras, the town of Spišská Belá, the association of municipalities of the Tokay region, the village of Pila, and the town of Skalica (Tab. 1).

Table 1. Cycling routes by region. Source: authors

Self-governing region	Number of administrators	Number of routes	Length [km]
Bratislava Region	12	41	749
Trnava Region	7	56	804
Nitra Region	6	21	387
Trencin Region	6	64	1,178
Zilina Region	6	112	2,200
Banska Bystrica Region	11	106	2,140
Presov Region	13	76	1,350
Kosice Region	13	54	1,291
Total	**74**	**530**	**10,099**

Cycling routes are divided as follows:
- by the nature of the terrain: road and mountain biking,
- according to the importance expressed in colour: red - blue - green - yellow,
- by intensity: recreation - sport - expert.

The marking for cycling tourism is regulated by the Slovak Technical Standard - STN 01 8028 - Cycling marking. The most of the cycling trails in Slovakia are marked with red colour (2,926 km), followed by cycling routes marked with blue colour (2,896 km), then green trails (2,588 km) and yellow marked cycling trails (1,220 km). Suitable trails for road bikes are about 47%, for mountain bikes about 53% of the total length of marked cycling routes.

3 Analysis of cycling traffic in the city of Košice

Cycling infrastructure in the city of Košice does not achieve the required quality. The number of cycling paths is deplorably low and the infrastructure for cycling around the city center is missing or insufficient. Košice has a limited range of cycling paths along the Hornad River (from the city part Nad jazerom to the city center), then on the north-south route on the pavements Južná - Hlavná - Komenského and the cycling paths on the sidewalks in the suburban part West. Dargovskych hrdinov housing estate does not have enough access for bicyclists. There are no bicycle lanes in the residential areas, and bicyclist there use mainly sidewalks.

Recreational paths are not well connected to urban communications. Crossings between paths and communications are often dangerous, light-controlled crosswalk usually causes pedestrians a long delay. The proportion of cycling routes is about 2%, but it is growing rapidly. The current arrangement of sidewalks with enable movement of bicycles will face problems and conflicts between the both groups (pedestrians and bicyclists). This problem has been already in the Hlavná Street. The total length of urban cycling routes in the city is 22,767 m (Fig. 2) and their distribution is the following.

City cycling routes in the main traffic area:

- cycling on-road lane - 211 m,
- cycling on-road route - 692 m,
- cycling on-street sharrow - 0 m,
- crossing for cyclists - 477 m.

City cycling routes outside the main traffic area:

- separate cycle path - 6,070 m,
- the shared path for pedestrians and bikers with mixed traffic - 72 m,
- the shared path for pedestrians and bikers with separate traffic - 15,245 m.

The cycle traffic share in Košice in the total transport work division is only 3 percent (Survey - BICY, Cities and Regions for cycling). In accordance with the vision of the National Cycling and Cycling Tourism Development Strategy in the Slovak Republic, which foresees a 10 percent share of bicycle transport in the total transport work division by the year 2020, the current state of affairs is not adequate. In order to achieve the desired status, it is necessary to start building a safe traffic area for cyclists and setting up attractive cycling routes. Currently, a cycling network of 26 km in the city and 146.8 km of cycling routes in the Municipal Forests is being built in Košice. Košice has ranked second in the length of urban cycle routes after the first Bratislava. Taking into account the number of inhabitants supposed to use cycle routes or the length of cycling trails related to the size of the city, Košice has come 7th and 6th out of ten cities. In Košice, segregation of cycling infrastructure prevails. Most urban cycling routes (up to 96%) are led outside the main transport area along with separate walkways or shared paths for pedestrians and cyclists. The remaining infrastructure consists of on-road cycle lanes and on-road cycle routes, which is less than 1 km in length.

4 Proposal of bicycle stands location in Košice

Proposal of the station locations within the bike-sharing system was based on experience from a similar system in New York [13], according to which the main criteria for station placement should be the capacity of stations, their relatively high density and also proximity to public transport. An important aspect is also the feedback from the community, which has not been taken into account in this proposal since it is only the initial design of station placement. Stations suggested for the bike-sharing system are close to the objects with the high-frequency of people such as schools, dormitories, department stores, shopping centers and various socially important buildings.

Altogether, 30 public bike-sharing stations were proposed (Fig. 3). Based on the proportional index analysis, the 15 most suitable stations were selected from this proposal.

Fig. 2. Highlighting of city cycle routes in Košice

Fig. 3. Proposal of bike-sharing stations location

The four criteria with the following weights were defined to calculate the most suitable position of the stations (Table 2).

Table 2. The weighting of the criteria. Source: authors

Criteria	Weight of criterion
Criterion 1 - proximity to public transport stops	**0.2**
Criterion 2 - density of people	**0.3**
Criterion 3 - purpose of the station	**0.4**
Criterion 4 - proximity to cycling trails	**0.1**
Total of weights	**1.0**

Individual stations were assigned a number from 1 to 30. Subsequently, individual stations were rated according to criteria on a scale of 1 to 10 (Table 3 and 4).

Table 3. Station rating – part 1. Source: authors

Criteria/Stations	1	2	3	4	5	6	7	8	9	10	11	12	13	14	15
Criterion 1	7	5	10	6	5	5	9	7	2	3	10	3	3	9	6
Criterion 2	7	4	7	5	10	3	7	4	7	7	7	7	8	9	7
Criterion 3	6	4	8	2	8	3	9	4	3	2	9	2	2	8	10
Criterion 4	8	3	8	8	5	8	10	6	4	4	4	3	3	5	4

Table 4. Station rating – part 2. Source: authors

Criteria/Stations	16	17	18	19	20	21	22	23	24	25	26	27	28	29	30
Criterion 1	9	7	4	6	5	4	4	5	5	6	4	4	3	9	3
Criterion 2	10	6	5	9	10	4	3	4	9	8	4	1	10	6	5
Criterion 3	10	9	2	9	8	3	3	2	8	8	3	3	9	9	2
Criterion 4	2	2	9	8	7	7	5	3	6	5	5	3	8	3	6

The individual station rating was then multiplied by the criterion weight. Using the ratio index analysis 15 stations with the highest rank in the total utility of the station U_m were selected (Fig. 4). The evaluated stations were divided into two groups. The group with a total U_m less than 5.8 was evaluated as unsatisfactory and the group with a total U_m greater than 5.8 was evaluated as satisfactory.

Appropriate location of stations (Fig. 5): Steel Arena, Festival Square, Technical University, Angels Arena, Továrenská, Aupark Košice, Ondavská, Staničné Square, City Swimming Pool, Dominican Square, Pavol Jozef Šafárik University, Poštová, Tesco Store, Košice State Theater, Jakab Palace.

Inappropriate location of stations: Štúrova - Fejova, Štúrova - Rastislavova, Timonova, Vojenská, Zvonárska, Bočná, Hrnčiarska, Kováčska, Strojárenská, Kuzmány, Bačíkova, Kasárny námestie, Zimná, Mlynárska, Františkánska.

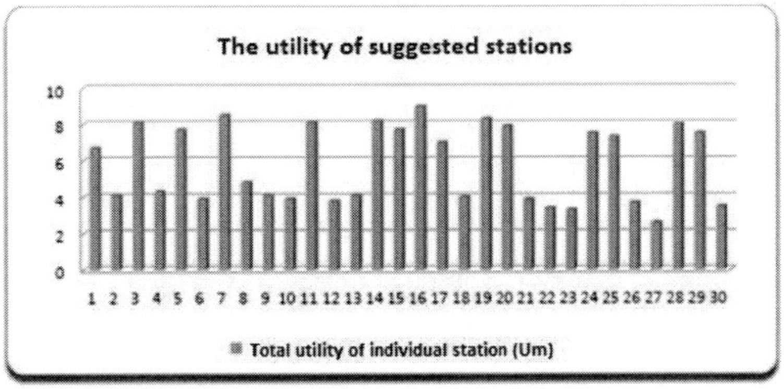

Fig. 4. The usefulness of individual stations. Source: authors

Fig. 5. Proposal of top 15 rated public bike-sharing stations. Source: authors

5 Conclusion

The idea to involve the form of recreational and leisure activities in public transport infrastructure is currently a topical theme in all major cities [14]. In the Slovak Republic, there is currently no official large-scale public bicycle system. There are only a few low-cost projects that run on volunteering and membership fees.

In the paper, the proposal for the publicly bike-sharing system creation in the city of Košice was demonstrated on the selection of appropriate locations for the placement of cycling stands and racks for citizens. When selecting suitable locations, criteria such as proximity to public transport, connection to cycling paths for comfortable and safe engagement of cyclists into urban transport were taken into account. Furthermore, the purpose of the cycle station was solved, whether it is designed for short-term or long-term parking and the number of people who are supposed to use the station. These aspects are further developing the capacity design of the stations and its technical solution and then the proposal for charging this service for the inhabitants of the city of Košice. This proposal is a contribution to addressing congestion in the city center by car transport, including the lack of capacity for car parking. With the development of cycle traffic in the context of the popularity of the using this service by the citizens of the city, an improvement in the traffic situation in Kosice city can be expected.

This work is a part of the project VEGA 1/0063/16.

This paper is supported by the research project "From horse-drawn railway to intermodal transport" within Visegrad Fund.

References

1. P. Apasnore, K. Ismail, A. Kassim, Accident Analysis and Prevention **106**, 141-148 (2017)
2. P. Czech, K. Turoń, R. Urbańczyk, *TSTP 2017 – 14th Scientific and Technical Conference on Transport Systems Theory and Practice* **631** (2018)
3. S. Cairns, F. Behrendt, D. Raffo, C. Beaumont, C. Kiefer, Transportation Research Part A: Policy and Practice **103**, 327-342 (2017)
4. D. de Waard, F. Westerhuis, D. Joling, S. Weiland, R. Stadtbäumer, L. Kaltofen, Ergonomics **60**, 1283-1296 (2017)
5. A. Pratelli, M. Petri, A. Farina, and M. Lupi, *TSTP 2017 – 14th Scientific and Technical Conference on Transport Systems Theory and Practice* **631** (2018)
6. Y.Q. Wang, F. Xing, L. Zhang, *7th International Conference on Green Intelligent Transportation System and Safety 2016* **419** (2018)
7. J.J. Lin, N.L. Wang, C.M. Feng, International Journal of Sustainable Transportation **11**, 633-641 (2017)
8. J. Mattson, R. Godavarthy, Sustainable Cities and Society **34**, 174-182 (2017)
9. G. Sun, J. Zacharias, Sustainable Cities and Society **35**, 323-330 (2017)
10. S. Zhang, Y. Ji, D. Sheng, J. Zhou, *7th International Conference on Green Intelligent Transportation System and Safety 2016* **419** (2018)
11. A. Ruda, L. Floková, I. Ivan, J. Horák , *Inspektor T. (eds) Dynamics in GIscience, GIS OSTRAVA 2017, Lecture Notes in Geoinformation and Cartography*, 383-388 (2018)
12. G. Fedorko, V. Molnár, J. Strohmandl, M. Vasil', *Transport Means – 19th International Conference on Transport Means*, 466-469 (Kaunas, Lithuania, 2015)
13. E. Jaffe, *The Methodology of Bike-Share Station Placement in New York City*, Available online: https://www.citylab.com/transportation/2011/10/how-new-york-city-will-choose-its-bike-share-stations/248/ (2011)
14. A. Kuptcova, P. Průša, G. Fedorko, V. Molnár, Transport Problems **11**, 21-31 (2016)

Low-carbon Scenarios Development for Modal Shift in the Chemical Industry

Barbara Ocicka[1,*]

[1]Warsaw School of Economics, Collegium of Business Administration, Department of Logistics, Al. Niepodległości 162, 02-554 Warsaw, Poland

Abstract. Supply chain managers have to deal with the performance requirement to significantly reduce CO_2 emissions in searching for excellence in green business processes management. The purpose of this article is to examine the perspectives on low-carbon scenarios development for modal shift in the chemical industry. The author outlines main research findings from the Interreg Central Europe ChemMultimodal project realised by 14 partners from 7 countries, among others by Department of Logistics at Warsaw School of Economics in Poland. The project idea is focused on analysing the potential and growth opportunities for multimodal transport usage in chemical supply chain management. Firstly, the objectives, current status and methodology of the project are explained. Then, the results of the research carried out among chemical and logistics companies operating in Poland are discussed. Furthermore, there is recognised that transport modal shift decisions determine changes in supply chain configurations that might be supported by planning and management tools. Consequently, the elements of the ChemMultimodal toolbox are outlined and its potential significance for low-carbon scenarios development is highlighted. As a result, both theoretical and practical implications of the research findings are indicated.

1 Introduction

Sustainability is one of the most important sources of competitive advantage of companies and supply chains in today's highly competitive and turbulent global business environment. Consequently, the phenomenon of best practices in logistics and supply chain management has gained strategic importance as a response to the rising problems and challenges, like e.g. limited energy supply, natural resources scarcity or global greenhouse gas (GHG) emissions at an alarming rate. The scope of best practices assessment is based on the necessity of their effects analysis in three categories: economic, social and environmental [1-3]. In this context, much of the recent logistics and supply chain management literature has focused special attention on the sustainability of transport management and operations, addressing especially the question of freight modal shift for greener logistics [4-6].

There is reported, that transport is responsible for 23 per cent of greenhouse gas emissions in the European Union [7]. A lot of initiatives and projects within European

* Corresponding author: barbara.ocicka@sgh.waw.pl

Union and other countries have taken into account to promote modal shift for decreasing emissions [8]. For instance, the flagship EU initiative is the Trans-European Network (TEN-T) programme, that aims to slow down the rate of increase of CO_2 emissions from freight transport operations through a combination of modal shift and more efficient operations [9]. However, there is a wide gap between anticipated real GHG emissions (CO_2 equivalents) and the ambitious EU purpose to reduce carbon emissions from transportation by 60 per cent by the year 2050 [10]. The just mentioned facts confirm the need to develop green transport practices and the necessity of their dissemination and promotion between companies. Furthermore, a decision support tools and frameworks for companies wanting to reduce their carbon footprint are rightly receiving a great deal of attention [11].

Facing the problem in an European context, there is a need to develop more projects devoted to particular regions and industries, considering their supply chain strategies and configurations. One of the initiatives contributing to the EU macro regional strategies is the Interreg Central Europe programme, that improves significantly capacities for regional development in innovation, CO_2 reduction, the protection of natural and cultural resources as well as transport and mobility. One of the initiatives co-funded by the Interreg CE programme is the ChemMultimodal project focused on 2 priorities: to increase multimodal transport by 10 per cent and to reduce CO_2 emissions by 5 per cent in the chemical industry.

The purpose of this article is to examine the perspectives on low-carbon scenarios development for modal shift in the chemical industry. It begins with a brief presentation of the ChemMultimodal project objectives, current status and methodology. In the next section, the results of the research carried out among chemical and logistics companies operating in Poland are discussed. Afterwards, the ChemMultimodal toolbox is described and its significance for low-carbon scenarios development is highlighted. Finally, the conclusions and implications of the research, the limitations of the work and recommendations for future research in the field of multimodal transport of chemical goods, are presented.

2 The ChemMultimodal project objectives and methodology

The main objective of the ChemMultimodal project is the promotion of multimodal transport of chemical goods by coordination and facilitation of cooperation between chemical companies, specialized logistics service providers (LSPs), terminal operators and public authorities in chemical regions in Central Europe. Its duration is limited to 3 years (from June 2016 till May 2019) and its budget amounts to 2.388.840 Euro. There are 14 project partners engaged from 7 countries: Austria, Czech Republic, Germany, Hungary, Italy, Poland and Slovakia. The specific objectives of the project take into consideration to: improve sustainability of chemical logistics, increase multimodal transport of chemical goods and facilitate cooperation of multimodal stakeholders. The achievement of the just mentioned specific goals should lead to increase multimodal transport by 10 per cent and to reduce CO_2 footprint by 5 per cent in the chemical supply chains reached in project lifetime. The ChemMultimodal scope has gained a noticeable significance, because the chemical industry is an important economic sector responsible for 8 per cent of freight transport in European countries.

2.1 The ChemMultimodal current status

The scope of the ChemMultimodal project encompasses 4 work packages (Table 1).

Table 1. ChemMultimodal current status. Source: [12]

Work package	WP description
WP1	• Analysis & Tool Development • **Output**: Tool for promotion of modal shift of chemical goods from road to intermodal transport
WP2	• Pilot Testing • **Output**: Pilot Projects and implementation in each country
WP3	• Capacity Building • **Output**: 6 Training Seminars (15 participants each) in DE, AT, HU, PL, IT, SK-CZ
WP4	• Strategy and Action Plan • **Output**: Transnational Strategy (1) and Action Plans (7)

The purpose of the first stage of the project (WP1) was to analyse multimodal transport of chemical goods in the countries and regions of project partners, while focusing on identifying multimodal transport importance, main routes, barriers and opportunities to improve the usage of multimodal transport in chemical logistics [13]. The results of the surveys were the starting point to develop the project tool, that should be helpful to promote and increase multimodal transport usage. The second part of work (WP2) is dedicated to pilot testing. The pilot projects will be realized in cooperation with chemical companies and logistics service providers in each country. The engaged entities will switch the so far used road routes to an intermodal road-rail solution. Next, the project partners will analyse and compare the reported data regarding the logistics performance of both transport processes, including: distance, volume, costs, service quality (transit time and punctuality) and CO_2 emissions. In consequence, the potential of a shift from road to intermodal road-rail transport, without jeopardizing logistics performance, will be identified. Afterwards, the knowledge generated from the pilot action will be shared and transferred to other companies and LSPs via 6 training seminars. The key purpose of the WP3 is capacity building with engagement of trained companies. Finally, the project team will develop one transnational strategy as well as action plans for each partner country (WP4). The current status of the project is the beginning of the WP2: pilot testing.

2.2 The ChemMultimodal methodology

In the WP1, a questionnaire-based survey was carried out among 21 entities between August and September 2016, including 12 chemical companies (producers and distributors) and 9 logistics companies (LSPs, carriers, rail and port operators) operating in Poland. Both groups are very important stakeholders of the ChemMultimodal initiative. The production or distribution companies as shippers make decisions on the purchase of transport and logistics companies play the key role in designing and providing environmentally sustainable solutions for supply chains [14]. Both project partners from Poland – Warsaw School of Economics and Polish Chamber of Chemical Industry were actively engaged.

The WP2 is focused on pilot testing. In this phase, the developed toolbox will be tested throughout the partner regions and countries in 7 pilot projects. Each pilot should work together with at least 5 companies with the objective to test a shift from road to intermodal road-rail transport in supply chains. The chemical and logistics companies engaged in the pilots will fulfil an evaluation questionnaire and report the original and tested routes. As a result, the collected case studies will be analysed to gain knowledge about the potential of new supply chain configurations to increase multimodal transport. Each project partner will also organise workshops with companies participating in the pilot stage.

The same methodology of the project throughout the partner regions and countries allows to successfully benchmark and transfer practices, which are worth following nationally and internationally.

3 Perspectives on multimodal shift in the chemical sector

The companies surveyed in 2016 evaluated the main advantages of multimodal transport on the scale from 1 – *not important* to 5 – *very important*. The outcomes of the evaluation are as follows: cost decrease (4.27), safety in terms of dangerous and hazardous goods transport (4.07), lowering CO_2 emission (3.50), time shortening (3.48), supporting CSR strategy (3.34) [13]. The respondents evaluated also the main disadvantages of multimodal transport, among others taking into consideration: lack of railway connections (4.13), lack of intermodal terminals (3.91), transit time (3.88), costs (3.76), low flexibility of transport management (3.64), extended planning and organizing forwarding time (3.31) and sector's regulations (2.46) [13].

In searching for the possibilities to shift road to multimodal transport, the chemical and logistics companies indicated the arguments encouraging them to use and offer multimodal transportation (Table 2). On the demand side, shorter transit time (5.0) and lower costs (4.7) are the main drivers, that could motivate the chemical companies to use multimodal transport. On the supply side, logistics companies may offer more competitive prices in case of large shipments and long-distance transport. The achievement of shorter transit time is more challenging because it depends strongly on external factors, like e.g. lack of direct railway connections or low interoperability within rail transport across Europe [13].

Table 2. Motivation (arguments) for multimodal transport. Source: [13]

Motivation (arguments) 1-*not important*, 5-*very important*	AVG. motivation for chemical companies	AVG. arguments for logistics companies
Cost	4.7	5.0
Safety	3.8	4.8
Capacity	2.8	3.8
Mode availability	3.3	
CSR strategy	2.3	3.8
Lowering CO_2 emissions	2.8	3.3
Transit time	5.0	1.0
Sector's regulations	3.0	

The respondents recognised the modal shift potential in any route characterised as follows: long enough to offer sufficient transport time, offered on a regular and long-term basis, offering backhaul opportunities. The chemical companies are especially interested in the development of the potential multimodal routes between Poland and such countries as: Czech Republic, Germany, Italy, Kazakhstan, Romania, Spain and Turkey.

It should be underlined that the perspectives on multimodal shift in the chemical sector are created by internal and external initiatives and incentives. On the one hand, there is the necessity for closer collaboration between chemical companies or development of sharing services by logistics companies to achieve higher availability of shipment consolidation and better effectiveness of transport solutions. It is also time to change habits of transport planners and enhance sustainability awareness among managers as well as operational workers to manage transport operations in the more environmentally-friendly way and measure key performance indicators in this area. Supply chain and logistics managers may use different frameworks and decision supporting tools to improve their practices in multimodal transport management and reduce the companies' carbon footprint [15]. On the other hand, the external incentives to promote multimodal transport are highlighted in the table 3.

Table 3. External incentives to promote multimodal transport. Source: [16]

Incentives	Average
Improvement of railway connections (frequency and punctuality)	4.50
Lowering costs of access to multimodal infrastructure	4.25
Consistency of transport time schedules	4.25
Better flexibility of transport management	4.13
Improvement quality of intermodal terminals	3.75
Increasing number of intermodal terminals	3.56
Sector's regulations (framework conditions)	2.50

4 The ChemMultimodal toolbox and its role in low-carbon scenarios development

4.1 Elements of the ChemMultimodal toolbox

The ChemMultimodal toolbox has been developed in close cooperation with logistics and chemical companies to promote modal shift from unimodal to multimodal chemical goods transport. It consists of four integrated pillars, namely (1) consulting services, (2) planning guidelines, (3) IT Visualization, (4) CO_2 Calculator (Fig. 1).

(1) **Consulting Services** Data base with stakeholders contacts Marketing platform	(2) **Planning Guidelines** Backbone of Toolbox National, legal regulations to be found here
(3) **IT Visualization** Intermodal Links Platform Visualize intermodal routes (frequency, available LSPs, information of pre- and post-haulage)	(4) **CO_2 Calculator** Calculation of CO_2 emission based on EN 16258 and McKinnon concept Create awareness for CO_2 savings

Fig. 1. Toolbox elements. Source: [12]

The first element called *consulting services* is focused on the partners' efforts to develop an approach to communicate and discuss the following issues with representatives of chemical and logistics companies: current transport patterns, existing potentials and possible actions to encourage modal shift and promote multimodal solutions. The second element known as *planning guidelines* is a backbone of the toolbox, because it provides national transport regulations, including e.g. loading and driving restrictions. The third element is a practical *IT visualization* established as an Intermodal Links Platform. It plays the role of the planner, that enables to visualize intermodal routes according to the requirements of the user, to present the list of available transportation offers of LSPs (including main, pre- and post-haulage routes) and contacts details. The fourth element is

 DOI: 10.1051/matecconf/201713400043

the *CO_2 calculator*, which gives useful feedback on CO_2 emissions with regards to different transport scenarios and helps in choosing the most environmentally-friendly option (Fig. 2).

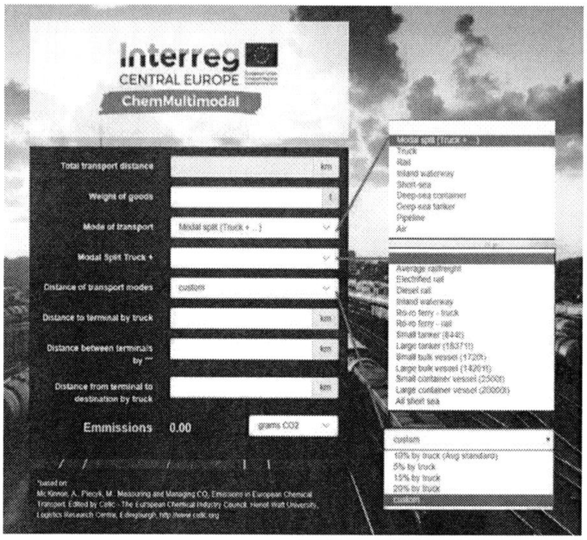

Fig. 2. CO2 Calculator for chemical transports. Source: [17]

It should be kept in mind that *consulting services* provide sufficient knowledge about the demand and needs of the chemical companies and LSPs using the three other elements (Fig. 3).

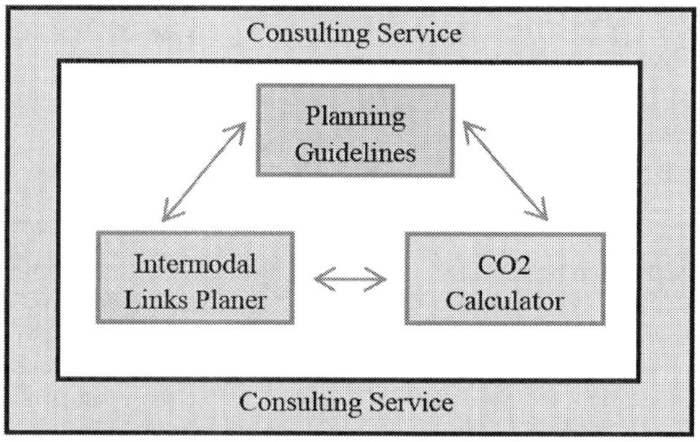

Fig. 3. Links between the toolbox elements. Source: [12]

4.2 Significance of the toolbox

The toolbox has a potential of a control tower for multimodal transportation planning and management, that supports chemical companies and LSPs in strategic and operational-level issues. The key purpose of its development and continuous improvement is to contribute significantly to optimization of the utilisation of multimodal transport solutions in the chemical industry. The Intermodal Links Platform creates a lot of cooperation possibilities between chemical companies and LSPs, accordingly on demand and supply side. There is not only the possibility to offer and buy full loads services. The planner offers as well the possibility to share transport vehicles and combine loadings ensuring the achievement of the highest possible filling rates and transport costs reductions. It allows in the best possible way, that every shipment in Europe might be planned and executed in a multimodal manner, including road service only for short pre-haulage and post-haulage distances to or from the transhipment terminals. According to shipper perspectives, the ChemMultimodal toolbox is useful on the most of the levels of low-carbon scenarios development in transport management.

5 Conclusion and future research

In summarizing the research findings in the ChemMultimodal project, switching to an intermodal road-rail solution is potentially worth considering, in particular, because of costs and safety reasons, but difficult to implement as it may jeopardize logistics performance with regards to transit time or flexibility. However, the strong need for low-carbon scenarios development in supply chain management requires to plan and manage modal shift in different industries. Logistics and supply chain managers need concepts and tools supporting and encouraging them to a change from unimodal truck to multimodal transport. The ChemMultimodal project delivers the toolbox, that allows to plan various multimodal scenarios and measure CO_2 emission as an indicator of environmental performance. It will be tested and further developed by the use of empirical data. In searching for low-carbon scenarios in the chemical industry, the future research will be focused on the empirical analysis of case studies of companies that will conduct modal shift in supply chains and verify its economic and environmental effects. This study has without a doubt several limitations, like: concentration on the single industry, limited number of pilots testing the scenarios for modal shift and difficult benchmarking analysis.

The article is a result of ChemMultimodal Project implementation and is co-financed by Interreg Central Europe Programme.

Interreg
CENTRAL EUROPE
European Union
European Regional Development Fund
ChemMultimodal

Scientific work financed from funding for science in the years 2016-2019 granted for the implementation of the co-financed international project.

This paper is supported by the research project "From horse-drawn railway to intermodal transport" within Visegrad Fund.

References

1. J. Elkington, *Cannibals with Forks: The Triple Bottom Line of the 21st Century Business* (Capstone Publishing, 1999)
2. C.R. Carter, D. S. Rogers, IJPDLM **38**, 360-387 (2008)
3. K. Rutkowski, *Best Practices in Logistics and Supply Chain Management. The Case of Poland* (OW SGH, Warsaw, Poland, 2009)
4. F. Eng-Larsson, Ch. Kohn, IJPDLM **42**, 36-59 (2012)
5. H. Pålsson, O. Johansson, Benchmarking: An International Journal **23**, 674-703 (2016)

6. H.A. von der Gracht, I. L. Darkow, IJLM **27**, 142-166 (2016)
7. European Commission, *A European Strategy for Low-Emission Mobility* (Brussels, Belgium, 2016)
8. R. Konings, H. Priemus, P. Nijkamp, *The Future of Intermodal Freight Transport. Operations, Design and Policy* (Edward Elgar Publishing, Cheltenham, UK, 2008)
9. A. McKinnon, M. Browne, A. Whiteing, M. Piecyk, *Green Logistics: Improving the Environmental Sustainability of Logistics* (Kogan Page, London, UK, 2015)
10. European Commission, *Roadmap to single European transport area – towards a competitive and resource efficient transport system* (Brussels, Belgium, 2011)
11. K. Liljestrand, M. Christopher, D. Andersson, IJLM **26**, 296-312 (2015)
12. N. Moritz, *Status quo. Toolbox. Future steps* (2017)
13. M. Cichosz, K. Nowicka, A. Pluta-Zaremba, M. Susik, *Report on analysis of multimodal transport of chemical goods for the identification of modal shift* (2016)
14. A. Bask, M. Rajahonka, IJPDLM **47**, 560-602 (2017)
15. M. Cichosz, K. Nowicka, *Narzędzia wsparcia zarządzania transportem multimodalnym – Projekt ChemMultimodal* (2017)
16. M. Cichosz, K. Nowicka, A. Pluta-Zaremba, Przemysł Chemiczny **7**, 1435-1439 (2017)
17. Interreg Central Europe, *CO2 Calculator for Chemical Transports,* Available online: https://ifsl50.mb.uni-magdeburg.de/chemmultimodal/ (2017)

Practical Use of the Braking Attributes Measurements Results

Ján Ondruš[1,*], *and Eduard* Kolla[2]

[1]University of Žilina, Department of Road and Urban Transport, Univerzitná 8215/1, 010 26 Žilina, Slovak Republic
[2]Institute of Forensic Research and Education of University of Žilina, Ulica 1. mája 32, 010 01 Žilina, Slovak Republic

Abstract. This contribution deals with issues of braking the passenger car. The measurement of braking deceleration of the vehicle Kia Cee′d 1,6 16 V was carried out by an optical device Correvit system. The measurement was carried out on the airport of the village of Rosina located close to Zilina. 10 drivers of different age, praxis, and kilometers driven participated in the measurement. The measured process was the vehicle full braking with the service brake of the initial speed of approximately 50 $km.h^{-1}$. Each of the drivers had 10 attempts. In the closure of this contribution the results of the performed measurements, their evaluation and comparison are presented. Practical result from the contribution is mainly the measurement set of braking deceleration of the respective vehicle during intensive braking.

1 Introduction

The braking deceleration value during intensive braking belongs to the important input data of the accident analysis. Only a small number of measurements made with the vehicles on common roads is published in the relevant literature. In this article the authors present their measurements results [1].

2 The experimental vehicle

A passenger car Kia Cee′d 1,6 16 V, production year 2008, belonging to the laboratories of the Department of Road and Urban Transport of the University of Žilina, was used as an experimental vehicle. During the trials the vehicle was loaded with two persons (a driver and a passenger who operated the measuring device). The experimental vehicle is shown in Fig. 1.

Tires Michelin Energy had dimension 185/65/R15 88 H. The front tires tread depth was 6.1 mm and the rear tires tread depth was 6.3 mm. At the time of the measurement the tires were pumped up to reach the compression specified by the producer, which was 220 kPa. The vehicle was equipped with the disc brakes on all wheels.

* Corresponding author: jan.ondrus@fpedas.uniza.sk

Fig. 1. The experimental vehicle. Source: authors

3 Measuring equipment

As the measuring device and recorder during the braking deceleration was used Correvit produced by the Corssys Datron. It is an optical device for measuring the braking deceleration process. This device consists of 4 main parts – microwave sensor, braking sensor, evaluation unit and control panel (fig. 2). The microwave sensor consists of sensor head and sensor electronics and accommodates additional connections, such as an interface for flow-measurement systems (for consumption tests), or trigger inputs for light barriers or brake switches, providing exceptional testing power and flexibility. The microwave sensor accommodates a precise, reliable and contactless measurement of the speed and length, using a technology based on a Doppler Effect [2-4]. To further identify additional data were, among others, used the devices for initial speed measuring - light barriers, cones, measuring tape, camera, camcorder, touch thermometer, slide calliper and notebook.

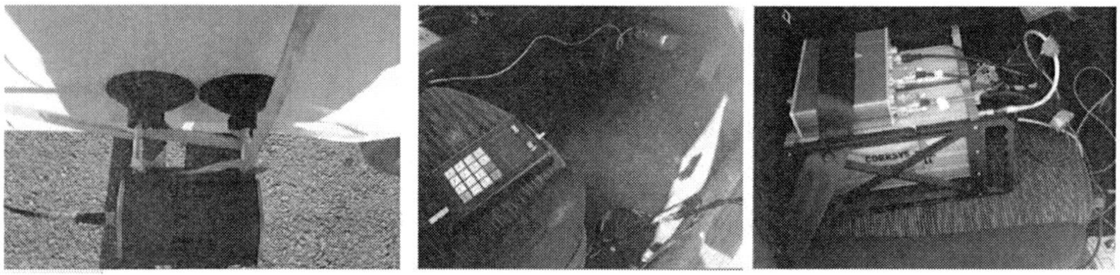

Fig. 2. Correvit – microwave sensor, evaluation unit, braking sensor and control panel. Source: Authors

The unit perceives the relative movement between itself and the test surface using a planar antenna, which projects two radar beams at 45° angles. Upon striking the test surface, the beams are reflected back to the sensor antenna. The resulting double frequency (equal to the difference of sent and received frequencies) is directly speed-proportional. The two-beam planar system increases accuracy by automatically compensating for mounting and pitch-angle errors. The gained signal is converted to the desired dimension via an onboard RISC processor and then sent to the corresponding outputs. Featuring an effective operating range of 300 mm to 1,200 mm, the Microstar Sensor can be used in applications demanding larger standoff distances without loss of accuracy [3, 4].

4 Conditions and methodology of the measurement

The experiment was carried out on 22. 09. 2015 at the Rosina airfield, located close to the town of Zilina. The surface conditions were: dry asphalt runway with the estimated adhesion factor of range 0.8 – 0.9 and the runway temperature was of 31.4 °C. The temperature of ambient air was 24 °C with very low speed wind so we could ignore its impact on the measurement. Prior to the experiment the calibration and test measurements to determine the necessary measurement characteristics (location of the braking manoeuvre, installation and calibration of the system, etc.) were carried out. The length of the experimental track was approximately 500 m. A particular measurement inaccuracy was caused by the surface gradient estimated at 2 %.

The experiment was carried out by 10 independent drivers of a different age, praxis length and kilometres driven. Out of these 10 drivers, 8 were men and 2 were women. Each of the drivers had 10 attempts and the conditions were equal for all of them. Table 1 shows the overview of the individual drivers.

Table 1. Basic information about the drivers. Source: authors

Driver	Gender	Age	Praxis length in years	Approximate no. of kilometres driven
1.	Man	24	6	9,000
2.	Man	41	23	50,000
3.	Man	38	20	65,000
4.	Man	45	27	85,000
5.	Man	37	17	70,000
6.	Woman	25	4	3,000
7.	Man	33	15	60,000
8.	Man	29	11	100,000
9.	Woman	27	9	20,000
10.	Man	36	18	150,000

After installation of the measuring device on the vehicle, the requested input data were set on the Correvit. System sampling rate was set to 200 Hz, i.e. we had 200 data per second available.

A phenomenon in question was a full braking of the vehicle with the vehicle service brake, while the vehicle was equipped with ABS. The test driver drove the vehicle up to the desired speed (50 $km.h^{-1}$). With the activation of the speed limiter fitted to a vehicle was achieved, that the vehicle did not exceed the speed. In about half of the departure runway the passenger activated a microwave sensor, and from that moment all important data were recorded. At the time of consistent desired speed of the vehicle the driver rapidly pressed the service brake pedal. This phenomenon was observed in the measuring system Correvit due to the location of the brake pedal sensor on the service brake, which was connected with this system. All other data were recorded in the memory of the evaluation unit and were available in a complete dataset of the program CeCallWin Pro. Braking was done to stop the vehicle till zero speed. In this state, when the vehicle was completely at rest, the measuring deactivated itself. After a short pause the next measurement was repeated.

5 Processing of measured data

The programme CeCallWin Pro 1.09.001 was used for the data evaluation. When used with the CeCalWin Pro Software, the Microstar Sensor functions as a complete data acquisition and evaluation system. The software functions enable test parameters and definitions to be permanently saved, along with online displays and evaluations, e.g. charts and plots. All measured data can be saved and evaluated off-line.

Data obtained directly from an external memory of the evaluation unit are stored in the program CeCallWin Pro in the ADF format. With this software, the data are exported to TXT format (fig. 3), and then imported into Excel file [5].

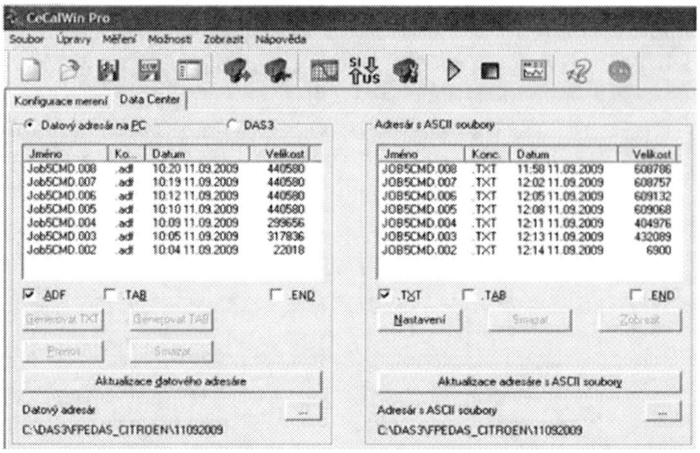

Fig. 3. Data export from ADF format to TXT format. Source: [5]

These data are clearly established in tabular form with the required parameters and it is possible to use them for future work. Data which were exported to the XLS file were: time, distance, speed, acceleration, respectively deceleration (each 0.005 seconds). Another important data point was an automatic indication of the commencement of braking, meaning the moment when a driver sharply pressed the brake pedal (in the Fig. 4 marked in yellow). Considering that a part of the experiment was the pursuit of the vehicle movement during the full braking deceleration, the start was set at this time point where I have added 0.2 seconds due to the technical specifications of the braking system (fully braking onset time + technical delays of brakes) and a time delay of a brake sensor (in the Fig. 4 marked in orange).

The end of the action was determined by the recorded stopping distance in Excel. The track has not changed, but measuring system still reported the movement of the vehicle. It means that because of the influence of centrifugal forces of the vehicle it was still swinging on the spot for a while (in the Fig. 4 marked in green). The individual kinematic data of the selected attempt are shown in Fig. 4.

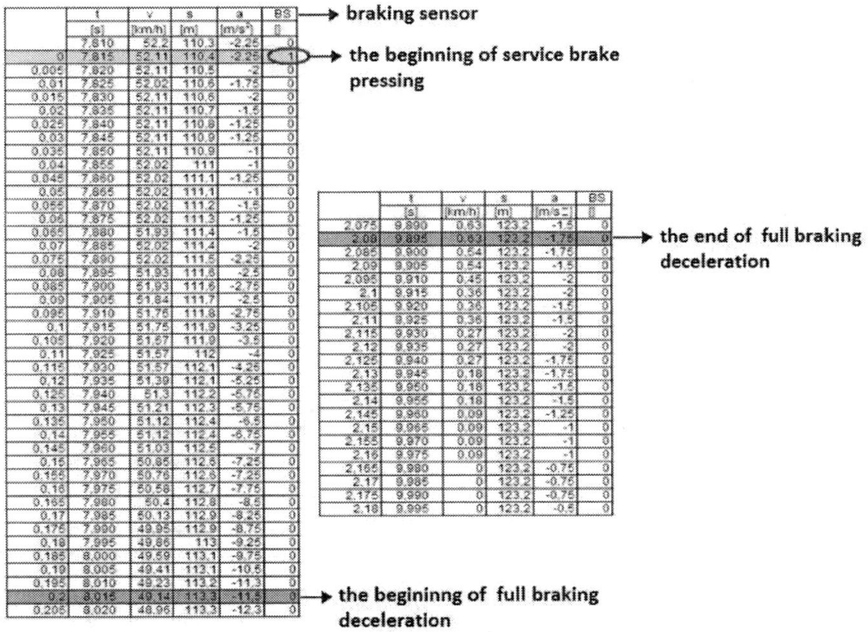

	t	v	s	a	BS
	[s]	[km/h]	[m]	[m/s²]	[]
	7.810	52.2	110.3	-2.25	0
0	7.815	52.11	110.4	-2.25	1
0.005	7.820	52.11	110.5	-2	0
0.01	7.825	52.02	110.6	-1.75	0
0.015	7.830	52.11	110.6	-2	0
0.02	7.835	52.11	110.7	-1.5	0
0.025	7.840	52.11	110.8	-1.25	0
0.03	7.845	52.11	110.9	-1.25	0
0.035	7.850	52.11	110.9	-1	0
0.04	7.855	52.02	111	-1	0
0.045	7.860	52.02	111.1	-1.25	0
0.05	7.865	52.02	111.1	-1	0
0.055	7.870	52.02	111.2	-1.5	0
0.06	7.875	52.02	111.3	-1.25	0
0.065	7.880	51.93	111.4	-1.5	0
0.07	7.885	52.02	111.4	-2	0
0.075	7.890	52.02	111.5	-2.25	0
0.08	7.895	51.93	111.6	-2.5	0
0.085	7.900	51.93	111.6	-2.75	0
0.09	7.905	51.84	111.7	-2.5	0
0.095	7.910	51.75	111.8	-2.75	0
0.1	7.915	51.75	111.9	-3.25	0
0.105	7.920	51.57	111.9	-3.5	0
0.11	7.925	51.57	112	-4	0
0.115	7.930	51.57	112.1	-4.25	0
0.12	7.935	51.39	112.1	-5.25	0
0.125	7.940	51.3	112.2	-5.75	0
0.13	7.945	51.21	112.3	-5.75	0
0.135	7.950	51.12	112.4	-6.5	0
0.14	7.955	51.12	112.4	-6.75	0
0.145	7.960	51.03	112.5	-7	0
0.15	7.965	50.85	112.6	-7.25	0
0.155	7.970	50.76	112.6	-7.25	0
0.16	7.975	50.58	112.7	-7.75	0
0.165	7.980	50.4	112.8	-8.5	0
0.17	7.985	50.13	112.9	-8.25	0
0.175	7.990	49.95	112.9	-8.75	0
0.18	7.995	49.86	113	-9.25	0
0.185	8.000	49.59	113.1	-9.75	0
0.19	8.005	49.41	113.1	-10.5	0
0.195	8.010	49.23	113.2	-11.3	0
0.2	8.015	49.14	113.3	-11.5	0
0.205	8.020	48.96	113.3	-12.3	0

	t	v	s	a	BS
	[s]	[km/h]	[m]	[m/s²]	[]
2.075	9.890	0.63	123.2	-1.5	0
2.08	9.895	0.63	123.2	-1.75	0
2.085	9.900	0.54	123.2	-1.75	0
2.09	9.905	0.54	123.2	-1.5	0
2.095	9.910	0.45	123.2	-2	0
2.1	9.915	0.36	123.2	-2	0
2.105	9.920	0.36	123.2	-1.5	0
2.11	9.925	0.36	123.2	-1.5	0
2.115	9.930	0.27	123.2	-2	0
2.12	9.935	0.27	123.2	-2	0
2.125	9.940	0.27	123.2	-1.75	0
2.13	9.945	0.18	123.2	-1.75	0
2.135	9.950	0.18	123.2	-1.5	0
2.14	9.955	0.18	123.2	-1.5	0
2.145	9.960	0.09	123.2	-1.25	0
2.15	9.965	0.09	123.2	-1	0
2.155	9.970	0.09	123.2	-1	0
2.16	9.975	0.09	123.2	-1	0
2.165	9.980	0	123.2	-0.75	0
2.17	9.985	0	123.2	-0.75	0
2.175	9.990	0	123.2	-0.75	0
2.18	9.995	0	123.2	-0.5	0

Fig. 4. Measured data view in MS Excel. Source: authors

The mean of fully braking deceleration in this case was calculated from the measured values of the stopping distance and speed using the relationship [6]:

$$b = \frac{v_1^2}{2.s} \tag{1}$$

Where: b - mean of fully braking deceleration [$m.s^{-2}$],
v_1 - the speed of the vehicle at the beginning of fully braking deceleration [$m.s^{-1}$],
s - distance travelled by the vehicle during the fully braking deceleration [m].

Relation (1) expresses the relationship to calculate the acceleration (deceleration) evenly accelerated (decelerated) movement.

6 Measured data evaluation

During each attempt the speed at the beginning of full braking deceleration, calculated mean of full braking deceleration, braking distance and time were evaluated. Some measurement attempts were not evaluated due to measuring equipment failure. This means that in some cases the measuring equipment did not measure the end of the measurement or the vehicle braking distance. Out of the 100 attempts 12 measurements were incorrect.

Due to the restricted range of this article we do not present the individual measurement results for each driver, but only their final average values. The highest average braking deceleration was reached by the driver no. 10, precisely 9.61 $m.s^{-2}$ (the driver with the highest number of kilometres driven – 150,000). On the other hand, the lowest braking deceleration was reached by the driver no. 3, precisely 9.07 $m.s^{-2}$ (the driver with 65,000 km driven).

The average value of all average braking decelerations of the individual drivers was 9.38 $m.s^{-2}$. Furthermore, the minimum and the maximum values of the reached braking deceleration, and the standard deviation were calculated for each of the drivers (tab. 2).

Table 2. Mean of full braking deceleration of the individual drivers. Source: Authors´ calculations

Driver	No. of incorrect measure-ments	Min. value of braking deceleration [m.s^{-2}]	Max. value of braking deceleration [m.s^{-2}]	Average braking deceleration of individual drivers [m.s^{-2}]	Standard deviation of individual drivers [m.s^{-2}]	Average value of braking deceleration [m.s^{-2}]
1.	3	8.94	10.01	9.34	0.374	9.38
2.	2	9.10	10.60	9.44	0.489	
3.	0	7.68	9.85	9.07	0.597	
4.	3	9.02	9.70	9.46	0.227	
5.	1	7.69	10.22	9.54	0.689	
6.	1	9.05	9.77	9.35	0.211	
7.	1	8.77	9.78	9.46	0.298	
8.	0	8.64	9.82	9.25	0.316	
9.	1	8.50	10.11	9.32	0.494	
10.	0	9.11	10.08	9.61	0.296	

Based on the above stated, we can observe that there are no major deviations among the calculated braking deceleration values. Also taking into account that some values were bigger or smaller that the calculated average, there was no major difference among the values. This was also statistically proved by the value of the standard deviation of the average braking deceleration, which is 0.146 m.s^{-2}.

7 Conclusion

The result of the measurements is, that despite of the different age, length of the praxis and number of kilometres driven of the individual drivers, their final braking deceleration was only slightly different. The results were also not influenced by the gender of the driver. During the measurements we calculated the braking distance and the full braking deceleration from the moment of the full braking impact commencement. It is necessary to point out, that we did not consider the driver´s reaction time, which is a factor essentially influencing the time and distance of the vehicle stopping. It would certainly be interesting to make a greater number of trials.

An important contribution of this experiment is the practical measurement of the braking behaviour of passenger car Kia Cee´d, in particular focus on the braking distances and average full braking deceleration. This value is very important for the Road Traffic Experts when elaborating the Forensic Expert's Reports of traffic accidents, where it represents the core entry value of braking behaviour of the respective vehicle; or alternatively in other areas, where the measured results can help to achieve the solution [7].

This contribution is the result of the project implementation:
2/KCMD/2017
Centre of excellence for systems and services of intelligent transport II, ITMS 26220120050 supported by the Research & Development Operational Programme funded by the ERDF.

This paper is supported by the research project "From horse-drawn railway to intermodal transport" within Visegrad Fund.

References

1. P. Hockicko, J. Ondruš, *NTF 2012: proceedings of the conference,* 214-217 (Brno: University of Technology Czech Republic, 2012)
2. DAS 3, *Corrsys Datron*, Available online: http://www.techtarget.com.br/portal/sites/default/files/cds-d_DAS-3_e.pdf (2007)
3. AB Dynamics, *Corrsys-Datron Optical Sensors*, Available online: http://www.abd-mf-jp.com/item/corrsys/index_e.html (2017)
4. Correvit, Corrsys *Datron Correvit SFII P User Manual Page 21,* Available online: https://www.manualslib.com/manual/912858/Corrsys-Datron-Correvit-Sfii-0P.html?page=21#manual (2008)
5. *Software CeCallWin Pro 1.09.001*
6. J. Vrábel, B. Šarkan, T. Skrúcaný, *Automotive safety 2016: problemy bezpieczeństwa w pojazdach samochodowych: X international science-technical conference*, 383-388 (Kielce – Ameliówka, Wydawnictwo Politechniki Świetokrzyskiej, Poland, 2016)
7. J. Vrábel, et. al., Machines, technologies, materials: international virtual journal **8**, 6 (2014)

MATEC Web of Conferences **134**, 00045 (2017) DOI: 10.1051/matecconf/201713400045
LOGI 2017

Synergies of the Liberalization of the Railway Transport Market

Michal Panák[1], *Eva* Nedeliaková[1], *Borna* Abramović[2,*], and *Denis* Šipuš[2]

[1]University of Žilina, Faculty of Operation and Economics of Transport and Communications, Department of Railway Transport, Univerzitná 8215/1, 01026 Žilina, Slovakia
[2]University of Zagreb, Faculty of Transport and Traffic Sciences, Department of Railway Transport, 10000 Zagreb, Croatia

Abstract. The liberalization of transport market brings various effects, which in the context of the assessment of synergies can affect in different ways the company, a customer as such, and transport undertakings operating on the railway transport market as well. This paper provides an innovative perspective on the relationship between liberalization and synergy, defining new types of synergies that have not yet been monitored in the conditions of railway transport. This approach is interesting because of the possibility of assessing the operation of railway undertakings in the open transport market. The eminent is characteristic of integration type synergies and emergence type synergies in relation to railway transport, as well as the breakdown of synergies in relation to the customer and carrier.

1 Introduction

In the nineties of the last century, the opening of borders and the associated emergence of new opportunities had an impact on transport. After such changing conditions it was the road transport that adapted in the best way possible. A great quantity of freight transport was moved to the road, and the public passenger transport underwent a similar change, with passengers starting to use individual road transport. Because of this development, the traffic load on roads increased significantly. The opening of the transport market and the lengthy preparation of conditions on the liberalization of railway transport market caused further decline in the share of railway transport on the transport market.

One of the possibilities of the global transport market and promotion of the improvement of railway transport services is liberalization, i.e. opening of the transport market, entry of carriers on the market on non-discriminatory terms.

Specification of measures to achieve interoperability and security on the entire continent is evident from the draft legislation in 2002, and the progress of work on the implementation of the adopted EU legislation. In addition to liberalization, the emphasis is

* Corresponding author: babramovic@fpz.hr

on safety and technical uniformity. This means that Member States must restrict adoption of new national safety regulations in order to harmonize them. Another objective is the interoperability aimed at an optimal level of technical harmonization.

The first task was to ensure the EU legislative and institutional coordination in the development of European standards within CEN, CENELEC, ETSI and UIC AEIF and their implementation into national systems of EU countries. A parallel task is to coordinate the preparation of technical specifications for interoperability of high-speed and conventional railways along with the implementation into the national environment. At present, the role of ensuring interoperability at European level is provided by the European Railway Agency.

The basic objective of adopted EU legislation is therefore enlarging market share using market forces to ensure the competitiveness of the rail sector. Railway transport services are the most difficult area to liberalize compared to other modes of transport, because infrastructural and technical standards are very complex, specific and vary largely between Member States. Apart from a few high-speed passenger trains, cross-border trains generally have to change locomotives and crews at the borders, which results in delays and higher costs. This is one of the main reasons for the low competitiveness of rail transport. It is of utmost importance to address the issue of monitoring synergies of liberalization of the transport market, since these effects may not always have a positive character. At the same time, however, it should be added that entry of new carriers on the transport market could lead to an increase in the supply and quality of railway transport services.

2 History and legislative regulation

The initial tendency to commence reforms of transport market in Europe was Directive 91/440/EC, which demanded accounting departments in historical railway undertakings. The directive has allowed access to national infrastructure within an international transport, which manager of infrastructure was obligated to charge [1]. This directive was followed by another one, Directive 95/18/EC concerning licensing transport undertakers, and then subsequent Directive 95/19/EC concerning railway infrastructure allocation and infrastructure charges.

Because the reform of European railway market was progressing slowly, four other directives were released to speed it up:

- 12/2001/EC – increased independence of railway undertakers, especially infrastructure managers from national budgets of state members. The Directive demanded strict separation of freight transport accounting from personal transport accounting, and within the personal transport avoiding cross-financing. This Directive is well-known for the definition of "Trans European Rail Freight Network". [2] Free access to services was to be included in this network as from 2003. A complete liberalisation of international railway freight transport was planned for 2008.
- 13/2001/EC – established common access to licensing railway undertakers and required mutual acceptance of licences within state members.
- 14/2001/EC – according to this Directive, an infrastructure manager had to introduce surcharge to marginal costs of charging infrastructure, by means of free and non-discrimination politics.

These three aforementioned directives were part of the so-called "First Railway Package", also referred to as Infrastructure package. Member States were obliged to incorporate the directives of this package into their national legislations until 2003.

First package was followed by "Second Railway Package" adopted in 2004 incorporated into national legislation by April 2006 [1, 3]. A part of its measures was the directive about railway safety, and a significant fact was the establishment of "European Railways Agency". Second Railway package changed Directive 96/48/EC about international operation of high-speed railway system and its part has become Directive 16/2001/EC, which discussed about interoperability in conventional transport.

"Third Railway package" tackles issues about compensations in case of non-compliance with contractual requirements on service quality, about passengers' rights and obligations and in a sense of liberalisation, the main part is about personal railway transport.

Part of "Fourth Railway package" includes the following directives, with a transposition deadline on or before until 16 of June 2019 [1, 4]:

- Directive EP and the Council of the EU 2016/797 of 11 May 2016 on interoperability of railway systems in EU.
- Directive EP and the Council of the EU 2016/798 of 11 May 2016 on railway safety.

Another major change in legislation is the Proposal for regulation of the Committee relating to fees paid by railway undertakings for EUAR. [1] This proposal adjusts charges and payments for services of EUAR resulting from the Regulation 2016/796 of the EP and the EU Council of, calculation methods, payment terms, and conditions.

Liberalisation of international railway transport is a key question when it comes to support of railway transport as an ecological way of transportation, which is why it is so important to monitor the impact of it on society, individuals and transport undertakings [1, 4, 5]. In the research, we monitored and defined different types of synergies that can emerge on the transport market regarding the new conditions in case of a new transporter entering the market.

3 Synergy and its relation to liberalization

The main reason for synergies in railway transport is their growing importance, and if required - operative interventions to railway operations that affect the quality of services. In these cases, there are many dynamic factors that affect the quality negatively, which evolve constantly over time. In most cases, it goes about unsatisfactory communication between infrastructure managers and carriers, which has a significant impact on the quality of services provided. This situation then ultimately affects satisfaction and the needs of existing customers of railway undertakings, as well as decisions of potential customers about the use of railway transport for the future.

Due to a very wide range of tasks and cooperative collaborations of two or more subsystems (external business partners, internal organizational units, teams, etc.), synergy in its management applications is very widespread with diverse content. Therefore, it is necessary to establish appropriate classification effects arising from their acts [6]. For the classification of the type of synergy that can be a source of inspiration for modern management, synergies are defined as a cooperation between manager infrastructure and carriers according to these prominent authors:

A) Peter A. Corning [7]

- Synergy of functional complementarities - are the cases synergy that accrues from accouplement accessories. Nowadays, the emphasis is placed on the effective cooperation within the meaning of the communication between manager of the railway infrastructure and the carriers whose services for customers each represent complements, i.e. they are linked, unable to exist without each other.
- Synergy of emergent phenomena - present in the cases of qualitative changes that occur due to mutual internal interaction of partial subsystems. In the implementation of the quality management system of railway undertakings, it is about an effort to improve

services. Qualitative changes are necessary for the effective functioning of railway undertakings on the transport market. Their implementation is difficult in terms of time and people factor, but they bring significant effects in logistics processes. Moreover, it is usually difficult to convince railway infrastructure managers and employees that change is necessary.

- Synergy of a combination of labour - the case of this synergy refers to the creation of conditions for the collection of the same or similar activities, so that individual activities are not conducted twice and so that the division of labour would stay fair.

B) Igor H. Ansoff [8]

- Sales synergy - arises when different services use shared distribution channels or selling points. In the case of railway passenger transport, it is the sale of different types of travel documents on the whole railway network and transport alone.
- Operating synergy - it results from the economical use of facilities and personnel of railway undertakings.
- Investment synergy - this synergy category is the result of improved recovery of resources. In railway undertakings, it refers to a more effective management of logistics processes (management of wagons, management of drive railway vehicles, management of invest resources to maintenance and repairs of railway infrastructure, and etc.).
- Managerial synergy - it occurs when the new railway undertakings recover existing well-established management knowledge.

C) Robert S. Kaplan and David P. Norton [9]

- Financial synergy - it is the integrated management of logistics processes in the value chain of business activities of the railway undertakings.
- Customer synergy - creation and implementation of joint offers to customers. In relation with transport services, it comes to getting and keeping satisfied customers.
- Synergy of internal processes - includes practical and effective changes in logistics processes railway undertakings for providing better quality services.
- Learning and growth synergy - association of innovative technologies in the field of logistics processes and customer services.

D) Michael Goold and Andrew Campbell [10]

- Integration type synergies
- Emergence type synergies

Integration type synergies belong to the synergy whose effects can be quantitatively evaluated. They arise out of the cooperation between minimally two partial processes of different levels of management railway undertakings. They are processes that can be objectively monitored using, for example, statistical methods of quality management. Projecting and using the integration type synergies and their effects are expected in the needed time period as it is characterized in the following Figure 1.

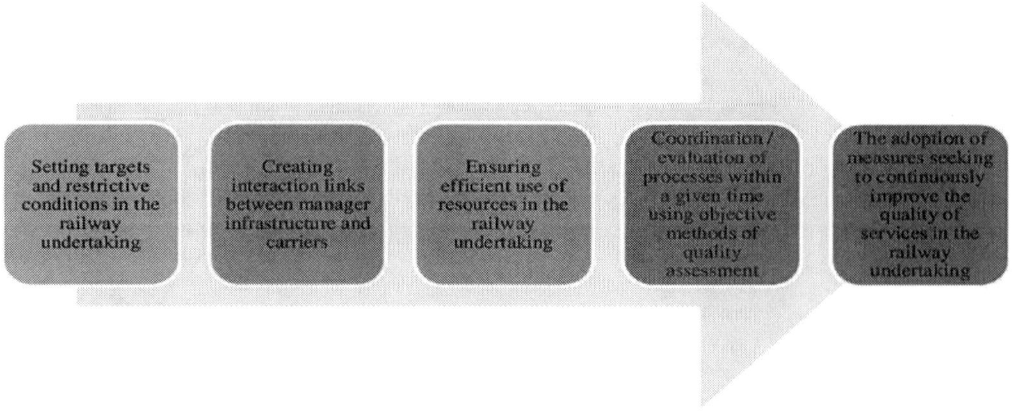

Fig. 1. Projection and use of integration type synergies and their effects in a time sequence. Source: authors

The basic idea for the formation of models synergies of integration type is a suitable distribution (allocation) or redistribution (reallocation) of activities and considered resources subsystems, and their use in favour of functioning of systems that create these subsystems. Contrario sensu are the emergence type synergies, in which arising effects are completely or largely qualitatively different from resources subsystems. [9] In one of their publications, Goold and Campbell report that "the results of emergence effects are manifested in the features, in the abilities to behave and in the results system of managerial work as a whole".

There are various types of models closely related to this synergy. One of them is Model of Critical Success Factors [5, 10, 11]. For modification processes with focus on the railway transport, see the following Figure 2.

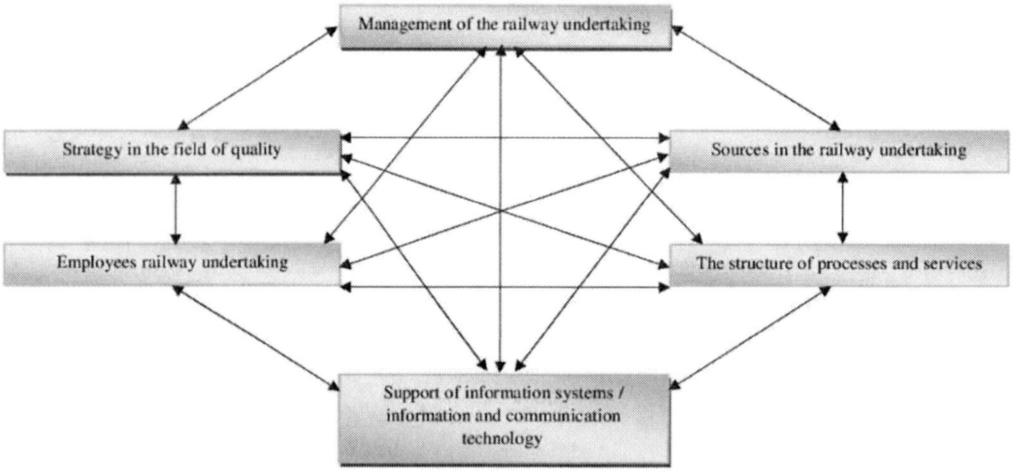

Fig. 2. Model Critical Success Factors railway undertaking and their book bands on the internal synergies. Source: authors

The purpose of the Model of Critical Success Factors is to focus attention of railway infrastructure managers on those aspects of their work that are essential for a more efficient provision of services. As far as the outputs are concerned, there are suitable selected arrangements of indicator activities for this undertaking. These indicators are characterized by prosperity, stability, competitive position, and others. The functioning of individual subprocesses is modified depending on their interactions with each other [5]. This interaction can vary in time, and emerging synergies may change over time, such as changing the quality of services.

These synergies significantly affect the performance of transport undertakings on the liberalized transport market, but also the customer service.

4 Basic types of synergies on the transport market

As already mentioned, synergy can be understood as a cooperation between infrastructure manager and carriers, or carriers to carrier, and customer to carriers, in terms of one mode of transport or a combination of several. At the same time, synergies can also work from outside business environment. These can be both positive and negative. As can be seen in Figures 3 and 4, synergies of liberalization can be distinguished from those that act on the customer and those that act on the carrier.

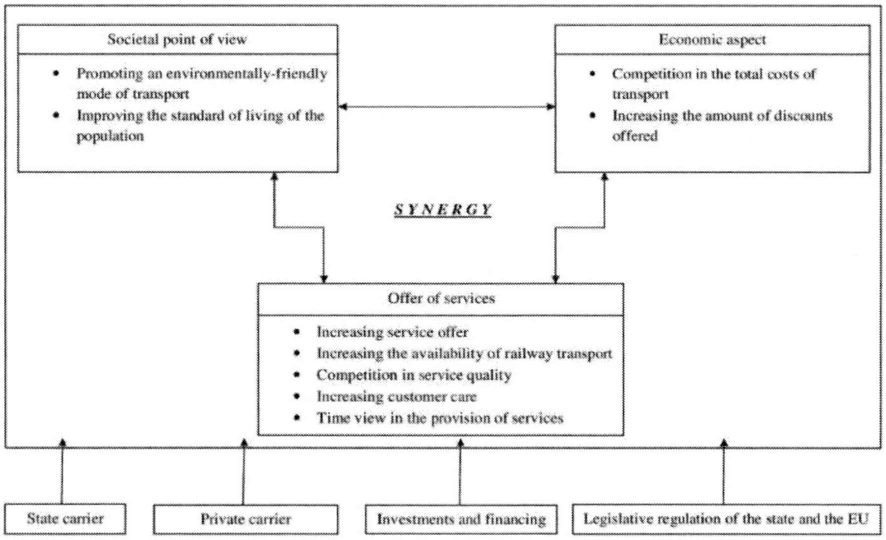

Fig. 3. Synergies of liberalization in connection with customer. Source: authors

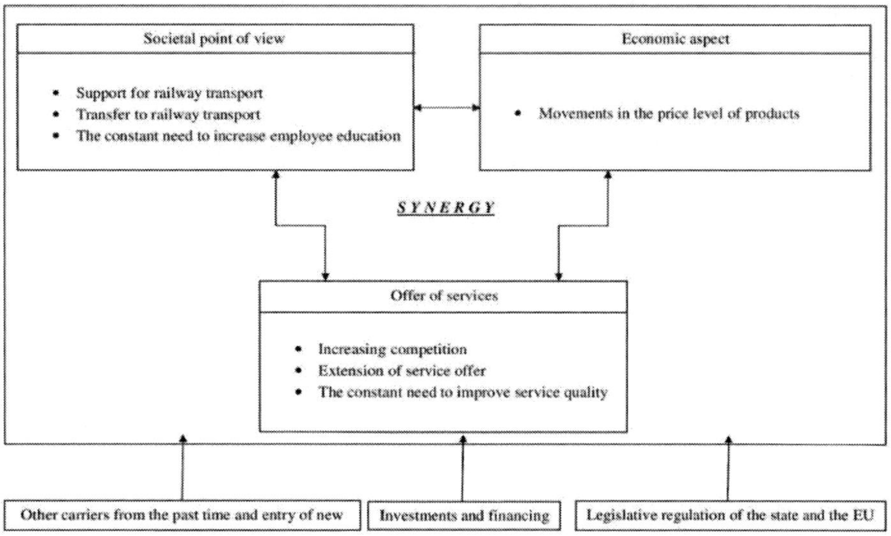

Fig. 4. Synergies of liberalization in connection with carrier. Source: authors

These synergies are closely linked to the competitiveness of railway transport more than other modes of transport. To create competitiveness in the railways, the EU has taken the path of separating rail infrastructure from rail services, although the success of such initiatives has varied under the terms of specific Member States. The current ability of service providers to take advantage of such rules depends on the willingness of States to help them [12]. Training of crew and certification for equipment can be a major obstacle for market access. For example, it took France and Germany years to mutually certify their respective high-speed trains for operation on the Paris-Frankfurt railway, which became operational in 2007. Also, the result of the White Paper on transport are several policy initiatives including encouraging the liberalization and harmonization of Europe's railway systems.

The European Union supports increasing the competitiveness of rail passenger services in the interests of the environment, and enhancing the quality of provided services. The revitalisation of rail transport on the European transport market is a challenging tasks, particularly because of inconsistencies in terms of national networks, the technical incompatibilities of rail systems, but also due to legislative restrictions [6]. The governments of some countries block the opening of passenger transport networks so as not to allow foreign companies access to their markets. A bright spot that determines the steps to liberalization of international passenger transport was the third railway package, which defined the year 2010 as the main objective of liberalization.

5 Conclusion

The opening up of railway sector to competition is given by the EU's approach in this area, which has gradually adopted legislation determining the direction and pace of liberalization railway sector binding for Member States.

Railway transport in the past constituted the basis of the transport system of individual European states, but it began to lose importance with the development of air and road transport. The fact that railways failed to respond to competitive pressure, in particular on road transport, resulted from two main reasons. The first is that railways were organized on a national principle - networks were built as national, taking into account national and regional requirements and therefore there was no interoperability. The second problem was that railways were large undertakings historically owned, controlled and regulated by the state, so they were not efficient and flexible enough to respond to changes in customer demand for transport. For these reasons, railways were unable to capture the potential of international cooperation and foreign trade growth, losing their overall position in the field of transport, particularly in favour of road transport.

The research into two major international projects has highlighted the clear importance of monitoring synergies in relation to liberalization. New synergies have been defined in relation to the customer and carrier, even when a new carrier enters the transport market. Research underlines that assessment of synergies significantly affects supply of services, quality of services, and the constant need to increase employee training. Synergies in relation to liberalization are one of the decisive factors that can affect the transfer of railway transport and promote the ecological aspect of transport. This research is also innovative in that it addresses the idea of raising the standard of living of the population.

This research was financially supported by the project KEGA 010ŽU-4/2017 "New methods of teaching quality management in the study program Railway transport with a focus on optimization of extraordinary events in terms of customer orientation" - Faculty of Operation and Economics of Transport and Communications, University of Zilina and the PROM – PRO - project "Adapting the railway system in the integrated passenger transport (ARSIPT) " solved at Faculty of Transport and Traffic Sciences, University of Zagreb.

This paper is supported by the research project "From horse-drawn railway to intermodal transport" within Visegrad Fund.

References

1. A. Tomová, *Models of structural reform of railways* (University of Žilina/EDIS - university press, Žilina, Slovak Republic, 2010)
2. B. Boskovic, M. Bugarinovic, Transport Policy **41**, 50-59 (2015), DOI: 10.1016/j.tranpol.2015.03.009
3. J. Kintler, *Managing and modelling of financial risks – 7th international scientific conference*, 358-366 (VŠB – Technical University of Ostrava, Czech Republic, 2014)
4. M. Jandova, T. Paleta, *European Regulation of the Railway Transport – the 1st Railway Package Recast and the 4th Railway Package Proposal* (2013), DOI: 10.5817/CZ.MUNI.P210-6425-5
5. E. Nedeliakova, M. Panak, *BEM2015 – International Scientific Conference: Business Economics and Management*, 172-179 (Procedia Economics and Finance **34**, 2015), DOI: 10.1016/S2212-5671(15)01616-0
6. E. Nedeliakova, B. Abramović, M. Panak, *Support for new approach within rail freight transport leading to creation of favourable conditions in transport market* (University of Žilina, Slovak Republic, 2015)
7. P.A. Corning, *Nature's Magic. Synergy in Evolution and the Fate of Humankind* (Cambridge University Press, Cambridge, 2003)
8. I.H. Ansoff, *Implanting Strategic Management* (Prentice Hall, Englewood Cliffs, 1990)
9. R.S. Kaplan, D.P. Norton, *The Execution Premium. Linking Strategy to Operations for Competitive Advantage* (Harvard Business Press, Boston, 2008)
10. M. Goold, A. Campbell, *Desperately Seeking Synergy* (Harvard Business Review, Cambridge, 1998)
11. E. Nedeliakova, M. Panak, J. Ponicky, R. Sousek, *Proceedings of the 2016 International Conference on Engineering Science and Managament*, 195-198 (ER-Advances in Engineering Research **62**, 2016)
12. V. Stefancova, E. Nedeliakova, C. Lopez-Escolano, *12th international scientific conference of young scientists on sustainable, modern and safe transport*, 834-839 (Procedia Engineering **192**, 2017), DOI: 10.1016/j.proeng.2017.06.144

Comparison of Transport Services in Rural Municipalities of South Bohemia

Petra Pártlová[1,*], and *Veronika* Humlerová[2]

[1]Institute of Technology and Business in České Budějovice, Faculty of Corporate Strategy, Department of Management, Okružní 517/10, 370 01 České Budějovice, Czech Republic
[2]Institute of Technology and Business in České Budějovice, Faculty of Corporate Strategy, Department of Tourism and Marketing, Okružní 517/10, 370 01 České Budějovice, Czech Republic

Abstract. Transport services and location are two of the most important key factors of regional development, especially in rural areas; in particular as in the Czech Republic, more than 90% people live in such areas. A location well related to the developing areas, population and economic activities density along the most important traffic flows seems to be the factor influencing all parts of the economical potential. It is in the rural area, where depopulation is currently occurring, however, the exposure to traffic is often neglected. The paper deals with the transport accessibility of the area according to the municipal action groups in the South Bohemian Region. The authors analyze the network of transport services in rural areas and propose measures that could be an opportunity to innovate efficiently and complete transport infrastructure in rural municipalities, including involvement of local actors.

1 Literature review

Rural areas in the Czech Republic and their population deal with a number of difficulties. After the post-totality reformation in 1989, there has been an increase in spatial imbalances in the distribution of population, but also to deepening the differences in socio-economic maturity of the regions [1-3]. Progressive activities, skilled labour, and human capital have increasingly concentrated in dynamically forming metropolitan areas, while rural area has been stagnant [4-5].

Many rural settlements have lost their services and facilities, reducing the frequency of public transport. The municipal infrastructure is inadequate, job opportunities are limited, and social and cultural activity is diminishing.

The outflow of younger and more educated people away from the area is a possible way how the population deals with the situation. Another possibility how to deal with the problems could be to join in activities at local and micro-regional levels. Support should therefore be given to the activation of local elites and the influx of actors from the outside [4]. If we want to deal with the problems of rural areas, i.e. to develop the area, we need to involve the factors important for the development of rural areas (generally in the

* Corresponding author: vachalova.p@seznam.cz

development of the territory). Binek et al. [6] defines endogenous and exogenous development factors. Exogenous factories are a framework of factors and tools that influence rural development, but individual actors cannot influence them directly. On the other hand, endogenous factors are directly linked to the activities of rural users. Such factors include population migration, lack of transport service, population activity, such as the development of the region through social networking of actors. Regarding rural development, there are different forms of working together and different organization that dealing with the problems of the municipalities. The organizations include: Spolek pro obnovu venkova (the Association for Rural Reconstruction), SMO ČR (the Union of Towns and Municipalities of the Czech Republic), Národní síť zdravých měst (the Healthy Cities of the Czech Republic), Národní observatoř venkova (National Rural Observatory) and the National Network of Local Action Groups in the Czech Republic [7-9]. Local action groups are formed in the EU under the LEADER method [7]. By Larsson [10], the original purpose of this Community initiative was to develop innovative ideas for a rural development model that would be useful in other areas too. Ray [11] describes LEADER as a program that helps the socio-economic viability of the area and supports innovative approaches to rural development at local level through the use of local resources, both physical and human. A well-working LAG, which wants to use the potential of its territory as much as possible, mainly applies the features of networking and cooperation [12-14]. It uses new approaches and procedures to guide the local development of the area [15]. Local action groups do not have the objective of generating profits, but of creating a synergy effect. LAGs are managed by people well aware of the area. They can exploit the potential of the area and the municipality.

The basic transport service is able to provide transport for all days of the week. The Czech Republic is one of the densest transport networks in Europe, especially rail and road, and in terms of density the transport infrastructure is comparable to the EU, but it lags significantly in qualitative parameters [16-17]. The same situation is in the region of South Bohemia. The transport is ensured by passenger and express trains of national and regional lines in the region and between regions and public bus lines operating in the region, and public bus lines operating on the territory of several regions [18-21]. One of the main tasks of regional or local policy in relation to transport services is undoubtedly the support of transport infrastructure. Related to this, it is necessary to note that although the current network of roads and motorway is quite dense s in the Czech Republic with a length of about 55,000 km, it is largely unsatisfactory in terms of traceability, surface quality and status of objects. Similar situation is reported for the rail transport, used for transport of goods and passengers too. The Czech Republic reports a high railway network density – 9,602 km (in 2003), but it lags behind the electrification of these lines and the number of two and multi-track lines (the operating length of these lines is 1,845 km). The biggest problems include neglected technical condition and obsolete communications and security devices [17].

The authors will attempt to map the situation in the South Bohemian region based on its transport infrastructure analysis and outline several problem areas including their possible solution. They also perform a detailed analysis of grant titles from national and European sources, including both the previous programming period and current programming period of 201-2020. The paper analyses the success of the Czech Republic in submission of project applications, obtaining subsidy titles and finishing projects. It further maps new possibilities and compares proposed and achieved input, output and global indicators to the previous programming period.

2 Materials and Methods

There are currently 17 LAGs in the region of South Bohemia. These 17 LAGs occupy a large part of the territory of the South Bohemian Region by their size. Similarly, this is also the case with the number of inhabitants living permanently in the territory of South Bohemian LAGs

There are 601 municipalities covering an area of 9,402 km^2 in the region of South Bohemia integrated into the area of LAGs. The compare with the total, the whole region covers the area of 10,057 km^2 with a total of 623 municipalities (in 2014). It means that the LAGs cover almost the whole area with the large majority of a population. The whole region has a population of 636,707. The area covered by LAGs has a population of 455,626. The difference does not include tows with a population of more than 25,000.

In South Bohemia, it means that the towns of České Budějovice, Písek and Tábor are outside the area of LAGs (with a total population of 157,831 in 2014). The difference was also adjusted by deduction of the municipalities of the LAG called Krajina srdce (Landscape of my Heart) which are outside the region of South Bohemia and adding the municipalities of Jemnicko LAG in the region (a population of 23,250). This means that LAGs cover more than 90% within the South Bohemia in the area, the number of municipalities and population.

Two types of indicators were used and analysed in the paper. The first is the transport services indicator and the second is the indicator of transport accessibility.

To analyse transport services, the method by [23] was used. The aim of the evaluation is to determine the number of bus lines. The selected criteria are the distance between the closest stop (in km) of individual municipalities from their catchment centres and the area of the unit. Furthermore, the time spent on the road in minutes to overcome a certain distance between the transport points. And the last pointer is the number of lines in the specified time period. The length of the line is shown in kilometres. The time period is shown is days (classified as week days, Saturdays and Sundays). The transport service uses weekdays from 6 to 20 o'clock. The resulting coefficient is as follows:

$$CO = di/ti*60*H/hi \tag{1}$$

where:
CO - Coefficient of transport services
Di - distance (in km) between the target points
ti - driving time (in min)
hi - the number of hours spent in period
H – the number of lines

The indicator of transport accessibility was the second indicator used in the analysis of South Bohemian LAGs and their transport services. The indicator measures the distance of the higher classes of roads and expresses the degree of transport accessibility of the municipality from other territories. The calculation was done by network analysis on motorways, expressways and roads. It takes into account different types of traffic in the Czech Republic only.

3 Discussion

Traffic service and its level are determined mainly by the efficiency of transport infrastructure, security, wide coverage and accessibility to all social groups. One of the decisive criteria for the sustainable development of the territory is its competitiveness among other territorial units.

It is possible to use several methods to analyze regions in terms of transport serviceability and to compare the achieved standard between individual regions, lower territorial units, such as micro-regions. The basic criteria for assessing transport serviceability can be the number of public transport lines (week days and weekends) and the availability of 1st class roads and higher roads and the availability of railway stations (in kilometres).

The region of South Bohemia, following the region of Vysočina and Plzeň, is the third worse region regarding transport service. This is also due to its area of the region (10,057 km^2), which is the 12.8%, meaning it is the second biggest region of the Czech Republic, following the Central Bohemian Region. On the other hand, it is the fourth region regarding the population and population density (63.3 of people per km^2; or even 45.3 of people per km^2 if calculated with the area of LAGs only).

As shown in figure 1, there are three regions with the weakest transport service - LAG Lužnice, LAG Blatensko and LAG Třeboňsko. These sub-regions are a part of a superior region. LAG Třeboňsko with the number of bus lines on week days/ at the weekends (29/7) is a part of the region of České Budějovice. The closest LAG in the region is LAG Hlubocko –Lišovsko with the number of bus lines on week days/ at the weekends (56 /17).

LAG Lužnice with the number of bus lines on week days/ at the weekends (28/5) is a part of Tábor area. The closest LAG is Lužnicko the quite large number of bus lines on week days/ at the weekends (90/30).

The last LAG Blatensko has a low number of bus lines. It is a part of Strakonice and Písek area – LAG Strakonice with 33/4 bus lines and MAS Písecká Brána (Písek´s Gate) with 34/6 bus lines. Both of them have quite a poor transport service, as shown in the table. They both are peripheral regions in the South Bohemia, quite distant from the regional city of České Budějovice.

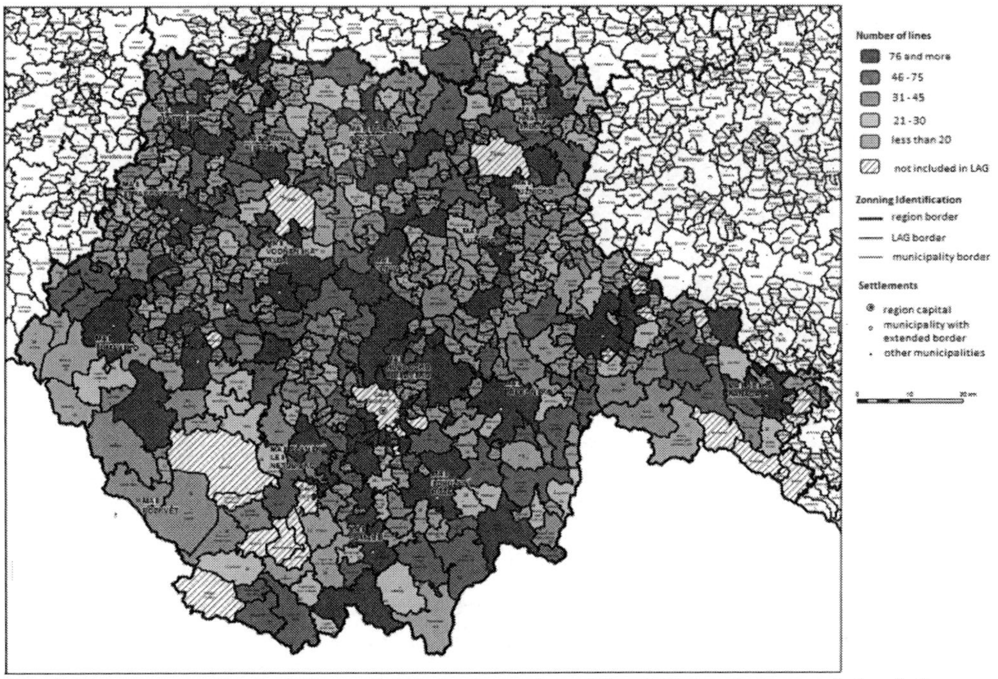

Fig. 1. Transport service in different LAGs in the region of South Bohemia (on week days). Source: authors

Transport accessibility from other areas is the second indicator of the analysis. At first, the indicator taking different regions, with the exception of Prague, into account was analysed.

Accessibility of roads in relation to other areas is 6.4 km. The region of South Bohemia and its LAGs are at the 7th place compared to all the regions of the Czech Republic.The following table shows the accessibility of the area by main roads and railway stations (km) in different LAGs of the region of South Bohemia.

If we analyse the rating of different LAGs in the region of South Bohemia, the worst accessibility is reported by the following LAGS: LAG Česká Kanada, LAG Sdružení Růže a LAG Šumava. These are border areas and it is well-known that these areas have been neglected from the point of view of road transport infrastructure. It is interesting that a bad situation is also reported by LAG Vltava. It is due to the fact, that the main roads are either in the direction to Tábor and Písek and municipalities of the Vltava LAG are right between these two regions (see figure 2).

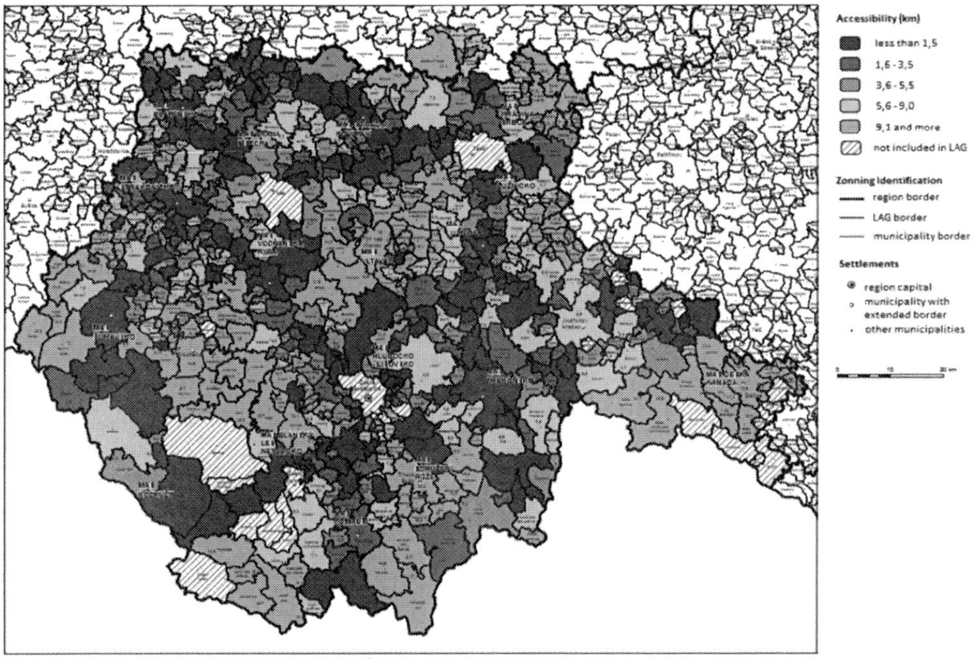

Fig. 2. Accessibility of the area by 1st class roads and main roads and railway stations (km) in different LAGs of the region of South Bohemia, Source: authors

Similar situation was reported to railway accessibility. In such comparison, the region of South Bohemia was the third worst; with the average distance from a municipality to a railway station of 6.3 km. Karlovy Vary was the second best region (following Prague) with the distance of less than 4.3 km.

Analysing different regions of the region of South Bohemia, the worst LAGs are the same as reported by road accessibility: LAG Sdružení Růže and LAG Vltava. Bad accessibility is also reported by LAGs Hlubocko-Lišovsko and Krajina Srdce. It is due to the fact, that those are neighbouring areas to České Budějovice and Tábor.

A possibility to improve the traffic infrastructure is to use grant aids. Grand aids of the European Union were chosen for a detailed analysis as their volume is larger for municipalities compared to the national budget. The municipalities in the South Bohemian region surprisingly benefited from operational programs of the programming period 2007-

2013 in 2014 and 2015. The type of the operational program is another surprise. It could be supposed that the largest aid is received through Transport OP; however, it is not true.

Beneficiaries of subsidies from OP Transport were the state and its institutions and organizations (Správa železniční dopravní cesty - the Railway Network of the Czech Republic), trade companies and contributory organizations (Ředitelství silnic a dálnic - the Road and Motorway Directorate), the capital city of Prague, Ministry of Transport and organizational units of the state (such as Státní fond dopravní infrastruktury - State Fund of Transport Infrastructure and Ředitelství vodních cest ČR - Directorate of Waterways of the Czech Republic).

The municipalities received transport aids from the Regional Operational Programme, and its axis 14.1. Within this axe, it was possible to receive support for the construction, modernization, repair and reconstruction of roads. In total, the ROP of the programming period in 2007-2015 supported 81 projects related to road repairs of 60 applicants in the region of South Bohemia.

Some municipalities, mostly larger towns, were multiple recipients of subsidies. The average amount for the project was 9,259 CZK per capita, with a median of 4,197 CZK per inhabitants.

The analysis does not include the towns of České Budějovice, Písek and Tábor, as they did not receive any subsidy from the ROP Jihozápad in the period. The biggest town receiving subsidies for roads from the ROP was Strakonice with a population of 23,020. The smallest village was Vrbice with a population of 57.

As a clarification, it is necessary to notice, that not all the municipalities are owners of the roads that pass through them. It is often an organization of ŘSD of the Czech Republic – the Road and Motorway Directorate. As only the owner is eligible to apply and receive subsidies, the municipalities often need to wait until their road becomes a priority for ŘSD of the Czech Republic.

4 Conclusion

Although the road network in the region, with some shortcomings, the availability of municipalities and production facilities is present, its quality does not correspond to the increasing demands on the capacity and quality of passenger and freight transport.

The density of the road network in terms of the area of the area is one of the lowest in the Czech Republic, which puts great demands on maintenance and it is obvious that in many regions the quality of these roads deteriorates. That is why the priority objective was defined for this period, namely the gradual improvement of the parameters of existing roads, including relocations, bypasses of towns and municipalities, and the extension of substandard parts.

A substantial part of the border is the state border with Austria (Upper and Lower Austria) and the Federal Republic of Germany (Bavaria) with a total length of 334 km. The cross-border nature of the region provides opportunities for effective cross-border cooperation in the area of production and services, together with strengthening social and cultural cooperation in these areas in order to avoid the negative impact of the depopulation of these areas. In these areas, the roads of the lower classes predominate. From this point of view, the South Bohemian Region has an insufficient number of quality roads across the border of the Czech Republic and it is a relatively heavy burden for the municipalities in these areas.

The level of railway transport is mostly characterized by its infrastructure and the fittings of railway rolling stocks. The structure of the railway network roughly corresponds to the needs of the region, the infrastructure does not. It significantly limits the capacity and speed of some railways, together with the rolling stock which is quite obsolete, although it has passed some modernization recently. Recently, the situation has improved slightly in both directions due to ongoing investments.

A detailed analysis of transport services of municipalities in the area of LAGs of the region of South Bohemia showed the following facts:

- Accessibility of remote areas and municipalities with a low population is a positive fact.
- There is at least 2/3 lines on a week day;
- There are more direct connections to the regional capital and the regional centres; compared to less stops in peripheral areas and areas outside the most important roads;
- Limited service in off-peak times on week days, due to low population in some areas;
- Absence of an integrated regional system, lack of interconnection between road and rail transport;
- Bad state of traffic communications, necessity of financing from European sources;
- Long distance to railway stations, far from settlements;
- A lack of railways, especially the absence of regional rail transport of some peripheral areas of LAG Strakonice;
- Outflow of passengers due to poor quality, the need to accelerate regional transport.

This paper is supported by the research project "From horse-drawn railway to intermodal transport" within Visegrad Fund.

References

1. J. Blažek, P. Csank, Czech Sociological Review **43**, 5, 945-965 (2007)
2. J. Blažek, P. Netrdová, European Urban and Regional Studies **19**, 1, 42-61 (2012)
3. V. Jančák, T. Havlíček, P. Chromý, M. Marada, Geography **113**, 3, 269-284 (2008)
4. M. Ouředníček, P. Špačková, M. Feřtrová, Czech Sociological Review **47**, 4, 777-803 (2011)
5. M. Hampl, M. Marada, Geography **121**, 4, 566-589 (2016)
6. J. Binek, H. Svobodová, J. Holeček, I. Galvasová, K. Chabičovská, *Synergies in rural areas - Actors and rural development tools*, 96 p. (Brno: GaREP Publishing, Czech Republic, 2009)
7. M. Woods, *Rural* (Routledge: Taylor&Francis Book, 2011)
8. K. Gorlach, M. Lošťák, P. H. Money, Journal of Rural Studies **24**, 2, 161-171 (2008)
9. J. Binek, I. Galvasová, H. Svobodová, K. Chabičovská, J. Holeček, *Approaches to strengthen cooperation and synergies in rural micro-regions. Methodology with certificate No. 1/2012 on the recognition of applied certified methodology in accordance with the conditions of "Methodology of evaluation of research and development results"*, 20 p. (Brno: GaREP, spol. s r.o., Czech Republic, 2012)
10. L. Larsson, *EU in the village: LEADER II, governance and rural development, In: Dargan, L., M. Shucksmith, 2008. LEADER and Innovation,* Sociologia Ruralis **48**, 3, 274-290 (2002)
11. C. Ray, Journal of Rural Studies **24**, 422-431, (2008)
12. A. Faludi, Regional Studies **40**, 6, 667-678 (2006)
13. R.D. Putnam, Journal of Democracy **6**, 1, 65-78 (1995)

14. R. Newbery, G. Bosworth, *The character of rural business relations. In: Kasabov E (ed.) Rural Cooperation in Europe: In Search of the "Relational Rurals"*, 29-48 (UK: Palgrave Macmillan, 2014)
15. J. Coleman, American Journal of Sociology **94**, 1, 95-120 (1988)
16. P. Hagget, *Geography – a Global Synthesis,* 833 p. (Prentice Hall: London, UK, 2001)
17. T. Hudeček, *Acceptability and Impacts of its Change in the Czech Republic in the Transition Period: Relationship to the Settlement System* (Dissertation work, Comenius University, Prague, Czech Republic, 2008)
18. R. Buehler, Journal of Transport Geography **19**, 4, 644-657 (2011)
19. L.Ch. Baum, Journal of Urban Economics **66**, 151-163 (2009)
20. M. Hampl, et. al., Journal of economics **56**, 7, 696-711 (2008)
21. M. J. Dargay, Transportation Researches Part E **38**, 351-366 (2002)
22. *Jikord, Transport accessibility of the South Bohemian Region,* Available online: http://www.jikord.cz/dopravni-obsluznost-regionu/ (2017)
23. P. Pova, Perner´s contact **1** (2009)
24. *Structural funds. Information about drawing subsidies,* Available online: https://www.strukturalni-fondy.cz/cs/Informace-o-cerpani (2017)

Localization Strategies of the Czech Companies on the Basis of Industrial-logistics Parks in the Sverdlovsk Region

Andrey Plakhin[1,*], *Rudolf* Kampf[2], *Ekaterina* Ogorodnikova[1], and *Alexandr* Kokovikhin[3]

[1]Ural state university of economics, Management Department, 620144 Ekaterinburg, Russia
[2]Institute of Technology and Business in České Budějovice, Faculty of Technology, 370 01 České Budějovice, Czech Republic
[3]Ural state university of economics, Institute of Management and Informatics, 620144, Ekaterinburg, Russia

Abstract. The Czech Republic is traditionally a strategic partner of the Sverdlovsk region, the foreign trade turnover in 2015 exceeded $ 134 million. This circumstance makes it possible to formulate a scientific hypothesis on strengthening in the long term international economic relations between our countries and increasing the degree of mutual penetration of economic entities. One of the most probable scenarios can be assumed, on the one hand, to create joint ventures both in the Czech Republic and the Russian Federation, on the other, Czech companies will focus on the opportunities that modern industrial-logistics parks provide to localize their productions. Thus, the purpose of this study is to study the theoretical, methodological and applied aspects of the strategic location of Czech companies on the sites of industrial-logistics parks in the Sverdlovsk Region. The article analyses the potential for localization of Czech companies in the Sverdlovsk Region as a whole, identifies areas for possible strategic interaction across economic sectors, explores the parametric characteristics of the industrial-logistics parks in the Sverdlovsk region, presents their rating and assesses the level of competitiveness.

1 Introduction

Localization of foreign companies' production is one of the most urgent areas of the import substitution strategy, which ensures the exchange of technologies. Today in many strategic industries the share of import consumption is estimated at more than 80% that makes a potential threat for both national security and the competitiveness of the Russian economy as a whole [1-3]. Accordingly, the urgent task remains to find the optimal strategy for placing industries localized in the Sverdlovsk region.

[1] Corresponding author : apla@usue.ru

2 Analysis of the localization potential of the Czech Republic companies in the Sverdlovsk Region

For several decades, the Czech Republic remains a foreign trade partner of the Sverdlovsk region, in 2015 the turnover exceeded 134 million dollars. In the structure of the Sverdlovsk region exports to the Czech Republic metals predominate and products made from them, import from the Czech Republic to the Sverdlovsk region is mainly composed of the high redistribution products: engineering products, pumps and electrical machines.

Among the key importers are the following enterprises: FGUP "Kombinat Elektrokhimpribor" (spare parts for gas cutting equipment), LLC "Unimatic" (machine tool equipment), OJSC "NPK Uralvagonzavod" (equipment for modernization), CJSC "The Ural Turbine Works" (components for the production of turbines), LLC "UMK Pumori" (auxiliary tools), LLC "Delcam-Ural" (metal cutting equipment, engineering).

Further in Fig. 1, the main indicators of foreign trade of the Sverdlovsk region and the Czech Republic in 2011-2015 are presented.

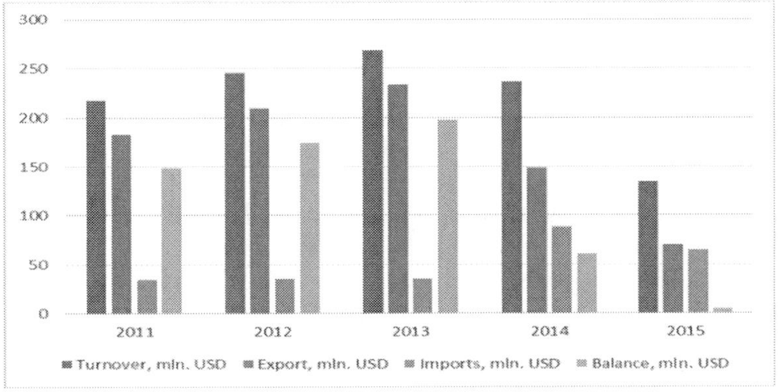

Fig. 1. Indicators of foreign trade of Sverdlovsk region with the Czech Republic in 2015. Source: [4]

In 2015, from the Sverdlovsk region to the Czech Republic metals and goods from them were mainly supplied (37%), including ferrous metals and products made from them (20%) (ferroalloys, semi-finished products, pipes, reduction of the supply volumes by more than 7 times), aluminum (14%) (raw, profiles, sheets), copper (3%) (reduction by 2 times). The share of machine-building products is 6%, including mechanical equipment (4%) (bearings), electrical equipment (2%) (transformers), parts of railway locomotives and wagons. Shipments of the aircraft parts were actually ceased. Also, mineral products were supplied (1%) (refined products), wood (5%) (plywood). The share of the undeciphered group of goods is (51%).

From the Czech Republic to the Sverdlovsk region machine-building products were mainly supplied (88%), including mechanical equipment (69%) (pumps, refrigeration equipment, processing centers for metal processing, machine tools, increase by 1.5 times in supplies of this group of products), electrical machines (11%), railway equipment (6%) (locomotives and their parts, significant growth), optical devices (6%), shipments of the aircraft parts are stopped. The share of products from ferrous metals is 2% (significant decrease), chemical products (2%) (plastics), mineral products (1%) (products of the fuel distillation), other goods - 2% (ceramics).

Cooperation between the two regions is not limited to foreign trade operations. On the territory of the Sverdlovsk region joint projects are being implemented at various stages of implementation, that allows to talk about the localization of individual industries.

Different facts allow to draw a conclusion about the urgency of the development of a localization strategy for Czech producers in the Sverdlovsk Region. Localization of production leads to the costs reduction due to savings on transaction costs (reduction in costs for concluding and maintaining contracts for the supply and maintenance of the equipment), transportation costs and taxes [5]. Localization issues often concern existing institutional barriers, including problems that contribute to the difficulties in accessing resources, problems related to raising capital, using infrastructure facilities, et al. The solution of this problem is possible by placing companies in industrial-logistic parks.

3 Characteristics of industrial-logistics parks in the Sverdlovsk Region

A typical model of an industrial-logistic park allows to solve a whole range of tasks by attracting innovative enterprises, that leads to an increase in the number of jobs; supporting startups, developing high-tech enterprises; developing an environment where local and international firms can interact with knowledge creation centers; acting as innovative hubs, facilitating interactive learning and commercializing the results of scientific researches; developing local entrepreneurial potential, et al.

In general, the territory of the Sverdlovsk region includes sixteen industrial-logistic park projects, including the project of Special Economic Zone "Titanium Valley" [6].

Consider and study the available information for each park and then compare the industrial-logistic parks of the Sverdlovsk region on the main characteristics that determine the decision of the resident to choose a site.

In the Sverdlovsk region 16 industrial-logistic parks are declared. Most of the industrial-logistic parks are created in the greenfield format (13 parks), the brownfield format is typical for the three parks of the Sverdlovsk region.

All industrial-logistic parks are significantly different in their key characteristics that determine the decision of a potential investor to choose a site for the construction of the facility [7, 12-15].

The following criteria have the most significant influence on the choice of a site for a certain industrial-logistic park:

- price of the land plot;
- provision of engineering infrastructure;
- distance from Ekaterinburg;
- status of the project (completion of creation the industrial-logistic park);
- the possibility of the flexible choice of the size and location of the land.

The characteristics of industrial-logistic parks in the Sverdlovsk region are represented according to the above criteria.

3.1 Status of the project and provision of engineering infrastructure

Only five parks out of the total number are operational or are at the final stage of implementation. These number includes the SEZ "Titanium Valley", HimPark Tagil, PRO-BUSINESS-PARK, "Berezovsky" and Uralmash. The other parks are in the process of elaboration of the concept for development and town-planning documentation.

Currently, the other industrial-logistic parks are in the following stages of project implementation:

- concept development: Muranitny;
- concepts are developed and park planning projects are developed or completed for the following parks: Magnitka, Novosverdlovsky, Uralsky, Novouralsky.

At the construction stage of the infrastructure are: Bogoslovsky, Titanium Valley.

DOI: 10.1051/matecconf/201713400047

Many industrial-logistic parks will require a change in the category of land or the type of permitted use of the land. Thus, some land parcels refer to agricultural land or forest fund lands. For some the issues of transferring rights to land plots to management companies have not been resolved. Thus, even with intensive study, the solution of these problems will require considerable time. It takes from 1 to 2 years before the emergence of the real possibility of locating residents in the territory of these parks. These parks are Uralsky, Novosverdlovsky, Isetsky and Muranitny.

Depending on the stage of project implementation, the provision of industrial-logistic parks with engineering infrastructure is different. In the Berezovsky Industrial-logistic park, the power supply capacity is only 1 MW, which makes it possible to place a limited number of residents on its territory (only logistics facilities without refrigeration equipment and small manufacturing). Only "PRO-BUSINESS-PARK", Uralmash and HimPark Tagil are provided with sufficient resources. The rest of the parks are not provided with engineering infrastructure at all.

In addition to the availability of spare capacity and ready networks, an important factor is also the state of the engineering infrastructure and the costs of its maintenance [8].The infrastructure of the PRO-BUSINESS PARK was built in 2014-2015 and with a high degree of probability would not cause problems with its operation. Whereas the construction of engineering networks of Uralmash and HimPark Tagil was carried out in the 60-70s of the XX century. In this regard, their technical condition requires additional expertise and may necessitate substantial investment by management companies in their reconstruction. Meanwhile, for the purposes of this study this factor can be neglected. The amount of modernization costs will not significantly exceed the cost of building new networks in the planned greenfield parks. While the modernization of the infrastructure of these parks can be carried out in parallel with the operation of existing production facilities of the residents.

3.2 Location next to the growing city and logistics accessibility

The most demanded industrial-logistic parks are located in the area of developing settlements. In this regard, today a greater number of Industrial-logistic Parks are planned to be located near one of the largest transport and logistics hubs in the country, important industrial centers.

The higher competitiveness of industrial-logistic parks, created in dynamically developing cities, is caused by a combination of obvious reasons [9]. First of all, a dynamically developing settlement generates businesses that demand for industrial and commercial real estate. In addition, the population of a large or growing city has the necessary competencies and is able to meet the demand for skilled labor for new and existing industries.

In the Sverdlovsk region these centers may be primarily Ekaterinburg and satellite towns, Nizhny Tagil, Kamensk-Uralsky.

3.3 Logistics accessibility

Logistical accessibility is defined in terms of the standard costs for transport communication between different points.

Currently, the transport connection between the Sverdlovsk region and neighboring regions is carried out through the regional center (the city of Ekaterinburg). There are 6 federal highways, 7 main railway lines, and the largest international airport outside the two capitals. Taking into account this circumstance, Industrial-logistic parks located in the

Ekaterinburg area along the Ekaterinburg ring road will have obvious transport accessibility.

At present, the geography of the Industrial-logistic Parks of the Sverdlovsk Region is the following: four industrial-logistic parks are located within the city of Ekaterinburg, six parks are located within 50 km from the city, four parks are located in the range from 51 to 150 km, two industrial-logistic parks are located at the distance of more than 150 km.

Special attention in terms of this criterion requires the industrial-logistic park "Bogoslovsky". This Industrial-logistic Park is located in the city of Krasnoturyinsk at a distance of 390-450km from both Ekaterinburg and the regional centers of neighboring regions. The key task for the initiators of this project is obviously the identification of such residents, in the cost of which transport costs can not constitute a significant share.

3.4 The presence of a branch specialization and anchor resident, around which a cluster can form

Industry specialization is the division of labor in the material (industry, agriculture, transport, construction, et al.) and non-material (science, education, trade, medicine, et al.) industries. Anchor residents increase the attractiveness of the Industrial-logistic Park for potential investors, provide demand for scientific and engineering personnel, set work standards, create effective demand for technology companies and also become the basis for the emergence of spin-off companies. Thus, production and cooperation chains of enterprises can form around the anchor residents, that creates necessary conditions for the formation of viable clusters on the basis of an industrial-logistic park [10, 12, 13]. At the same time, it should be noted that specialization is not a strict restriction, on the territory of these parks there may be located production unrelated with the key residents industries. However, the availability of the possibility of forming a cluster makes it possible to generate ready-made investment proposals for companies in the target industries.

4 Assessment of the industrial-logistics parks competitiveness in the Sverdlovsk Region

In order to determine the competitiveness of industrial-logistic parks located in the Sverdlovsk region, their multi-attributive assessment was performed on the totality of the above criteria, which were significant for the residents.

The quantitative assessment of industrial-logistic parks for each criterion was carried out on the basis of the method of expert estimations. For this purpose, a survey of the business community representatives and employees of public authorities was conducted. Respondents were asked to assess the level of industrial-logistic parks attractiveness according to the certain parameters on a scale from 0 to 10. The cumulative assessment by criterion was determined as the arithmetic average of the assessments of all experts divided by 10. The sum of scores for each indicator shows the final coefficient of competitiveness.

A composite matrix was developed to visualize the findings, that allows in a visual form to group industrial-logistic parks of the Sverdlovsk Region in terms of their competitiveness. The horizontal axis shows the attractiveness for residents, expressed by the coefficients of competitiveness. The vertical axis reflects the stage of implementation: the status of the park, the degree of infrastructure readiness and the prospects for its creation in the near future, the time frame for creating the park (Fig. 2).

 DOI: 10.1051/matecconf/201713400047

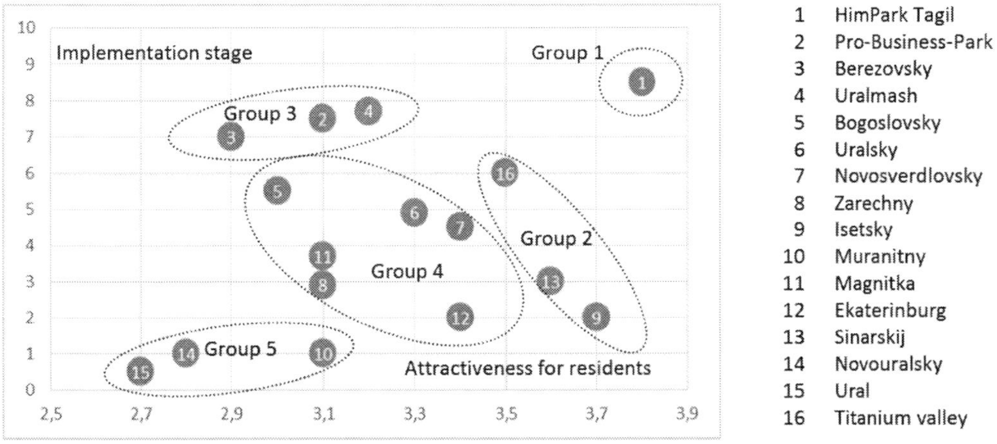

Fig. 2. Grouping of industrial-logistic parks of the Sverdlovsk region by competitiveness. Source: compiled by the author

As a result, five groups of industrial-logistic parks in the Sverdlovsk region are identified.

Group 1. This group is characterized by the highest attractiveness for residents. Currently, this group includes only one industrial-logistic park, this is the operating park "HimPark Tagil". Chemical park is an industrial-logistic complex, consisting of chemical production, trade and service enterprises. The main objective of the project is the development of local business by attracting external investments and technologies of the park residents, creating favorable conditions for the development of chemical production in a single infrastructure complex.

Group 2 includes the second level parks in terms of attractiveness, which are characterized by favorable geographical position and subject to successful implementation, they will be provided with the necessary infrastructure in the near future. This group includes the Titanium Valley, Sinarskij, Isetsky. The most competitive platform among the parks being implemented is the Titanium Valley. This site also provides residents with extended benefits for federal and regional taxes as a special economic zone. Among the projected parks Isetskiy and Sinarskij parks occupy a favorable position. The specific sectoral focus of the economic zone is titanium and related production, which do not have competitors on the territory of Russia. Isetsky Park specializes on knowledge-intensive industries and is ready to place about 25 production sites with the creation of a cluster of one direction industries. Moreover, the infrastructure potential of the site will allow to expand the territory of the par in the long term. Specialization of the Sinarskij Park is engineering and metallurgy. These industrial-logistic parks have a small number of analogues due to the industry specialization and the availability of an anchor resident.

Group 3 includes existing industrial-logistic parks, on the territory of which it is possible to place production in the shortest possible time. Meanwhile, the parks of this group have significant drawbacks. In particular, the areas available for accommodation of residents are very small, the prices for plots are extremely high. This group includes industrial-logistic parks (in order of attractiveness reduction): Uralmash, "PRO-BUSINESS PARK", Berezovsky. These parks are located in the city of Ekaterinburg and therefore they have excellent logistical accessibility. Parks do not have a specific specialization, that is why there are many alternatives. Industrial-logistic Park Berezovsky occupies a weaker position among the existing parks. The park has good logistical accessibility, but it is poorly provided with engineering infrastructure, therefore it has a large number of analogues.

Group 4. The largest is the group of implemented and projected parks, whose competitive advantages are not obvious. It should be noted that all the parks of this group are at very early stages of creation. The dynamics of the implementation of most of the group's projects is purely declarative. Many of the parks included in this group are located in remote settlements from the regional center. They do not have an obvious and attractive anchor resident. Currently, the objective parameters of the group parks are also poorly estimated by investors. In this regard, the success of the projects to create industrial-logistic parks will largely depend on the ability to find a niche in the commercial and industrial real estate market. Implemented parks with average indexes are: Magnitka, Bogoslovsky, Uralsky, Novosverdlovsky, Zarechny, "Ekaterinburg". Industrial-logistic parks are located next to the growing populated areas, have good logistical accessibility, but they have quite a lot of alternatives.

Group 5. The last group includes projected parks with low attractiveness for the residents. These parks are: Novouralsky, Ural, Muranitny. Industrial-logistic parks have stopped on the development of concept and planning projects, the time constraints of infrastructure construction and commissioning of facilities are not known. Thus, it is possible to recommend implementation of the localization strategy for the companies of the Czech Republic in the industrial-logistic park of the first group - "HimPark Tagil".

According to the results of the study, it should be noted that the supply of industrial-logistic parks in the Sverdlovsk region is limited [11]. In fact, it is possible to talk only about 4-5 really operating industrial-logistic parks.

Platforms in the regional center, undoubtedly, would be more attractive for attracting residents. These are industrial-logistic parks Uralmash, "PRO-BUSINESS PARK". However, according to experts, at present land plots in their territory are over and the offer is very limited. This circumstance determines the high price of land for investors and the inability to flexibly choose the configuration and location of the site for construction.

There is no substitute for such sites in the short term. Despite two or even three years of work on the problem of creating industrial-logistic parks in the Sverdlovsk Region, this process is far from being completed. All worthy alternatives to the existing industrial-logistic parks are still at the initial stages of implementation.

In the medium term, subject to the successful implementation, the competition for HimPark Tagil may be formed by prospective industrial-logistic parks. First of all, it is necessary to focus on the sites in the Ekaterinburg area: Novosverdlovsky, Uralsky, Ekaterinburg. Also, the second-tier platforms may be interesting: Magnitka (Pervouralsk), Isetsky (Sredneuralsk). But in relation to these sites, HimPark Tagil also has a relative advantage in the form of a clear specialization in chemical production and the ability to locate objects with high hazard classes.

Thus, despite the large number of industrial-logistic parks declared on the territory of the Sverdlovsk region by the end of 2015, the creation of the industrial-logistic park "HimPark Tagil" is actual in the current conditions.

Undoubted competitive advantage of the industrial-logistic park "HimParkTagil" is also the presence of "anchor residents" in the form of the chemical industry enterprises: OJSC "Uralchimplast", OJSC "Ural Plant Of Polymer Materials", CJSC "UralMethanolGroup". This circumstance makes it possible to form a vertically integrated chemical cluster on the territory of this industrial-logistic park.

"HimPark Tagil" is potentially required and quite competitive in comparison with other parks, because of the following advantages:

- the possibility of placing objects of chemical industries, industries with a high level of danger;

- despite chemical production, environmental safety remains;
- a relatively low price for renting land and premises;
- provision of engineering infrastructure with high capacities.

This paper is supported by the research project "From horse-drawn railway to intermodal transport" within Visegrad Fund.

References

1. I.V. Mukhanova, Russian State University for the Humanities Bulletin, Series: Economy, Management Law **3**, 55-64 (2016)
2. E.V. Mezentseva, *Proceedings of Southwest state University* **3**, 66, 102-111 (2016)
3. N.V. Zuboreva, Electronic Scientific Journal **13**, 10-3, 47-51 (2016)
4. The Ministry of international and foreign economic relations of the Sverdlovsk region materials, *The cooperation of the Sverdlovsk region with the Czech Republic*
5. V.Yu. Markova, D.G. Shuvalova, Economic systems management: electronic scientific journal **58**, 10, 61 (2013)
6. Investment portal of the Sverdlovsk region, Available online: http://invest.midural.ru/
7. A.E. Plakhin, A.B. Stavrova, T.V. Kochergina, *Proceedings of the Baikal state University* **26**, 3, 400-410 (2016)
8. A.E. Plakhin, The manager **4**, 50, 72-78 (2014)
9. E.S. Ogorodnikova, M.M. Sidorenko, Agri-food policy in Russia **11**, 23, 97-99 (2013)
10. A.E. Plakhin, D.S. Mironov, *ECONOMY, SOCIETY, PEOPLE: THEORY, METHODOLOGY, REALITY, Compilation of scientific publications: in 2 parts*, 26-30 (The Ministry of Education and Science of the Russian Federation, Ural State University of Economics, Ekaterinburg, Russian Federation, 2015)
11. A.Yu. Kokovikhin, E.S. Ogorodnikova, D. Williams, A.Ye. Plakhin, Ekonomika regiona [Economy of Region] **13**, 1, 80–92 (2017)
12. E.De Simone, M. D'Uva, Social Indicators Research **133**, 3, 1031-1045 (2017)
13. I. Siskos, Van Wassenhove, N. Luk, Journal of Industrial Ecology **21**, 4, 802-814 (2017)
14. S. Zheng, W. Sun; J. Wu, et al., Journal of Urban Economics **100**, 80-103 (2017)
15. B.G. Hwang, L. Zhu; J.S.H. Tan, Journal of Cleaner Production **153**, 1, 209-219 (2017)

MATEC Web of Conferences **134**, 00048 (2017) DOI: 10.1051/matecconf/201713400048
LOGI 2017

Harmonization of Market Conditions in Provision of Freight Forwarding

Miloš Poliak[1*], and *Patrícia* Šimurková[2]

[1]University of Žilina, Faculty of Operation and Economics of Transport and Communications, Department of Road and Urban Transport, Univerzitná 1, 010 26 Žilina, Slovak Republic

Abstract The paper deals with the issue of different market access for the transport procurement in the EU in context of open market of providing forwarding services. It highlights different requirements of market access in freight forwarding. The article proposes conditions of market access unification in freight forwarding based on analysis of the existing situation. It also refers to the different responsibility of freight forwarder depending on membership in the national associations of freight forwarders. The contribution analyses the limitation of liability in freight forwarding and shows differences in contract of freight forwarding for particular shipments by model examples. There is also pointed to the need for the development of uniform freight forwarding conditions in the EU.

1 Introduction

Freight forwarders play an important role in transport chain. They provide professional services in national and international transport. The basic task of every freight forwarder is procuring of transport and providing of advisory service for all modes of transport. Freight forwarders seek to ensure non-tariff discounts on behalf of customer. They check invoices for transportation outputs and the other expenditures, carry out all kinds of discounts and their direct debits and give it to the customer according to previously agreed principles. Forwarders also ensure claim damages or may, in agreement with the customer handle complaints directly [1]. Another important task is drawing up the calculation of transportation outputs, recommendation of the optimal transport route and determination of the optimal tariff declaration for transported goods. A freight forwarder operates a collection service, carries out national collection and distribution of shipments, and ensures completion, distribution of general cargo, storing, signing and repairing of packaging. They are also responsible for providing various benefits for all modes of transports for example the issuing of documents. A freight forwarder can carry out, in agreement with customer, customs clearance of consignments and can represent customer in customs procedures. Last but not least, he can provide legal aid to the customer in defending his interests in disputes with his partners [2].

* Corresponding author: milos.poliak@fpedas.uniza.sk

It follows that the role of freight forwarder is not easy and the freight forwarder has a high responsibility for his actions in these tasks. When ordering services, the customer expects to receive quality services which can be provided only by specialist.

Whereas the market access conditions in freight forwarding are significantly different in the EU, the aim of this paper is to highlight the importance of position of freight forwarder and define differences in market access. The aim of the paper is also suggest a way of harmonizing market access for freight forwarding in the EU.

The freight forwarding is not adjusted uniformly in the EU by common legislation in light of forwarding contract, for example like CMR Convention in the road transport. Given that the aim of this paper is also to point out different responsibilities of freight forwarders who conclude forwarding contracts under the national law and to propose a way of uniting responsibility [3, 4].

2 Market access in freight forwarding in the selected countries

Freight forwarding has unmistakable role in the providing of transport and logistics services in countries with developed market economy. Freight forwarding is providing transport or other activities related to the carriage under his own name, on his own responsibility and on behalf of third.

As the conditions of market access are not harmonized for freight forwarding in the EU, the aim of this paper is identify whether freight forwarders from the other selected countries have comparable conditions of market access. Access to the freight forwarding business is governed by national law. Because of this it is necessary to identify the specific national legislation of EU Member States by analysing market access in freight forwarding. Based on the analysis of national law which is available in [5] it is possible to divide the countries according to Table 1.

Table 1. Comparison of requirements for market access in freight forwarding. Source: [5, 6]

Country	Required experience	Required education	Financial guarantee	Unregulated trade
Belgium	yes	yes	yes	
Bulgaria				yes
Croatia				yes
Czech Republic				yes
France	yes	yes		
Germany				yes
Netherlands				yes
Poland	yes		yes	
Portugal	yes	yes	yes	
Austria	yes	yes		
Slovakia	yes	yes		
Spain	yes			
Sweden				yes
Italy	yes	yes	yes	
Great Britain				yes

In the position of freight forwarder may be the company or a person who has no professional training or professional representation in seven of analysed countries, as Bulgaria, Croatia, Czech Republic, Germany, Netherlands, Sweden and the Great Britain.

In these countries is the freight forwarding unregulated trade, it means any person who meets the general business conditions, can make business in freight forwarding. In the other analysed countries are different conditions.

The mandatory conditions for obtaining a business license in Belgium are experiences of the last six years and at least:

• five continuous years in a company holding a license,
• three years, if was met a 3-year training,
• two years, if was met a 3-year training or if the applicant holds a university diploma
• six months if the applicant passes a professional examination in forwarding or if has own certificate of professional competence in transport (additional test for specific faculties of freight forwarding).

Another condition is a financial guarantee in amount of 12 394.68 € for the license, the integrity which is a proof of good moral conduct of public authorities and it is showing that the applicant did not commit any crime. The other duties include, for example, a certificate of professional experiences which is issued by person, who is assuring the daily management of the company.

In France, activity of freight forwarder is regulated by several conditions. The first condition is the registration in business register and second condition is the registration in a register of freight forwarder. The registration is conditional and is required qualification for profession and integrity. Requirement of registration is professional competence that is certified by the "Prefect of the regions". Conditions to achieve professional qualifications are the ownership level III diploma "Diplôme de niveau III (Bac + 2)" and also major in transport. There it is also required at least 200 hours of management experience and successful completion of a special written examination. The applicant must demonstrate five consecutive working-years in a transport company but no more than three years before filing the application. The other condition is an integrity, which is defined by clean criminal record. A professional qualification is certified for those who have mastered either a written test, have a sufficient expertise or demonstrate related diplomas [5].

Freight forwarding services in Poland can be provided only to entrepreneurs who meet the prescribed conditions of required competence. Professional competence in Poland, which is proved by certificate of professional competence, is identical to the certificate of professional competence for the road transport. The certificate of professional competence must be at least one person from the management company or a person who is working in a company that is responsible for operating road transport operations. Entrepreneur who wants to provide services related to the acquisition or brokering transport, must demonstrate a minimum share capital 50 000 €.

In Portugal, the professional competence is a mandatory condition for obtaining a business license in freight forwarding, which is obtained through a special examination by National Committee for Transport. National Committee for Transport is the authority in the field of freight forwarding. The applicant must also meet financial guarantee which is demonstrated by capital in amount of 50 000 € and a liability insurance in the amount of 100 000 €. Another important condition is to prove integrity. It means that the applicant has no criminal record for managers, board members, and technical directors [5].

In Austria, the applicant must meet the general conditions of business and professional competence. A technically competent person is one who has successfully completed the listed schools (transport, economic and law) and has one year of experience or passes the exam and has two years of experience in the field. The exam is organized annually by the Economic Chamber. The test consists of two parts. Written part of the exam includes as correspondence, drawing up documents, payments and lending, calculating performance, billing the customer and internal accounting evidence according to the special requirements of accounting. The oral part consists of 13 headings.

Requirements for a freight forwarder in the Slovak Republic are established by Act no. 455/1991 Coll. On Trade Licensing. This act requires meeting criteria of specific conditions (in the form of professional competence) in addition to general business conditions by the applicant for freight forwarding. According to the Act no. 455/1991 Coll. on Trade Licensing is professionally qualified the applicant who has completed secondary education and has at least two years' experience in the field or a university degree in the field of study, and at least one year experience in the field. There are only general business conditions required in the case of transportation brokerage business by Trade Licensing. Brokering is classified as unregulated trade in the Slovak republic [7, 8].

3 The responsibility of freight forwarder for the transportation procurement

The freight forwarding business is regulated by Act no. 513/1991 Coll. Commercial Code and General Conditions of Consignment in the Slovak Republic. If it is based on the Act. 513/1991 Coll. Commercial Code (freight forwarders should not use any forwarding conditions) then each freight forwarder who breaches his obligations from the contractual relation contract is obligated to compensate damage caused to the other – to the customer. Except that the freight forwarder proves that the breach is caused by circumstances excluding his responsibility. These circumstances include for example floods, fires and the like. The responsibility is not excluded by an obstacle which occurred during the time when the obliged party was already in delay with the performance of its obligation or arose out if its economic situation. The effects excluding liability are limited to the period of time during which exists the obstacle, with which such effects are connected. It means that the freight forwarder, who does not have a particular responsibility in forwarding the modified conditions is liable under law no. 513/1991 Coll. Commercial Code to the full damage [9].

The scope of the responsibilities may be adjusted by General Conditions of Consignment. General Freight Forwarder's conditions are issued by Association of Logistics and Freight Forwarding of the Slovak Republic (hereinafter referred to as ALFF SR) and determinate responsibility of freight forwarder. Forwarding conditions ALFF SR are applicable to trade between the customers and freight forwarders in national and international freight forwarding, in case they become part of the contract of freight forwarding.

If the freight forwarder is responsible for damage, his duty to compensate is limited by General Freight Forwarder's conditions ALFF SR as follows:

- in case of damage which is occurred on the consignment during carriage by means of transportation or at the handling operations connected with carriage (at loading, transport, loading, unloading), the amount of damage is limited to 8.33 XDR per 1 kg gross weight of the goods damaged, destroyed or lost, the maximum amount being 20 000 XDR per one case of damage,
- if the damage is caused by late delivery of the consignment, the freight forwarder's duty to compensate the damage is limited by the agreed amount of remuneration to the carrier.
- in case of storage the responsibility of the freight forwarder for the goods lost, damaged or destroyed is limited to the amount of:
- 3.925 XDR per 1 kg of gross weight of the goods damaged, lost or destroyed; however, the maximum amount being 3 925 XDR per one case of damage (one consignment),
- 19 625 XDR, if damage is caused to the depositor (the customer) consists of the difference between required and actual state of the stored goods,

- in cases of other damages the freight forwarder's duty to compensate the damage is limited to the amount of 20 000 XDR per one case of damage.

This scope of responsibilities of freight forwarder is in force only if the reference for General Freight Forwarder's conditions ALFF SR is part of the contract of freight forwarding. If the contract is included with freight forwarder in another country, it is possible that part of the freight forwarding contract are forwarding conditions of other associations, which are setting different responsibilities. In the Table 2 is processed the overview of limitations of freight forwarders on the basis of general freight forwarding associations in the analysed countries. Table 2 shows the limitation of damaged or destroyed shipment per kilogram (if XDR value is converted also into euro in General Freight Forwarder's conditions) and per one case of damage.

Table 2. The limitation of responsibility of freight forwarder per kilogram of lost, damaged or destroyed shipment and per one case of damage. Source: [5, 6]

Country	XDR/kg	€/kg	XDR	€
Slovakia	8.33	9.36	20 000	22 480
Germany		5	unlimited	
Croatia	unlimited	unlimited	unlimited	unlimited
Austria	-	1.09	-	1 090.09
Belgium	-	5	-	25 000
Italy	8.33	9.36	unlimited	unlimited
Romania	4	4,49	10 000	11 240
Estonia	8.33	9.36	50 000	56 200
Netherlands	4	4.49	10 000	11 240
Slovenia	5	5.62	25 000	28 100
Poland	2	2.25	50 000	56 200
Czech Republic	8.33	9.36	20 000	22 480
Norway, Finland, Denmark, Sweden	8.33	9.36	50 000	56 200

The responsibility of freight forwarder is limited according to forwarding conditions and it can be in the amount from 1.09 €/kg (Austria) to unlimited responsibility of freight forwarder (in Croatia). In the most of analysed countries is the responsibility of freight forwarder limited to the amount of 8.33 XDR per kilogram of damage or destroyed of goods as in the Slovak Republic (Slovakia, Italy, Estonia, the Czech Republic, Finland, Norway, Denmark and Sweden).

The responsibility of freight forwarder is also limited by the maximum responsibility in the case of damage, which is bounded from 1 090.09 € per case of damage in Austria to unlimited liability (in Croatia, Italy). In the most analysed countries is the maximum responsibility of the freight forwarder in amount of 50 000 XDR pre one case of damage, which corresponds to the value of 56 200 €.

4 Proposal for amendment of market access for freight forwarding business

Freight forwarder is responsible for a consignment by forwarding contract because transportation of shipment is carried out according to instructions of freight forwarder. Freight forwarder also appoints a carrier [10]. If there are requirements of professional competence and financial competence for a carrier directly by Regulation (EC) No.

1071/2009 across the EU, freight forwarder should also demonstrate some professional competence. To enable market access in freight forwarding the content of professional competence should be given by internationally recognized course - Freight Forwarding Expert. Freight forwarding should be provided only by person who successfully completes the exam before the professional authority or demonstrates completed knowledge of the subjects which are subject for the freight forwarding course - Freight forwarding expert. These are the subjects: Introduction to freight forwarding, Maritime transport, Maritime containers, Multimodal transport, Air transport, Road transport; Rail transport; National and international transport by inland waterway; Customs procedures; Logistics, warehousing and distribution; Insurance; Dangerous goods; Safety and security; Information and Communication Technologies in freight forwarding.

The differences in EU market access should be deleted by adoption of rules. Freight forwarders, who are operating for the procurement of transport across the whole market, would have to meet the same requirements for market access.

In the analyse has been analysing a current state and it has been found that the forwarding contract is established accordance with national law and there is no unified responsibility of the carrier like, as in the road transport (CMR Convention), rail transport (COTIF Convention) or waterways transport (e.g. The Hague-Visby rules). Within the EU, or CLECAT, a convention should be accrued or at least the forwarding conditions that harmonize the level of freight forwarder´s responsibility for damage. If we based on the average values, the responsibility would be limited to the level of:

- 5.10 € / kg gross mass damaged, lost or destroyed goods
- 23 000 € per shipment.

The resulting values of responsibility will be result of the negotiations of EU members or CLECAT but regardless the boundary is united responsibility very important.

6 Conclusion

In the EU, it is not a position of market access in freight forwarding uniformly regulated. Each country has its own specific conditions, which the applicant must meet. As the freight forwarder can provide his services in the EU, a potential customer receives a variety quality of services from freight forwarders [11, 12].

It follows from the above that, the requirements of market access in freight forwarding is needed to harmonize. It is because of that the freight forwarders do not provide their services in local markets but provides transport services through the EU market. If this were to happen, the market would have served the same quality freight forwarder and customers should be ensured that they get the service to the extent and quality they expect.

Within the procurement of transport, there is also the problem of different responsibilities of freight forwarders, depending on the conditions under which they operate forwarding. For the customer who is not specialized in transport, it can be unclear. As in the case of the CMR Convention for road transport it would be appropriate to harmonize the responsibility of freight forwarders operating on the Community market. The analysis points out significant differences in limitation of responsibility of freight forwarder. It should be noted that other differences are in defined circumstances, which exclude the liability of freight forwarder. Also the responsibility of freight forwarder is affected by the national rules governing trade relations in the country.

This work was supported by the Slovak Research and Development Agency under the contract No. SK-PL-2015-0019 and VEGA no. 1/0143/17 POLIAK, M.: Increasing the competitiveness of Slovak carriers providing road transport services in the common market of the European Union.

This paper is supported by the research project "From horse-drawn railway to intermodal transport" within Visegrad Fund.

References

1. J. Gnap, et al., *Freight Forwarding (Zasielateľstvo)*, 2nd Edition (University of Zilina EDIS – Publishing, Slovak republic, 2011)
2. M. Poliak, *Business in Freight Forwarding and Transport (Podnikanie v zasielateľstve a doprave)* (University of Zilina EDIS – Publishing, Slovak republic, 2011)
3. Z. Říha, J. Tichý, *The 19th International Scientific Conference*, 388-391 (Kauno technologijos universitetas, 2015)
4. B. Caplan, The Economic Journal **112**, 433-458 (2002)
5. T. Račeková, *Proposal to unify market access conditions for freight forwarders in EU*, (Diploma Thesis, M. Poliak - supervisor, University of Zilina, Slovak Republic, 2016)
6. D. Beňová,: *Proposal of common freight forwarding conditions for CLECAT members*, (Diploma Thesis, M. Poliak - supervisor, University of Zilina, Slovak Republic, 2016)
7. O. Cyprich, V. Konečný, K. Kilianova, PROMET - Traffic&Transportation **25**, 6, 533-541 (2013)
8. M. Poliak, Journal of Economics **61**, 2, 206-220 (2013)
9. M. Poliak, S. Semanova, A. Poliakova, *Risk allocation in transport public service contracts*, Ekonomski Pregled **66**, 4, 384-403 (2015)
10. O. Stopka, L. Černá, V. Zitrický, Nase more **63**, 3, 189-194 (2016), DOI: 10.17818/NM/2016/SI21
11. I. Šimková, V. Knečný, S. Liščák, O. Stopka, Transport Problems **10**, 113-124 (2015)
12. M. Poliak, A. Poliaková, *Relation of social legislation in road transport on driver's work quality* (TST, Wroclaw, Poland, 2015)

Life Cycle Cost Calculation at the Transport Company in the Supply of Production of Wooden Houses – Case Study

Marek Potkány[1,*], *Miloš* Hitka[1], and *Lucia* Krajčírová[1]

[1]Technical University in Zvolen, T. G. Masaryka 24, 960 53 Zvolen, Slovakia

Abstract. A correct information manager's decision-maker database is a very important element that substantially affects its success. This article presents the potential of using the methodology of life cycle cost calculation in the conditions of a transport company that focuses on the logistic supply of wood-housing producers. The problem is presented through a case study and addresses the decision-making aspect of the decision about acquisition of the transport vehicle. This decision uses time value indicators, inflation rates, average rate of profitability of industry and life cycle costs. Due to the short life cycle of the analyzed period, it was not necessary to consider the ergonomic requirements resulting from the trend of anthropometric dimensions growth.

1 Introduction

The business activity is accompanied by a number of decisions - both operative and strategic level. One of the strategic tools that can be used in this area is Life Cycle Cost calculation. Managers in their activities must also deal with various strategic decisions. Life Cycle Cost (LCC) Calculation is a method that is one with strategic decision-making tools.
The main goal of this article is to present the use of LCC methodology in the decision making of the transport company, the object of which is the supply of building materials to the producers of wooden houses in Slovakia.

2 Theoretical backgrounds

Many authors of studies, like Popesko and Papadaki, but also Šoljaková, Wagner and Petřík are concerned with the calculation of life cycle costs at the practical conditions. In their work, they are associated with the performance of the enterprise and talking about it as a concept of economic accounting [1-4]. According to the author, it is suitable for complete project management, control and planning and evaluation of investment projects. This calculation shows all costs and revenues of the project during its economic life cycle. For corporate governance, LCC is an important source of information. By using this method, the company knows how much the cost has actually been incurred at all stages of the life

* Corresponding author: potkany@tuzvo.sk

cycle, from acquisitions, through operating to retirement. The main task of the LCC is to optimize all costs over the entire economic life of the project without losing its overall performance.

According to Šoljaková, which states, that it is also considering the costs in the pre-production phase and the stages that are connected with the end of the production cycle [2]. Crespo Márquez et al state, that a life cycle cost analysis calculates the cost of an asset for its entire life span (Figure 1).

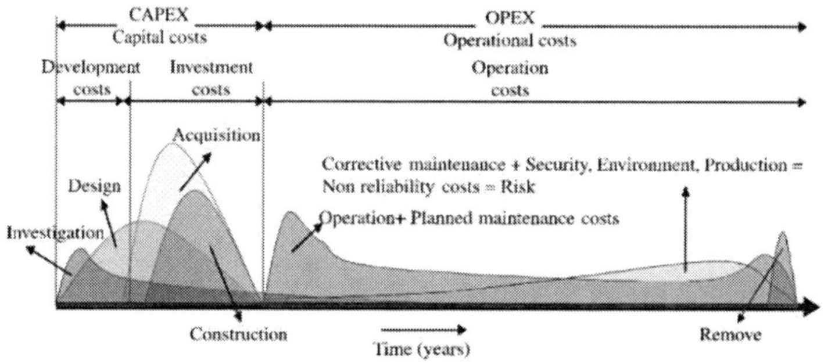

Fig. 1. Life cycle cost analysis. Source: [5]

According to Šoljaková, which states, that the LCC can be expressed as a calculation of the partial or the entire evaluated period. One of the parameters that are the basis for its compilation is the exact definition of the life cycle length [2]. For the LCC application, besides the cycle length itself, it is necessary to know the level of the discount rate, the rate of price growth and the interest rate. According to Zhang and Yuan the standard discount rate is determined by the interest rate and the rate of inflation [6].They state in their work, that life cycle of steel bridge deck pavement is often cannot reach the designed working life, taking 20 years as life cycle in the consideration of actual use situation of steel bridge deck pavement. Standard discount rate is decided by interest rate and inflation rate, discount rate (I) = (interest rate inflation rate) / (1 + the inflation rate). The relevant studies have shown that discount rate is between 3% and 5% usually.)

LCC finds its application in various transport areas [7], construction [8], the automotive industry [9]. It is linked to the area of environment and construction [10].

According to Štefko and Reinprecht, the wooden house is a construction work, the main supporting structure of which is made of wood, wood elements and wood-based materials. It is a wider concept [11]. It is not just the wooden structure itself. This, however, always forms the supporting part of the wooden house.A frame construction system is simple and does not require a great deal of mechanisation. The object is completely fitted on the building site [12]. A level surface is made on the building site (production-assembly plate). on which the building walls are made. The prepared walls are erected and fitted into the object foundations. The walls covered with wide area materials provide stability and strength and they can also be reinforced using diagonals embedded into the wall mullions.

3 Methods

Petković et. al. states that at the decision about implementing of investment project, the terms net present value (NPV) and net future value (NFV) are used [13]. According to the authors, net present value is defined by the sum of discounted values of individual concepts in the flow, where each term is discounted to a common reference date. According to Rossi et. al. the net future value measures the profitability at the end of the planning horizon compared to what might have been gained by investing elsewhere [14]. The discount rate is used for discounting the cashflow to the present. The NPV of the estimated cost over the planning horizon is the discounted value of the NFV to the present.

According to Rossi et. al. this method is very often used in the conditions of various industries to decide on the feasibility of projects. It often uses uncertain information and needs to be aware of the cost risk that can influence the analysis [15]. Spickova and Myskova state, that LCC method calculates both present and future value of discounted cash flow [16]. Authors writing that Net Present Value is the basic tool for comparison of the money values in the different times. The LCC application itself can cause some problems. These may occur mainly according to in the process of unification of individual stages of the investment project [17]. If a project is judged by discount techniques, is the solution of transforming all cash flows into real cash flows [18, 19]. Armendariz-Lopez et. al. used in their study to take account of the value of money in time for determining the payback period [20].

This paper takes into account the currency value over time as an economic indicator for the financial evaluation of projects, in order to consider the payback period. The net present value requires the net cash flow. The resulting value is deducted from the initial net investment.

Fourie and Tendayi argue that life-cycle cost analysis basically takes into account all the costs that will be incurred sometime in the future, so the cost of a particular decision-making place must be avoided [7]. The decisive point is known as the present value. The following information is used to the methodology of LCC for decision-making database:

- a project lifespan,
- dicount rate,
- capital equipment cost for the maintenance and operational calculations.

Heralová state, that there are several approaches to the LCC application [21]. One is a way of using NPV. In principle, it is necessary to calculate the initial cost of acquisition. Another is to use a stochastic approach where inputs are considered as random variables with assigned probability density functions. There is also the possibility of using a deterministic approach. This is based on an expert estimate of input values from past periods supplemented by a sensitivity analysis.For the conditions of our case study, the total lifecycle costs will be calculated according to the formula (1).

$$LCC = acquisition\ price + operating\ cost \times RBF \pm \left(RP \times \frac{1}{(1+r)^n}\right) \tag{1}$$

where:
LCC – the sum of the life cycle cost at the present value,
RBF – discount factor for operating costs,
RP – residual price,
r – discount rate,
n – the length of the analysis period.

RBF is a discount factor (Rentenbarwertfaktor), which works with the discount rate indicator (r) and the length of the analyzed period (n). By the literature source [22] it can be determined by the formula (2).

$$RBF = \frac{(1+r)^n - 1}{(1+r)^n \times r} \tag{2}$$

For the determination of the RBF factor, we used the interest rate calculated by the relationship reported by Petřík [4], which is also reported in study Baum et. al. [23], where we took into account the level of inflation (i) and the average rate of profitability of industry (p) determined by formula 3.

$$discount\ rate\ (r) = [(1+p) \div (1+i) - 1] \times 100 \quad (3)$$

We used the following relationships to determine the cost of each cost item. We have taken into account fuel coststhe average consumption (AC) of each vehicle for the previous period, the expectancy distances ($\sum$km) and diesel price for current period/liter (DP/l) (formula 4).

$$fuel\ cost = \frac{\left(\frac{AC}{100} km \times \sum km\right)}{100} \times DP \quad (4)$$

The engine oil cost, we take into accountof oil change interval (I - 50,000 km), oil tank volume (O_v -38 l) and actual engine oil price (EOP - €/ l) (formula 5). Cost of tires depends from price of tires (PT) in their number (n) and tires change payments TCH. Also from tires life ($T_{l/km}$) and the number of kilometers driven ($\sum$km) (formula 6).

$$engine\ oil\ costs = \frac{(O_v \times EOP)}{I} \quad (5)$$

$$cost\ of\ tires = \frac{(PT \times n) + TCH}{T_{l/km}} \times \sum km \quad (6)$$

Maintenance costs include the costs associated with regular replacement of tires, oils, filters, lubricants, spare parts, unsupervised inspections and additional costs. On the acquisition of new mean of transport, these costs foresee lower level versus used mean of transport. The operating costs included insurance cost, motor vehicle tax, technical and emission control and highway toll. In the first year, the cost of registering a motor vehicle was also included in their total amount. For operational costs, we consider the rental office cost, as well as parking cost of vehicles because our company rents. This group also included administrative costs, energy and telecommunication costs. Compensation costs (liquidation cost) were also included in the life cycle cost calculation. A new vehicle is considered to sell it at a residual price of 20% of the entry price. The used vehicles are considered the cost of liquidation in the amount of 3,000 €. In the framework of the analysis, the residual price shows a positive share in relation to the cost of life cycle and disposal costs shows the negative share to the cost of life cycle.

4 Application of LCC – case study

The subject of our case study is a transport company that focuses on logistic supply to the Slovak producers of wooden houses. The most commonly transported materials are spruce / fir wood, OSB boards, insulating materials, material needed to make perimeter walls of column construction, roof construction, etc. The most commonly transported materials are spruce/fir sawn timber, OSB boards, insulating materials, material needed to make perimeter walls of column construction, roof construction, etc. Decision-making role is based on an assessment of alternatives to the acquisition of new means of transport. The options are as follows:

- 1st alternative: acquisition of new semi-trailer unit in acquisition price 92,000 €,
- 2nd alternative: acquisition of used semi-trailer unit in acquisition price 51,000 €.

In our case, the lifetime of both vehicles was set for a period of 6 years. Since the cost item depreciation does not represent an expense for the enterprise, we did not count the operating costs.

In our forecast, we calculate the mileage of 110,000 km/year. The average fuel consumption for the new semi-trailer is 27 l/100 km, and 35 l/100 km. Wage costs represent the sum of the gross wage (3.5 €/hour) and also the statutory social security and health insurance company contributions. We expect the driver to work on average 200-220 hours per month.

Total operating costs in the first year represent81,780.97 € at the 1st alternative and 93,458.97 € at the 2nd alternative. In the next years, they are achieving values81,621.97 € a 93,299.97 €.

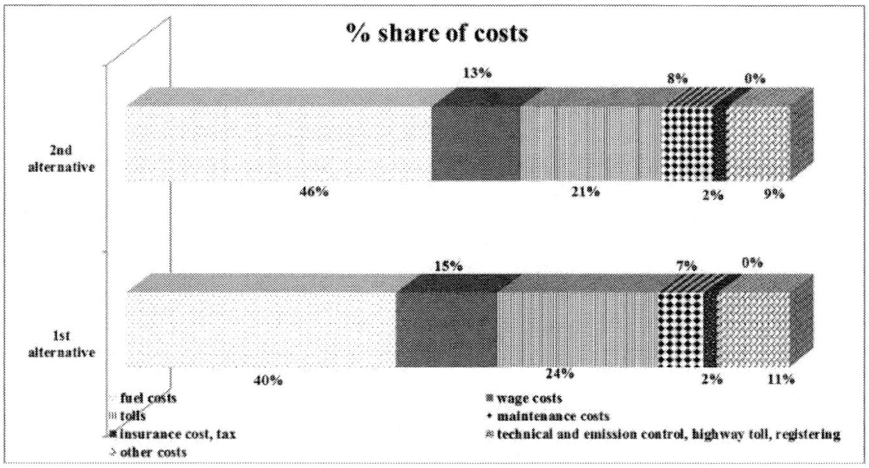

Fig. 2. The share of individual groups of costs for each alternative. Source: authors

The share of costs (Figure 2) is roughly the same level in both cases. The largest share of operating costs are expenses in connection with the procurement of fuel, wages and tolls.

For the determination of life cycle costs, the following data were used in the methodology:

- inflation rate i = 1.1%
- average rate of profitability of industry p = 4.3%
- RBF factor ≐ 5.388
- 1st alternativeresidual price -20% from acquisition price
- 2nd alternative liquidation cost + 3,000 €

Based on the results, we can state that by using the relationship (1) we have determined the value of the lifecycle cost for the new vehicle (1st alternative) at over 516,000 €. For 2nd alternative, using the LCC methodology, we received a cost level 556,000 €.

$$LCC1^{st}\text{ alternative} = 92{,}000 + 81{,}621.97 \times 5.388 - 18{,}400 \times (1.03165)^{-6} \doteq 516{,}500\ €$$
$$LCC2^{nd}\text{ alternative} = 51{,}000 + 93{,}299.97 \times 5.388 + 3{,}000 \times (1.03165)^{-6} \doteq 556{,}200\ €$$

Thus, we can conclude that the initial higher investment in the acquisition of new vehicle is disadvantageous for the company just beginning. Changing the situation and reducing the total cost of the first alternative will occur after the first year of potential use of the vehicles. These conclusions are presented in Figure 3. Our recommendations are, despite the higher acquisition price, use of the possibility of using 1st alternative.

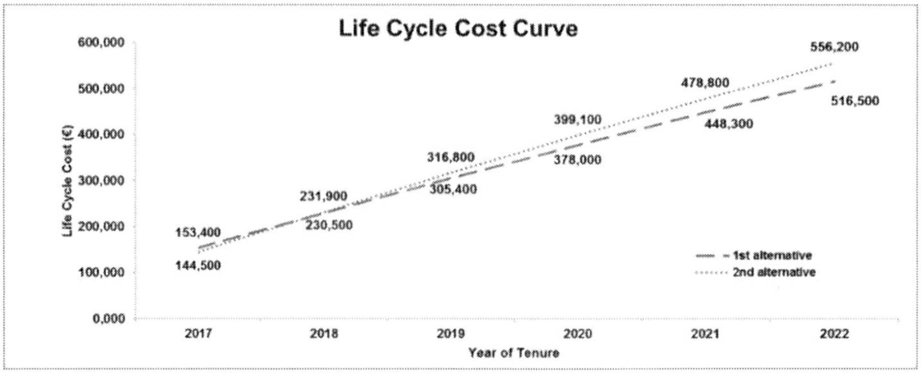

Fig. 3. LifeCycleCostCurve. Source: authors

5 Conclusion

Life Cycle Cost calculation is an important tool for managerial decision-making. Its importance is based on the cost quantification across the lifecycle of the cost object at the applied time value of money, which considers the level of inflation rate and the current interest rate. The aim of this calculation is to provide the user important information for the right decision in terms of strategy. For decisions of that type, information is important, which should be accompanied by the minimum level of allocated costs. In this case study, we present an approach that is based on selecting the most appropriate alternative to acquisitions of the transport vehicle at the conditions of assumed purchase price level, the 6-year life span, the inflation rate 1.1% and the average rate of profitability of industry4.3 %. The presented approach is based on the business reporting approach in terms of capturing all real costs and revenues and their managerial clarity.

The paper is a partial result of the grant scientific project VEGA No.1/0320/17 Economic and Social Context of European 20/20/20 Targets from the Viewpoint of Economy Low-energy Houses and project APVV 16-02-97 Updating of anthropometric database of Slovak population.

This paper is supported by the research project "From horse-drawn railway to intermodal transport" within Visegrad Fund.

References

1. B. Popesko, Š. Papadaki, *Moderní metody řízení nákladů. Jak dosáhnout efektivního vynakládání nákladů a jejich snížení*, 264 p. (2016)
2. L. Šoljaková, *Strategicky zaměřené manažerské účetnictví*, 206 p. (2009)
3. J. Wagner, *Měření výkonnosti. Jak měřit, vyhodnocovat a využívat informace o podnikové výkonnosti*, 256 p. (2009)
4. T. Petřík, *Procesní a hodnotové řízení firem a organizací – nákladová technika a komplexní manažerská metoda: ABC/ABM (Activity Based Costing/Activity Based Management)*, 911 p. (2007)
5. A. Crespo Márquez, P. Moreu de León, J. F. Gómez Fernández, C. Parra Márquez, M. López Campos, Journal of Quality in Maintenance Engineering **15,** 167 (2009)
6. D. Zhang, F. Ye, J. Yuan, Procedia Social and Behavioral Sciences **96**, 785 (2013)

7. C.J. Fourie, T.G. Tendayi, South African Journal of Industrial Engineering **27**, 142 (2016)
8. K.K. Lawania, W.K. Biswas, Clean Technologies and Environmental Policy **18**, 2419 (2016)
9. S. Kara, W. Li, N. Sadjiva, Procedia CIRP **61**, 767 (2017)
10. B. Grzyl, E. Miszewska-Urbańska, M. Apollo, E3S Web of Conferences **17** (2017)
11. J. Štefko, L. Reinprecht, *Dřevěné stavby, konstrukce, ochrana a údržba*, 196 p. (2004)
12. J. Štefko, R. Soyka, *Wooden buildings*, 124 p. (2013)
13. D. Petković, S. Shamshirband, A. Kamsin, M. Lee, O. Anicic, V. Nikolić, Revewable and Sustainable Energy Reviews **57**, 1270 (2016)
14. B. Rossi, S. Marquart, G. Rossi, Journal of Environmental Management **197**, 41 (2017)
15. L. Kirkwood, E. Shehab, P. Baguley, A. Starr, Procedia CIRP **38**, 245 (2015)
16. M. Spickova, R. Myskova, Procedia - Economics and Finance **34**, 337 (2015)
17. B. Král et al. *Manažérské účetnictví 3*, 660 p. (2012)
18. F. Nemec, S. Lorincová, M. Hitka, L. Turínská, Nase more **62**, 131 (2015)
19. M. Pavlíčková, K. Teplická, Acta Montanistica Slovaca **19**, 59 (2014)
20. J.F. Armendariz-Lopez, A. Luna-Leon, M.E. Gonzalez-Trevizo, A.P. Arena-Granados, G. Bojorquez-Morales, Renewable Energy **87**, 564 (2016)
21. R.S. Heralová, Procedia Engineering **85,** 454 (2014)
22. Lebenszykluskostenrechnung, Available online: https://www.f08.th-koeln.de/imperia/md/content/personen/professoren/schmieder_matthias/controlling/loesung_lebenszykluskostenrechnung.pdf (2013)
23. H.G. Baum, A.G. Coenenberg, E. Günther, *Betriebliche Umweltökonomie in Fällen. Band I: Anwendung betriebswirtschaftlicher Instrumente*, Available online: https://books.google.sk/books?id=hNedCgAAQBAJ&pg=PA256&lpg=PA256&dq=Lebenszykluskosten+Break-Even-Time&source=bl&ots=y6ondKquN0&sig=2P5CUGpRAcBFoxJQjSATTiNY8Ds&hl=sk&sa=X&ved=0ahUKEwjtq6KHssfKAhVIw3IKHfGaCQwQ6AEIGjAA#v=onepage&q=LCC&f=false (2017)

Optimization of Processes in a Freight Forwarding Company Using a Simulation Model

Michal Sedláček[1,*]

[1]College of Logistics, Přerov, Department of magister studies, Palackého 1381/25, 750 02 Přerov, Czech Republic

Abstract. This paper discusses the possibilities of optimizing processes in forwarding companies that provide services in international truck transport. The computer simulation model was used to optimize logistics processes in the real distribution company. Software SIMUL8 was chosen as a development environment for its possibility of dynamic discrete simulation of company processes. This model will help us to analyze bottlenecks in company processes and find optimal solutions using simulation experiments. Optimization is focused on the company's warehouse, where a sequence of vehicles loading and unloading operations and emerging downtime are explored.

1 Introduction

Simulation models can be used to optimize both external and internal business processes. Procedures and methods to optimize business processes are many [1, 2] and one of them is the use of computer modeling. This method can help us solve logistics problems such as minimizing inventory in stock, optimal utilization of handling equipment, shortening shipping routes and the introduction of lean production [3, 4]. The simulations allow us to mimic processes that show shortcomings and bottlenecks. In the corporate process, we have to fix the problem and then propose its solution. The simulation is a cost-effective option for each company in terms of both financial and time-consuming. The company does not need to invest in the physical change of the process immediately, but after the simulation experiments and obtaining the necessary input data. Some methods, which use simulation models in solving logistic processes, were presented in [5-7].

Based on simulation experiments, we can easily identify bottlenecks at a particular point in the logistics process. If we find such a place, we can easily find solutions in the form of combinations to have the most effective impact on the whole simulation. The result of the simulation model should be to propose optimization of a particular process to solve the problem.

The paper focuses on solving the specific problem of eliminating the downtime of trucks in the forwarding company. This issue was solved in a specific forwarding company within the scope of the diploma thesis [8] at the College of logistics. The simulation model is used to obtain data for the decision-making process to eliminate truck downtime.

[*]Corresponding author: michal.sedlacek@vslg.cz

2 Computer simulation

Based on simulation experiments, we can easily identify bottlenecks at a particular point in the logistics process. If we find such a place, we can easily find solutions in the form of combinations to have the most effective impact on the whole simulation. The result of the simulation model should be to propose optimization of a particular process to solve the problem.

2.1 Discrete simulation

The discrete simulation model [9, 10] can be defined as a method of studying complicated probability dynamic systems through experiments in a computer model. Simulate means to mimic the operation of a real process using computer simulation. The complexity of the model is only a relative concept, which means that the model of a real process, cannot be solved by analytical steps, that will lead to a precise solution. With the whole model, we perform various experiments and observe the behavior of each activity. This is a statistical attempt, from which it follows that the result of the simulation is an estimate, not an exact meaning. Unlike a laboratory experiment, the simulation experiment is performed using a computer. The simulation as a scientific and practical field is closely related to the development of computer technology. Over time, simulation programs have been developed to simplify the production of simulation models in general. Thanks to these programs, we can easily perform simulation experiments and analysis of results.

Discrete simulation models are characterized, that all the state variables take only discrete values and change over time in jump steps [11-14]. Discrete dynamical system is a mathematical structure designed by main three components:

- Intervals in which all the possible values of the state variables lays,
- Function defined on specific interval,
- Differential equation which represents the "movement" equation of the system.

We quantify this change in terms of an updating function, which takes the current state of the system as an input, and specifies what the state of the system will be after some interval of time, an interval which corresponds to the specific updating function we are using. In general, an updating function is written:

$$m_{t+1} = f(m_t), \tag{1}$$

Where m_t is the current state of the system, and m_{t+1} is the state of the system after one interval of time has passed. By iterating the above calculation multiple times, we can find subsequent states of the system. Equation can be written in the form:

$$f\,{}^{\circ}f(m_t) = f(f(m_t)) = f(m_{t+1}) = m_{t+2}, \tag{2}$$

This equation describes the changing states of the system in dependence on time. This equation is also used as a theoretical basis for understanding discrete time steps in simulation model that we created.

Simulation also serves to link the theoretical knowledge of probability, statistics, operational research, system theory and informatics. We can use it both to solve theoretical problems and to solve practical problems. Application of the simulation model may be for:

- analysis of the dynamic behavior of a complicated or projected enterprise system through computer simulation,

- analysis of the sensitivity of the acquired solution to change the various parameters of the model,
- comparison of different system layout variants according to specified parameters for optimization of processes,
- replacing experiments with a business process that cannot be done in practice for some reason, but only by experiment using simulation.

Simulating means imitating the whole process just by computer simulation. The aim of the simulation is to solve specific situations or problems.

2.2 Input preconditions of the simulation model

To create the simulation model, a warehouse was selected, which is part of the entire logistic process of the company. In the warehouse the consignments are brought in during the week and the shipment ends at the end of the week and the consignments leave the warehouse.

The problem of downtime is solved in the warehouse. This problem occurs regularly every Friday and especially in traffic peak. At that moment a certain number of pick-up vehicles arrives and the warehouses are overwhelmed. This is mainly the capacity of warehouses and the number of loading ramps. Gradual steps have been compiled in simulation models using SIMUL8 simulation software to get the most accurate results. Data from these models were used for the final evaluation.

2.3 Description of the simulation model

The simulation model (Fig. 1) thoroughly examines the process of collecting consignments, especially the time-out of deliveries. The whole process works according to reality and the data entered into the simulation copies the real state of the system. The entire simulated process starts with the shipment of shipments from customers to the company terminal. This process is carried out using smaller vehicles. These vehicles are waiting for unloading, followed by unloading the pallets into the intermediate storage. As soon as the pallets are loaded into the intermediate storage, they are weighed into the warehouse shelves according to the destination. At this point, when all scheduled shipments are ready to be loaded, trucks arrive in to individual destinations. After loading all trucks, the model successfully ends with loading and leaving the last truck.

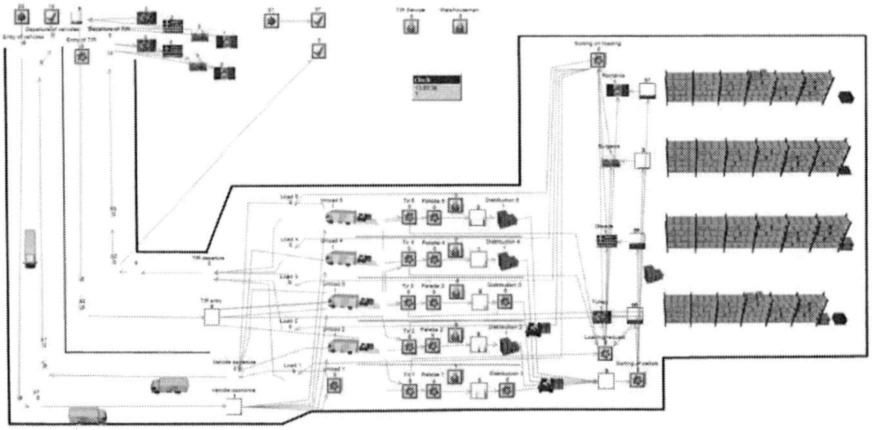

Fig. 1. Simulation model of the freight forward company. Source: authors

The simulation begins at 7:00 AM when the warehouse opens and the warehouse staff is ready to go. This is a simulated expedition day and this was Friday. The simulation should end at 8:00 PM by leaving the warehouse and leaving the tenth truck. These data were determined based on the opening hours of the warehouse and the practical experience. However, this time is not fixed. Since this is a collection service, warehouse keepers are sometimes in the warehouse for longer because consignments are delayed or waiting for trucks.

3 Creation of a simulation model

The simulation model is composed of the following basic elements, which are the main building of the entire simulation project:

- Truck (Vehicle),
- Truck (TIR)
- Pallet,
- Warehouseman,
- Truck (TIR) service,
- Destination.

The basic elements of the model are truck, pallet and TIR. Vehicle arrivals are sorted by time. The first batch of trucks arrives in the warehouse is between 7:00 AM- 12:00 AM 8 cars in the interval of 37,5 minutes. In the time span, that lasts from 12:00 AMto 3:00 PM, 45 trucks arrive at the warehouse in the interval of 4 minutes. In the time span between 3:00 PM and 5:00 PM, only 7 trucks arrive at the warehouse in the interval of 17 minutes. In the simulation model, we investigate the time tip in the warehouse to find out the possible downtime in truck arrivals and try to design the optimal solution.

The warehouse has only 5 ramps and the warehouse capacity is limited. Friday is a critical day, and most of the week's raid is taking place. Using label action function in simulation model, we distinguish the destinations for trucks deliveries. There are three pallets in each. Shipments are sorted by destination destinations. These are Greece, Turkey, Romania and Bulgaria.

The inventory called "vehicle downtime" is the most important element in the simulation model. Penalties from carriers are set at 500 CZK per hour if the truck waited without access to the ramp. Based on this data, a rate of 8.33 CZK was set for every minute of the car's downtime. This rate is listed in the stack, so we'll figure out how much money the downtime of trucks will cost. Manipulation with pallets takes approximately 20 minutes. This time was determined based on measurements. Other activities, such as ramp arrival, pallet unloading, pallet marking and subsequent confirmation of documents by the carrier are included. After unloading the pallets into the intermediate warehouse, the pallets are unloaded into the final warehouse shelf. The pallets are sorted in front of the shelves by the pallet sorting.

In the simulation model there are 10 warehouses, the morning group is in the warehouse from 7:00 AMtill 3:30 PM, the afternoon group comes at 12:00 AMdue to the tip in the warehouse. From 3:30 PM, the warehouse is limited to 5 warehouses. An important point in the simulation is the arrival of the truck (TIR) to be loaded at 4:00 PM. All trucks are divided into four destinations: Greece, Turkey, Romania and Bulgaria. Each of the international trucks will pick up 33 pallets and the warehouseman will load the truck. Loading the truck takes about 66 minutes. Subsequently, the truck leaves the ramp and departs to a destination that has been pre-defined using the "Destination" label in simulation model.

 DOI: 10.1051/matecconf/201713400050

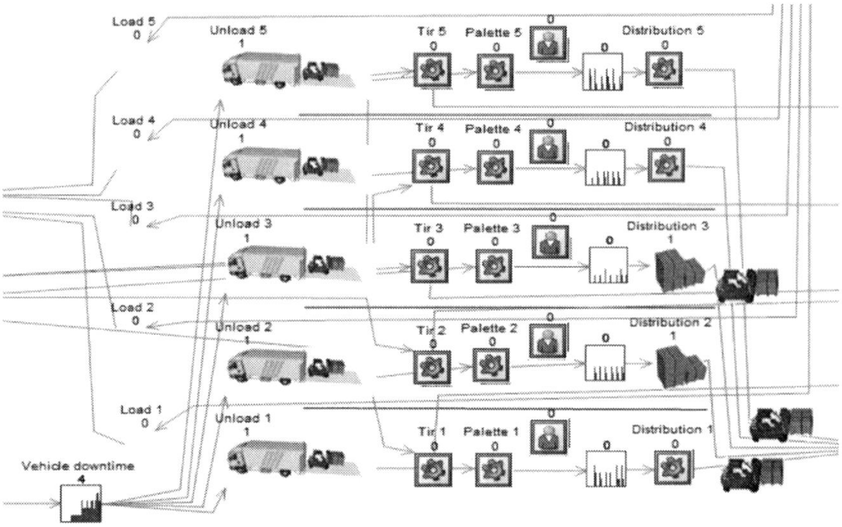

Fig.2. Part of the simulation model – problematic situation of the truck downtime. Source: authors

The result of the simulation is the apparent downtime of the trucks (vehicles) waiting to be unloaded listed in the vehicle downtime stack (Fig. 2). Another important feature is all served trucks and TIR. The sequence of all activities in the model is correctly defined and the final result is a functional simulation model corresponding to the real state of warehouse.

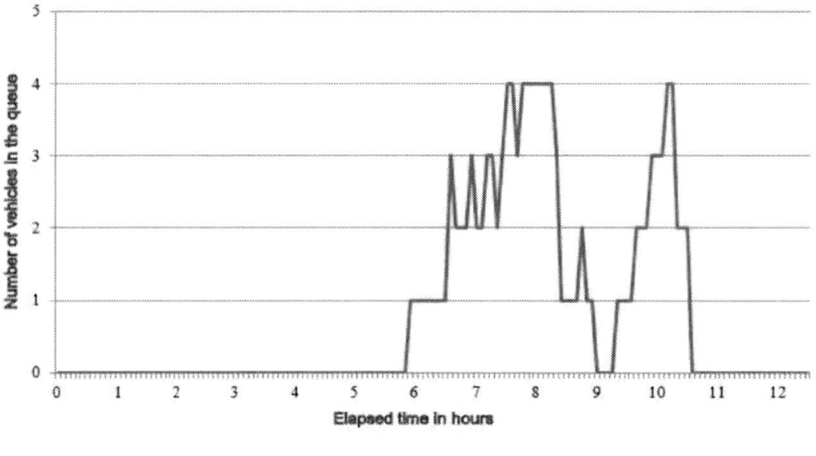

Fig. 3. The number of vehicles waiting in the queue (initial condition). Source: authors

4 Simulation experiments

A simulated model that matches the reality of the company will provide us with the current state of the system. We will focus mainly on arriving trucks to be unloaded and on trucks to be loaded to eliminate downtime, which is a financial burden of the company. The result of the simulation experiment is illustrated in Figure 3, which shows, that after 6 hours a queue of pick-up vehicles is being created, when there are 4 trucks waiting in the queue. This is a problematic situation, in terms of financial problem, it can become a company 4915, CZK each day. In simulation experiments, we will focus on eliminating this problematic downtime.

4.1 Simulation experiment 1

The first option to improve downtime is to add one warehouseman. He will help speed up the entire landing process, including TIR loading. An extra warehouseman is available from 12:00 AM. At the same time, there will come an extra warehouseman at 3:30 PM. The results of the simulation are shown in Figure 4.

The results show acceleration of the whole process and reduction of downtime. In total, the company would pay 500,-CZK for a day. In real terms, it is also necessary to count the possible financial costs for additional warehouseman, if the company would not solve it from its personnel resources, but by external staff.

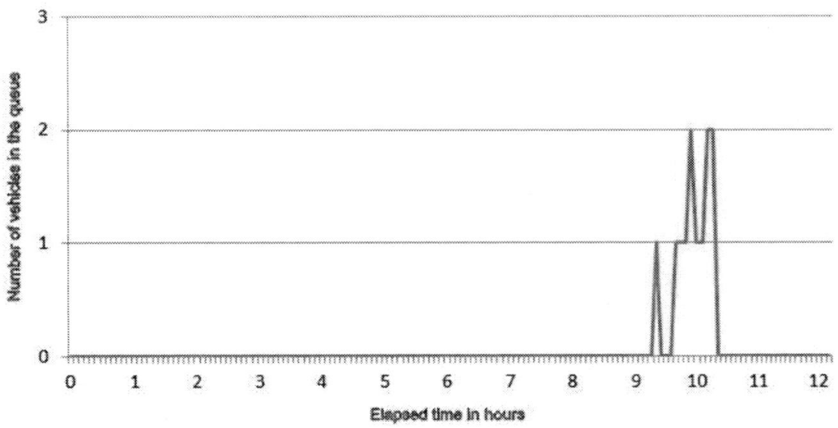

Fig.4. The number of vehicles waiting in the queue (Experiment 1). Source: authors

4.2 Simulation experiment 2

The second option is to shift the truck's arrival time to an earlier hour at 3:00 PM. However, this option will show clear deficiencies in the capacity of ramps and labor, which are considerably overloaded and disproportionately prolonged downtime. Generated graph on Figure 5 demonstrates this situation.

If we start loading trucks earlier, then higher downtime will occur. At this time, the simulation generated a loss of 1870,-CZK. The displacement of the loading time at 5:00 PM will also not solve the situation, the loss caused by the waiting truck donation is 3540,-CZK.

 DOI: 10.1051/matecconf/201713400050

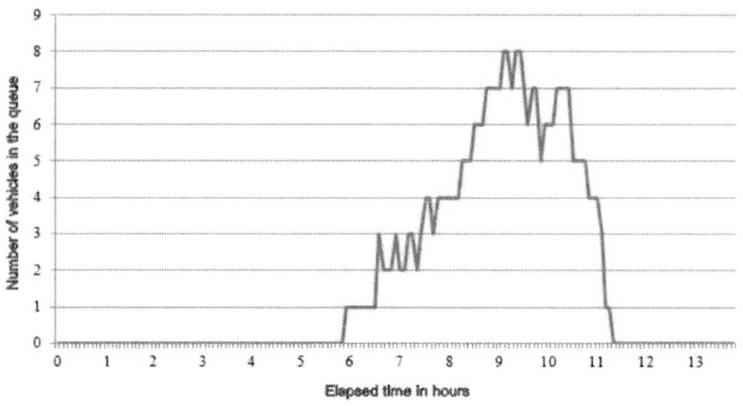

Fig. 1. The number of vehicles waiting in the queue (Experiment 2). Source: authors

4.3 Simulation experiment 3

Possible optimal solution to the current situation is a variation with the loading time at 5:00 PM and the addition of one warehouseman for loading. The result of this option is the zero time-lag of the trucks, but the possible additional labor costs. The last option is to add another loading ramp that could accelerate the entire loading process and reduce truck downtime. Adding the sixth ramp could help the current situation, but unfortunately it does not seem like a real deal in today's business conditions. There is no room for expansion in the warehouse, and the owner would not allow it either. This option can only be created in the simulation model.

Table 1. Costs of truck waiting times according to simulation experiments. Source: authors

	Arrival time of TIR	Total waiting time for all trucks	Cost of truck downtime
Experiment 1	4:00 PM	60 min	500,- CZK
Experiment 2.1	3:00 PM	225 min	1870,- CZK
Experiment 2.2	5:00 PM	425 min	3540,- CZK

5 Conclusion

The simulation model, which was constructed and on which experiments were performed, provided data to assess the suitability of the proposed optimization variant. The problem was related to the downtime of vehicles in the warehouse. The experiments consisted of a change in the number of workers in the warehouse and the arrival time of the lorries to load the goods. As an optimal solution, the option was chosen to change the time of arrival of trucks and to add one labor force to the warehouse. This option is optimal especially if the trucking company uses its own resources - unused workforce.

Possibilities of using of the simulation models in praxis are large. Simulation models can be used with advantage in dealing with different logistical problems. As example can be the models of production logistics, distribution logistics, reverse logistics, transportation logistics and logistics services [4, 5, 15]. In other fields simulation models are used to solve problems of crisis management and obtaining input data for solving various problems of human activities [6, 7, 16]. Simulation tools occupy an important place in improving the efficiency of logistics processes of companies, among them we can mention the so-called digital factory [13, 17] and production information systems based on an internet of things [3].

The aim of this paper is to demonstrate the specific use of simulation models for solving the truck downtime in freight forwarding company. This problem was solved in a manufacturing enterprise and subsequently realized. Simulation model was served as a basis for strategic decisions of the company management. The paper describes three variants of solving the truck downtime, describes the assembly simulation experiments, execution and evaluation of the different options according to the simulation results. Simulation models have been used in solving the specific logistical problem in the real company.

This paper is supported by the research project "From horse-drawn railway to intermodal transport" within Visegrad Fund.

References

1. G. Fedorko, Z. Čujan, *12th International Conference on Industrial Logistics*, 167-175 (Island Brac, Croatia, 2014)
2. G. Fedorko, V. Molnár, J. Strohmandl, M. Vasiľ, *Transport Means 2015*, 466-469 (Kaunas University of Technology, Lithuania, 2015)
3. T. Martoch, O. Kodym, *SGEM 2013 – International Multidisciplinary Scientific GeoConference Surveying Geology and Mining Ecology Management*, 207-212 (Albena, Bulgaria, 2013)
4. M. Andrejiová, A. Pavlisková, N. Husáková, *Carpathian Logist. Congr.*, 1-6 (Tanger LTD. Ostrava, Czech Republic, 2012)
5. O. Kodym, L. Kavka, M. Sedláček, *SGEM 2015 – International Multidisciplinary Scientific GeoConference Surveying Geology and Mining Ecology Management*, 183-190 (Albena, Bulgaria, 2015)
6. G. Fedorko, N. Husáková, G. Dudáš, Acta Montan. Slovaca **15**, 14-22 (2010)
7. V. Molnár, *Carpath. Logis. Congr.*, 1-6 (Tanger LTD: Ostrava, Czech Republic, 2012)
8. J. Vyšín, *Optimalizace procesů ve spediční společnosti s využitím simulačních modelů* (Diploma thesis, College of logistics, Přerov, Czech Republic, 2017)
9. M. Dlouhý, *Simulace podnikových procesů* (Brno: C. Press, Czech Republic, 2007)
10. I. Křivý, E. Kindler, *Simulace a modelování* (Ostrava: OU, Czech Republic, 2001)
11. R. Bartošíková, J. Bilíková J. Strohmandl, V. Šefčík, P. Taraba, *24th International Business Information Management Association Conference*, 1479-1483 (Milano, Italy, 2014)
12. S. Li, X. Wang, International Journal of Control and Automation **8**, 2, 71-82 (2015)
13. H. Neradilova, G. Fedorko, Acta Logistica Moravica **1**, 36-43 (2015)
14. SIMUL8, *Corporation. Simul8: manual and simulation guide* (Herndon, VA: The Corporation, 2001)
15. A. Torok, A. Torok, F. Heinitz, Transport and Telecommunications **15**, 4, 292-298 (2014)
16. M. Čičak, T. Mlinarić, B. Abramović, Promet - Traffic - Traffico **16**, 2, 63-69 (2004)
17. J. Nikolic, I. Ristovic, D. Vasovic, Journal of Environmental Protection and Ecology **16**, 3, 1082-1090 (2015)

Efficiency Assessment of Fuel Borne Catalyst

Marie Sejkorová[1,*], *Branislav* Šarkan[2], and *Jan* Verner[1]

[1]University of Pardubice, Faculty of Transport Engineering, Department of Transport Means and Diagnostics, Studentská 95, 532 10 Pardubice, Czech Republic
[2] University of Žilina, Faculty of Operation and economics of Transport and Communications, Department of Road and Urban Transport, Univerzitná 1, 010 26, Žilina, Slovakia

Abstract. Fuel additives are offered to motorists at petrol stations and on the internet. Cerium or iron in the form of ferrocene is often used as a catalytically active metal in these additives. Producers of these fuels borne catalyst declare that their use will improve certain properties of engine. Therefore, in this paper are presented the result of testing and effects of these additives on the smoke and the engine noise. The results do not confirmed the positive effect of these fuel additives on the researched properties of engine.

1 Introduction

Thanks to the invention of the combustion engine, transport accelerated significantly and at the same time it has been enabled to transport people and goods for longer distances. Transport has therefore entered a new dimension of breaking the barriers. Nowadays, transport not only enables meeting people, connecting cultures and increase of economic and social functions but it has also become a significant factor influencing the environment. In 2012, the total share of emissions produced by the transport sector, including international transport, was 24%. Transport is therefore the only sphere of the European Union (EU) where the emission has increased since 1990 [1]. Transport, especially road transport, is also the main producer of noise when considering the amount of inhabitants exposed to the noise levels exceeding hygienic limits. Approximately 210 million people in the EU are exposed to excessive noise caused by road transport. A study on the risk of the influence of excessive decibels on human health, which nearly a half of Europeans is exposed to, was published [2]. Unwanted excessive noise may aside from bad sleeping lead to e.g. development or worsening of lifestyle diseases such as high blood pressure or diabetes. According to World Health Organization noise above 55 decibels means serious discomfort and may set off development of many diseases and other disorders.

One of the possible ways how to reach an improvement in the area of road transport emission production is to use fuel catalytic converters. Fuel catalytic converters are organometallic compounds dissolved in organic carrier fluids. Their main contribution is supporting of diesel particulate filter (DPF) system functions. Fuel catalytic converters and diesel burn together in the combustion chamber and respective oxides develop from the metals present. These inorganic particles are subsequently diffused and are collected

* Corresponding author: Marie.Sejkorova@upce.cz

on the DPF as a homogenous compound made of carbonaceous particles and metal oxide particles. Metal oxide particles which are diffused in the carbonaceous particles decrease the temperature of their combustion, i.e. DPF regeneration temperature. The advantage of these catalytic converters is a better combustion and therefore a lower development of hydrocarbons, carbon monoxide, and solid parts. The disadvantage may be the fact that their catalytic potential manifests during the fuel oxidation in the fuel tank itself [3]. A lot of studies reported the results of the metal oxide additives, e.g. aluminum oxides [4], copper oxides, manganese oxides or cerium oxides [5] in diesel. These compounds may lower e.g. the soot production thanks to strengthening the contact between the catalytic converter and soot [6]. Cerium (Ce) or iron (Fe) in the ferrocene form ($(C_5H_5)_2Fe$) are frequently used by catalytically active metals [7-9].

Ferrocene is an organometallic compound which used to be added into automotive gasoline for increasing the octane number. Ferrocene has never reached such popularity as lead antidetonators and some studies proved that iron oxides developing during ferrocene combustion increase the wear of the engine because of their abrasive effect, especially at cylinder walls and piston rings [4]. Based on the experimental studies, some researchers reached the conclusion that the addition of ferrocene prevents soot production. It has been proved that the addition of iron containing compounds increases the speed of soot oxidation and decreases the total soot amount by combustion [10]. Iron oxide particles function as a nucleation centers for soot production and these are subsequently oxidized by the $Fe_xO_y + C \rightarrow CO + Fe_xO_{y-1}$ reaction [7].

Adding ferrocene into the motor diesel has an impact on the process of soot production and the molecular structure and nano-structure of soot. According to the data [11], it appears that the carbon structure of soot changes dramatically in favor of aliphatic structures. With regards to the fact that the motor diesel and ferrocene compound produces soot with a less graphitic yet more aliphatic structure, it may be one of the ways for increasing the soot oxidation. Nevertheless, with regards to respiratory toxicity of iron oxides [11] which may be present in soot and ash, subsequent filtration of exhaust fumes is necessary in order to prevent the release of toxic substances into the air.

Cerium oxide is an extraordinary catalytic converter thanks to its ability to alter between two oxidation states, i.e. CeO_2 (+4) and Ce_2O_3 (+3) using a relatively low energy reaction [12]. At the same time, cerium oxide lowers emissions of HC, soot, and NO_x in the exhaust fumes of combustion engines because of this unusual transformation ability. Cerium oxide is theoretically and experimentally proved redox catalyst used for decreasing harmful emissions of diesel engines.

Fuel catalyst converters may be in a liquid or solid form. These additives, which are added directly into the fuel tank, may be e.g. in the form of tablets. Manufacturers of these products claim reduction of pollutants in exhaust gases, decreasing of fuel consumption, increasing of engine performance, and a cleaning effect for the fuel system and combustion chamber.

Zhnag et al. [13] were researching the influence of cerium oxide based nanoparticles additive dosage of Envirox in diesel, engine performance, and emissions of polluting substances. Using this additive led to an improvement of the fuel effectivity of up to 5.6%, CO_2 emission reduction of 10.6%, and reduction of solid particles of 24.4%, but also an increase in ultrafine particles of 32%, and a NO_x increase of 9.3%. Sajeevan and Sajith [14] and Sajith et al. [15] state that when using cerium oxide based nanoparticles as an additive into motor diesel, an average reduction of H_xC_y emissions of 45% was found [15]. As a part of the study [3], the liquid product Envirox underwent and X-ray fluorescence (XRF) analysis which did not prove the presence of cerium nor any other metal and it was subsequently stated that it was only a hydrocarbon product.

2 Materials and methods

The "Power Cab Tab" product by Aquel Bohemia, s.r.o. distributor was selected to be the fuel conditioner tested. This product in the form of tablet is widely known among the public and it is the best-selling fuel conditioner. This producer also claims this product causes noticeable differences in the behavior of the engine after a relatively short time of using this product.

The application of the tablets was always done immediately before refueling as stated by the producer of the fuel conditioner. The reason for this application is to prevent the tablet from getting stuck e.g. in the filler pipe of the fuel tank. One whole tablet is meant for 40 l of automotive gasoline or diesel. In total, 9 tablets were used and 346.5 l of fuel were taken in. The vehicle used in the experiment was a Škoda, Octavia II Combi 4x4 car. For the car parameters, see Tab 1. The table shows the initial state of kilometers or the state of the mileage until the first measuring (measuring without the use of fuel conditioner) and the subsequent state which means the state of the counter when the second measuring was conducted, i.e. after the fuel conditioner application. The distance driven between the first and the second measuring is 4,476 kilometers.

Table 1. Basic vehicle parameters. Source: authors

Vehicles	Škoda Octavia II Combi 4x4
Engine	diesel engine, 1.9 TDi, 77 kW, typ BXE
Fuel	diesel
Year of production	2008
Type of emission system	Controlled with OBD, oxidation catalyst, without DPF
Initial kilometres [km]	282 028
Total distance [km]	4 476

The measuring of smoke emission was conducted in a specialized laboratory for measuring emissions. The measuring itself and its evaluation was conducted using the equipment shown in Tab 2. The period during which the tablets were being applied was 3 months. The producer claims that an improvement of the parameters measured should manifest during this period.

Table 2. Used equipment. Source: authors

Place of measurement	ZDAR, a.s. (Technical inspection Žďár nad Sázavou)
Used equipment for measurement emission	opacimeter MAHA MDO - 2; Software DEKRA Emissions (2016.0.3)
Used equipment for measurement sound	sound level meter VOLTCRAFT 322 Datalog
The time difference first and second measurements	3 month

 DOI: 10.1051/matecconf/201713400051

The measuring was conducted according to the methodical procedure of emissions measuring approved by the Ministry of Transport of the Czech Republic. Before the smoke emission measuring itself, an engine diagnosis with the inclusion of the error log which was without any entry. The value of the corrected absorption coefficient, which is given by the producer and which must be subsequently followed by the vehicle, was generated automatically by an on-line measuring tool. It is the maximum value possible which guarantees fulfillment of the emission requirements. In this case it was 1.00 m^{-1}.

The noise measuring was conducted at the same time as the smoke emission measuring for the sake of following the identical conditions of the experiment.

Initially, the first measuring was conducted for finding the noise values of the engine using diesel without additives and subsequently second measuring after the application of fuel with fuel conditioner added during the testing period in real running conditions, i.e. for three months.

The microphone was placed at the height of 1.1 m and in the middle in front of the vehicle with an open hood in the distance of 0.2 m. There were not any reverberation obstacles in the immediate surrounding area. The measuring is shown in Fig. 1. All measuring parameters are shown in Tab 3.

Fig. 1. Measurement of engine noise. Source: authors

Table 3. Results from measurement of noise. Source: authors

	1st measurement	2nd measurement
used equipment for measurement sound	sound level meter VOLTCRAFT 322 Datalog	
used weighting filter	weighting filter A	
the height of the microphone from the floor [m]	1.1	
distance of microphone from vehicle in top view [m]	0.2	
microphone placement relative to vehicle	front part, at the center of the vehicle with the bonnet open	
distance of the nearest object [m]	2.0	
distance of other objects [m]	> 6.5	

3 Results and discussion

Value of 1.11 m^{-1} was obtained by the measuring of the smoke emission conducted at the technical inspection of the ZDAR, a.s., Žďár nad Sázavou company at the beginning of the experiment, i.e. without the use of fuel conditioners, as seen in an excerpt from the protocol shown in Tab 4.

Table 4. 1st measurement (before using the conditioner fuel). Source: authors

results check of the readiness code		set	
		Mil: off	
the result of check of the engine control unit		without error	
engine speeds	predefined speed	measured	
idling	800-1000	830	
overrun	4650-5050	4850	
corrected absorption coefficient by the label [m^{-1}]		1.00	
value of smoke [m^{-1}]		1.11	
smokiness range of values of four consecutive measurements [m^{-1}]		permissible	0.25
		measured	0.00

After three-month-long using of Power Car Tab fuel conditioner when the second measuring was conducted, the value of 1.08 m^{-1} was obtained (see Tab 5). With regards to the fact that 1.00 m^{-1} is the maximum limit allowed, the vehicle could not be approved as capable of operation on the roads of the Czech Republic in case of emission measuring and being a subject to a subsequent vehicle inspection.

Table 5. 2nd measurement (after using the fuel conditioner). Source: authors

results check of the readiness code		set	
		Mil: off	
the result of check of the engine control unit		without error	
engine speeds	predefined speed	measured	
idling	800-1000	830	
overrun	4650-5050	4860	
corrected absorption coefficient by the label [m^{-1}]		1.00	
value of smoke [m^{-1}]		1.08	
smokiness range of values of four consecutive measurements [m^{-1}]		permissible	0.25
		measured	0.00

In both cases presented, diesel was refueled at the same authorized company which is supposed to sell fuel following the requirements given by the Czech legislation. Using fuel conditioners in the form of tablets reduced the smoke emission by 0.03 m^{-1}; value of 1.00 m^{-1}, which is the limit for compliance of the emission limits, was not reached.

The difference of the average measured noise values of the engine with idle speed between the first and the second measuring is 0.3 dB. In the case of overrun speed, the difference was 0.4 dB, as shown in Tab 6. The measuring accuracy of the noise meter used is ± 1.5 dB. The results measured do not meant conclusive reduction of engine noise claimed by the tablet producer.

Table 6. Results of noise measurement. Source: authors

	1st measurement	2nd measurement
Max noise at idling speed for 30 seconds [dB]	72.1	71.7
	71.9	71.8
	72.0	71.6
The average value of max noise at idling speed for 30 seconds [dB]	72	71.7
Max noise at overrun engine speeds for 5 s [dB]	101.9	101.6
	101.7	101.4
	102.1	101.5
The average value of max noise at overrun engine speeds for 5 seconds [dB]	101.9	101.5

4 Summary and conclusion

Using the fuel based catalyst converter is one of the ways how to reduce pollutant emissions from combustion engines. Fuel additives are additives into the fluid hydrocarbon fuel which are supposed to influence their quality or improve combustion. The producers of fuel conditioners claim the products improve engine characteristics such as higher performance, lower fuel consumption; production of pollutant emissions lowered by 90%, noise reduction and higher engine protection. The question is if it is not only a business stunt of the producers of the additives. Because of this, emission measuring and noise measuring of a Škoda Octavia car with and without fuel conditioner was conducted. The fuel conditioner selected was Power Car Tab.

Using of the aforementioned conditioner lasted for the period of three months and with range of 4,476 km. All requirements for unambiguous recognition of the functionality of the additives defined by the producer were therefore met. However, the differences of emission and noise measuring before and after the application of fuel conditioner did not confirm the producer's positive claims; the smoke emission was reduced only by 0.03 m^{-1}, the noise values decreased only by 0.3 dB, or 0.4 dB, respectively.

The results of this thesis are in accordance with the research results [3] which observed the changes in oxidative stability of fuels during using of fuel conditioner and it also did not prove any positive effect of fuel conditioner application.

This study is partly funded by the University of Pardubice (No. SGS_2017_009).

This paper is supported by the research project "From horse-drawn railway to intermodal transport" within Visegrad Fund.

References

1. Evropská agentura pro životní prostředí, *Evropské životní prostředí – stav a výhled*, 212 (2015)
2. LC Den Boer, A. Schroten, CE Delft **70** (2007)
3. J. Černý, P. Šimáček, *Úsporné tablety do paliva – obchod s lidskou důvěřivostí, Kniha přednášek mezinárodní konference Tribotechnika v provozu a údržbě 2016*, 9-14 (VŠB – Technical University of Ostrava, Czech Republic, 2016)
4. M.J. Kao, B.F. Lin, T.T. Tsung, *18th Internal Combustion Engine Symposium* (Jeju, Korea, 2005)
5. A.M.V. Selvan, R.B. Anand, M. Udayakumar, Fuel **130**, 160-167 (2014), DOI: 10.1016/j.fuel.2014.04.034
6. J. Lahaye, et al, Combustion and Flame **104**, 1-2, 199-207 (1996), DOI: 10.1016/0010-2180(95)00176-X
7. CH. HU, et al, Journal of the Energy Institute (2016)
8. A.C. Sajeevan, V. Sajith, Fuel **183**, 155-163 (2016), DOI: 10.1016/j.fuel.2016.06.048
9. K. Tian, Z.S. Li, S. Staude, B. Li, Z.W. Sun, A. Lantz, M. Aldén, B. Atakan, *Proceedings of the Combustion Institute* **32**, 1, 445-452 (2009)
10. J. Zhang, C.M. Megaridis, Combustion and Flame **105**, 4, 528-540 (1996)
11. Z.H. Zhang, R. Balasubramanian, Environmental Science & Technology **51**, 8, 4248-4258 (2017), DOI: 10.1021/acs.est.7b00920
12. A. Trovarelli (ed.), *Catalysis by ceria and related materials* (World Scientific, 2002)
13. J. Zhang, et al, Environmental science & technology **47**, 22, 13077-13085 (2013), DOI: 10.1021/es402140u
14. A.C. Sajeevan, V. Sajith, Journal of Engineering **2013**, 1-9 (2013), DOI: 10.1155/2013/589382
15. V. Sajith, C.B. Sobhan, G.P. Peterson, Advances in Mechanical Engineering **2010**, 1-6 (2010), DOI: 10.1155/2010/581407

Price for the Allocation of Railway Infrastructure Capacity as a Tool for the Improvement of Train Transport Planning

Jaromír Široký[1,*]

[1]University of Pardubice, Faculty of Transport Engineering, 532 10 Pardubice, Czech Republic

Abstract. The introducing of legal background for capacity allocation in the Czech Republic. Author analysis of price for capacity allocation in the Czech Republic and in Europe. The proposal of the capacity allocation price which should lead the Railway undertaking's path request to the suitable product. The author summarizes the legislative background of the railway infrastructure capacity allocation, and then compared the levying of charges for the allocation of railway infrastructure capacity in the Czech Republic as well as in the neighbouring states – Germany, Poland, Austria and Slovakia.

1 Introduction

Economic development has a bearing on an increased transport demand. Large amounts of goods need to be transferred from producers to consumers, which require quality transport services. Along with the increasing transport demand, there is a growing supply of carriers on the transport market. It is crucial for the state to adopt certain regulatory actions in order to preserve market equilibrium and protect the market. Issuing of licences and certificates to carriers is one of these regulatory actions; the allocation of railway infrastructure capacity is another one. In view of the geographical, demographic, economic and urbanistic make-up of Europe, most states depend on the import and export of their products and goods. Uniform conditions of network access and the provision of facilities and services need to be set for the opening transport markets. The uniform conditions for the EU member states are approved and made public by the European Parliament and the Council in their regulations and directives. A regulation enters into force and becomes legally applicable in all member states just as it was approved and made public. Directives need to be implemented in the legislation of individual member states. Directive 2001/14/EC of the European Parliament and of the Council of 26th February 2001 considers the allocation of railway infrastructure capacity, the levying of charges for the use of railway infrastructure and the issuing of safety certification. This directive was amended by the Directive 2012/34/EU of the European Parliament and of the Council of 21st November 2012 establishing a single European railway area.

* Corresponding author: jaromir.siroky@upce.cz

2 The legislative background of the allocation of railway infrastructure capacity

In the Czech Republic, the conditions for the allocation of railway infrastructure capacity are defined by Act No. 266/1994 coll. on railways, which governs the relationships between the railway infrastructure capacity allocation body (railway manager) and the carrier. The process of railway infrastructure capacity allocation must be non-discriminatory for all carriers who meet the conditions for access to railway infrastructure. Správa železniční dopravní cesty (hereinafter only referred to as „SŽDC") is the railway infrastructure capacity allocation body for the national railway and the state-owned regional railways [1].

Upon the accession of the Czech Republic to the European Union (EU) on 1st May 2004, the need arose to adapt the legislation of the Czech Republic. Act No. 103/2004 Coll. of 11th February 2004 amending Act No. 266/1994 Coll. on railways, as amended, came into force [2].

The amendment to the Act No. 266/1994 Coll. implemented the Directive 2001/14/EC of the European Parliament and of the Council of 26th February 2001 on the allocation of railway infrastructure capacity and the levying of charges for the use of railway infrastructure and safety certification.

In 2012, the Directive 2001/14/EC was amended by the Directive 2012/34/EU of the European Parliament and of the Council of 21st November 2012 establishing a single European railway area [3].

The Directive 2012/34/EU† of the European Parliament and of the Council states that the EU member states must ensure that the schemes for levying charges for the use of railway infrastructure and for the allocation of railway infrastructure capacity are governed by the principles given in the Directive 2012/34/EU [3] and make it possible for the infrastructure managers to trade the available infrastructure capacity and ensure its optimal effective use. The Directive 2012/34/EU [3] defines the content of the minimum access package, it defines what belongs to the railway access to equipment and services, it defines which services belong among basic services and which are provided as ancillary services. These services must be provided by all the railway managers in the EU to all the carriers meeting the conditions of access to the railway network.

The allocation of railway infrastructure capacity in the Czech Republic is a part of the minimum range of services in compliance with the Decree No. 76/2017 Coll. on the content and scope of services provided to the carrier by the railway manager and by the service facility manager. The railway to be used for the train run represents another part of the minimum scope of services provided, i.e. ensuring the railway operation (operation management) and ensuring the operability of the railway (maintenance and repairs of the infrastructure).

For railway lines where SŽDC is the railway infrastructure capacity allocation body, the calculation of the price for the allocation of railway infrastructure capacity depends on:

- The duration of the period from the submission of the application and the allocation of railway infrastructure capacity and the requested first day of its drawing,
- The relationship of the submitted application for the allocation of railway infrastructure capacity and the term of the establishing of the annual working timetable or its planned changes,
- The processing difficulty of the application.

For railway lines where SŽDC is the railway infrastructure capacity allocation body, the price for the use of a railway for a train run depends on:

† Directive 2012/34/EU of the European Parliament and of the Council of 21st November 2012 establishing a single European railway area

 DOI: 10.1051/matecconf/201713400052

- The length and parameters of the railway in question,
- Type of transport (passenger, freight),
- The parameters of the train and on the type of price applied – basic price, promotion price, increased price.

Each railway infrastructure capacity allocation body shall draw up their rail network statement and make it public no later than 12 months before the beginning of effectiveness of the working timetable in a way which enables remote access. The network statement shall set non-discriminatory rules for the allocation and withdrawing of railway infrastructure capacity, for access to the railway, for its use and for the calculation of the price for this use [2].

3 The analysis of the system setting the process for the allocation of railway infrastructure capacity in the Czech Republic and neighbouring states

In this chapter, the author shall deal with the charging scheme for the allocation of railway infrastructure capacity and the charging scheme for the use of railway infrastructure in the Czech Republic and in the neighbouring states, i.e. in Germany, Poland, Austria and Slovakia. Charges are levied on the minimum access package; some states levy a charge on it. Some states levy charges on it by parts, such as the Czech Republic, which levies charges for the allocation of railway infrastructure capacity and charges for the use of railway infrastructure separately.

The idea of calculating a single charge for the use of the railway infrastructure in the whole EU originated as early as in 2003 as the PARTNER project (Path Allocation Re-engineering of Timetable Networks for European Railways), which was opened on 1st October 2003. Currently, the project is no longer taking place. It has been replaced by project Christine.

Each railway manager from any EU country issues their Network Statement which must have a unified structure so that it's easier to understand for the applicants, The Network Statement must contain basic information, such as conditions of access to the railway network, the description of infrastructure, the description of the process of application and allocation of the railway infrastructure capacity, dispute resolution in case of insufficient railway infrastructure capacity, services provided to the applicants and carriers (minimum access package and railway line access to facilities and services for carriers), the price for the use of the railway and for the services provided. So far, a unified charging scheme has not been implemented on the European network due to various legal and legislative reasons on the part of the member states. The carriers would agree with the unified charging mechanism, as it would substantially simplify the reporting of their performances and services used and their recording. Furthermore, all the charges would be unified and the carrier would be able to calculate the price of transit through all the states in advance.

As part of their minimum access packages, the individual railway managers define the content of the minimum access package and which services the carriers are entitled to and how they are broken down.

In the Czech Republic, there are several railway infrastructure capacity allocation bodies. The levying of charges, which is subject to price regulation in the Czech Republic, is a part of the process of allocation of railway infrastructure capacity. The levying of charges depends on the length of the railway in question and the total amount of eligible costs of the allocation of the railway infrastructure capacity [4].

The railway managers and railway infrastructure capacity allocation bodies charge the applicants for the use of the national and regional railway infrastructure:

1. The prices including the economically substantiated costs directly incurred on the operation of railway transport, i.e. the following:
   - The price of the capacity allocation body for allocating the railway infrastructure capacity,
   - The price of the railway manager for the use of the railway for the train run,
   - The price of the railway manager for ensuring railway network access of carriers to service facilities
2. Prices for additional services.

The minimum scope of services provided by the railway manager to the entitled carriers are laid down by Decree No. 76/2017 Coll. on the content and scope of services provided to the carrier by the railway manager and by the service facility manager.

Prices for regulated services stated in paragraph 1) are subject to rectification as per the effective assessment of the Ministry of Finance, which is made public in the Official price bulletin of the Ministry of Finance – „The use of state and regional railway infrastructure".

The table (Table No.1) summarizes the charging for the allocation of capacity by the individual capacity allocation bodies in the Czech Republic.

Table 1. The Charging for the allocation of railway infrastructure capacity in the Czech Republic. Source: Source: author, based on [5-12]

Railway infrastructure capacity allocation body	Charging for the allocation of railway infrastructure capacity		
AŽD Praha			NO
Czech railways			NO
Jindřichohradecké místní dráhy			NO
Moravian-Silesian Region			NO
KŽC	YES	For the train path / per day / different rates for regional timetable and ad hoc	
Inter-municipal cooperative of the Desná valley	YES	For the train path / per day / progressive rates	
SŽDC	ANO	For the train path / per day / per km of planned path / different rates for regional timetable and ad hoc	

The SŽDC, a member of RNE, is the most significant railway infrastructure capacity allocation body.

The maximum timespan of the allocated railway infrastructure capacity is 20 hrs (the period from the departure from the first point of the network until the arrival to the last point in the network of SŽDC). For railway lines belonging to the European Rail Network for Competitive Freight (ERNCF) as per the Regulation 913/2010/EU‡, the capacity may also be allocated by the Corridor-OSS (C-OSS) on behalf of SŽDC. The conditions and procedures for the allocation of railway infrastructure capacity by C-OSS are made public for the individual corridors in the Corridor Information Document (CID) [13].

The company DB Netz AG (railway manager in Germany), a member of RNE (Rail Net Europe), is the most significant railway manager in Germany. The same applies to the company PKP Polskie Linie Kolejowe (PKP PLK) in Poland (railway manager in Poland),

‡ Regulation 913/2010/EU of 22nd September 2010 concerning a European rail network for competitive freight (ERNCF)

to the company ÖBB-Infrastruktur AG in Austria (railway manager in Austria) and to the company Železnice Slovenskej republiky (ŽSR) in Slovakia (railway manager in Slovakia).

The Table (Tab. No. 2) summarizes the levying of charges in the given states and shows the differences in charging. Only basic parameters are given. For example, the parameter of the environmental impact of the traction unit applied by ÖBB-Infrastruktur AG isn´t considered. The parameter considering the use of the tilt body applied by DB Netz AG isn´t considered either [14].

The most similar model of charging for the railway infrastructure to the one used by SŽDC is the model used by ŽSR, where charges are also levied according to actual parameters of trains (also considered by ÖBB-Infrastruktur AG), considering the type of railway, the weight of the train and an independent traction unit on electrified railway lines. To the contrary, neither the speed of the train, nor the railway lines with overloaded infrastructure ae considered. Furthermore, SŽDC uses a different price for passenger and freight transport as opposed to ŽSR.

DB Netz Ag and PKP PLK levy charges according to the planned train parameters. The type of railway is considered by all the stated railway managers. DB Netze, PKP PLK and ŽSR do not have a different price for passenger and freight transport. The rate according to the weight of the train is applied by PKP PLK (gradually), by ŽSR and by ÖBB-Infrastruktur AG for freight transport. DB Netz AG has a single rate for freight of up to 3000 t [15].

An independent traction unit on electrified railways isn´t considered by DB Netz AG, PKP PLK and for the passenger transport for ÖBB-Infrastruktur AG (they do take it into account for freight transport). Apart from ŽSR, the consumption of traction energy isn´t considered by DB Netz AG, PKP PLK, ÖBB-Infrastruktur AG. The speed of the train is considered by DB Netz AG, where there is a surcharge for slow trains (under 50 km/h) on high-speed tracks and by ÖBB-Infrastruktur AG, where there is a surcharge for trains with a speed of more than 160 km/h. Railways with overloaded infrastructure are considered by DB Netz AG and ÖBB-Infrastruktur AG.

Table 2. Comparison of charges levied for the infrastructure in the Czech Republic and neighbouring states. Source: author, based on [5, 13-21]

	Payments as per planned or real parameters	**The type of railway considered.**	**Different prices for passenger and freight transport**	**The weight of the train considered**	**Independent traction unit on an electrified railway considered**	**Traction energy consumption considered**	**The speed of the train considered**	**Railways with overloaded infrastructure considered**
SŽDC	real	yes	yes	Gross tonne-kilometres	yes	no	no	no
DB	planned	yes	yes	Rate per train km up to 3000 t is the same	no	no	Yes – for slow trains on high-speed railways	yes
PLK	planned	yes	no	Rate per train km by weight	no	no	no	no
ÖBB	real	yes	yes	Gross tonne-kilometres only for freight transport	Yes, for freight transport	no	Surcharge for speed of above 160 km/h	yes
ŽSR	real	yes	no	Gross tonne-kilometres	yes	yes	no	no

4 Conclusion

In each part of the present contribution, the author summarised the legislative background of the railway infrastructure capacity allocation, then compared the levying of charges for the allocation of railway infrastructure capacity in the Czech Republic as well as in the neighbouring states – Germany, Poland, Austria and Slovakia. It follows from the comparison that the price for the allocation of the railway infrastructure capacity is accounted for and charged independently only by SŽDC. The railway managers of the neighbouring infrastructures, such as DB Netze AG, PKP PLK, ÖBB-Infrastruktur AG or ŽSR, do not charge for the allocation of the railway infrastructure capacity independently, but they include it in the minimum access package. However, all these cooperating railway managers apply sanctions or reservation charges for the cancellation of the allocated capacity of the railway infrastructure or for the failure to use it as a regulatory measure.

This paper is supported by the research project "From horse-drawn railway to intermodal transport" within Visegrad Fund.

References

1. J. Široký, et al., *Transport Technology and Control* (Tribun EU: Brno, Czech Republic, 2014)
2. *Act No. 266/1994 coll. on railways* (2016)
3. *Directive 2012/34/EU of the European Parliament and of the Council* (2012)
4. H. Saláková, *Monitoring of economic factors by allocation of railway infrastructure capacity* (University of Pardubice, Czech Republic, 2015)
5. Railway Infrastructure Administration. *Statement on national and regional railway*, Available online: http://www.szdc.cz (2017)
6. SŽDC, *The Network Statement on nationwide and regional rail networks* (2017)
7. SŽDC, *The Network Statement on regional rail networks*, (AŽD Prague, Czech Republic, 2017), Available online: http://www.azd.cz/admin/files/Dokumenty/pdf/Kompletni-zneni-Prohlaseni-o-draze-2017-Cizkovice-Obrnice-Kopidlno-Dolni-Bousov.pdf
8. *The Network Statement on nationwide and regional rail networks of the Czech railways* (2017)
9. *The Network Statement on regional rail networks Jindřichohradecké místní dráhy*, Available online: http://docplayer.cz/1333471-Prohlaseni-o-draze-regionalni.html (2016)
10. *The Network Statement on regional rail networks Moravskoslezský kraj,* Available online: http://www.msk.cz/assets/doprava/prohlaseniodraze_2017.pdf (2016)
11. KŽC, *The Network Statement on regional rail networks*, Available online: http://www.kzc.cz/muzejni-zeleznice-senovka/ prohlaseni-o-draze (2016)
12. *The Network Statement on rail Inter-municipal cooperative of the Desná valley*, Available online: http://www.sart.cz/wp-content/uploads/Prohlasení-o-draze-aktualizace-2016.pdf (2016)
13. *Regulation 913/2010/EU of concerning a European rail network for competitive freight* (ERNCF) (2010)

14. DB NETZE, *Network Statement*, Available online: http://fahrweg.dbnetze.com/fahrwegen/network_access/network_statement/2017.html (2016)
15. S. Bahri, Y. Huseyin, U. Yasin, A. Guneri, G. Bahadir, E. Turan, The Scientific World Journal **2014** (2014), DOI: 10.1155/2014/630320
16. PKP PLK, *Network Statement*, Available online: http://en.plk_sa.pl/files/public/user_upload/pdf/Reg_przydzielania_tras/Regulamin_2016_2017/Network_Statement_2016-2017_v11.pdf (2016)
17. ÖBB Infrastruktur AG, *Network Statement,* Available online: http://www.oebb.at/infrastruktur/en/Network_Access//NetworkStatement/2017.jsp (2016)
18. ÖBB Infrastruktur AG, *Product_Catalogue_2017_Train_Path.pdf,* Available online: http://www.oebb.at/infrastruktur/en/Network_Access/Product_Train_Path.jsp (2016)
19. Li, Rui, R. Roberto, Journal of Construction Engineering and Management **143**, 6 (2017)
20. T. Dollevoet, D. Huisman, L. Kroon et al., Computers & Operations Research **143**, 203-217 (2017)
21. N. Stojadinovic, B. Boskovic, M. Bugarinovic, *Transport Means – 20th International Scientific Conference on Transport Means Location*, 735-740 (Juodkrante, Lithuania, 2016)

Methods of Bargaining in Logistics Processes

Jasmina Pašagić Škrinjar[1,*], *Borna* Abramović[1], and *Martina* Furdić[1]

[1]University of Zagreb, Faculty of Transport and Traffic Sciences, Vukelićeva 4, 10000 Zagreb, Croatia

Abstract. This paper outlines the methods of bargaining taken from game theory. The methods and types of bargaining are defined with the aim of establishing a cooperative relationship among the participants to mutual benefit. The methods in the paper are explained under various bargaining conditions and through instances of their application. The paper also examines the application of game theory and methods of bargaining within different logistics processes.

1 Introduction

Game theory should be understood as means of modelling real-world issues with the aim of ensuring easier decision-making and taking into consideration all actions and their quantification. A model of a game should be encompassing enough so that it can be considered a valid approximation of the actual market situation that is being analysed, but at the same time simple enough so that it is technically suitable for a more specific analysis. There is no algorithm for balancing the two requirements, but a solution is proposed for each specific situation.

Game theory models are based upon the mathematical knowledge and intuition of a researcher about what is relevant for the specific situation, and what is not. A successful application of such models in the analysis of certain logistical and transport issues requires a thorough understanding of the issue being analysed. A concrete analysis demands a multidisciplinary approach, which means it should involve experts from many fields (e.g. mathematicians, logistical process expert, and others). The lack of a team approach might result in deviations in determining optimal strategies.

Game results are commonly expressed in numbers. A larger number indicates that the certain strategy results has a higher score, which in logistics may refer to higher profit, faster delivery, safer transport, a more long-term business, and competitive advantage.

Bargaining games are a separate class of games popular in economic and other analyses, business processes, political action, logistical and transport process, and in daily life.

Person A wants to sell a used forklift for no less than 10,000 €, while Person B wants to purchase such a forklift, ready to pay up to 13,000 €. If the seller sells the forklift for a price within that range, he makes a profit. If the buyer gets the forklift for the price within his range, he has also benefited. There is, therefore, a space in which both parties are involved in an exchange and thereby ensure mutual benefit.

* Corresponding author: jpasagic@fpz.hr

Different variables impact the process of negotiation, divided into external and internal. The effect of the external and internal options was defined by Muthoo [1] on the example of marital relationships, but this paper will analyse an example in logistics.

Companies A and B are logistical companies. Company A has a more complete fleet of vehicles and better storages, but a smaller storage capacity than Company B. Furthermore, Company A has a better marketing service and position on the local market. It cooperates with logistical Company B in terms of leaving a part of its business to it and occasionally using its storage facilities (outsourcing). Company B provides services for large international department stores and holds a stronger position on the international market. Both companies take to the external market since it increases their chance of ensuring work and gaining higher profit. In their negotiations, both logistical companies use their respective advantages to achieve a more positive outcome from themselves.

2 Literature review

The application of negotiation games has been in used for a number of years in various aspects of business, and the following part will examine the instances of the use of game negotiations in logistics processes. In his work, Reyes [2] applies Shapley's concept on solving the unloading / loading issues in order to maintain stable conditions within the logistics network. Sucky [3] is concerned with supply chain management from the perspective of inventory management. For a buyer to be motivated to order in quantities that are favourable to the supplier, the paper analyses the supplier's ability to introduce a cooperative purchasing policy for the buyer. The research in this paper offers several negotiation models dependent on alternative supplier strategies. With the help of these models, a cooperative relationship between the buyer and supplier can be established to mutual benefit.

The research presented in another paper written by Sucky [4] provides a bargaining model with asymmetric information about the buyers cost structure, assuming that the buyer has the power to impose his individual optimal policy. Plambeck and Taylor [5, 6] analyse two similar games between two firms that have an option of pooling their capacity and investments to maximize the total value. In the first stage, firms choose investment into an effort that affects the market size. In the second stage, firms bargain over the division of the market and profits. Esmaeili, Aryanezhad and Zeephongsekul [7] proposed several seller–buyer supply chain models, which incorporate both cost factors, as well as elements of competition and cooperation between seller and buyer. They modelled the relationships between seller and buyer on non-cooperative and cooperative games, and compared the results between different models.

Suh and Park [8] analyse the delay loss cost per ton and per day between the charterer and the ship owner, by determining the equilibrium charterage in bargaining with asymmetric impatience. They used Nash equilibrium theory, which aims at reducing the negotiation time. The analysis made by Wang, Guo and Efstathiou [9] explains non-cooperative behaviour in a two-echelon decentralized supply chain, composed of one supplier and n retailers. They built the approximate decision model of base stock level for sufficient supply from the supplier, in which the suppliers reactions are not considered, and its non-cooperative behaviour is obtained. For insufficient supply from the supplier, a much more complicated non-cooperative behaviour is obtained, and they find that the competition occurred between all the retailers, as well as the supplier. In order to guarantee optimal cooperation in the system, they designed several Nash equilibrium contracts in echelon inventory games and local inventory games. Kwak, Kim and Moon [10] also explored bargaining between suppliers and buyers. They developed two different models: one for the situation where the supplier has a superior bargaining power over the buyer, and the other for the reverse situation. For each model, they derived a method to find the best strategy of each agent through analysis based on the game-theoretic approach. The solution they found using that method was verified to be the Nash Equilibrium of each model.

Game theory also finds some interesting problem solution Pašagić Škrinjar, Abramović and Brnjac [11] for planning of urban transport, especially problem of introducing the new bus line. Then Koryagin [12] again use Game theory for urban planning for the Travel Demand Management. Catalano amd Migliore [13] have use game theory for solving the design of logistic terminals. Liu, Wilson and Luo [14] use cooperative game to solve the impact of Panama Canal expansion on the container-shipping market. Babu and Mohan [15] use an integrated approach to evaluating sustainability in supply chains using evolutionary game theory.

3 Bargaining methods

The most well-known bargaining solution was put forward by John Nash in 1950 [16]. The author describes the bargaining problem.

Rational individuals *A* and *B* willingly enter negotiations. They are familiar with all the outcomes of these negotiations, to which they have attributed various benefits, based on their respective expectations. The negotiations can be concluded in many different ways, and the set of possible agreements represents the permissible area *D*, as seen in Figure 1.

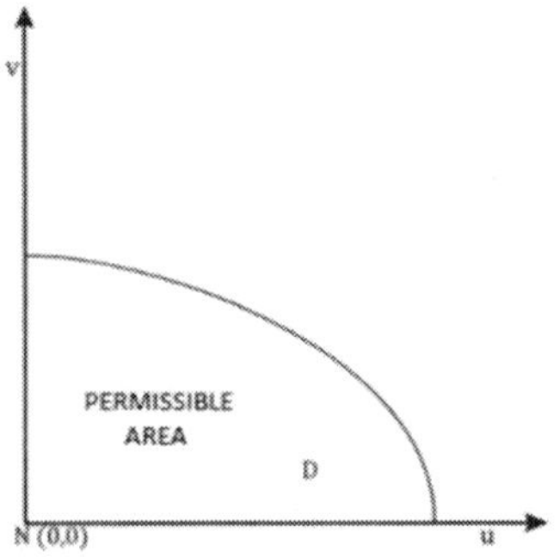

Fig. 1. Permissible area in a bargaining situation. Source: authors

The coordinates u and v in Figure 1 represent the winnings of Player *A*, i.e. Player *B*. The permissible area also contains the failure point (status quo, disagreement point) $N(0, 0)$, the negative outcome of bargaining. Disagreement point N is not necessarily located in the origin of the coordinate system. It can be located within the area *D*, denoted by (u_1, v_1) in $N(u_1, v_1)$. Other points in the permissible area denote the possible agreements which are differently acceptable for both players.

Since the players are perfectly rational, they will not observe all possible outcomes, but rather the ones that are Pareto efficient.

In a Pareto efficient outcome neither individual can make their preference criterion better off without making any other individual's preference criterion worse off. The set of all Pareto efficient allocations (points) comprises negotiating set P within the permissible area, $P \subset D$, as illustrated in Figure 2.

The allocations of the negotiating set are payoff pairs (u, v), whereby individuals prefer any point in the set P. Since both individuals wish to reach an agreement, they are willing to give up on for them most favourable gains in the boundary points of the negotiating set.

A successful conclusion of negotiation refers to only cooperative behaviour that improves the negotiating outcome.

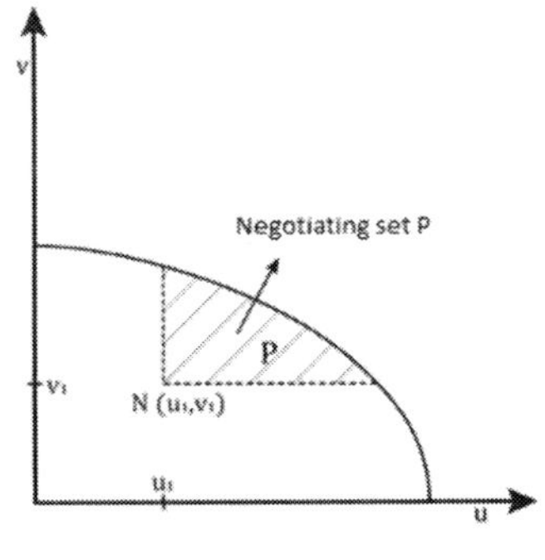

Fig. 2. Negotiating set *P*. Source: authors

In case of an agreement, individuals achieve benefits which are equal $(u - u_1)$ for the first and $(v - v_1)$ the second individual. If the negotiating set contains only one point, the problem becomes trivial and the successful conclusion follows immediately. Therefore, let us assume that the set *P* contains several points and let us set the problem of allocating the agreement point (u^*, v^*).

Nash proposes four axioms which should be satisfied in each negotiating solution:

Pareto-optimum (PO): The accepted allocation must be a unique Pareto optimal solution.

Symmetry (S): If the permissible area is symmetric, then the allocation of (a, b) includes the allocation of (b, a). If $u_1 = v_1$ for the disagreement point, then the same is true for the agreement point, $u^* = v^*$.

Invariant to equivalent utility representations (EP): Assume that the payoff functions of the two individuals, $u(\cdot)$ and $v(\cdot)$, based on which the agreement point was allocated: (u^*, v^*). If we apply positive linear transformation and transform their payoff functions into new functions:

$u(\cdot) = a_1\, u(\cdot) + b_1$, where a_1 and b_1 are positive constants,

$v'(\cdot) = a_2\, v(\cdot) + b_2$, where a_2 and b_2 are positive constants,

then the new agreement point represents the linear transformation of the preceding point, $(a_1u^* + b_1,\ a_2v^* + b_2)$. This agreement point remains the same, now expressed in payoffs measured by the new scales.

Independence of irrelevant alternatives (IA): Let two problems in negotiating have the same conflict points and let that negotiating set of the 2 is within the negotiating set of 1. If the agreement point of 1 (u_1^*, v_1^*) is located within the negotiation set of 2, then it is at the same time the agreement point for 2 $(u_1^*, v_1^*) = (u_2^*, v_2^*)$.

Based on these four conditions, Nash proved the following theorem:

If conditions *PO*, *S*, *EP*, and *IA* have been met, then the agreement point (u^*, v^*) is the solution of the problem:

$$max\ (u - u_1)(v - v_1) \tag{1}$$

in the negotiating set, if there are points that prove $u > u_1$ and $v > v_1$.

The obtained solution is referred to as Nash's bargaining solution and is the only one that satisfies all the four axioms.

Nash's bargaining solution can be applied in various forms of bargaining, some of which include symmetric games, asymmetric bargaining games, different scales for determining payoffs, various risk forms in bargaining, and many others.

3.1 Example 1: Asymmetric bargaining games

The first asymmetry concerns the starting positions of the players. Assume that the disagreement point N is not $(0, 0)$ and that Player A or Player B can count on a definite profit on the side, in case the negotiation fails. If Player A has a guaranteed amount of 250€, and the profit of Player B depends exclusively on the outcome of the observed negotiation, the disagreement point N is $(250, 0)$ and it affects the final outcome. It cannot be expected that the players divide the money into equal parts, as seen in Figure 3.

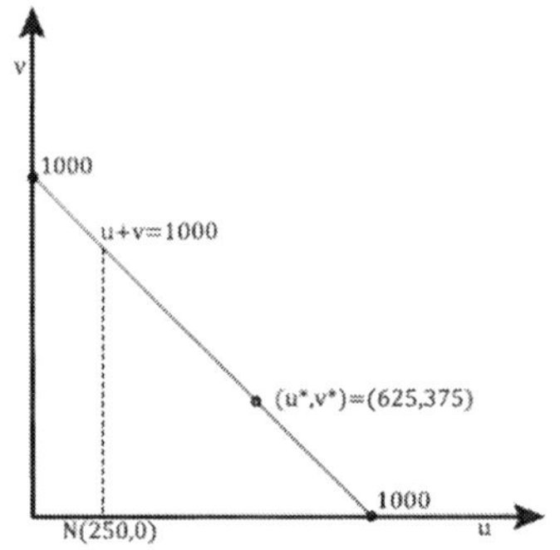

Fig. 3. Asymmetric bargaining games. Source: authors

Assuming that players are still neutral towards risk, which makes their payoffs equal to financial outcomes $u = x$, $v = y$, therefore the efficient distribution is $u + v = x + y = 1000$. The bargaining solution is the solution of the problem:

$$max\ (x - n_1)\ (y - n_2) \tag{2}$$

that is,

$$max\ (x - 250)\ (y - 0) = max\ (xy - 250y) \tag{3}$$

$$\text{for } y = 1000 - x \tag{4}$$

in the expression above:

$$max\ (1250x - x2 - 250000) \tag{5}$$

A zero first derivative of $1250 - 2x = 0$, is $x = 625$, and $y = 375$. The result $(625, 375)$ shows that the concluded agreement is asymmetrical in favour of the player that enters bargaining with a more favourable starting position [17].

3.2 Example 2: Negotiator's attitude towards risk

So far, it has been assumed that negotiators are not prone to taking risk, which meant that in the case of symmetric disagreement points $N = (0, 0)$ the available sum was divided into equal parts. The negotiator's attitude towards risk affects the bargaining outcome, because it shapes the payoff functions of the negotiator. Let us assume that Player A is still neutral towards risk, and let $u = x$, while Player B shows a slight animosity towards risk, so that the sums y are added payoffs which are $v = y^{4/5}$, i.e. $y = v^{5/4}$. By substituting these payoffs for lines $x + y = 1000$, there follows a curve $u + v^{5/4} = 1000$, as seen in Figure 4.

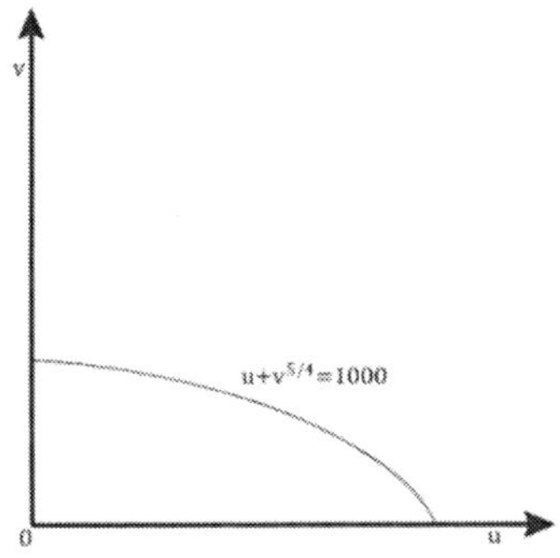

Fig. 4. Negotiator's attitude towards risk. Source: authors

Substituting $v = (1000 - u)^{4/5}$ into max $(u - n_1)(v - n_2)$ it follows: $max\ u\ (1000 - u)^{4/5}$ let first derivative be 0:

$$(1000 - u)^{4/5} + (4/5)\ u\ (1000 - u)^{4/5 - 1}(- 1) = 0 \quad (6)$$

$$(1000 - u)^{4/5} = (4/5)\ u\ (1000 - u)^{-1/5} \quad (7)$$

and multiplying the first and the second equality by expression $(1000 - u)^{1/5}$, it follows:

$$1000 - u = (4/5)u \quad (8)$$

after the calculation we determine the bargaining result for an individual who is neutral towards risk, $u = x = 555.5$, which is followed immediately by $y = 1000 - 555.5 = 444.5$ €.

The solution of the bargaining game $(555.5;\ 444.5)$ shows that the different attitude of the negotiator towards risk conditions unequal distribution. When the negotiator shows animosity towards risk, the fear of failure makes him bargain unfavourably away from equal distribution of income. The higher his animosity towards risk, the more unfavourable the distribution of income. If Player A is neutral, and $u = x$, and Player B shows animosity towards risk so that his payoff function equals $v = y^{1/2}$, the solution of the bargaining is $(666.6;\ 333.3)$.

The inequalities of distribution are even more evident if Player A is prone to risk, and Player B against it. The final outcome is then substantially asymmetrical in favour of Player A [17].

Although logical, such result evokes much criticism, even contestation of Nash's solution. These solutions favour individuals prone to risk, which is why certain authors state that it encourages players to falsely demonstrate individual preferences. It is in the interest of rational individuals who are neutral or show animosity towards risk to conceal their preferences and enter negotiations as players prone to risk. Nash's solution only makes sense if there's an assumption that it is based on real and not false payoff functions of both negotiators.

The initial assumption about perfectly rational and intellectually equal partners is what separates game theory from practice. The theory completely disregards the individuality of players, even though knowledge, intelligence and character traits affect the final outcome of a negotiation in real-world situations. The theoretical solution is based solely on the specificity of conflict situation.

3.3 Example of negotiating in channels of distribution

This example is based on building the basic bargaining model in which a manufacturer (with utility u_M) sells to a retailer (who is similar to a news vendor with utility u_R). The players negotiate on the terms of trade, and effectively the utility allocations. The retailer faces a random demand, the density of which is $f(\varepsilon)$. He has to place an order of Q units in anticipation of the demand. He incurs purchasing, holding, and shortage costs, and gains revenue proportional to the quantity he sells. We will denote his revenue per unit as s, the other costs per unit as w, h and v, respectively. The manufacturer faces a production cost per unit, denoted as c, and sells to the retailer at a wholesale price w. The profit, given a demand realization ε, of the retailer and the manufacturer as $\pi_R(w,Q,\varepsilon)$ and $\pi_M(w,Q,\varepsilon)$, respectively. With a slight abuse of notation, we denote the expected profits as $E[\pi_R(w,Q,\varepsilon)] = \pi_R$ *and* $E[\pi_M (w,Q,\varepsilon)] = \pi_M$, by simply dropping ε from the arguments. For the sake of exposition, we describe the construction of the game using wholesale price contracts. Recall that a NB game requires us to identify a feasible set of payoffs and a disagreement point that are pre-determined and are independent of the negotiations. To do so, let us first assume that the supplier and the retailer negotiate on (w,Q). The feasible set is constructed as follows: we first define:

$$\Delta = \{(E(u_M(\pi_M(w,Q,\ \varepsilon))),\ E(u_R(\pi_R(w,Q,\ \varepsilon)))) : (w,Q) \in T\} \tag{9}$$

for some suitable compact set $T \subseteq R$. The set T is assumed to be a rectangular set. This is be the case, for instance, when both the wholesale price and the purchasing quantity are allowed to take values in a closed interval in the positive real axis. We let Ω be the convex hull of Δ and we henceforth refer to Ω as the feasible set.

The disagreement outcomes have several interpretations, depending on the context. They represent a certain level of expected utility that each player requires as a minimum level of acceptance to participate in the trade. We set this to zero. In a setting where players are risk neutral, this is without loss of generality, i.e. the structure of the results trivially change. When players are risk averse, they need to be a bit more careful. However, the results for the system above (a manufacturer – retailer chain) in this paper essentially remain the same.

When players are risk-neutral and negotiate on both w and Q, a few observations can be made. First, the bargaining solution of the game will not be a randomized outcome. This is because risk neutrality implies $\Delta = \Omega$. Furthermore, note that the bargaining solution necessarily requires Pareto optimality as a condition. This immediately ensures that when the players are risk-neutral, the negotiated quantity is always the coordinating quantity, Q_c. This simplifies the analysis and removes double marginalization effects. The negotiated value of w allocates the channel surplus between the two players. This simple observation continues to hold when more complex contracts are considered (i.e., buyback, revenue sharing, etc.), as long as one assumes that all parameters of the contract are simultaneously negotiated [18].

4 Conclusion

Game theory does not adopt the role of an advisor, nor does it give directions to players on how to win. The main goal of the theory is to identify and model the essence of conflict which is specific in a certain situation. The initial assumption about perfectly rational and intellectually equal players is what separates theory from practice. The former completely disregards individual impacts of players, whilst in real-world negotiating circumstances the knowledge, intelligence and character traits of the negotiator indeed determine the final outcomes. The theoretical solution, i.e. solution proposal is determined solely on the specificity of the conflicting situation.

Since theory observes only the perfectly rational individuals that always select from a set of optimal solutions, the Pareto-optimum is accepted theoretically. In real-world negotiation circumstances, the rationality condition is infringed, so final outcomes of numerous negotiations do not represent the Pareto efficient solutions. Decision-makers often disregard the assumptions of a perfectly rational decision-making, while the complexity of negotiating situations makes it impossible to precisely determine the Pareto optimal solutions set. The symmetricity condition in a bargaining game should be ensured so that negotiation is a fair game. If the players are in opposite and completely symmetrical positions, then a deal is fair when both obtain equal benefits. The condition understands symmetricity of results expressed in benefits, which does not implicitly state that they are equal in monetary or some other expression.

The paper is supported by the PROM – PRO research project "Adapting the railway system in the integrated passenger transport (ARSIPT)", that is solved at Faculty of Transport and Traffic Sciences, University of Zagreb.

This paper is supported by the research project "From horse-drawn railway to intermodal transport" within Visegrad Fund.

References

1. A. Muthoo, World Economics **1**, 2, 145-166 (2010)
2. P. Reyes, Applied Mathematics and Computation **168**, 2, 1419-1431 (2005), DOI: 10.1016/j.amc.2004.10.030
3. E. Sucky, International Journal of Production Economics **93-94**, 253-262 (2005), DOI: 10.1016/j.ijpe.2004.06.025
4. E. Sucky, European Journal of Operational Research **171**, 2, 516-535 (2006), DOI: 10.1016/j.ejor.2004.08.039
5. E.L. Plambeck, T.A. Taylor, Management Science **53**, 12, (2001)
6. E.L. Plambeck, T.A. Taylor, Management Science **51**, 1, (2005)
7. M. Esmaeili, M.B. Aryanezhad, P. Zeephongsekul, European Journal of Operational Research **195**, 2, 442-448 (2009), DOI: 10.1016/j.ejor.2008.02.026
8. S. Suh, N. Park, The Asian Journal of Shipping and Logistics **26**, 2, 247-261 (2010), DOI: 10.1016/S2092-5212(10)80004-5
9. H. Wang, M. Guo, J. Efstathiou, European Journal of Operational Research **157**, 2, 372-388 (2004), DOI: 10.1016/S0377-2217(03)00233-9
10. T.C. Kwak, J.S. Kim, C. Moon, Computers & Industrial Engineering **51**, 2, 219-228 (2006), DOI: 10.1016/j.cie.2006.02.004
11. J. Pašagić Škrinjar, B. Abramović, N. Brnjac, Tehnicki Vjesnik – Technical Gazette **22**, 6, 1617-1621 (2015)
12. M. Koryagin, Periodica Polytechnica Transportation Engineering, (to be publish)
13. M. Catalano, M. Migliore, Journal of Transport Geography **41**, 63-73 (2014), DOI: 10.1016/j.jtrangeo.2014.08.010
14. Q. Liu, W.W. Wilson, M. Luo, Maritime Policy & Management **43**, 2 (2016), 209-221, DOI: 10.1080/03088839.2015.1131863
15. S. Babu, U. Mohan, Computers & Operations Research **89**, 269-283 (2018), DOI: 10.1016/j.cor.2017.01.008
16. J.F. Nash, Jr., Econometrica **18**, 2, 155-162 (1950)
17. D. Pavličić, *Teorija odlučivanja* (Ekonomski fakultet, Belgrade, 2010)
18. M. Nagarajan, G. Sosic, European Journal of Operational Research **187**, 3, 719-745 (2008), DOI: 10.1016/j.ejor.2006.05.045

MATEC Web of Conferences **134**, 00054 (2017) DOI: 10.1051/matecconf/201713400054
LOGI 2017

Impact of Cargo Distribution on the Vehicle Flatback on Braking Distance in Road Freight Transport

Tomáš Skrúcaný [1,*], *Ján* Vrábel [1] , *Martin* Kendra [1], and *Patrik* Kažimír [1]

[1]University of Žilina, Faculty of Operation and Economics of Transport and Communications, Univerzitná 8215/1, 010 26 Žilina, Slovakia

Abstract. Vehicle centre of gravity influences directly the breaking characteristics. It dues through the actual each axle load and the actual gross weight of the vehicle. This fact is significant by full braking maneuver. Full braking measurements with different operational vehicle conditions - cargo weight and its location on the vehicle – were done to evaluate this influence. The medium sized good vehicle (category N2) was breaking on a wet asphalt surface. The value of the braking deceleration MFDD and the braking distance was measured by vehicle decelerometer. The results compare different axle- and vehicle load and show the area for future related measurements.

1 Introduction

Braking plays a key role in critical situations like the early stop in front of an obstacle or its avoidance. In road freight transport, in which trucks weigh tens of tonnes, it prevents causing damage on the vehicle, carried goods, property and what the most important, it can save lives. Having the knowledge of the variables that affect braking can thus help us to avoid these undesirable circumstances.

From the point of view of road safety, the most important is the track needed to stop a vehicle. It is composed of several parts.

The driver response time t_r is different for each person, it is influenced by driver's concentration, physical and mental state. The lowest response time of 0.6 - 0.7 s is for the driver who is concentrated; the highest time of up to 2 seconds is for a tired driver.

During the delay time of brake effect t_d, it happens to set off of backlash in braking system mechanism. It takes values of 0.05 s for hydraulic braking systems and 0.15 s for pneumatic braking systems.

Throughout the time to develop fully brake effect t_{df}, the increasing pressure in the transfer medium and the permissible deformation of the brake system parts occur. It takes values of 0.15 s for hydraulic braking systems and 0.2 s for pneumatic braking systems.

The time of fully brake effect t_f flows from the moment the brake reaches its full braking effect until the vehicle stops.

* Corresponding authors: tomas.skrucany@fpedas.uniza.sk

 DOI: 10.1051/matecconf/201713400054

However, the actual deceleration of the vehicle during braking is more complicated than the ideal one because of the changes in the brakes temperature, vehicle speed and tire slip ratio.

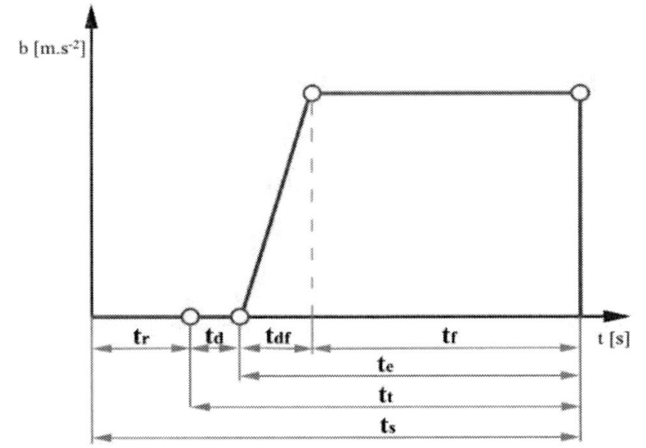

Fig. 1. Braking time process. Source: authors

Legend:
t_r – driver response time (s)
t_d – brake effect delay time (s)
t_{df} – time to develop fully braking effect (s)
t_f – time of fully braking effect (s)
t_e – effective braking time (s)
t_t – total braking time (s)
t_s – time to stop vehicle (s)

2 Measurement procedure

To find out the answers a good vehicle of the N2 category, the MAN TGL 8.180, on which the decelerometer measured brake characteristics during the braking tests, was used.

Fig. 2. MAN TGL 8.180. Source: authors

The tests were divided into four situations with a different load distribution on the vehicle's loading surface in the following order: from the first situation - highest front axle load to the fourth position - highest rear axle load. The differently placed cargo caused a different axle load and this was measured by the portable weighing system. In addition to the four different axle loads, the impact of the initial vehicle speed at 40 $km.h^{-1}$, 65 $km.h^{-1}$, 80 $km.h^{-1}$ was also investigated. Overall, 32 brake tests were planned.

The aim was to find out how mean fully developed deceleration called also MFDD ($m.s^{-2}$) and braking distance (m) would change. The decelerometer mounted at the front of the load compartment calculated these values automatically after each braking.

Fig. 3. The cargo distribution on the loading surface in the individual situations. Source: authors

Based on the knowledge of maximum allowable axle load, the maximum pay-mass and the dimensions of the car used, a load distribution diagram was developed. The load in specific situations is also shown on the diagram.

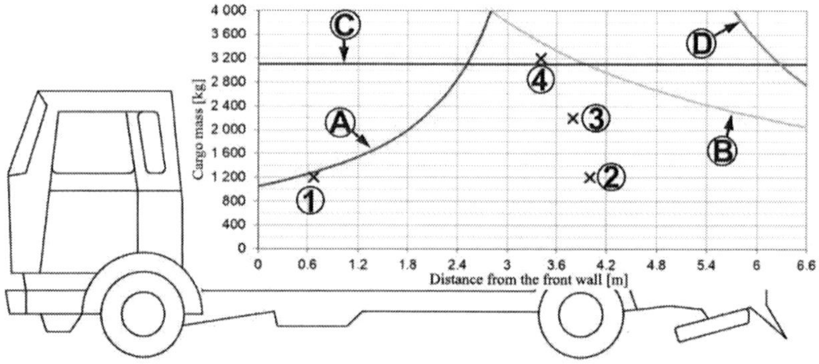

Fig. 4. Load distribution diagram of MAN TGL 8.180. Source: authors

Legend:
A – maximum permissible front axle load
B – maximum permissible rear axle load
C – maximum pay-mass
D – minimum front axle load
1 – load in 1st situation
2 – load in 2nd situation
3 – load in 3rd situation
4 – load in 4th situation

When loading, the maximum pay-mass and the maximum permissible axle load must not be exceeded, while respecting the minimum permissible axle loads. For category N vehicles under Council Directive 96/53/EC the weight borne by the driving axle must not be less than 25 % of the total laden weight of the vehicle or vehicle combination. By the Government Regulation of Slovak Republic 349/2009 Z.z. which is adopting Council Directive 96/53/EC, also the weight on the steering axle must not fall below 20 % of the gross vehicle weight. The cargo is loaded correctly if its centre of gravity on the diagram is in the area bounded by the curves A, B, C, D. The curve of the minimum permissible rear axle load is not shown in the diagram. It limits this area only in specific vehicles, because the rear axle relief can cause only the load placed in front of the front axle.

On the load distribution diagram of the used car, the area of the correct load centre of gravity location is bounded by curves A, B, C. In the first case, the load was placed at the maximum permissible front axle load limit. Load in the 2nd and 3rd situation is in the area bounded by the curves, meaning that the vehicle is in both cases in an ideal operating load. In the fourth situation, the largest total laden weight of the vehicle was slightly exceeded, as it can be seen on the diagram where the 4th point is just above the maximum pay-mass line.

3 Results

For the measurement purposes, the braking sections have been selected from the time to develop fully brake effect to the stopped vehicle. The Driver response time and brake effect delay time were not taken into account as the weight and load distribution are not related to them. During the tests, these parameters were measured: MFDD ($m.s^{-2}$), initial speed ($km.h^{-1}$), braking distance (m) and braking time (s).

Table 1. Average measured values. Source: authors

	Initial speed ($km.h^{-1}$)	**MFDD ($m.s^{-2}$)**	**Brake distance (m)**	**Braking time (s)**
1. situation 39/99/47 *	40	5.51	12.68	2.23
	64	5.18	33.27	3.60
	84	4.63	58.22	5.06
2. situation 39/79/63 *	40	5.98	11.51	2.06
	64	5.19	31.31	3.49
	82	5.54	48.86	4.39
3. situation 71/80/81 *	41	5.97	12.28	2.17
	64	5.63	29.91	3.36
	82	5.76	48.32	4.27
4. situation 103/86/95 *	42	5.61	13.41	2.31
	65	5.89	30.50	3.33
	81	6.10	44.78	3.94

*Percentage of used load in order: payload of the vehicle / maximum permissible front axle load / maximum permissible rear axle load;

Table data was used to determine the dependence between axle load and MFDD - the quantity characteristic of the braking deceleration.

 DOI: 10.1051/matecconf/201713400054

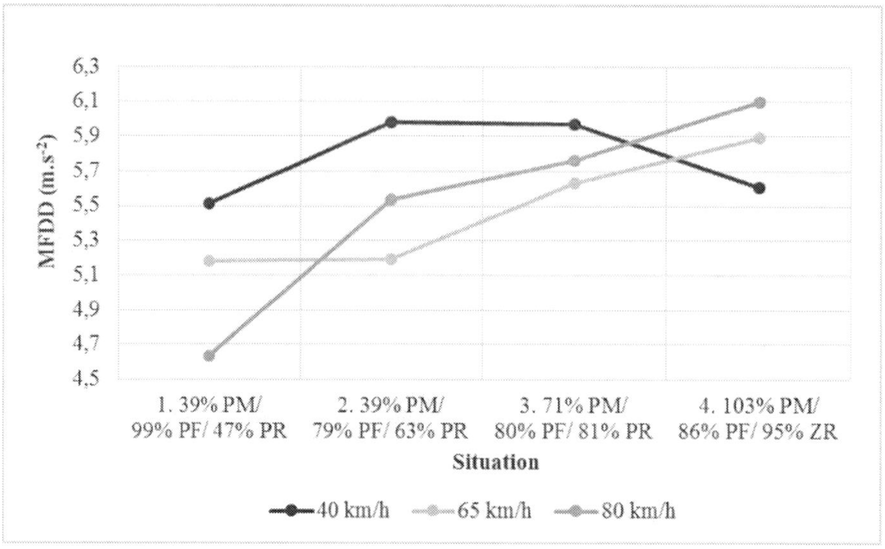

Fig. 5. Dependence of MFDD on axle load. Source: authors

Legend:
PM – pay-mass of the vehicle
PF – maximum permissible front axle load
PR – maximum permissible rear axle load

There are large differences in MFDD in individual initial speeds. Spread of braking decelerations by initial speed was the highest in the 1st situation – 16 %. On the contrary, the lowest spread of deceleration was recorded in the 3rd case - only 6 %. It is true that with an increase in the initial speed, the change in MFDD is also increasing with a various load. Lower initial speeds are probably an exception; in this situation it is the speed of 40 km.h^{-1} where no clear dependence of braking deceleration on the change of axle load has been confirmed. However, it has been found out, that at higher initial speeds the braking deceleration is also increasing with increasing of the rear axle load.

While at the maximum front axle load, the MFDD is the highest at the lowest initial speed and the lowest at the highest initial speed, at the maximum rear axle load the situation is the opposite one.

By averaging the MFDD of all initial speeds and selecting one MFDD for each situation, we draw further conclusions.

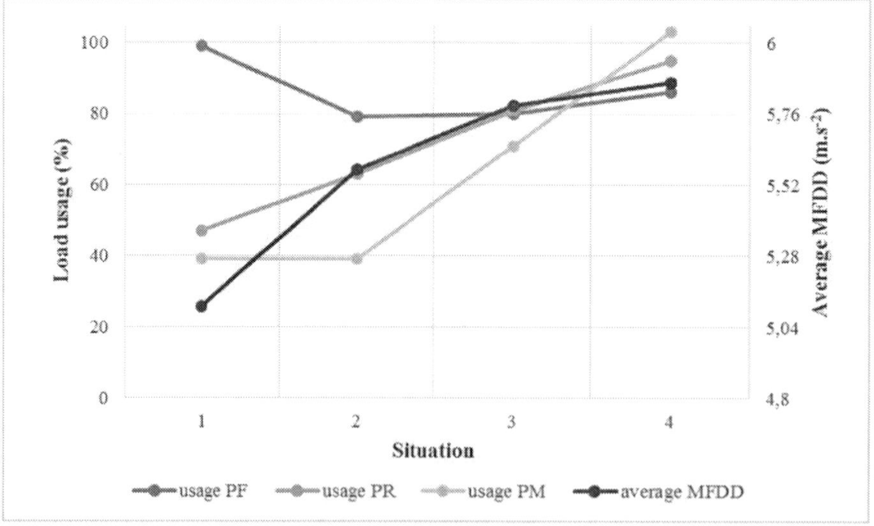

Fig. 6. Influence of load usage on MFDD. Source: authors

Better deceleration is achieved with the axles of the vehicle loaded more evenly. This fact is confirmed by the third and fourth situation with the highest achieved results, with an ideal load distribution (80 % front axle load, 81 % rear axle load) in the 3rd case. In particular, by comparing the first and the second case with the same total vehicle weight, but with a cargo located diametrically differently, we gain the notion that the more evenly the load is on the axles, the higher the braking deceleration is. In the 4th situation the MFDD was higher than in the third situation but the weight distribution was less uniform than in the third case. Based on that, it can be concluded that the increase of the MFDD depends also on the weight above the rear axle.

In contrast to the highest MFDD values in the 3rd and the 4th situation, the lowest braking deceleration was in the 1st case where the mass was concentrated mainly on the front steering axle and it was even higher during braking due to the torque changing axle load induced by the inertia of the centre of gravity. In this context, the role of the height of the centre of gravity is also worth mentioning, since the higher the centre of gravity is, the greater the torque changing axle load is created. During braking, this torque causes an increase of mass on front axle and a drop of mass on the rear axle. When placing a cargo on a loading surface, it is therefore necessary to consider the braking that may result into this overload and it is also important not to place the cargo in the way that the load caused by it would be brought to the maximum permissible front axle load.

In the case of the maximum front axle load, the braking deceleration was decreasing with the increasing initial speed. This means that the faster the vehicle was at the start of the braking, the greater the deceleration drop occurred.

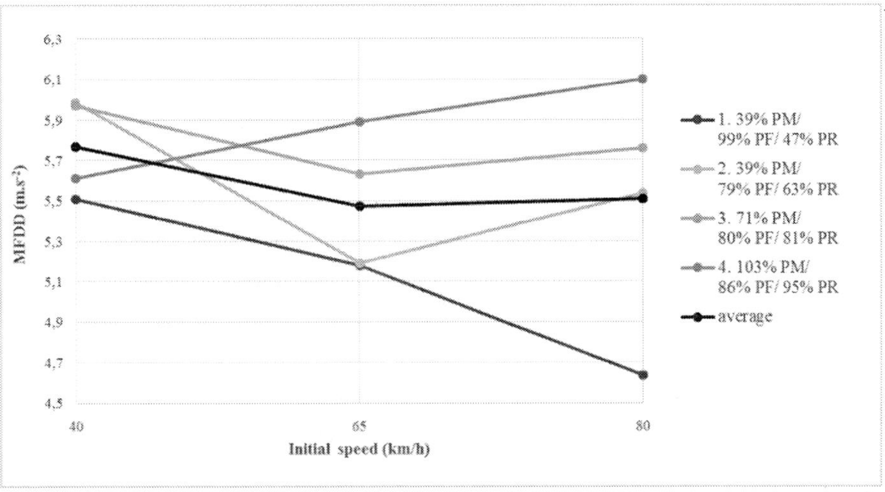

Fig. 7. Dependence of MFDD on initial speed. Source: authors

In situations no. 2 and 3, it is possible to see a similar decrease in deceleration, but only to an initial speed of 65 $km.h^{-1}$. The decrease is smaller the greater the mass on the rear axle is. In the subsequent increase of the initial speed the deceleration slows. The unambiguous braking deceleration was recorded at the maximum rear axle load when the deceleration raised with increasing of initial speed. From the above findings it follows that the braking deceleration chiefly growth with the additional rear axle load, with the exception of the low initial speeds, when an unknown measurement error could also occurred. The exception does not apply to the equal load of both axles where a high braking deceleration was achieved at all initial speeds.

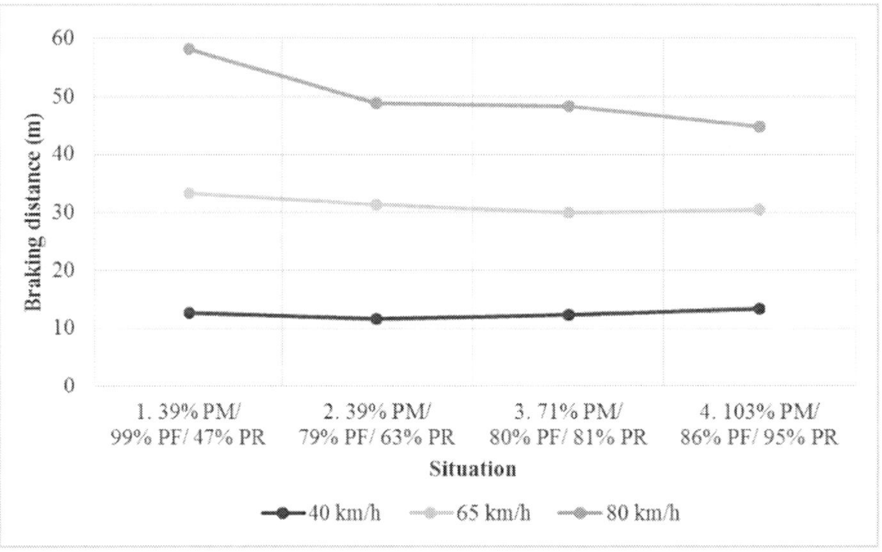

Fig. 8. Dependence of braking distance on axle load. Source: authors

At the lowest initial speed, the shortest braking distance was when the vehicle was loaded with just 1 tonne cargo placed directly above the drive axle. At an average initial speed of 65 $km.h^{-1}$, the braking distance is reduced with a more uniform load on each axle. At the highest initial speed, the braking deceleration increases with increasing rear axle load, the difference being almost 15 m when comparing the braking distance with the maximum front axle load and the maximum rear axle load. In general, it is possible to conclude that the path needed to stop the vehicle is the shorter the greater the load on the rear axle. This fact also confirms the comparison of the first and second situations when the vehicle had an identical gross weight but the load focused in one case on the front and the second on the rear axle.

4 Conclusions

The results reveal that not only the gross weight of the vehicle, but also the cargo location on the vehicle's loading surface influences the braking process. It has been found that higher load on the rear axle (as opposed to higher front axle load) favourably influences the braking deceleration of the car, helping to reduce the braking distance. The braking distance difference was greater the faster the vehicle was at the moment when deceleration began raise. These facts accompany the peculiarities at a low initial speed. The same use of maximum permissible axle load had a much clearer effect on braking deceleration. In a case of ideal axle load was achieved high braking deceleration at each initial speed. It could be inferred that when placing the cargo on the loading surface of the vehicle, it is important to place it so that all axles of the vehicle are evenly loaded, making it much more convenient to place the load on the rear axle than on the front.

Surprising results were obtained when investigating the effect of the initial speed. At the lowest initial speed was achieved highest braking deceleration with the centre of gravity closer to the front axle, and the lowest braking deceleration was reached by the vehicle with the most rear axle loaded. Exactly the opposite situation occurred at the highest initial speed when the centre of gravity approaching of the front axle adversely affected the braking deceleration.

This paper is supported by the research project "From horse-drawn railway to intermodal transport" within Visegrad Fund.

References

1. Council Directive 96/53/EC, Available online: http://eur-lex.europa.eu/legal-content/EN/TXT/HTML/?uri=CELEX:01996L0053-20150526&from=SK (1996)
2. Goverment Regulation of Slovak Republic 349/2009 Z.z., Available online: https://www.slov-lex.sk/pravne-predpisy/SK/ZZ/2009/349/ (2009)
3. J. Mašek, L. Černá, J. Čamaj, *Horizonty železničnej dopravy – International conference*, 159-173 (Strečno, Slovakia, 2014)
4. J. Gašparík, J. Grenčík, P. Meško, V. Zitrický, V. Ľupták, *Mechanika v železničnej doprave*, 278 p. (University of Žilina, Slovakia, 2016)
5. J. Vrabel; J. Jagelcak; J. Zamecnik, J. Caban, *Transbaltica 2017 – 10th International Scientific Conference: Transportation Science and Technology*, 89-99 (Procedia Engineering **187**, Vilnius, Lithuania, 2017), DOI: 10.1016/j.proeng.2017.04.354
6. B. Sarkan; O. Stopka; J. Gnap, J. Caban, *Transbaltica 2017 – 10th International Scientific Conference: Transportation Science and Technology*, 775-782 (Procedia Engineering **187**, Vilnius, Lithuania, 2017), DOI: 10.1016/j.proeng.2017.04.437
7. J. Lizbetin; O. Stopka; F. Nemec, *Transport Means 2016 – 20th International Scientific Conference on Transport Means*, 151-155 (Juodkrante, Lithuania, 2016)
8. P. Kohút, R. Jurina, J. Ondruš, *25th Annual Congress of EVU: proceedings: 20-22*, 263-275 (2016)
9. V. Rievaj, F. Synák, Scientific journal of Silesian university of technology: Series transport = Zeszyty naukowe Politechniki Śląskiej: Transport **94**, 187-197 (2017)
10. D. Barta, M. Mruzek, D. Kalincak, *BulTrans: anniversary scientific conference on aeronautics, automotive and railway engineering and technologies*, 54 -57 (Sofia, Bulgaria, 2013)
11. J. Grenčík, P. Volna, M. Mikolajčík, Autobusy: technika, eksploatacja, systemy transportowe **6,** 881-885 (2016)
12. D. Földes, C. Csiszár, *ICTTE 2016 – International Conference for Traffic and Transport Engineering*, 533-540 (Belgrade, Serbia, 2016)
13. T. Figlus, M. Stańczyk, Metalurgija **55**, 75-78 (2016)

Proposal for Streamlining the Railway Infrastructure Capacity on the Specific Track Section in the Context of Establishing an Integrated Transport System

Ondrej Stopka[1,*], *Mária* Chovancová[1], and *Rudolf* Kampf[1]

[1]Institute of Technology and Business in České Budějovice, Faculty of Technology, Okružní 517/10, 370 01 České Budějovice, Czech Republic

Abstract. This paper is processed based on assessing the current capacity condition of the specific railway track section which connects two cities. Particularly, it proposes a set of possible measures in order to increase the permeable railway tracks performance (permeability) for the purpose of establishing an integrated transport system. In detail, it examines the capacity solutions by the construction of a double-track liner for smooth crossing and a second track on a single-track railway section.

1 Introduction

For a long time, individual regions of the Slovak Republic have been trying to find effective ways to harmonize, and especially, make public passenger transport more attractive, and thus attract new customers. Providing the railway infrastructure represents a prerequisite to achieve a high-quality public passenger transport which could be considered as the skeleton of the entire transport system, not only in the national, but also in the regional conception.

Thus, integrated transport system (ITS) becomes an alternative for individual car (car passenger) transport and passengers have more options to choose how to reach their journey destination more comfortably.

When considering the increase of the railway infrastructure capacity, the required range of railway transport plays a major role in the specific railway section. All the adjustments and upgrades affecting the railway track over decades must be reflected from the required range of railway transport. First of all, it is necessary to analyze the current state (condition) of infrastructure capacity to make it clear what changes can be expected in terms of number of trains.

The Košice - Prešov transport relation (railway track section) is included into the Košice and Prešov self-governing regions. Individual self-governing regions have some intentions that they would like to apply to this section in order to improve transport serviceability between these cities and adjacent cities and municipalities [1-3].

* Corresponding author: stopka@mail.vstecb.cz

 DOI: 10.1051/matecconf/201713400055

2 Transport policy and the "Masterplan" for the development of railway infrastructure by 2020

The Strategy for the Development of Transport Infrastructure of the Slovak Republic by 2020 and the Strategy for the Development of Public Passenger Transport of the Slovak Republic by 2020 identified especially the capacity and safety limitations of railway infrastructure and inappropriate equipment of carriers' fleet as the main constraints of this situation in the Slovak Republic. This results in limited speed of cargo and passengers' transportation and reduced reliability of this transport mode [4].

In 2013, the Ministry of Transport, Construction and Regional Development of the Slovak Republic drafted a transport plan aimed at improving the quality of transport infrastructure and the quality of provided transport services under the title "Masterplan" [4].

In regard to transport policy and the current state in the field of railway transport, the main issues and the resulting needs are defined as:
- balance between transport offer and transport demand,
- balance between infrastructure offer and transport demand,
- establishing the conditions for the proper railways operation.

The realization of these visions will be done through the implementation (fulfillment) of the partial strategic objectives including, as follows [4]:
- quality and competitive railway passenger transport,
- high-quality and competitive railway freight transport,
- modern and safe railway infrastructure,
- efficient organization and planning the railways development,
- economically sustainable railways.

3 Traffic flows of the Košice integrated transport system

In the Kysak - Prešov railway track section, the early-morning heavy traffic flow is divided into two sections, one part directs into Prešov station and second part into Lipany. Transport demand in Lipany - Prešov section is spread almost evenly throughout the day. The backflow is spread over two time periods in the afternoon and evening, and the traffic rush-hour is repeated in reverse direction. Passengers' traffic flows per day are processed separately for each direction in Table 1. It contains a statistical overview of passengers' flows from Košice to Prešov and Table 2 summarizes flows from Prešov to Košice.

Table 1. Passengers' traffic flows – direction into Prešov (in 2015). Source: [5]

Track number	Track direction	Railway track section	Number of passengers
188	Košice - Prešov	Košice – Ťahanovce	2,945
		Ťahanovce – Kostoľany n. Hornádom	2,982
		Kostoľany n. Hornádom – Trebejov	2,440
		Trebejov – Kysak	2,372
		Kysak – Ličartovce	1,519
		Ličartovce – Drienovská n. Ves	1,562
		Drienovské n. Ves – Kendice	1,596
		Kendice – Haniská pri Prešove	1,602
		Haniská pri Prešove – Prešov	1,751

Table 2. Passengers' traffic flows – direction into Košice (in 2015). Source: [5]

Track number	Track direction	Railway track section	Number of passengers
188	Prešov - Košice	Prešov – Haniská pri Prešove	1,590
		Haniská pri Prešove – Kendice	1,588
		Kendice – Drienovská n. Ves	1,498
		Drienovská n. Ves – Ličartovce	1,438
		Ličartovce – Kysak	1,446
		Kysak – Trebejov	1,872
		Trebejov – Kostoľany n. Hornádom	1,900
		Kostoľany n. Hornádom – Ťahanovce	2,227
		Ťahanovce – Košice	2,125

4 Railway infrastructure between the cities of Prešov and Košice

The railway track section connecting the cities of Prešov and Košice is 33 km long, for which the streamlining is proposed, consists of two tracks on a network of the Železnice Slovenskej republiky (ŽSR; Railways of the Slovak Republic). These include the track designated by the number 188 (Košice - Prešov - Plaveč) and the track 180 (Košice - Poprad - Tatry - Žilina), which has a common numbering with the track 188 in the Košice - Kysak section; according to the numbering in the book timetable.

The traffic management is decentralized, which means that the management is provided by the transport staff directly in railways operating buildings. There is also a remote (dispatcher) controlled track between the cities of Kysak and Ličartovce on the assessed infrastructure section. All the operating buildings are constantly occupied without the transport service lockout [6-9].

In the Košice - Kysak section, the railway infrastructure connecting the cities of Prešov and Košice is built as a double-track, 16 km long, and in the Kysak - Prešov section, this infrastructure is built as a single-track and 17 km long. According to the ŽSR railway tracks categorization, the whole Košice - Prešov section is included among tracks of 1st category.

Fig. 1 depicts a track section scheme showing the stations and stops located on this section as well as the branch (turning) tracks from the railway stations Košice, Kysak and Prešov.

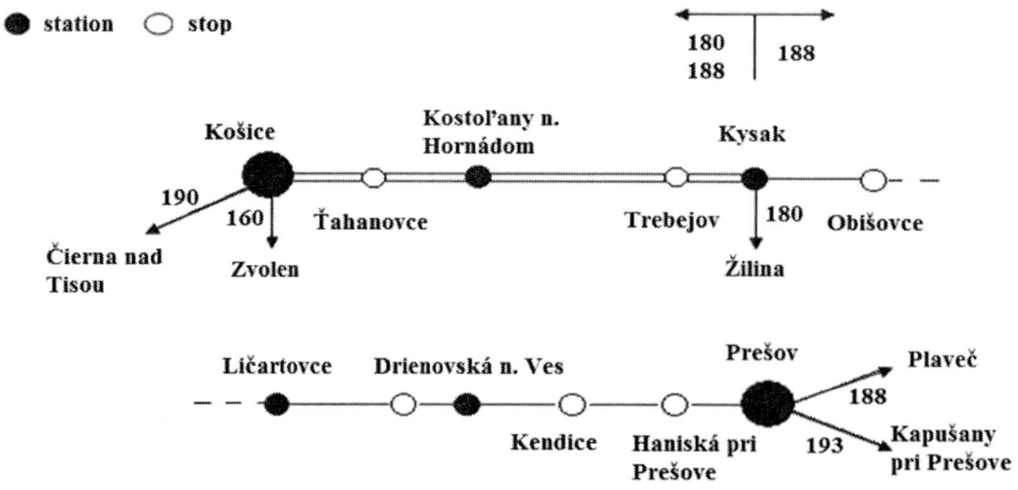

Fig. 1. Košice - Prešov track section scheme. Source: authors

 DOI: 10.1051/matecconf/201713400055

5 Results

Based on the analysis of passengers' flows and identification of a possible maximum range of transport performance offer within regional passenger transport, the railway infrastructure capacity options between the cities of Košice and Prešov, in the context of implementation of interval transport within the planned ITS, are assessed.

On the basis of the knowledge of a set of possible measures for increasing the permeable performance of railway tracks and the knowledge of the railway infrastructure condition in the assessed track section, further, the following alternative measures will be developed:

- construction of a double-track liner for smooth crossing,
- construction of a second track on a single-track railway section.

5.1 Railway tracks permeability indicators in the current state

For the purpose of the permeability calculation, identified facts by analyzing the elements of the train graph schedule diagram (TGS) in 2015, such as interstation section occupation times, station and track intervals, and so on. Table 3summarizes the numbers of operated connections and the capacity for the Prešov - Košice section.

Table 3. Numbers of operated connections within the TGS (in 2015). Source: [10]

Track number	Track section	Train type	Number of connections							
			Regular – in TGS		As needed – in TGS		Free capacity		Capacity	
			E	O	E	O	E	O	E	O
105 A	Košice - Kysak	Passenger	47	48	2	1	79	74	173	171
		Freight	24	25	21	23				
107 A	Prešov - Kysak	Passenger	21	20	0	0	35		96	
		Freight	5	5	5	5				

Explanations: E – even train direction; O – odd train direction; TGS - train graph schedule diagram

Table 4 contains calculated parameters characterizing the current state of the Prešov – Košice track section permeability.

Table 4. Prešov - Košice track section permeability (in 2015). Source: authors

Track section	Direction	Total number of trains	T_{occ}	T_{gap}	N_{add}	N_{ar}	D_o	C_p	t_l
			t_{occ}	t_{gap}	p	t_{fix}	b		
Prešov - Kysak	E	26	1,165	275	39	35	0.43	52.9	52
	AN	5	11.65	2.75	96	0	15.6		x
	O	25	x	x	x	x	x	x	x
	AN	5	x	x	x	x	x		x

Explanations: E – even train direction; O – odd train direction; T_{occ} – total occupation time of the interstation section by all trains [d]; t_{occ} – occupation time per one train [d]; T_{gap} – total gaps time [d]; t_{gap} – gaps time per one train [d]; N_{add} – number of additional routes; p – practical permeability [t/d]; N_{ar} – number of added routes; t_{fix} – time of fixed operations per one train [d]; D_o - occupation degree; b – backups time [d]; C_p – coefficient of the permeability utilization; t_l – lockouts time [d]

Table 5 contains calculated parameters characterizing the current state of the Košice - Kysak track section permeability.

Table 5. Košice-Kysaktrack section permeability (in 2015). Source: authors

Track section	Direction	Total number of trains	T_{occ}	T_{gap}	N_{add}	N_{ar}	D_o	C_p	t_l
			t_{occ}	t_{gap}	p	t_{fix}	b		
Košice - Kysak	E	71	1,269	171	87	79	0.36	40.9	60
	AN	23	7.01	0.94	173	0	12.4		x
	O	73	1,262	178	83	74	0.38	42.8	60
	AN	24	7.09	1	171	0	11.8		x

5.2 Calculation of the Kysak - Prešov track section permeability after the construction of a double-track liner for smooth crossing

In the case of insertion of a double-track section into a single-track line on the Kysak - Prešov section, the permeability will be favorably influenced by the fact that trains in the double-track section will be able to cross during the movement (smooth crossing).

In order to meet this goal, it is necessary to determine the optimum length of a double-track liner (section) with the corresponding safety device (see equation 1) [11-14]:

$$L_k = \frac{(\tau_{pv}^A+\tau_{pv}^B)*(v_1*v_2)}{(v_1+v_2)*0.06} = \frac{(2+2)*(40*44)}{(40+44)*0.06} = 1{,}396.83 \text{ m} \quad (1)$$

where: L_k – the minimum distance of the double-track liner for smooth crossing [m]; v_1 – speed of the first train [km/h]; v_2 – speed of the second train [km/h]; τ_{pv}^A – interval of successive entrances in the railway operating building A; τ_{pv}^B – interval of successive entrances in the railway operating building B.

From the above calculation, it is clear that the length of a double-track liner for smooth crossing must be at least 1397 m. After the performed analysis of the track section, it would be most efficient to build a double-track liner in a constraining section of the single-track line between Drienovská nová Ves station and Haniská pri Prešove station. The distance between these two points is 5,100 m, which meets the condition of the minimum distance of a double-track liner for smooth crossing.

Table 6summarizes calculated parameters characterizing the Košice - Kysak track section permeability after the construction of a double-track liner.

Table 6. Košice-Kysaktrack section permeability after the construction of a double-track liner (in 2015). Source: authors

Track section	Direction	Total number of trains	T_{occ}	T_{gap}	N_{add}	N_{ar}	Cp	t_l
			t_{occ}	t_{gap}	p	t_{fix}		
Košice - Kysak	E	71	379	1,000	x	0.27	44.85	60
	AN	23	6.22	16.4	136	12.4		x
	O	73	x	x	x	x	x	x
	AN	24	x	x	x	x		x

From above calculations, it is obvious that, after the construction of a double-track liner for smooth crossing in the given track section, its capacity has a potential to significantly increase from the original 96 trains per day to 136 trains per day. At present, 61 train connections are operated on this track section, which means that other 75 train connections can be deployed after implementing this measure [11, 12].

5.3 Calculation of the Košice - Prešov track section permeability after the construction of a second railway track

Construction of new railway tracks, or double-tracking railway sections, has a high positive effect in terms of increasing the capacity of the assessed section. For such construction measures, the permeability is increased by more than 100%. When decision-making such as double-tracking a railway section, it is necessary to equip the entire section not only with a new railway track but also a new railway safety device, which has a positive effect on the permeability of the track section as well [14-17].

Similarly to all operating devices, it is necessary to determine the practical permeability of the railway tracks separately (see equation 2 for even direction and equation 3 for odd direction) [18-20]:

$$n_E = \frac{T_{occ}-(T_E+T_{ois})}{t_{occ}+t_{gapp}} = \frac{1{,}440-(0+0)}{(4+4)} = 180 \text{ t/d} \quad (2)$$

$$n_O = \frac{T_{occ}-(T_E+T_{ois})}{t_{occ}+t_{gapp}} = \frac{1{,}440-(0+0)}{(4+4)} = 180\text{t/d} \quad (3)$$

where: n_E – practical permeability for even direction [trains per day = t/d]; n_O – practical permeability for odd direction [t/d]; T_E – total time needed for regular scheduled examinations [d]; T_{ois} – occupation time of an inter-station section by regular siding, operating and service trains which are not included in the number of trains [d]; t_{occ} – average occupation time per one train [d]; t_{gap} – required gaps time per one train [d].

After calculations, it is clear that the practical permeability is for even and odd directions of 180 trains per day [21].

6 Discussion

After the made calculations, it is necessary to evaluate the individual design measures. Fig. 2 shows comparison of the practical permeability of the examined railway track section for the current state (without applied measures) and the proposed measures variants; a double-track liner and a complete double-tracking [22-24].

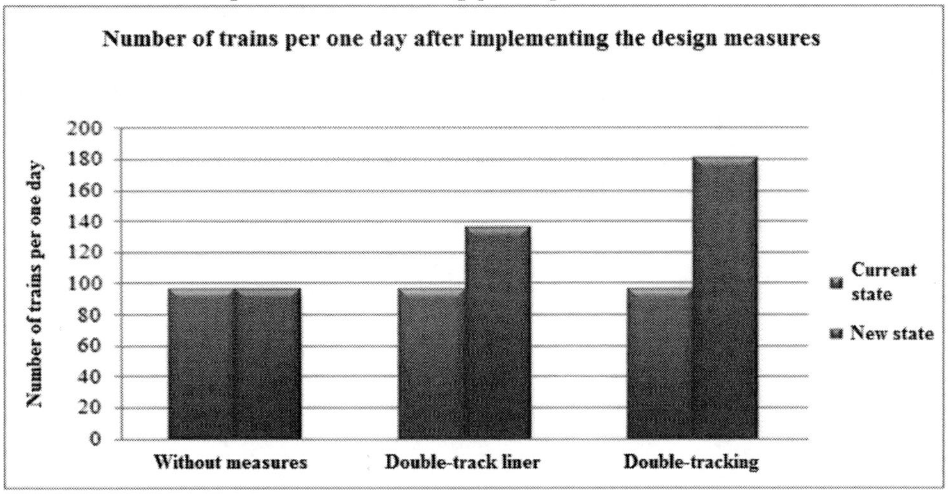

Fig. 2. Comparison of the original permeability and permeability after modernization. Source: authors

Following the section upgrade, i.e. the construction of a double-track liner for smooth crossing, the section capacity would increase from the original 96 trains per day to 136 trains per day.

Double-tracking the whole section from the Prešov station to Kysak station represents next design measure. This most demanding measure in terms of investments also has the most favorable impact on the whole section capacity, while the permeability would increase from the original 96 trains per day in both directions to 180 trains per day in both the even and the odd direction [25-27].

6 Conclusion

As mentioned above, when considering the increase of the railway infrastructure permeable performance from an integrated transport system establishment point of view, the required range of connections within railway transport plays a major role in particular railway section. All the changes affecting the railway track must be based on this railway transport range.

Košice – Kysak – Prešov railway connection represents a link of two important settlements, the centers of regional departments and creates significant interconnection within the functional unit. On this interconnection, considerable demands on the quality and capacity of the railway infrastructure, especially in terms of its utilization for the creation of a primary system within an integrated transport system.

Among the potential regional railway tracks, where significant investments could be made available, the track between the cities of Prešov and Košice can be included. It has a high passengers transportation potential between cities and a strong current traffic flow.

This paper is supported by the research project "From horse-drawn railway to intermodal transport" within Visegrad Fund.

References

1. G.C. Li, B.R. Li, M.Y. Ju, Z.J. Zhang, *WCTR – 14th World Conference on Transport Research*, 3402-3415 (Transportation Research Procedia **25**, Shanghai, China, 2017), DOI: 10.1016/j.trpro.2017.05.231
2. V. Molnar, *CLC 2013 – Carpathian Logistics Congress*, 561-566 (Krakow, Poland, 2014)
3. J. Ližbetin, L. Černá, M. Ľoch, Nase More **62**, 3, 147-152 (2015), DOI: 10.17818/NM/2015/SI11
4. Ministry of Transport, Construction and Regional Development of the Slovak Republic, *Operačný program Integrovaná infraštruktúra* 2014 – 2020, *5th revised draft* (Bratislava, Slovak Republic, 2013)
5. Ministry of Transport, Construction and Regional Development of the Slovak Republic, *Strategický plán rozvoja dopravnej infraštruktúry SR do roku 2020, 1st phase* (Bratislava, Slovak Republic, 2013)
6. D. Michniak, Quaestiones Geographicae **35**, 4, 107-120 (2016), DOI: 10.1515/-quageo-2016-0039
7. B. Abramovic, V. Zitricky, V. Biskup, European Transport Research Review **8**, 4 (2016), DOI: 10.1007/s12544-016-0215-7

8. K. Borsekova, M. Sokolowicz, *21st IntSciConf of PGV Network - For Stronger Europe Renovation of the Economic, Entrepreneural, Political, Territorial and Educational Structures*, 179-194 (Banská Bystrica, Slovak Republic, 2016)
9. M. Nadanyiova, *Transport Means – 18th International Scientific Conference on Transport Means*, 120-124 (Kaunas, Lithuania, 2014)
10. *ŽSR – Railways of the Slovak Republic, Book of the railway tracks permeability of the ŽSR for the TGS 2015/2016* (Bratislava, Slovak Republic, 2016)
11. R.L. Parsons, A.J. Rahman, J. Han, T.E. Glavinich, Transportation research Record **2448**, 133-141 (2014), DOI: 10.3141/2448-16
12. Y.G. Zhang, Q.F. Zeng, D.Y. Lei, X.Y. Wang, Discrete Dynamics in Nature and Society, Article Number: 2319437 (2016), DOI: 10.1155/2016/2319437
13. B. Abramovic, V. Zitricky, P. Mesko, LOGI – Scientific Journal on Transport and Logistics **8**, 1, 1-10 (2017), DOI:10.1515/logi-2017-0001
14. R.W.B. Stell, IEEE Vehicular Technology Magazine **6**, 3, 99-105 (2011), DOI: 10.1109/MVT.2011.942537
15. J. Gasparik, V. Gaborova, V. Luptak, *Transport Means – 20th International Scientific Conference on Transport Means*, 245-249 (Juodkrante, Lithuania, 2016)
16. M. Sol-Sanchez, F. Moreno-Navarro, M.C. Rubio-Gamez, Materials **7**, 8, 5903-5919 (2014), DOI: 10.3390/ma7085903
17. S.Y. Feng, L.M. Wei, C.Y. He, Q. He, Journal of Central South University **21**, 7, 2921-2927 (2014), DOI: 10.1007/s11771-014-2258-z
18. B. Poliaková, I. Kubasáková, Autobusy: technika, eksploatacja, systemy transportowe **5**, 104-110 (2014)
19. G. Fedorko, H. Neradilova, M. Sutak, V. Molnar, *Transport Means –20th International Scientific Conference on Transport Means*, 169-174 (Juodkrante, Lithuania, 2016)
20. R. Francesco, M. Gabriele, R. Stefano, European Transport Research Review **8**, 4 (2016), DOI: 10.1007/s12544-016-0216-6
21. J. Gasparik, V. Luptak, P. Mesko, *ICTTE – 3rd International Conference on Traffic and Transport Engineering*, 388-392 (Belgrade, Serbia, 2016)
22. J. Jagelčák, I. Kubasáková, NaseMore **61**, 5-6, 106-116 (2014)
23. J. Gašparík, L. Pečený, P. Blaho, *Regulovaná a neregulovaná konkurence na kolejích (Regulated and unregulated competition on rails)*, 40-48 (Telč, Brno: Masaryk University, Czech Republic, 2013)
24. V. Lendel, M. Varmus, Logi – Scientific Journal on Transport and Logistics **3**, 2, 21-34 (2012)
25. J. Zamecnik, J. Jagelcak Communications: scientific letters of the University of Žilina **17**, 4, 21-27 (2015)
26. L.X. Yang, X.S. Zhou, Z.Y. Gao, Omega-International Journal of Management Science **48**, 75-93 (2014), DOI: 10.1016/j.omega.2013.11.004
27. Z. Cujan, G. Fedorko, Open Engineering **6**, 426-431 (2016), DOI: 10.1515/eng-2016-0057

Virtual Reality as a Tool for Large Scale Systems

Vladimír Strakoš[1], *Libor* Kavka[1,*], and *Filip* Beneš[2]

[1]College of Logistics, Přerov, Palackého 1381/25,75002 Přerov, Czech Republic
[2]VŠB – Technical University of Ostrava, Institute of Economics and Control Systems, 17. listopadu 15/2172, 708 33 Ostrava, Czech Republic

Abstract. Virtual reality (VR) is a wonderful opportunity of information technology which enables the movement of people in the virtual world. The virtual world, into which a person dives, allows them to forget for some time that they are not in the real world, but right now they are living in an entirely different world with unexpected properties. This idea is the grounds for putting forth a proposal to use the virtual world, which is not a copy of the real world, but it is created from purposefully modified 3D images of technical means and properties of these means. Such a world can then be used to control processes, especially when the space of a virtual world is created from quantities obtained in the real world. The huge importance of creating a virtual world from quantities that we have long been storing lies in the fact that one glance, one flight over the surface of such a world is sufficient to get an overview of the behavior of ultra-large-scale systems with a number of monitored quantities. It is the only way to assess, evaluate and control systems presented by a large amount of regularly monitored quantities, which are mutually connected with material and information bonds.

1 Introduction

We perceive the outside world with our senses and behave according to how the world around us affects us. Either we adapt to it or we affect it in such a way as to live our lives well and therefore to be satisfied with ourselves. Now suppose that the world around us is an area we are responsible for, and we are expected to manage it and maintain it in a very good condition. There are a number of effects on our world, against which we have to protect our area. In other words, we must control such a large-scale system. What does it take to control a large-scale system? The key is to know our area, to monitor it closely and regularly and if something is happening, to take measures against undesirable effects.

If we are in charge of a surface area of several km², a very acceptable way to maintain an overview of such a large area is to have an airplane or, ideally, a helicopter, and to fly over this area from time to time and check if something is happening. How do we know from a helicopter that something is wrong? Only by comparing an image which is stored in our memory and is not real with the image we are seeing. If there is something in a place

* Corresponding author: libor.kavka@vslg.cz

which previously was not there, then we land, we find out the details, evaluate the situation and decide what must be done. Then, we look for a person who will do what we want or what we need.

Now, suppose that we have a device to create a virtual, that is non-existent world in reality. The world, that is the surface of the area of interest, will be created from z coordinates which illustrate for example average salaries of people in a big city. Then, wearing glasses or surrounded by projected images, we find ourselves over the surface of such an area, which does not actually exist but it is created according to salaries of the inhabitants. Without being told, we do not know over what area we are, what surface it is below us. We will see elevations, hills, mountains, and we will start wondering what it is beneath us. If we learn the fact, e.g. average salaries of residents, then from our position we can see immediately where the rich live, where the middle class live and where the poor are concentrated. We can see the differences, big and small, the size of areas with high income and low income, etc. How do we know how many people with high income there are in an area – many or few people? For such information we put colour marking of population density on the surface we created and thus we get a significantly better overview of how many rich people and how many poor people there are. We will fly there to see both the rich and the poor and then, after returning to our office, we can contemplate how to use the information we gathered to improve the situation [1].

It can be reasonably assumed that during our flying over a landscape of lower and higher salaries improved by colour indicating the density of the population, we forget that we are over a landscape which is not real but created as virtual space. Only after landing we will look around in surprise to see where we are and recollect the world in which we lived for some time and in which we dealt with some critical situations. We can invite experts for the next trip so that they can monitor the whole situation with us and together decide what to do and where. It is an incredible opportunity to use virtual reality as a tool for our decisions on utterly real problems.

Consider, as an example, a nice picture of heat consumption in a particular town in Fig.1. The red lines represent the amount of flowing heat. Beneath the surface of these hills there is actually a hidden group of interesting data influencing the consumption of heat in these areas. We can assume, for example, that those tall formations in the back may represent the consumption of factories with energy-intensive production.

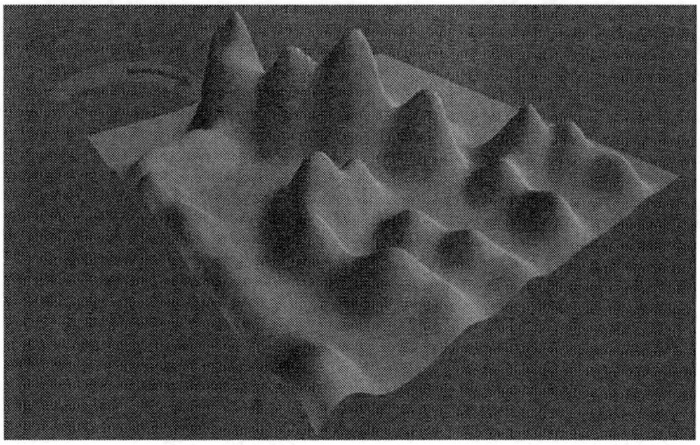

Fig. 1. Example of 3D space representing heat consumption in a mid-size city. Source: authors

On the other hand, in the front on the right there may be a group of houses with a large number of apartments. The valley in the middle may be the city centre surrounded by houses with relatively high consumption. A lot of questions can be derived from such a figure and after adding other layers we get very interesting and necessary information for the management of supply and distribution of heat; all this almost "at a glance" in a very short time. This is exactly what present day with a huge amount of information around us requires. Obviously, only an expert can read what is hidden beneath the surface. This is exactly the perfect example of what the authors have promoted all the time and the fact that the model and its properties is important only for experts and not for laymen. A layman can look at a lot of information just as they would look at an image while an expert spends a lot of time studying the model, but they gain a lot of hidden information. Often, it is information they cannot obtain in any other way [1].

From this perspective, our virtual landscape will have the feature of a large-scale model and on this model we can again simulate certain situations over our landscape, we can solve options to find the most favourable measure.

Does it sound complicated? Of course it is very difficult, but when we look, with great admiration, at the ever-expanding possibilities of information technology, we are astonished at the speed with which these options improve. It is very likely that even the creation of an artificial world only from the monitored quantities will be much easier than ever before, and such realization is close ahead. But in order to make use of virtual reality to control, we must perform several difficult follow-up operations.

First, we have to carry out the reduction of our real object onto a system with the individual elements of the system being later displayed as a set of points representing quantities which are available in the various subsystems and throughout the system as well. Then we will pre-think the appearance of the landscape which will be compiled from the selected quantities so that their connections are clear, and we will suggest their location on the surface of the 3D space that we will be creating. Then all the quantities from the real process will be displayed as z coordinates of the future 3D surface. Perhaps the most difficult task will be the fact that the area which we will look at must give us an idea of mutual connections of quantities in the real object. Because we can use colour coding of individual parts, as well as different textures, different characters and spatial objects that complete the displayed area, this is a very difficult task indeed. However, it can be also completed later. Perhaps the last thing to consider is the possibility of placing the elements on the surface, which will allow the control of some quantities through actuators and thus interfere in the real process.

At the end of this introduction, it must be pointed out that we are talking about an artificially created virtual world that is not a copy of the real world. This is an area of constantly improving and well-selling games and that is why we must think about how these great opportunities of information technology can help us in the control of large scale systems.

2 Real process and control

We virtually never see real large scale process, which we control, as a whole. Even if we are close to the process, we see only a part and even the view of that part gives us little information. Fig. 2 shows as an example data on the activity of the production process a real company.

Even if we had knowledge of what the individual numbers and characters represent in the real world of a production plant, an overview of the activities of the production process would be very hard to gain and it would be very difficult, if not impossible, to decide about control interventions in this process.

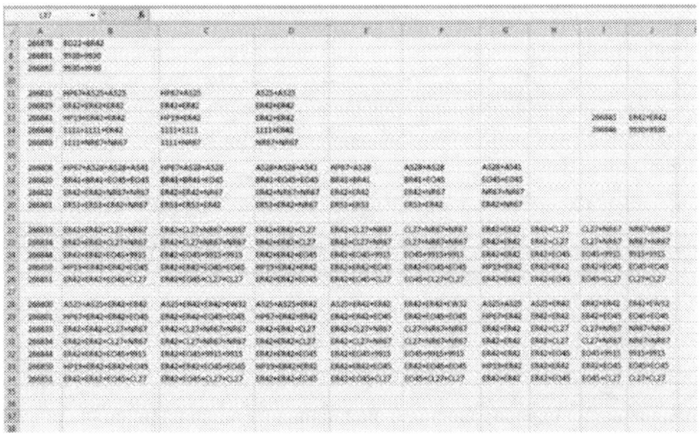

Fig. 2. Example of a screen containing a number of practical operational data, but it is almost impossible to understand this data in order to be able to make decisions. Source: authors

This is one of views which leads the authors to believe that it is much more advantageous to reduce the data into an image or preferably into 3D space, or the space of virtual reality. Consider now as an example the power plant boiler. It is a large object which takes up a lot of space and thus we can always see only a part of it, and even so, we can see its surface and what is essential for us is inside, and this is what we cannot see. Therefore we always reduce the controlled object to a system, we measure important quantities, and these we display on the screen area, for example using SCADA systems, and this is the principle of our view of the use of VR. For the control of – say the above-mentioned boiler, we do not need to see the boiler, we do not even need to know where the sensors are located and what type of sensors they are, but what we need to see is the mutual connections of individual quantities. We do not even need to see a diagram of the boiler, but we need to see the connection between the individual flows of energy, transfer points, the quantity of transferred energy, loss of energy, the amount of fuel which is being withdrawn, the amount of energy transferred to the turbine and the generator etc. Therefore we can view the quantities needed for the control by so far unused display and it is 3D space.

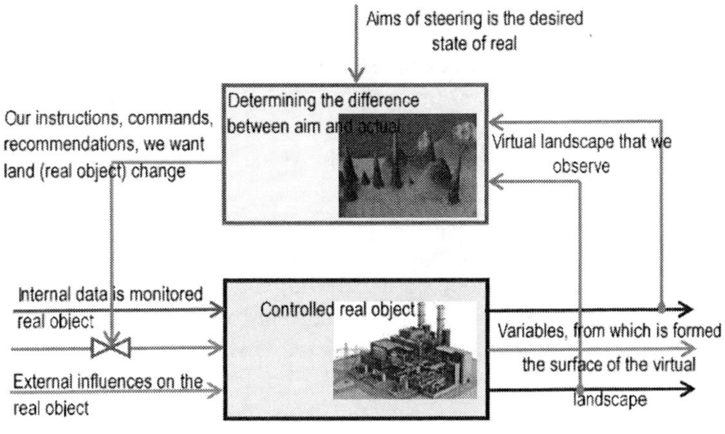

Fig. 3. General control scheme focused on the area of operational and medium-term object control using a 3D view. Source: authors

Suppose that we control the production process of an object whose information system monitors e.g. 3000 quantities. If we wanted to have an overview of all measured quantities, then we have practically only two ways to display them. One possibility is to create a 50x60 table of measured values. It will not fit in one screen. By grouping several screens we would be able to see the entire table e.g. as in Fig. 2. But then, we have a lot of almost chaotic numbers and it will definitely be hard for us to understand them. What is more, the screen will show us only a situation in one selected point. Computing, however, offers us a number of other ways to study the table, we can create graphs of various structures, we can monitor the development of selected quantities in time, we can use data mining and thus we get information that is hidden in the amount of data and that is not visible without statistics. We have a lot of possibilities, but only a reduction of the data into a properly assembled 3D image will enable us to look at all the data at once, including their mutual connections [2].

The VR environment allows us to display measured quantities in different ways. The easiest possibility, though not very suitable, is to select only one type of quantity that is measured in many places, such as the temperature of the media in some parts of the large scale equipment and to create the surface of the virtual landscape from these values [4].

Thus, we reached a significant fact, namely that we always work with data and information and even though in our case the controlled process is real, we still perceive it with our senses and we put the image of the real world in our brain. If we control such a process, all the decision-making is based on our perceptions and the signals we receive through our senses.

3 Transforming signals of a real object into the virtual world

The preparation of the virtual world for control partly follows Fig. 3. The important information for creating a "copy" of the real world is information about the real object. This is important, but for the control of a real object, e.g. a nuclear power plant, the size of cooling towers is not important, but what is important are their properties in relation to the functioning of the entire process. That is, how much heat and from what amount of coolant they are able remove at a certain temperature of the external environment. Their shape, material from which they are built etc. is not so important for the control , but only those values which are needed to fulfil their function, which can be water level, temperature in different locations, functions and state of pumps, resistance of piping systems, etc.

Within the scope of this article we cannot create a detailed analysis of a nuclear power plant in order to create 3D space of virtual reality, but all we want to do is to outline a possible procedure for creating 3D space. Do we really need to create space of virtual reality or is a 3D image sufficient? 3D image is enough, but if this space is put into a virtual reality environment, then we can move freely over this space, we can touch the actuating elements, we can perhaps even watch the space above us, watch the shadows from a certain direction of light, moving objects, etc. [2].

Consider now the quantities essential for creating the 3D space representing the operation of a nuclear power plant. We do not want to create a 3D object of a nuclear power plant because this would be a model of a real object, which would almost useless for control. We are not interested a copy of a real object, but we are interested in the properties and behaviour of a real object, not its shape and size. This is the fundamental difference we are talking about, it is the image suitable for process control what we have in mind.

Fig. 4 shows a beautiful scheme of the primary circuit of a nuclear power plant designed to simulate the plant operation. Such a scheme is very suitable for our thinking. It does not include details which are not significant for the assessment of the operation. But even in this part of the operation many other quantities which are necessary for the control of the primary circuit in real time are measured and checked. The use of virtual reality is not suitable for process control in real time. VR is designed more for the operational and medium-term control and for this the example in fig. 4 is ideal.

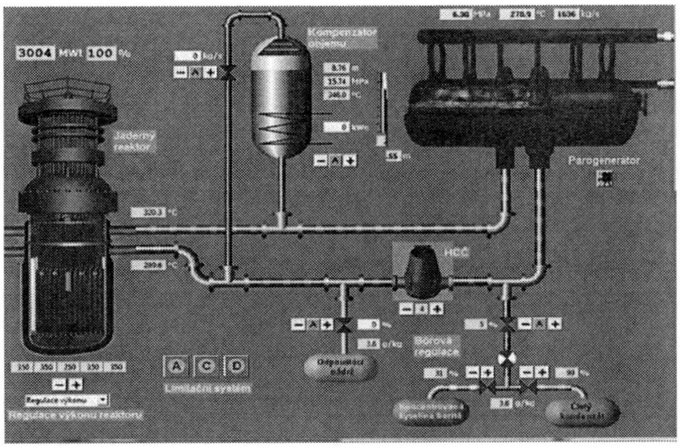

Fig. 4. Scheme of the primary circuit of a nuclear power plant with the most important data. Source: Simulation Program - Information Centre of the Temelín NPP

The reactor is a source of thermal energy, as we can see, a temperature of 320^{0}C. In this situation the reactor is used for 100%. The thermal energy is transferred through specially treated water in carefully isolated circuit to the steam generator. There the energy is transferred to the water in the secondary circuit, which thus has no direct link with the reactor and the primary circuit and radioactivity does not reach beyond the primary circuit. Volume compensator, which compensates for the change in the volume of water due to the change of heat, is integrated in the circuit. If part of the water in the primary circuit is lost, then condensate and boric acid is added to the primary circuit. For a less informed reader we point out that due to the high pressure in the primary circuit of 15.7 MPa, the liquid having a temperature of 350^{0}C is still in a liquid state. That is why a liquid flows in the primary circuit while in the secondary circuit, that is behind the steam generator, it is steam. Consider how this scheme will be transformed into 3D space, at the same time making all the important data clearly visible.

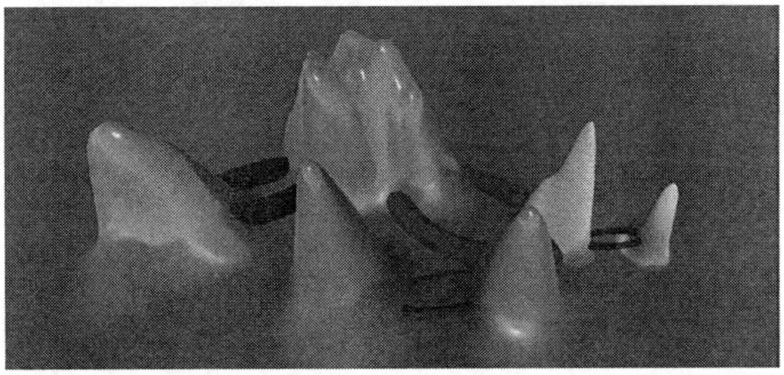

Fig. 5. Scheme of the main energy flows in a nuclear power plant, including several indicators which can be used to check the details in the model. Source: authors

Fig. 5 shows us that it is possible to create a 3D space which would represent, in our case, energy flows in a nuclear power plant, in many variants. It is difficult to say whether this view is the best one. It is certainly not the best, because there are endless possibilities, and therefore the thing is only to create a view that we, senior managers, will like. The thing is not only whether we like it or not, but whether it provides us with enough information that we need to control this process. We would like to point out again that this is not a real-time control, which the control room is in charge of, but it is partly operative control, but mainly it is medium-term control.

When looking at fig. 5, we will certainly think that it does not make any sense, that we can display this clearly on screen with commonly used tools designed for the construction of schemes of controlled process. After all, the diagram in Fig. 4 proves this, even if it does not show the whole scheme. When we look at both pictures, probably nobody can say which view is clearer.

Thus we come to the essence of the problem and this is namely how large the number of data necessary for the control at a certain control level must be so that it is worth creating a 3D view. Note the well-arranged space in Fig. 6; this is electric power distribution. This energy system covers a large area of supplied space and apart from that, it indicates a cross-border connection on the right where the energy flow implies significant financial means which "flow" between the connected countries. Additionally, they can sometimes really affect the stability of the whole energy system. The number of end customers, which corresponds to the number of terminal branches, is several million. Such a large number of branches in terms of medium-term control may not even be interesting for us. Consumption of the minor end-user is practically irrelevant in terms of controlling the whole unit. Consumption is, however, significant from a business perspective, because the customer must pay us for the consumed energy. We will focus our control only on larger areas and then display of the distribution of energy consumption in various locations of the monitored area will be very interesting and it will also be necessary for high-quality and economical controlling. 3D space, unlike common ways of display, is irreplaceable for displaying such a move of energy, its sources and places of its consumption and yet we have all the necessary data. Implementation of 3D space compiled from data on the movement of electrical energy can have an almost infinite number of variations [3].

Fig. 6. Part of 3D landscape for displaying the distribution of electrical energy. Source: authors

A simpler variant, which the authors tried to create in Figure 6, will suffice for now. It is loosely based on high voltage distribution systems in the Czech Republic. This means that only significant power plants and some routes of electric line are approximately recorded. This figure, unlike the previous one, is focused on the spatial distribution of the main elements of the entire system for supplying electrical energy to the population. The authors believe that even such a relatively simple representation of the distribution of electrical energy gives a better overview of the whole situation than any other. It will be even better when we make a full use of colour coding of the utilization of electric power distribution, the capacity of power plants, decreases in long distribution lines, cross-border flows, etc. The possibility of using the monitoring panel to "touch" any point in space gives such displays a very interesting overview of the entire system [3].

We can observe what capacity various sources use, what cross-border flows are and the direction of energy flows, how busy the individual line elements are, etc. It would be possible to process more detailed views on the state of switches, transformers, power factor correction etc. in the same way in substations. Of course, one can get all this information in detail on the screen, e.g. in a SCADA system or else, but for a senior manager who takes a decision about the situation for a certain period in advance, such a view of the controlled system is certainly more enjoyable and mainly it is free of unnecessary data.

Obviously, the success of such a display depends on the skill and imagination of the creator, and this is indeed one of the major problems with using information 3D space or for other utilization of the possibilities of virtual reality. Some scientific groups dealt with the possibilities of new technologies in control of the technological processes, such as the use of RFID technology, 3D models and use the virtual reality in process control [5].

3 Conclusions

The development of information technology brings challenging fast-growing possibilities, which we should take advantage of also for the benefit of the society. Specifically, the issues associated with controlling, as shown in Figure 3, is the part of our activity where we can use technology for the transmission and processing of information. The video game industry, which is constantly developing, is the driving force for the development of the system of representation of the space in which the games take place. These games challenge us to incorporate their possibilities also into the systems control. When considering the ever-increasing transport options for both short and long distances and the increasing speed of transportation, we come to the conclusion that the space for trading and thus also for controlling the movement of goods will continue to expand. It is the development of such technical means which is ideal for the presentation of data and information necessary for the control of movement in a large area. This is the main impulse which led the authors to the fact that it is necessary to prepare possibilities for using the virtual world space for controlling large-scale systems. But in fact, large-scale systems do not have to be large only in terms of space. Consider, for example, the distribution of blood in our body, where there are several billion passages and the failure in any of these, even in the relatively small and insignificant one somewhere in the middle of this network, causes this person major inconvenience, or perhaps even death. This naturally created large-scale supply system is, nevertheless, very well controlled, but it is a question of multi-agent systems and this is a different area of control, emerging again from the development of information technology.

This paper is supported by the research project "From horse-drawn railway to intermodal transport" within Visegrad Fund.

References

1. S. Aukstakalnis, Old Tappan, *Practical augmented reality: a guide to the technologies, applications and human factors for ar and vr, 1st edition*, (NJ: Pearson Education, 2016)
2. V. Kebo, O. Kodym, Virtualni realita a rizeni procesu (VSB-TU Ostrava, Czech republic, Montanex, 2011)
3. T. Parisi, *Learning virtual reality: developing immersive experiences and applications for desktop, web, and mobile, 1st edition,* (Sebastopol, CA: O'Reilly Media, Inc.2011)
4. A. Okita, De Luca, *Learning C# programming with Unity 3D*, (Boca Raton: CRC Press, 2015)
5. R. Danel, V. Kohut, O. Kodym, *13th International Multidisciplinary Scientific Geoconference, Information Systems in Coal Preparation Plants and the Use of New Technologies* (2013)

Spatial Econometric Analysis of the Hungarian Border Crossings

Zsombor Szabó[1,*], *Tibor* Sipos[1], and *Árpád* Török[1]

[1]Budapest University of Technology an Economics, Department of Transport Technology and Economics, H-1111 Budapest, Hungary

Abstract. The analysis on the border crossings, as the objects to decrease the separation effect, is really important for our country, to evaluate the connection of the neighbouring and other nearby countries and international economic centres with Hungary. In our country's life the cross-border partnerships have an important role because of historical and cultural causes. In recent article, our task is to analyse changes of the cross-border traffic, with the help of the spatial econometrics. The goal is, to show the effect of five outstanding areas on the crossing traffic, and to analyse the spatial connection between them.

1 Introduction

The analysis of the borders is exceptional nowadays. While in the European researches the main goal is to find the nature of the relations, in the rest of the world the focus is on the international cooperation's effect on the economy [1, 2]. Recent paper's exceptional goal is to describe the system of the relations between the borders, borderlands and border crossings specific parameters and the crossing traffic flow. Numerous papers show that there is a time-based correlation for the flow through the border gates, but now another approach is analysed [3, 4]. Numerous international papers have shown, that in case of rating the border crossings locations, the relations between them, or the connected traffic needs, it is advised to take into consideration the properties of the road network system which provide the border crossings (for example: type of the road, international road systems or special infrastructure elements like bridge or tunnel), and the relations with the neighbouring countries especially for the uncontrolled borders described by the international agreements [5, 6]. Exceptional problem that in Hungary the mean distance between the border crossings is 25 kilometres, while in Western Europe this number is only 5-6. Because of these facts recent papers main goal is to describe a tool which is available for the analysis of the goodness the recent and the planned border crossings. For the tool, the chosen method is the spatial econometrics, although some other methods are available, [7] because numerous papers shown, that this method can be used for answering some questions made by the transportation processes [8].

* Corresponding author: szabo.zsombor@mail.bme.hu

2 The used model

In our model the border crossings of Hungary are analysed by spatial econometric method. For this, the dependent variable will be the traffic flow on each border crossing in vehicle unit/day/two directions. In the analysed time in Hungary there was 103 border crossings, which locations are shown on the next map. (Fig. 1.)

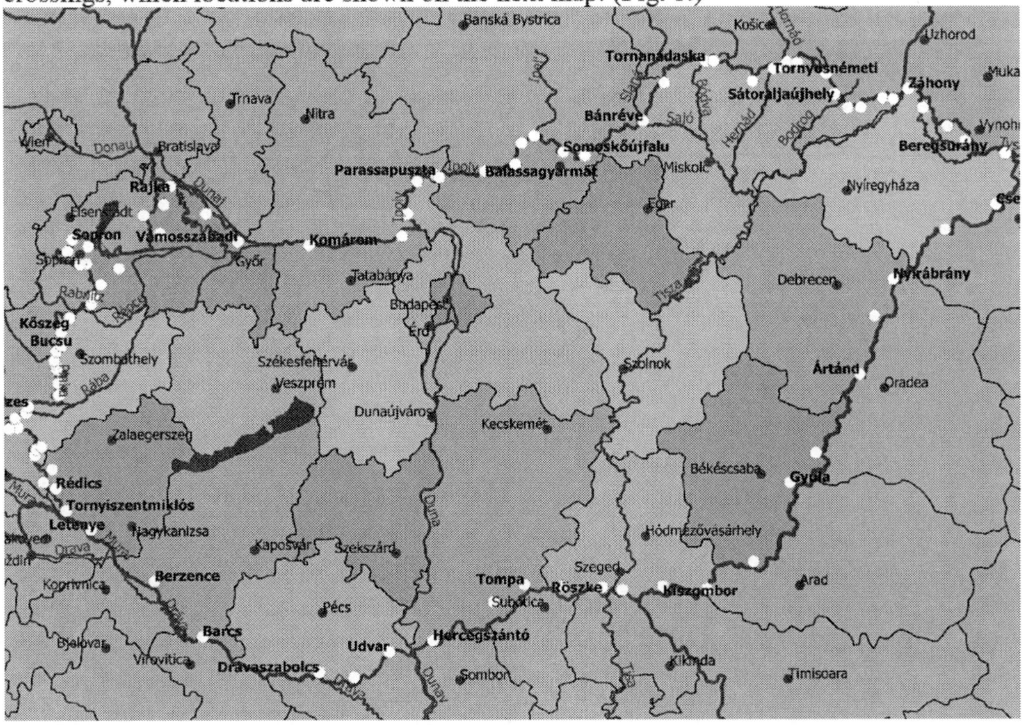

Fig. 1: The map of Hungary's border crossings. Source: authors

The analysis has five outstanding point of view which is need to be considered. These are the next five:

- the effect whether the border gate is on a bridge
- the type of the road
- the effect of the European International E-network
- the country where the border gate is opening
- the effect of the Schengen borders [9]

The analysis of the outstanding effects is in harmony with the past research papers like [5, 6].

After than several possible decision variables are eliminated due to multicollinearity issues, the remaining description variables can be organised in numerous groups. The first group is the territories demographic values. In the Hungarian side the populations of the nearest city, the processus (which stands for the NUTS-4 or LAU-1 level) and the county is used. This wide choice of the areas is need to be used because in Hungary 14 out of 19 counties lye near the border [10]. On the other side of the border only the population of the NUTS-3 territories will be used, because the nearest cities p-value is too high to use.

The second group is the rules of the border usage. In this sections five variables are need to be considered. The first is the number of the opening hours in a week, the second is the capacity of the border gate. The third is the total weight capacity for the road, the forth is the total weight capacity for the trucks, and the fifth is a dummy variable which is one if the buses are allowed to cross the border at the border gate.

The third group has only one variable which is the effect of the bridge. This is also a dummy variable. It is one, when the border crossing is also a bridge. The fourth group is the different country's dummy variables. In this case there are six variables where Ukraine is the base of the comparison.

The fifth, sixth and seventh groups are considering on the road network. They are dummy variables for the European International E-network, the type of the road in Hungary, and in the neighbouring country also. The last variable, which belongs also to the bridges is the width of the river.

The base model upon the previously shown decision variables, is shown in the next table. The . stands if the $p<0.1$, the * stands if $p<0.05$, the ** stands if $p<0.01$, and *** stands if $p<0.001$ [11-14]. The model seems to be quite good with $R^2=0.7757$, and the p-value of the F-statistics is $1.38*10^{-14}$.

Table 1. The result of the classic linear regression model. Source: authors, based on [11-14]

		Estimate	Std. Error	Pr(>|t|)	
Intercept		2280	4271	0.5951	
Hungarian population	County	3.11E-03	1.67E-03	0.0669	.
	Processus	-3.53E-03	8.28E-03	0.6714	
	City	1.66E-01	3.84E-02	4.64E-05	***
Neighbouring NUTS-3 population		2.13E-03	3.99E-03	0.5947	
Border crossing	Opening time	-12.49	14.88	0.4040	
	Capacity	-0.46	0.43	0.2840	
	Total weight capacity	14.47	19.81	0.4675	
	Truck weight capacity	1098	931.3	0.2422	
	Bus banning	-1483	1170	0.2089	
Bridge		-550	798	0.4928	
Country's dummies	Austria	5467	2004	0.0080	**
	Croatia	-1107	1914	0.5647	
	Romania	809.4	1863	0.6652	
	Serbia	-180.2	1899	0.9247	
	Slovakia	1628	2294	0.4801	
	Slovenia	4331	2199	0.0526	.
European international E-road network	A-class reference	4190	1592	0.0103	*
	A-class intermediate	-631	1348	0.6411	
	B-class	1921	1826	0.2961	
Hungarian road dummies	Motorway	8434	3458	0.0171	*
	Highway	4580	3797	0.2315	
	National road	2641	1282	0.0429	*
	Other road	828.8	955	0.3883	
Neighbouring road dummies	1st class national road	-2420	3123	0.4409	
	2nd class national road	-3717	3150	0.2418	
	Regional road	-5042	3231	0.1229	
	Local road	-5601	3223	0.0864	.
River width		4.16	3.73	0.2680	

3 Method

The method is the spatial econometrics based on Luc Anselin [15] and Varga Attila (in Hungarian) [16] works. There are two models which is preferable to use. One is the spatial lag model, where the spatial effects are taken into consideration as a spatial lag, like a time lag in the classic econometrics. The model's mathematical representation is the next (1) [16].

$$\mathbf{y}_{(N\times1)} = \rho\mathbf{W}_{(N\times N)}\mathbf{y}_{(N\times1)} + \mathbf{X}_{(N\times K)}\boldsymbol{\beta}_{(K\times1)} + \boldsymbol{\varepsilon}_{(N\times1)} \tag{1}$$

where:
- $\mathbf{y}$: vector of the dependent variables
- ρ: autoregressive parameter
- $\mathbf{W}$: weight matrix
- $\boldsymbol{\beta}$: the coefficient vector
- $\mathbf{X}$: the matrix of the independent variables
- $\boldsymbol{\varepsilon}$: vector of errors ($E(\varepsilon_i)=0, V(\varepsilon_i)=\sigma^2$)
- N: number of observations
- K: number of independent variables

Among the parameters only the ρ and $\boldsymbol{\beta}$ are unknown. The $\boldsymbol{\beta}$ can be calculated from the ρ. Because of the multicollinearity the ordinary least squares method cannot be used. For this, another estimator model should be used, which is the maximum likelihood estimation. Its formula is the next (2) [16].

$$L_C = \sum_{i=1}^{N}(1-\rho\omega_i) - \frac{N}{2}\ln\frac{(\mathbf{e}_0-\rho\mathbf{e}_L)^T(\mathbf{e}_0-\rho\mathbf{e}_L)}{N} \tag{2}$$

where:
- ω_i: eigen value of the $\mathbf{W}$
- L_c: the loglikelihood parameter
- $\mathbf{e}_0$, $\mathbf{e}_L$: the error vector of the linear regression models

The other model is the spatial error model, where the spatial effects are taken into consideration as an error. In this case the model's mathematical representation is the next (3-4) [16].

$$\mathbf{y}_{(N\times1)} = \mathbf{X}_{(N\times K)}\boldsymbol{\beta}_{(K\times1)} + \boldsymbol{\varepsilon}_{(N\times1)} \tag{3}$$

$$\boldsymbol{\varepsilon}_{(N\times1)} = \lambda\mathbf{W}_{(N\times N)}\boldsymbol{\varepsilon}_{(N\times1)} + \boldsymbol{\zeta}_{(N\times1)} \tag{4}$$

where:
- $\boldsymbol{\zeta}$: vector of spatial dependent errors
- λ: autoregressive error parameter

In order to decide which model should be used, the R project's LMtest order was chosen, which is in the spdep library. This function determines the Lagrange Multiplier test, and its robustness for both model, for a given weight matrix, and a basic linear regression model [11-14, 17].

4. Results

4.1. Choosing the proper model

Before introducing the results, an important question is need to be answered. In order to analyse the data six weight matrix are constructed. The type of the matrix can be binary (B) or row standardised (W) [13]. The assignment has also four types:
- All: all of the border crossings have effect on each other
- Neighbour: only the neighbouring border crossings have effect on each other
- Main road: only the border crossings have effect on each other which are between two border crossings with 2nd class, or higher road type.
- Country: only the border crossings have effect on each other, where the crossings points toward the same country.

The a parameter for the weight matrices is 1.94 based upon [18].

To choose the proper model, the next table is need to be taken into consideration [17]. Between the parameters, the first is the Moran I's value, and the second is its p-value. The remaining for rows shows the p-values of the Lagrange multiplier (LMerr, LMlag), and the Robust Lagrange multiplier tests (RLMerr, RLMlag), for the two models (where the err stands for the spatial error model, and the lag stands for the spatial lag model) [8-11, 14].

Table 2. Choosing the proper model. Source: authors, based on [11-14]

	0. model	1. model	2. model	3. model	4. model	5. model
Type of matrix	W	W	W	B	W	B
Assignment	All	Neighbour	Main road	Country	Country	Main road
Moran I	-0.1392	-0.1841	-0.1768	-0.0386	-0.1474	-0.0185
Moran I's p-value	0.3296	0.4276	0.3205	0.4534	0.3811	0.2924
LMerr p-value	0.0796	0.0981	0.0668	0.1494	0.0816	0.7544
LMlag p-value	0.6192	0.9014	0.7861	0.7901	0.6303	0.0594
RLMerr p-value	0.0175	0.0364	0.0188	0.0153	0.0172	0.0086
RLMlag p-value	0.0933	0.1982	0.1350	0.0492	0.0899	0.0013

4.2. Comparing the models

Based on Table 2. five models seem to be appropriate. Among this five models one is spatial lag model, for the 5th matrix (sar5), while the other for are spatial error models for matrices no. 0, 1, 2 and 4 (errorsalm0-errorsalm4). For the five model, the results are shown in Table 4. As it can be seen the error models have better p-values, but the goodness of the models can be seen in other table (Table 3.).

Table 3. Comparing the goodness of the five model. Source: authors, based on [11-14]

	sar5	errorsalm0	errorsalm1	errorsalm2	errorsalm4
R^2	0.7839	0.8286	0.7995	0.8118	0.8291
ρ/λ	0.0343	-0.6004	-0.2909	-0.4244	-0.5836
p-value	0.0820	1.04E-04	1.24E-02	1.95E-03	1.30E-04
L_c	-931.1929	-925.1766	-929.5786	-927.9058	-925.3812
AIC	1924.4	1912.4	1921.2	1917.8	1912.8

 DOI: 10.1051/matecconf/201713400057

Table 4. Results of the five model. Source: authors, based on [11-14]

		Spatial lag model for 5th matrix				Spatial error model for 0th matrix				Spatial error model for 1st matrix			
		Estimate	Std. Error	Pr(>|z|)		Estimate	Std. Error	Pr(>|z|)		Estimate	Std. Error	Pr(>|z|)	
Intercept		2461.30	3526.50	0.4852		2685.40	3232.80	0.4061		2454.50	3498.10	0.4829	
Hungarian population	County	2.53E-03	1.43E-03	0.0774	.	2.85E-03	1.10E-03	0.0096	**	3.00E-03	1.21E-03	0.0135	*
	Processus	-4.43E-03	6.88E-03	0.5192		-1.11E-02	5.17E-03	0.0319	*	-7.22E-03	5.78E-03	0.2116	
	City	1.57E-01	3.25E-02	1.39E-06	***	1.37E-01	2.82E-02	1.18E-06	***	1.49E-01	2.99E-02	6.44E-07	***
Neighbouring NUTS-3 population		1.73E-03	3.39E-03	0.6106		6.93E-03	2.71E-03	0.0105	*	5.05E-03	2.97E-03	0.0885	*
Border crossing	Opening time	-11.22	12.34	0.3633		-15.25	11.80	0.1962		-14.62	12.36	0.2371	
	Capacity	-0.42	0.36	0.2434		-0.70	0.33	0.0312	*	-0.60	0.35	0.0852	*
	Total weight capacity	11.53	16.54	0.4858		-7.14	14.05	0.6112		7.18	15.02	0.6328	
	Truck weight capacity	1267.70	779.93	0.1041		-1694.40	729.89	0.0203	*	-344.17	759.27	0.6503	
	Bus banning	-1511.70	974.45	0.1208		443.13	957.01	0.6433		-605.54	982.10	0.5375	
Bridge		-595.37	664.96	0.3706		-1121.30	614.71	0.0681	.	-969.80	626.45	0.1216	
Country's dummies	Austria	4474.60	1682.60	0.0078	**	7586.30	1306.90	6.44E-09	***	6815.80	1460.80	3.07E-06	***
	Croatia	-1440.70	1551.10	0.3530		-83.72	1256.60	0.9469		-486.16	1351.90	0.7191	
	Romania	515.42	1546.60	0.7389		297.00	1283.60	0.8170		555.32	1353.10	0.6815	
	Serbia	-309.49	1524.20	0.8391		1457.00	1185.20	0.2189		751.03	1328.70	0.5719	
	Slovakia	1339.80	1944.60	0.4908		682.90	1620.80	0.6735		1108.30	1707.60	0.5163	
	Slovenia	3473.10	1821.40	0.0566	.	7036.90	1395.90	4.63E-07	***	5908.70	1558.30	1.50E-04	***
European international E-road network	A-class reference	4317.30	1340.30	0.0013	**	3621.20	1122.70	0.0013	**	3713.30	1247.50	0.0029	**
	A-class intermediate	-473.07	1132.00	0.6760		-782.20	1041.50	0.4526		-724.15	1077.40	0.5015	
	B-class	1844.10	1506.70	0.2210		2628.40	1400.70	0.0606	.	2427.90	1527.90	0.1121	
Hungarian road dummies	Motorway	8067.10	2881.10	0.0051	**	10947.00	2488.40	1.09E-05	***	9901.50	2722.50	2.76E-04	***
	Highway	4874.80	3170.00	0.1241		4564.50	2787.90	0.1016		4320.00	3035.40	0.1547	
	National road	2422.90	1072.90	0.0239	*	2784.10	961.13	0.0038	**	2657.90	1007.00	0.0083	**
	Other road	642.26	805.12	0.4250		1062.70	691.26	0.1242		891.38	730.66	0.2225	
Neighbouring road dummies	1st class national road	-2324.10	2610.60	0.3733		-638.81	2403.70	0.7904		-1526.20	2548.40	0.5493	
	2nd class national road	-3445.50	2628.20	0.1899		-3525.00	2404.90	0.1427		-3705.30	2576.40	0.1504	
	Regional road	-4756.00	2699.00	0.0780	.	-6211.60	2452.20	0.0113	*	-5657.70	2641.60	0.0322	*
	Local road	-5317.30	2686.60	0.0478	*	-6600.00	2488.90	0.0080	**	-6192.40	2667.10	0.0202	*
River width		4.40	3.12	0.1580		5.17	2.72	0.0569	.	5.03	2.80	0.0724	.

		Spatial error model for 2nd matrix				Spatial error model for 4th matrix			
		Estimate	Std. Error	Pr(>|z|)		Estimate	Std. Error	Pr(>|z|)	
Intercept		2068.20	3415.20	0.5448		2941.90	3242.50	0.3643	
Hungarian population	County	3.03E-03	1.15E-03	0.0085	**	2.85E-03	1.09E-03	0.0091	**
	Processus	-8.33E-03	5.48E-03	0.1287		-1.10E-02	5.14E-03	0.0319	*
	City	1.49E-01	2.85E-02	1.85E-07	***	1.37E-01	2.82E-02	1.16E-06	***
Neighbouring NUTS-3 population		6.34E-03	2.82E-03	0.0243	*	6.83E-03	2.68E-03	0.0107	*
Border crossing	Opening time	-14.50	12.05	0.2289		-16.70	11.86	0.1593	
	Capacity	-0.64	0.34	0.0594	.	-0.73	0.33	0.0255	*
	Total weight capacity	4.16	14.82	0.7789		-6.93	14.08	0.6227	
	Truck weight capacity	-1170.60	763.91	0.1254		-1708.30	730.04	0.0193	*
	Bus banning	25.24	977.35	0.9794		478.98	956.51	0.6165	
Bridge		-1034.40	619.17	0.0948	.	-1147.00	614.04	0.0618	.
Country's dummies	Austria	7414.10	1376.20	7.15E-08	***	7622.50	1279.20	2.54E-09	***
	Croatia	-147.46	1259.20	0.9068		92.25	1199.80	0.9387	
	Romania	543.54	1281.20	0.6714		395.29	1226.40	0.7472	
	Serbia	1238.30	1244.60	0.3197		1501.50	1139.00	0.1874	
	Slovakia	902.14	1618.80	0.5773		789.02	1582.30	0.6180	
	Slovenia	6636.30	1462.70	5.70E-06	***	7111.80	1360.50	1.72E-07	***
European international E-road network	A-class reference	3370.60	1194.70	0.0048	**	3599.80	1113.30	0.0012	**
	A-class intermediate	-665.69	1052.00	0.5269		-817.29	1043.60	0.4335	
	B-class	2653.20	1507.40	0.0784	.	2689.30	1429.50	0.0599	.
Hungarian road dummies	Motorway	10494.00	2652.20	7.59E-05	***	10942.00	2485.40	1.07E-05	***
	Highway	4422.00	2965.60	0.1359		4614.70	2808.50	0.1004	
	National road	2683.40	990.41	0.0067	**	2789.30	963.35	0.0038	**
	Other road	965.48	713.59	0.1761		1035.50	689.46	0.1331	
Neighbouring road dummies	1st class national road	-1201.80	2492.00	0.6296		-651.23	2401.90	0.7863	
	2nd class national road	-3677.10	2510.00	0.1429		-3543.10	2399.80	0.1398	
	Regional road	-5971.50	2565.60	0.0199	*	-6244.50	2452.40	0.0109	*
	Local road	-6352.60	2601.10	0.0146	*	-6643.80	2489.80	0.0076	**
River width		5.28	2.75	0.0550	.	5.33	2.70	0.0481	*

As it can be seen the error models seem to better then the only one lag model. In the first row, the R^2 values can be seen. In the second row there are the proper autoregressive parameters, and in the third row their p-value. In the fourth and fifth row another qualifying parameters can be seen, the value of the log likelihood, and the AIC (Akaike Information Criteria) which can be calculated from the L_c value. There are more accurate information criterias, but because of the constant number of the independent variables, there is no need to use them. Among the error models the all assigned and the country assigned models seem to be the best.

In the Table 4, every negative estimates are signed by red. As the results are analysed some important facts can be seen. The bridge and the rivers effect have an opposite direction as it expected. This can be because the border can be presented as a legal separator.

At the road effects the Hungarian, and the neighbouring countries road systems provide the expected effects, but the international E-road system's numbers do not provide them, at the A-class intermediate roads. This can be because there are seven border crossings with this kind of E-road classification, and two of them are also A-class reference type too.

Analysed the effects of the neighbouring countries it can be seen, that the strongest relations are with the members of the European Union. The strongest relations are settled with Austria and Slovenia. After them comes Slovakia and Romania. Serbia's and Croatia's effects are not robust; they are changing from model to model. But usually their p-value is also too high.

Analysed the predicted and the real traffics the next statements can be settled. For one hand there are not general rules, that we can set up, like if the model predicts lesser flow then the real, then the border crossing is all right do not need to do anything. In general, two results can indicate the necessity of an upgrade. From one hand if the predicted values are higher, than the counted that can mean that its role can be more serious, but it needs an upgrade. But on the other hand if the predicted is more lesser than the real that indicates that for some reason it is overused, which can be solved by upgrade. So in every case it is need to choose which state is appropriate.

From the data, that can be seen, that in harmony with the long term infrastructure upgrade plans, the upgrades seem to necessary. This means the motorway upgrades (M15, M70), [19, 20] and the highway buildings (42, 44, 56, 8, 84, 41), but it can be seen, that new bridges should be necessary over the Danube (14, 13, 11326 and the ferry).

5. Conclusion

After the set up of the spatial econometric model the following properties can be stated. If the spatial lag model is taken into consideration, there is only one model which can be set up. This model has low autoregressive parameter value, with low p-value. The spatial error models bring better results.

To continue the research some additional fact is need to be taken into consideration. For example, in the future it is necessary to destruct the flow data for the different vehicle types. Another interesting direction is to understand better the European International E-road system. Beside this the international traffic is need to be taken into consideration in the future papers.

This paper is supported by the research project "From horse-drawn railway to intermodal transport" within Visegrad Fund.

References

1. M. Moniruzzaman, H. Maoh, W. Anderson, Transportation Research Part C: Emerging Technologies **63**, 182-194 (2016), DOI: 10.1016/j.trc.2015.12.004
2. H.F. Maoh, S.A. Khan, W.P. Anderson, Journal of Transport Geography **53**, 12-21 (2016), DOI: 10.1016/j.jtrangeo.2016.04.002
3. A. Jarašūniene, Transport **20**, 4, 165-171 (2005)
4. S. Opasanon, S. Kitthamkesorn, Transport Policy **48**, 1-12 (2016)

5. Japan International Cooperation Agency, *Oriental Consultants Global Co., LTD., and Padeco Co., LTD. Preparatory Survey on the Cross-Border Road Network Improvement Project – Final Report* 1: *Main Repor* (Dhaka, Bangladesh, Roads and Highways Department, Ministry of Road Transport and Bridges, The People's Republic of Bangladesh, 2016)
6. United Nations Economic and Social Commission for Asia and the Pacific. *The Road Networks connecting China, Kazakhstan, Mongolia, the Russian Federation, and the Korean Peninsula* (UNESCAP, 2001)
7. F. Gumz, Á. Török, Periodica Polytechnica Transportation Engineering **43**, 2, 92-97 (2015)
8. T. Sipos, Periodica Polytechnica Transportation Engineering **45**, 2, 101-105 (2017)
9. I. Trupac, Promet Traffic-Traffico **15**, 2, 105-115 (2003)
10. I. Illés, *Regionális gazdaságtan – Területfejlesztés*, 190-205 (Budapest, TYPOTEX, Hungary, 2008), (in Hungarian)
11. R. Bivand, G. Piras, Journal of Statistical Software **63**, 18, 1-36 (2015)
12. R.S. Bivand, J. Hauke, T. Kossowski, Geographical Analysis **45**, 2, 150-179 (2013)
13. Roger Bivand et. al., *Spatial Dependence: Weighting Schemes, Statistics and Models* (Bergen, Norges Handelshøyskole, Norway, 2017), Available online: https://cran.r-project.org/web/packages/spdep/spdep.pdf
14. R Core Team, *A language and environment for statistical computing* (Wien, R Foundation for Statistical Computing, Austria, 2017), Available online: https://www.r-project.org/
15. L. Anselin, *Spatial Econometrics: Methods and Models* (Dordrecht, Kluwer Academic Publishers, Netherlands, 1988)
16. A. Varga, Térökonometria, Statisztikai szemle **80**, 4, 354-370 (2002), (in Hungarian)
17. Ignacio Sarmiento-Barbieri, *An Introduction to Spatial Econometrics in R. Champaign*, (IL, University of Illinois, USA, 2016), Available online: http://www.econ.uiuc.edu/~lab/workshop/Spatial_in_R.html
18. P. Hagget, *Geography. A Global Synthesis* (Harlow, The United Kingdom, Pearson Education Limited, 2001)
19. F. Szalóki, *XV. European Transport Congress and X. Budapest International Road Congress*, 12-17 (Budapest, Hungary, 2017)
20. F. Szalóki, *XV. European Transport Congress and X. Budapest International Road Congress*, 18-24 (Budapest, Hungary, 2017)

Application of DEA Method to Evaluate Financial Health of Selected Transport Companies

Martin Telecký[1,*]

[1]University of South Bohemia in České Budějovice, Faculty of Economics, Department of Accounting and Finance, 37005 České Budějovice, Czech Republic

Abstract: The paper strives to apply the selected bankruptcy and creditworthy models on the transport companies operating under the conditions common in the Czech Republic. The paper aims to assess credibility of individual classification models which will be consequently verified by the DEA method. The task is to support or refute the hypothesis within the identification of the most effective unit, i.e. a transport company. The result is to determine the ranking of transport companies based on the results of the application of bankruptcy and creditworthy models and the application of the DEA method. Consequently, the hypothesis will be supported or refuted.

1 Introduction

The aim of the financial management is to obtain a complete overview of the financial situation of the company and the financial aspects. It is necessary to know more information to determine the basic objectives of the company in the future. To do so, it is necessary to identify the customers, suppliers, competitors, a quality of provided services, etc. Assessment of the company´s financial health is part of efficient decision-making. It is necessary to find out the status of financial doubts and which measures can be applied. The efficient management of the company requires the optimum level of the assets, debts, forms of financing the operating activity using the company´s own or external capital, the amounts of costs and revenue and the amounts of income and expenditure.

If the company is able to earn sufficient money and to comply with its obligation, it will not become insolvent and there will be no risk of bankruptcy. It means that the financially healthy company can report sufficient return and find the optimal sources of financial risk coverage. The financial risk is associated with the use of various forms of capital and the risk of financial insolvency.

The financial health of the company cannot be directly calculated but it can be determined upon the diagnostic methods using the financial analysis which is utilized by the general public and various economic entities with the aim to obtain information on the management and economic situation of the given company. Every economic entity focuses

[*] Corresponding author: telecm00@jcu.cz

on a different type of information. The company management is interested in the information from all points of view, namely, whether its company

- is liquid,
- is solvent,
- generates a profit,
- is taking on a risk in case of an investment and how big this risk is,
- background and assumptions for further development.

The financial analysis utilizes the diagnostic methods which evaluate the company's management from the viewpoint of the past, the presence and the expected future. The aim is to reveal the weak points which could entail serious problems in the future and to start to build-up the financial success and company advantages on the strong points [1-2].

2 Methods

Classification models distinguish between two main groups of models, namely:

- bankruptcy models,
- creditworthy models.

Bankruptcy models, also known as prediction models, were compiled and investigated using the detail analysis of individual financial indicators which were applied to the various types of samples of examined companies (accounting entities). The main goal of bankruptcy models is to warn against a possible bankruptcy of the company "in time". The main factors are such selected indicators which imply a threat to the financial health of the company.

Bankruptcy models differ from creditworthy ones. They were and are created upon the real data. **Creditworthy models**, on the other hand, take into consideration the theoretical knowledge and on some models they answer the question whether the company creates a certain value for the owner and the investors (financial performance of the company). The goal is to classify the companies by the degree of their financial health, from the very strong to very weak undertakings. Most creditworthy models are evaluated upon the amount of points of the relevant ratio indicators. The achieved score should illustrate a certain degree of financial health [3-4].

2.1 Bankruptcy Model Index IN 95

This bankruptcy model was created by the Neumaiers spouses who received the Nobel Prize. They compiled the so-called credibility index IN. The application was carried out on the basis of the Czech accounting reports under the conditions of the Czech economy. They, thus, considered both the Czech accounting reports and they relied on the economic situation of the country.

The index IN 95 for the "transport" industry has the following form:

$$\begin{gathered} 0.22 * \text{assets} / \text{external resources} \\ 0.11 * \text{earnings before interest and taxes (EBIT)} / \text{interest payable} \\ 8.33 * \text{earnings before interest and taxes (EBIT)} / \text{interest payable} \\ 0.52 * \text{revenues} / \text{assets} \\ 0.10 * \text{current assets} / (\text{short-term liabilities} + \text{short-term bank loans})) \\ -16.8 * \text{overdue liabilities} / \text{revenues} \end{gathered} \tag{1}$$

The IN 95 index is a specific bankruptcy model designed for the Czech conditions. It means that it does not have the determined market value of the company in any single indicator. The model is designed for those companies which do not trade at the exchange. It is generally well known that there is a high level of insolvency in the Czech Republic. Mr and Mrs Neumaier considered this factor in the "overdue liabilities / revenues" indicator. It represents a certain inability of companies to pay off their debts in time (the index value is reduced) [5-6].

IN 95 > 2
An ability to comply with the financial obligations without problems
1 < IN 95 < 2
Grey area
IN 95 < 1
Insufficient ability to comply with the financial obligations

2.2 Creditworthy Model IN 99

The IN 99 index ranks among the creditworthy models that are specifically designed from the perspective of an owner. The discrimination analysis was used both on the IN 95 index and the IN 99 index. Weights of individual indicators differ from the IN 95 index. The reason is enforcement of achieving a positive value of the economic profit.

*-0.017 * assets / external capital*
*4.573 * earnings before interest and taxes (EBIT) / total assets*
*0.481 * sales / total assets*
*0.015 * current assets / short-term liabilities*
(2)

The following information determines the final qualification of the company [5-8]:

IN 99 >2.07
The company creates a new value for the owner
1.42 < = IN 99 <2.07
It is supposed that the company still creates a value for the owner
1.089 < = IN 99 <1.42
It is not sure whether the value for the owner is created
0.684 < = IN 99 <1.089
The company rather does not create the value for the owner
IN 99 <0.684
The company does not create a value for the owner

2.3 Index IN 01

This model was created upon the discrimination analysis. The IN 01 index is a specific model which takes into account the previous two models, i.e. IN 95 and IN 99. It is characterized as a bankruptcy and creditworthy model. It consists of five main indicators. This model does not contain the market value of the capital. It means that the model is especially designed for the companies not trading at the stock exchange.

*0.13 * assets / external capital*
*0.04 * earnings before interest and taxes (EBIT) / interest payable*
*3.92 * earnings before interest and taxes (EBIT) / total assets*
*0.21 * sales / total assets*
*0.09 * current assets / short-term liabilities*

(3)

The following information determines the final qualification of the company [5-8]:

IN 01 >1.77
The company creates a value.
0.75 <= IN 01 <1.77
The company does not create a value, yet it is considered creditworthy.
IN 01 <0.75
The company goes into bankruptcy.

2.4 Index IN 05

The IN 01 was helpful in creating the recent model IN 05. The goal of this model is not only to determine whether the company is near the bankruptcy, but to determine also a value for the owners. The prerequisite of this model is that it should estimate the bankruptcy of the company with probability higher than 72 % [7-9].

*0.13 * assets / external capital*
*0.04 * earnings before interest and taxes (EBIT) / interest payable*
*3.97 * earnings before interest and taxes (EBIT) / total assets*
*0.21 * sales / total assets*
*0.09 * current assets / short-term liabilities*

(4)

IN 05 >1.6
A company creates a value.
0.9 < IN 05 < 1.6
A grey area of neutral results
IN 05 <0.9
A company destroys a value

We must not forget the fact that the indicator representing the interest coverage has very distorting effects in this model. These effects can result in the wrong rating of the company's financial health. Based on the authors' recommendation, it is apt to choose the maximum value 9. The reason is that the interest payable tends to be close to 0 and the interest coverage indicator can achieve extreme values also provided that the weight is only on the level of 0.04.

2.5 DEA Method

Thanks to its characteristics and functions, the DEA (Data Envelopment Analysis) method can analyse the economic, accounting and logistic issues in common practice. The result of the method is the information on the efficiency of individual activities (products) in the selected group in absolute terms and as a percentage. Unlike other methods, the DEA method is considered a new non-parametric method which evaluates the efficiency, performance or productivity of homogeneous production units. The main advantage is that the method allows us to evaluate efficiency of individual units (transport companies) individually within the chosen set of units. It shows which units are efficient and which are not efficient in performing their activity.

The DMU (Decision Making Unit) refers to an organizational unit which uses for its activity a certain number of inputs and outputs and represents the resulting values. The rule of selecting the lower number of inputs and the higher number of outputs applies. The unit´s efficiency is calculated upon the following formula:

$$\textbf{\textit{Efficiency}} = output/input \tag{5}$$

This formula represents the simplest applicable case.

The CCR model maximizes the rate of effectiveness of the evaluated unit as a quotient of weighted outputs and weighted inputs on the condition that the rates of effectiveness of other units are less than or equal to one. The input-oriented CCR model focuses on such an amount of inputs which are consequently evaluated by this model. The model recommends such changes so that the ineffective unit becomes the effective one. At the same time, the model foresees the constant revenue of scale; i.e. the change in the amount of inputs will be directly proportional to the change in the output amount. For each unit this model will set individual weights of inputs and outputs to maximize the technical efficiency coefficient. Certain conditions, however, must be met, namely:

- Weights cannot be negative
- When using this set of weights for all entities, no coefficient of the technical efficiency must be greater than one.

The input-oriented BCC model is the modification of the previous model CCR. This model already takes into consideration the variable revenue of scale, i.e. increasing, decreasing or constant revenue. It foresees the variable, in certain parts linear revenue of scale and can evaluate the efficiency of entities for the generally non-constant output of scale [10-12].

A basic sample of transport companies which provide the traffic services using the public line transport in the selected regions of the Czech Republic was chosen. The bankruptcy and creditworthy models created by the Neumaiers for the companies doing business in the Czech Republic were applied on the data for the year of 2016.

The input data for the DEA method application was as follows:

Table 1. Input data for the DEA method analysis in the Czech crowns. Source: author

	Assets	External capital	Short-term liabilities	Interest payable	Sales	EBIT
CSAD A	293891	114496	62849	980	334065	9435
CSAD B	363539	146533	44216	378	226158	13222
CSAD C	51529	49070	44138	0	927	379
CSAD D	317752	269646	146167	2990	257011	8984
CSAD E	192638	88319	34212	984	350643	15182
CSAD F	414345	198180	117971	2185	475161	31452
CSAD G	69812	64972	29427	1047	115781	1726
CSAD H	376913	62445	36889	257	591624	72088

Due to the sensitivity of economic data processing, the transport companies are marked with the letters. The author knows all economic and financial structures of individual transport companies.

3 Results

Table 2. Indices IN 95, IN 99, IN 01 and IN 05 based on the calculation using EBIT. Source: author

	IN 95 EBIT	IN 95 Overall ranking as per EBIT	IN 99 EBIT	IN 99 Overall ranking as per EBIT	IN 01 EBIT	IN 01 Overall ranking as per EBIT	IN 05 EBIT	IN 05 Overall ranking as per EBIT
CSAD A	2.58	5.	0.69	5.	1.1	5.	1.09	4.
CSAD B	5.28	2.	0.5	7.	2.23	2.	1.2	3.
CSAD C	0.31	8.	0.04	8.	0.27	8.	0.27	7.
CSAD D	1.44	7.	0.51	6.	0.57	7.	0.57	6.
CSAD E	4.54	3.	1.26	2.	1.76	3.	1.51	2.
CSAD F	3.71	4.	0.89	4.	1.41	4.	1.2	3.
CSAD G	1.88	6.	0.9	3.	0.67	6.	0.67	5.
CSAD H	35.49	1.	1.67	1.	13.35	1.	2.5	1.

Table 3. Results found upon the DEA method analysis when the BCC model is applied. Source: author

Unit name	Score	Efficient
CSAD A	66.10%	False
CSAD H	100.00%	True
CSAD C	100.00%	True
CSAD B	39.10%	False
CSAD G	91.10%	False
CSAD D	44.40%	False
CSAD E	100.00%	True
CSAD F	65.90%	False

The Table 2, which analyses the financial health of transport companies, lists the final values of individual bankruptcy and creditworthy models, including the final ranking from the best transport company to the worst one. The transport company CSAD H is a winner. It means that it has the best results of all transport companies under study. CSAD C, on the other hand, demonstrates the worst financial health in all classification models [13, 14].

The situation is, however, different after the DEA method is applied. The Table 3 shows which transport companies are rated as efficient and non-efficient. CSAD H can be considered an efficient unit. This statement is identical also as to the results of bankruptcy and creditworthy models which set the financial health as the best. The transport companies CSAD E and CSAD C are considered to be an efficient unit. According to the Table 2, CSAD E came second and third. CSAD C is almost last on the list. This implies that the bankruptcy and creditworthy models are not absolutely credible and fail to have a 100% explanatory power. Also other factors which affect the financial health of the company matter. An example is also the structure of the classification model itself which can more efficiently rate the financial health of the company [15]. The success of the models IN 95, IN 99, IN 01 and IN 05 is estimated with credibility of 57%, 81%, 52% and 62%, respectively.

4 Conclusions

In the past, the models evaluating the financial health of the company, basically, consisted of the indicators which characterized the properties and actions of the company in the financial sphere. As the time goes by, the technology develops and the business processes improve, the new models are updated (innovated) or compiled to better evaluate the financial health of companies. They consider the factors which were not reflected in the past or were not so important to be addressed by the users. They were not substantial in the past. At present, it is necessary to create such classification models which would evaluate the financial health of the company in the specific fields of business at the given territory and not to apply the versatile bankruptcy or creditworthy models only.

The paper is drawn-up within GAJU 075/2017/S.

This paper is supported by the research project "From horse-drawn railway to intermodal transport" within Visegrad Fund.

References

[1] J. Fibírová, L. Šoljaková, *Value Tools of Company Management and Performance Measurement,* 1st issue (Prague: ASPI, a.s., Czech Republic, 2005)

[2] E. Kislingerová, J. Hnilica, *Financial Analysis: Step by Step,* 1st issue, 137 p. (Prague: C. H. Beck, Czech Republic, 2005)

[3] R. Grünwald, J. Holečková, *Financial Analysis and Planning of the Company* (2002)

[4] E. Kislingerová et al, *Managerial Finance*, 3rd issue, 811 p. (Prague: C. H. Beck, Czech Republic, 2010)

[5] R. Grünwald, J. Holečková, *Financial Analysis and Planning of the Company,* 318 p., (Prague: Ekopress, Czech Republic, 2009)

[6] I. Neumaierová, I. Neumaier, *Company Performance and Market Value*, 1st issue, 216 p. (Prague: Grada Publishing, Czech Republic, 2002)

[7] Finanalysis, *Used Bankruptcy and Creditworthy Models*, Available online: http://www.finanalysis.cz/pouzite-bankrotni-modely.html (2014)

[8] M. Vochozka, *Methods of Comprehensive Evaluation of the Undertaking*, 248 p. (Prague: Grada Publishing, Czech Republic, 2011)

[9] M. Zikmund, *IN05 – Financial Analysis: Bankruptcy Index from the Czech Republic Which Works When Applied to the Czech Companies*, Available online: http://www.businessvize.cz/financni-analyza/in05-bankrotni-index-z-ceska-ktery-funguje-na-ceske-firmy (2011)

[10] P. Fiala, *Production System Modelling and Analysis*, 1st issue, 260 p. (Prague: Professional Publishing, Czech Republic, 2002)

[11] J. Jablonský, M. Dlouhý, *Efficiency Evaluation Models of Production Units*, 1st issue, 184 p. (Prague: Professional Publishing, Czech Republic, 2004)

[12] R. Hindls et al., *Statistics for the Economists*, 7th issue, 420 p. (Prague: Professional Publishing, Czech Republic, 2006)

[13] M. Poliak, Š. Semanová, P. Varian, Logi – Scientific Journal on Transport and Logistics **5**, 1, 51-60 (2014)

[14] R. Kampf, O. Stopka, I. Kubasakova, V. Zitricky, *WMCAUS 2016 – World Multidisciplinary Civil Engineering-Architecture-Urban Planning Symposium,* 1538-1544 (Procedia Engineering **161**, Prague, Czech Republic, 2016), DOI: 10.1016/j.proeng.2016.08.623

[15] H. Bínová, E. Březina, Logi – Scientific Journal on Transport and Logistics **6**, 1, 13-25 (2015)

Assessment of Citizens' and Employees' Mobility by Public Transport

Michal Turek[1,*], and *Kamil* Peterek[2]

[1]College of Logistics, Přerov, Palackého 1381/25, 750 02 Přerov, Czech Republic
[2]University of Žilina, Faculty of Safety Engineering, 1. Mája 32, 010 26 Žilina, Slovakia, external doctoral student

Abstract. The contribution analyzes the level of service by bus and rail transport in the Přerov area. Exploration is carried out through key indicators such as transport speed and number of connections. Based on these indicators, the public transport service is assessed between the statutory city of Přerov and the towns of Hranice, Kojetín, Lipník nad Bečvou, Potštát and Tovačov. Thereafter, there are proposed measures to increase the mobility of citizens and employees by public transport.

1 Introduction

Mobility is one of the basic needs of the population. The increasing need for relocation of citizens brings an increase in individual car traffic, to which must be offered an adequate alternative to public transport. To ensure successful public service provision, it must be adapted to the individual requirements of citizens who need to be transported to schools and school facilities, to public authorities, to employment, to healthcare facilities, and to satisfy cultural, recreational and social needs. The most significant group of passengers, however, represents the number of people transported outside of the morning transport peak 5:00 - 8:00 and the afternoon transport peak from 13:00 - 17:00 on working days, mainly those who are working in three-shift operation. Therefore, it is necessary to deal with public transport both throughout the day but also before and at the end of each shift [1, 2].

2 Solution process

In order to assess the service of the territory by public transport it is necessary to look at it from a passenger's point of view, which requires time availability of the area by the given speed of connection in a given period of time and sufficient number of connections to be carried out during the given period of time [3,4].

In terms of quantity, the problem of the service of the territory provided by public transport can be expressed by the speed of the connection, which is expressed by means of distance and travel time, the number of connections per hour, which is expressed

* Corresponding author: michal.turek@vslg.cz

by the number of connections during the reference period and the indicator of the public transport operator including the speeds of the connections and the number of connections. The procedure used was proposed by Pova [5, 6].

Indicator of the speed of connection:

$$v_{ij} = \frac{d_{ij}}{t_{ij}} \cdot 60 \text{ [km/h]} \tag{1}$$

where:
d_{ij} - the length of junction between the i-th place and j-th place [km],
t_{ij} - the time of travel between the i-th place and j-th place [h].

Indicator of the number of connections per hour:

$$n_{ij} = \frac{N_{ij}}{T_{ij}} \text{ [h}^{-1}\text{]} \tag{2}$$

where:
N_{ij} - the number of connections between i-th place and j-th place in the monitored time period [-],
T_{ij} - the time period in which the number of connections between the i-th place and j-th place is monitored [h].

Indicator of the service area:

$$o_{ij} = \frac{d_{ij} \cdot 60 \cdot N_{ij}}{t_{ij} \cdot T_{ij}} \text{ [km/h}^2\text{]} \tag{3}$$

where:
d_{ij} - the length of junction between the i-th place and j-th place [km],
N_{ij} - the number of connections between i-th place and j-th place in the monitored time period [-],
t_{ij} - the time of travel between the i-th place and j-th place [h],
T_{ij} - the time period in which the number of connections between the i-th place and j-th place is monitored [h].

3 Assessment of the mobility of citizens and employees

The application part will be focused on public transport service in the Přerov district, which is a major economic and industrial area. This is the assessment of the mobility of citizens between Hranice, Kojetín, Lipník nad Bečvou, Potštát and Tovačov, which consists of towns in the Přerov district and the statutory town of Přerov. The assessment of citizens' mobility will then be extended to assess the mobility of employees in the three-shift operation. Therefore, increased attention needs to be paid to this issue.

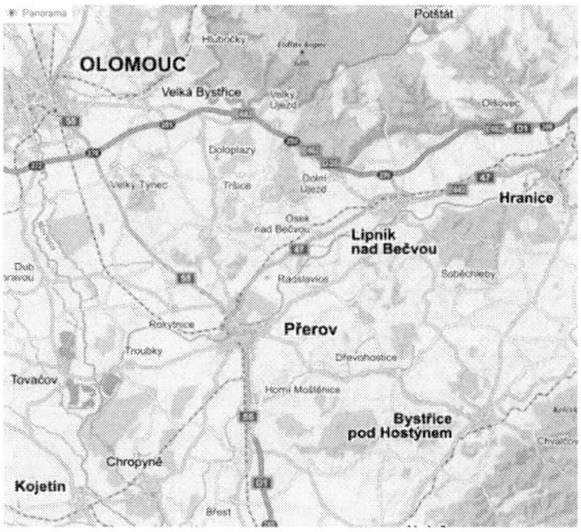

Fig. 1. Map with cities in the district Přerov. Source: [7]

3.1 Assessment of the mobility of citizens

The mobility of citizens in the Přerov region was examined on a working day during the period 4:00 - 24:00, when the active life of the inhabitants takes place. The input data for the mobility of citizens in the period 4:00-24:00 are listed in Table 1.

Table 1. The input data for the assessment of the mobility of citizens. Source: [8]

Session	Bus transportation			Rail transportation		
	Length of junction [km]	Time of travel [min]	Number of connections [-]	Length of junction [km]	Time of travel [min]	Number of connections [-]
Hranice - Přerov	31	55	6/6	29	20	41/42
Kojetín - Přerov	28	33	6/6	17	18	31/31
Lipník n. B. - Přerov	17	30	15/16	15	12	18/17
Potštát - Přerov	-	-	0/0	-	-	0/0
Tovačov - Přerov	13	21	22/23	-	-	0/0

On the basis of information on connections in the assessed sessions, the research was carried out using indicators of the speed of connections, the number of connections, the operation of the territory, which are clearly presented for bus and train transport in Table 2.

Table 2. Indicators for the assessment of the citizens' mobility. Source: authors

Session	Bus transportation			Rail transportation		
	Speed of connection [km/h]	Number of connections [h^{-1}]	Operation area [km/h^2]	Speed of connection [km/h]	Number of connections [h^{-1}]	Operation area [km/h^2]
Hranice - Přerov	34	0,3/0,3	10/10	87	2/1	174/87
Kojetín - Přerov	51	0,3/0,3	15/15	57	1,6/1,6	91/91
Lipník n. B. - Přerov	34	0,7/0,8	24/27	75	0,9/0,8	68/60
Potštát – Přerov	0	0/0	0/0	0	0/0	0/0
Tovačov - Přerov	37	1,1/1,2	41/44	0	0/0	0/0

Table 2 shows that, with the exception of transport in the Potštát - Přerov session, there is a good offer of train transport connections, supplemented by bus transport in the Hranice - Přerov, Kojetín - Přerov, Lipník n. B. - Přerov routes, which confirm the higher values of the territory and a bus service in the Tovačov - Přerov session.

3.2 Assessment of the mobility of employees

The mobility of employees in the Přerov region was gradually examined on a working day in the hours of 4:00 - 6:00 and 14:00 - 16:00, 12:00 - 14:00 and 22:00 - 24:00, 20:00 - 22:00 and 6:00 - 8:00 in which the transfer of employees to the three-shift operation takes place. The input data for examining the mobility of employees during the periods mentioned are shown in Table 3.

Table 3. The input data for the assessment of the mobility of employees. Source: [8]

Session	Bus transportation			Rail transportation		
	Length of junction [km]	Time of travel [min]	Number of connections [-]	Length of junction [km]	Time of travel [min]	Number of connections [-]
Hranice - Přerov	31	55	1/1	29	20	4/4
Kojetín - Přerov	28	33	1/1	17	18	1/3
Lipník n. B. - Přerov	17	30	2/3	15	12	2/2
Potštát - Přerov	-	-	0/0	-	-	0/0
Tovačov - Přerov	13	21	2/4	-	-	0/0
Hranice - Přerov	31	55	1/0	29	20	3/2
Kojetín - Přerov	28	33	1/1	17	18	3/1
Lipník n. B. - Přerov	17	30	2/2	15	12	1/1
Potštát - Přerov	-	-	0/0	-	-	0/0
Tovačov - Přerov	13	21	4/1	-	-	0/0
Hranice - Přerov	31	55	0/2	29	20	3/4
Kojetín - Přerov	28	33	1/1	17	18	2/3
Lipník n. B. - Přerov	17	30	2/2	15	12	1/3
Potštát - Přerov	-	-	0/0	-	-	0/0
Tovačov - Přerov	13	21	2/2	-	-	0/0

On the basis of information on connections in assessed sessions, the survey was carried out using indicators of the speed of connections, the number of connections and the operation of the area, which are clearly presented for bus and train transport in Table 4.

Table 4. Indicators for the assessment of the employees' mobility. Source: authors

Session	Bus transportation			Rail transportation		
	Speed of connection [km/h]	Number of connections [h^{-1}]	Operation area [km/h^2]	Speed of connection [km/h]	Number of connections [h^{-1}]	Operation area [km/h^2]
Hranice - Přerov	34	0.5/0.5	17/17	87	2/2	174/174
Kojetín - Přerov	51	0.5/0.5	26/26	57	0.5/1.5	29/85
Lipník n. B. - Přerov	34	1/1.5	34/51	75	1/1	75/75
Potštát - Přerov	0	0/0	0/0	0	0/0	0/0
Tovačov - Přerov	37	1/2	37/74	0	0/0	0/0
Hranice - Přerov	34	0.5/0	17/0	87	1.5/1	131/87
Kojetín - Přerov	51	0.5/0.5	26/26	57	1.5/0.5	86/29
Lipník n. B. - Přerov	34	1/1	34/34	75	0.5/0.5	38/38
Potštát - Přerov	0	0/0	0/0	0	0/0	0/0
Tovačov - Přerov	37	2/0.5	74/19	0	0/0	0/0
Hranice - Přerov	34	0/1	0/34	87	1.5/2	131/174
Kojetín - Přerov	51	0.5/0.5	26/26	57	1/1.5	57/85
Lipník n. B. - Přerov	34	1/1	34/34	75	0.5/1.5	38/113
Potštát - Přerov	0	0/0	0/0	0	0/0	0/0
Tovačov - Přerov	37	1/1	37/37	0	0/0	0/0

Table 4 shows that except the Hranice - Přerov session, there is a decrease in the quality of the connection offer in some periods. These are mainly train connections in the Kojetín - Přerov session for the morning shift, the afternoon shift and the night shift see the low values of the operator indicator of areas 29, 29 and 57 and in the Lipník n. B. - Přerov session for afternoon shift and afternoon shift, see the low values of the 38/38 operator indicator and the bus transport connections in the Tovačov - Přerov session for the morning shift, afternoon shift, night shift and night shift, see low values of the 37, 19, 37/37 traffic indicator.

4 Evaluation

All cities, except the city of Potštát, located on the northern outskirts of Přerov, have a public transport connection with the statutory city of Přerov (Table 5 and 6).

Public transport, which provides mobility for citizens and employees, shows a higher level of service to the area by train transport and a lower level of service to the area by bus transport, due to more favorable conditions on railway lines in the case of train transport and less suitable conditions on roads in the case of bus transport (Table 5 and 6).

An analysis based on the comparison of citizens' mobility throughout the day and the mobility of employees at partial intervals revealed that there were significant differences between the values of quality indicators of public transport (Table 5 and Table 6).

Table 5. Evaluation of bus transport in Přerov district. Source: authors

Session	Operation area 4:00-24:00 [km/h^2]	Operation area 04:00-06:00/14:00-16:00, 12:00-14:00/22:00-24:00, 20:00-22:00/06:00-08:00 [km/h^2]		
Hranice - Přerov	10/10	17/17	17/0	0/34
Kojetín - Přerov	15/15	26/26	26/26	26/26
Lipník n. B. - Přerov	24/27	34/51	34/34	34/34
Potštát - Přerov	0/0	0/0	0/0	0/0
Tovačov - Přerov	41/44	37/74	74/19	37/37

Table 6. Evaluation of the rail transport in Přerov district. Source: authors

Session	Operation area 4:00-24:00 [km/h^2]	Operation area 04:00-06:00/14:00-16:00, 12:00-14:00/22:00-24:00, 20:00-22:00/06:00-08:00 [km/h^2]		
Hranice - Přerov	174/87	174/174	131/87	131/174
Kojetín - Přerov	91/91	29/85	86/29	57/85
Lipník n. B. - Přerov	68/60	75/75	38/38	38/113
Potštát - Přerov	0/0	0/0	0/0	0/0
Tovačov - Přerov	0/0	0/0	0/0	0/0

5 Conclusion

During providing of the public transport from cities to statutory cities, it is necessary to take into account their location within the district, as the cities that are located in the peripheral parts of the district do not have to move the traffic streams to the statutory city but may go beyond the neighboring district.

To ensure high-quality mobility for citizens and public transport employees, it is necessary to deal with train and bus services separately, including the possibility of combining them, and by improving the supply and speed of bus transport connections.

In addition to ensuring high-quality mobility for both citizens and bus and train staff, qualitative indicators must be ensured both throughout the day and at partial intervals in which workers are transported in three-shift traffic.

This paper is supported by the research project "From horse-drawn railway to intermodal transport" within Visegrad Fund.

References

1. Z. Lokošová, Perner's Contacts **2**, 7, 64-69 (2007)
2. Act No. 194/2010 Coll., *On public transport services* (2010)
3. P. Drdla, *Personal transport of regional and supra-regional importance,* 411 p. (The University of Pardubice, Pardubice, Czech Republic, 2014)
4. L. Zelený, *Passenger Transport* (ASPI, Prague, Czech Republic, 2007)
5. B. Hejhalová, Transport and communications **6**, 1, 84-98 (2010).
6. P. Pova, Perner´s Contacts **4**, 164-174 (2009)
7. *Mapy.cz*, Available online: https://mapy.cz/zakladni?x=17.3322337&y=49.5327162&z=11&l=0 (2017)
8. IDOS, *Vlaky + Autobusy + MHD (všechna) - Vyhledání spojení*, Available online: https:// http://jizdnirady.idnes.cz/vlakyautobusymhdvse/spojeni/(2017)

Online Reputation in Automotive

Josef Vodák[1], *Milan* Kubina[1], *Jakub* Soviar[1,*], *Diana* Zraková[1], and *Patrik* Ferenc[1]

[1]University of Žilina, Faculty of Management Science and Informatics, Univerzitná 8215/1, 010 26 Žilina, Slovakia

Abstract. This paper deals with the issue of online reputation, namely the social networking profile of businesses. Selected companies in the automotive industry through social profiles communicate with their customers, the public and they trying to improve their name and the name of their products in the public eye. Online reputation analysis was carried out to determine the current situation on the territory of Slovakia. On the basis of the data found, measures were proposed to improve the current state and reputation of automotive companies. Recommendations suggested by the findings can be used on any market to improve the current state and increase the competitiveness of automotive companies.

1 Introduction

The current turbulent market development causes a number of business challenges and opens up the question: how to maintain a positive corporate reputation for the long term or the reputation of its brands and products? At present, companies are still underestimating the power of reputation and reputation management. Because of understatement the power of reputation a number of companies lost what they have been building for years - a good name and an image. In order to, the companies lost everything in a short time, just only one bad step is enough. Customers and the public perceive every bad step - disrespect or deceptiveness or other information after a disappointment are very fast spread further, which contributes to the destruction of the good name and the positive opinions of the general public in very short time. For company si very easy loss of reputation, but repair the damage of reputation takes much longer and it is represent a conscious process - a reputation management process. Reputation management is the process of managing how companies, brands, or people are perceived around the business. The business environment is represented by customers, business partners, traders, competitors, journalists and the media, potential employees, as well as co-workers, investors, various personal contacts and other stakeholders. The goal of most of them is to protect the reputation of the company in the offline world, but also on the Internet by trying to prevent problems that revolve around the company's reputation. Reputation, online reputation, and trusted branding should be a priority for all companies.

* Corresponding author: jakub.soviar@fri.uniza.sk

2 Reputation and online reputation

Reputation can be defined as a summary of the attitudes, expectations and judgments of entities that are relevant to the company, which implies that the views, attitudes, expectations and judgments of all stakeholders have a direct or mediated influence on the enterprise and its activities [1]. Gattiker argues that reputation represents a collective expression of the enterprise's perception by several stakeholders on the basis of the efforts made by the company [2]. Fombrun's reputation attributes to businesses the ability to bring value to the business in addition to multiple stakeholders. According to the author, the reputation represents a set of internal relationships with employees and external relations with all stakeholders, partners and competitors [3]. Barnett and Pollock define the reputation as follows: Reputation is a prediction of future behaviors, which is based on an assessment of how past behaviors met the expectations following the comparison of social status [4].

Reputation is related to company strategy and must be in line with marketing strategy [3, 5]. Reputation is a value for stakeholders that represents product quality assurance. It is also important in terms of marketability and competitiveness. Creating reputation consists in building the knowledge, orientation and confidence of the business in certain parts towards the public in order to maintain customer value. Reputation is a very important. It decides on the future behavior of stakeholders towards the business. Positive reputation enhances performance, helps to withstand risks, and helps buyers to help with buying decisions [6-9]. In order to promote positive reputation, it is necessary to keep in touch with our customers. Communication reflects corporate culture. If companies want to maintain a positive reputation and get loyal to customers, they must constantly communicate with them, in short, to know their entire audience [10]. For this reason, its necessary to do an audience survey, for example on social networks. Authors say, that individuals are focusing around the companies or brand for the obvious reason online. Customers concentrate on social networks around automobile compartments for the purpose of collecting and exchanging information, interacting with each other [11, 12]. Social networks enable organizations to communicate regularly with clients and the public about crisis events, among other things in crisis situations. The public has the opportunity to respond to emerging crisis situations through social networks and can remove the uncertainty they experience. If public reactions are negative, company should respond appropriately and respond truthfully, which is beneficial for its reputation [13].

Online reputation represents an image of how the public and stakeholders are perceived by an enterprise when searching in the online environment. Online reputation can include news articles, blog posts, social profiles, and all the information that can be found and retrieved by search engines. Vartiak also states that there are three steps in building the right online reputation and are as follows [14]:
- monitor and search for what is said online about the business and its products,
- analyze how the visible information affects the reputation of the company, the brand or its products,
- influence outcomes and eliminate negative feedback by participating in communication.

The overall reputation of the business depends on how the companies perceives individual stakeholders in both online and offline environments. The better the stakeholders are perceived by society, the more positive the company has in its overall reputation.

3 Survey

The survey in this article focused on the analysis of online reputation in automotive in Slovakia. The purpose of the survey was to compare the profiles on the selected social network in which the individual carmakers build their reputation. This survey should have as certain the current state of reputation of selected automotive companies among Slovak residents.

The Facebook social network has been selected, as it is one of the most popular in Slovakia. In analyzing company profiles on the social network, the following weighting factors were selected: number of fans / followers with assigned weight 4, the number of posts for the last month with assigned weight 3, average number of posts per week with assigned weight 2, management response time with weights of 4 and interaction with users (the number of responses to comments) weighing 3. The weights of the individual factors were on the scale from 1 to 5 (5 being the most important factor and 1 the least important factor). Table 1 lists the values which are then multiplied by the weights and the resulting values given in Table 2 are calculated.

However, the time for the response to the management was to be recalculated to maintain the conversion methodology as for the other factors. Message response time was Škoda one day, for Volkswagen 2 hours, Kii 3 minutes and Peugeot and Land Roveri one day. The resulting recalculation of response time is based on the assignment of penalty points. For every minute of the management response delay, the -1 value was deducted.

Table 1. Profile Analysis on Facebook Social Network. Source: authors

Company	Number of fans	Number of posts	Average number of posts per week	Management reaction time	User interaction
Škoda	59 802	23	4	-1 440	high (3 replies)
Volkswagen	19 365	24	4	-120	high (3 replies)
Kia	60 019	30	6	-3	high (3 replies)
Peugeot	10 825 350	27	5	-1 440	middle (2 replies)
Land Rover	14 016	20	3	-1 440	middle (2 replies)

In the above table, information on selected automotive companies is identified by analysis. For each company, results are shown for each factor. It can be seen that Peugeot has the most fans on its official facebook profile. This factor is assigned the highest weight 4. With this number of fans, the question arises as to whether it really is due to a great company activity or that fans are split into a number of officially verified facebook profiles of companies with global reach, while Peugeot does not have such official worldwide sites, respectively. are not easily viewable. Land Rover has the least fan, which may be due to the fact that this company is not a long-time producer on the Slovak market, so it is not even in the awareness of most Slovaks. However, this fact may change in the future.

For the second factor with a fairly high weight, it can be seen that the number of contributions is in all cases greater than 20. In all cases, except for Kia and Peugeot, the contributions to the social network were roughly the same. In terms of the number of posts per week, Kia is again winning, much like the reaction time factor. Kia, unlike other companies, uses a social network for its reputation. Kia responds to a comment or social networking from the public in minutes. Volkswagen is also relatively good in this area. It can give feedback within two hours. Other companies are neglecting feedback as they respond to contributions post-dated. The last factor of interacting with users is not very large. The number of responses to comments is roughly the same for all selected companies.

Table 2. Overall rating of individual companies. Source: authors

Company	Number of fans	Number of posts	Average number of posts per week	Management reaction time	Interaction	Sum	Company
Škoda	239 208	69	8	-5 760	9	233 534	Škoda
VW	77 460	72	16	-480	9	77 077	VW
Kia	240 076	90	12	-12	9	240 175	Kia
Peugeot	43 301 400	81	10	-5 760	6	43 295 737	Peugeot
Land Rover	56 064	60	6	-5 760	6	50 376	Land Rover

From the results shown in Table 2, a sovereign winner can be seen in the care of his reputation on the Facebook social network. The winner in this area is Peugeot, who leads the table with a big lead. If we do not think about this company, since it seems to have accumulated all customers on one social network, the best company would be Kia. This company excels in several factors. It has high interaction and the lowest response time for messages, it has the largest number of posts and fans. No serious deficiencies were found in analyzing facebook profiles of all carmakers. Several of the selected car manufacturers were mistaken for mileage information and also provided various information about the foundations. Škoda and Peugeot show the year of founding the whole company, not just the Slovak branch. Land Rover does not mention this information at all. Kia reports the creation of a facebook profile. More important information missing from most of the contact was at least the email address. If a contact customer wants a company and does not want to communicate through the social network only, they must rejoin the company's official website to find contact information. All of the analyzed profiles were almost completed, they were active and verified.

Based on the analysis of profiles on the social network, the following ranking was drawn: Peugeot first, second Kia, third Škoda, fourth Volkswagen and the last Land Rover.

From the analysis of individual companies and their efforts to maintain constant communication with customers, each of the rated companies uses the phenomenon of social networks to their advantage and their profiles are trying to regularly interact with the fans.

It is also possible to argue that despite the fact some of the carmakers have done some repairs, their reputation did not have a great impact on the territory of the Slovak Republic. Despite the scandal that has been published, customers continue to buy these brands and continue to support them and communicate with their facebook posts. As an example, it is worth mentioning Kia's fiasco in the US and Canada. In 2012, Kia mistakenly reported lower car consumption than it actually did. Customers perceived this situation as a big deception and misappropriation. The reputation of the company suffered a loss and also suffered as a drop in shares [15]. The second example is Volkswagen and the famous dieselgate case, when the carmaker falsified emissions tests not only for the Volkswagen brand, but also Škoda, Audi and Seat [16].

4 Conclusion

Due to the constantly increasing innovation potential and business digitization, it is necessary not only to invest time, money and energy in online reputation not only in the real world but also in the virtual world. It impacts on customers, the public, and other stakeholders. Online reputation encourages the need to invest free assets in products and services with a positive reputation. Since, at present, many companies do not pay much attention, its importance must be highlighted.

Based on the survey conducted by selected companies in the automotive field, it can be argued that, despite the publicly known and still current under-estimation of the strength and the importance of reputation, the state of online reputation in the selected area is above average.

Despite the positive current situation of selected companies, it is worth recalling and proposing measures for improvement. To improve online reputation, it is necessary to have a full profile on the social network so that the customer can learn as much information as possible in one place and, of course, cannot even be mistaken for the official website of the car offer. It is advisable to provide contact details on the social network so that the customer does not have to search for the necessary information when contacted.

Even on the basis of selected factors, it is clear that you need to respond promptly to all the comments and reactions of your customers, fans or the public. If the business is active and decent and engaging with the public, it will feel that the business is interested in it and it depends on it. It is important to create a confidential relationship between customers and fans. It is clear that the company does not always meet on the social network with positive reactions only. In this case, it is important not to delete the post, because for others it can act as concealment, an inability to cope with a negative situation, and it can give fans and customers a feeling of insecurity and deception.

For customers, it is important for businesses to be active and to learn from them. This implies that businesses should add more frequent contributions to the different events they participated in, news in the field of planned actions or, for example, price shares, but also, for example, news from the field of planned model series innovations and the like. Businesses should also encourage their fans to work, for example, trying to induce communication with the addition of photos of their "four-crowned pets" in a story, or the situation and the owner of the best photo for which the general public would vote would be rewarded by a little bit.

It follows that automotive companies are able to increase their reputation by busy, decent and active communication through official profiles on social networks that need to be maintained and must be taken care of by the company. Trust, loyalty, and truthfulness are incredibly important for customers, and businesses should strive to achieve the best and most trustworthy relationship with their fans and customers, as well as the public. Online marketing is also influenced by marketing and various marketing activities as outlined above. Businesses should be aware that it is sometimes better to take advantage of smaller and less expensive marketing activities distributed through social networks, especially because of low implementation costs and the ability to work on different segments at the same time. To increase online reputation, it is also important to act ethically, socially and professionally to the public. It is for this reason that the business should highlight all its activities in this area and publish it on its social networking profile, thereby also presenting customers with the willingness to contribute not only to improving their products but also to improving the global situation in the world.

This article is supported by the Slovak research and Development Agency under the contract No. APVV-15-0511.

This paper is supported by the research project "From horse-drawn railway to intermodal transport" within Visegrad Fund.

References

[1] J. Klučka, *Good name is difficult to get and easily lose (Dobré meno sa ťažko získava a ľahko stráca) (In Slovak)*, Available online: https://www.etrend.sk/podnikanie/dobre-meno-sa-tazko-ziskava-a-lahko-straca.html (2006)

[2] U.E. Gattiker, *Brand versus reputation*, Available online: http://commetrics.com/articles/branding-versus-reputation-jeff-bezos-richard-branson-josef-ackermann-and-pat-russo-to-the-rescue/ (2008)

[3] C.J. Fombrun, C. Van Riel, Corporate Reputation Review, 5-13 (1997)

[4] M.L. Barnett, T.G. Pollock, *The Oxford Handboook of Corporate Reputation*, 528 p. (OUP Oxford, UK, 2012)

[5] Š. Hittmár, *Management (Manažment) (In Slovak)*, 301 p., (EDIS, University of Žilina, Slovak Republic, 2006)

[6] M.L. Barnett, J.M. Jermier, B.A. Lafferty, Corporate Reputation Review **9**, 1, 26-38 (2006)

[7] R.J. Burke, G. Martin, C.L. Cooper, Corporate *Reputation: Managing Opportunities and Threats*, 356 p. (CRC Press, 2016)

[8] V.A. Knipp, *Trust the Brand – Corporate Reputation Management in Private Banking*, 148 p. (Diplomica Verlag, 2009)

[9] J. Vaculčiaková, *Reputation affects the value of the business (Reputácia ovplyvňuje hodnotu firmy) (In Slovak)*, Available online: http://www.etrend.sk/podnikanie/reputacia-ovplyvnuje-hodnotu-firmy.html (2003)

[10] R. Burkhardt, *Reputation Management in Small and Medium-sized Enterprises: Analysis and Evaluation of the Use of Reputation Management. A Survey of Small and Medium-sized Enterprises in Germany*, 122 p. (Diplomica Verlag, 2008)

[11] J. Gutiérrez-Cillán, C. Camarero-Izquierdo, R.S. José-Cabezudo, *How brand post content contributes to user's Facebook brand-page engagement. The experiential route of active participation*, 258-274 (2017)

[12] E.D. Raj, L.D.D. Babu, *An enhanced trust prediction strategy for online social networks using probabilistic reputation features*, 412-421 (2017)

[13] H. Crijns, V. Cauberghe, L. Hudders, A. Claeys, *How to deal with online consumer comments during a crisis? The impact of personalized organizational responses on organizational reputation* (2017)

[14] L. Vartiak, *Transcom Proceedings, Section 2: Economics and Management*, 270-276 (2015)

[15] B. Benčat, *Hyundai and Kia have made a hard hit on the consumer fiasco. (Hyundai a Kia tvrdo doplatili na fiasko so spotrebou) (In Slovak)*, Available online: https://www.etrend.sk/auto/hyundai-a-kia-tvrdo-doplatili-na-fiasko-so-spotrebou.html (2012)

[16] M. Kern, *Which cars are related to an emission scandal and what are their owners waiting for in the near future? (Akých áut sa týka emisný škandál a čo ich majiteľov v najbližšom čase čaká) (In Slovak)*, Available online: https://dennikn.sk/254341/akych-aut-sa-tyka-emisny-skandal-a-co-ich-majitelov-v-najblizsom-case-caka/ (2015)

Principles of Logistics Applied to Railway Passenger Transport

Martin Vojtek[1,*], *Martin* Kendra[1], *Vladislav* Zitrický[1], and *Jozef* Daniš[1]

[1]University of Žilina, Faculty of Operation and Economics of Transport and Communications, Department of Railway Transport, Univerzitná 8215/1, 010 26 Žilina, Slovakia

Abstract. Main challenge of logistics is delivering right assortment of products in exact amount, to exact place, in exact time, ecologically and for exact price. Logistics deals with freight transport but when the word 'products' is changed to 'passengers', then many principles can be applied to passenger transport. Railway passenger transport is the key part of passenger transport system, so it is necessary to optimize it on logistics philosophy at first.

1 Introduction

Logistics is defined as an interdisciplinary science, which deals with coordination, harmonisation, connection and optimization of materials, products and services flows including flows of information and financial resources. The main aim is to satisfy needs of the customer with importance of quality aspect and optimal costs [1-4]. Nowadays, logistics become more popular thanks to global trends such as widely opened global market, increasing influence of IT sector and ecological aspects, deregulation, standardization, technical revolution, international migration and industrial society. Majority of logistics research and practical technologies are used in manufacturing enterprises and freight transport but there is a possibility that some modern principles of logistics can be applied in passenger transport [5]. There are some ideas how to apply it specifically in railway passenger transport.

2 Railway station as a multi-directional cross-docking system

Classic cross-docking system is the logistics system, where commodities are delivered to the distribution centre and then all commodities are divided in required amount, structure and quality to the specific retail unit. The distribution centre is not used as a warehouse but it is used as a mixing centre [6-8]. Commodities from suppliers are delivered mostly by high-capacity mode of transport for instance railway transport. When some kind of commodity is combined with other commodities to one complete consignment according to customer requirement, than the consignment is sent to the specific customer mostly by freight road transport [9].

* Corresponding author: martin.vojtek@fpedas.uniza.sk

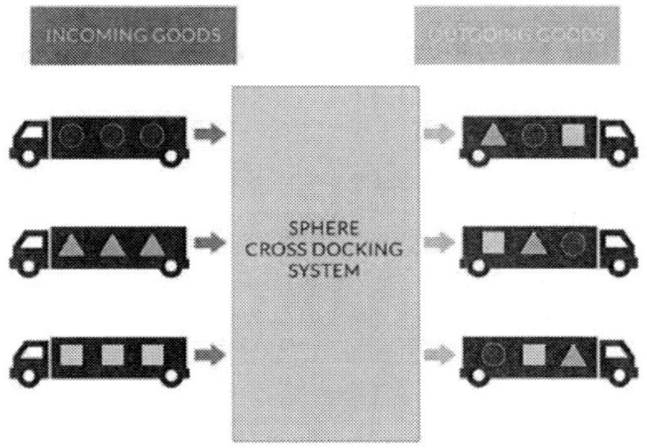

Fig. 1. Classic cross-docking system in freight transport. Source: www.sphereglobal.com.au

Railway passenger station is some kind of transport hub – a starting and finishing point for flows of passengers. Passengers have the opportunity to change the train type from long-haul train to regional train or contrariwise or simply enter or leave the system of railway transport. This is the reason, why it is multi-directional system, not one-way system, as the classic cross-docking system used in freight transport. Main functions of railway passenger stations are:

- ensuring safe, regular and fluent transport,
- ensuring technical and technological services for operators,
- ensuring operation of facilities for traveling public (waiting room, ticket store etc.).

Fig. 2. Zvolen passenger station with marked sectors. Source: authors, base on www.google.sk

In the picture, there are illustrated sectors of real passenger railway station. Purple sector (hall) represents input and output to the system from external systems. Red sector (platforms) represent input and output from internal system consisting of regional and long-haul trains. Green sector (underpass) connects all inputs and outputs and it represents the place, where flows of passengers are mixing up.

 DOI: 10.1051/matecconf/201713400061

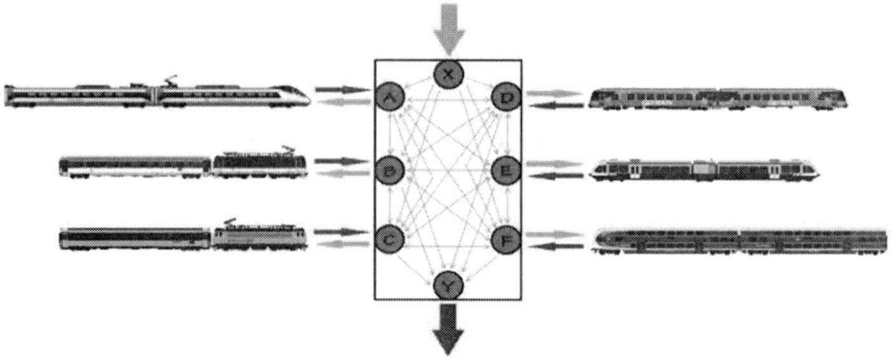

Fig. 3. Simple scheme of cross-docking system applied to railway passenger station. Source: authors

In the picture, there are illustrated flows of passengers between trains which arrive from different directions and depart to different destinations. Points A, B, C, D, E and F represent trains whereas point X represents input to the transport system and point Y represents output from the transport system where passengers use another mode of transport instead of railway transport. Classic cross-docking system is substituted by multi-directional cross-docking system and external input and output are added, but the main character of this logistics system is preserved, because passengers are mixed up and their routing can be changed according to their will and requirements.

3 Terminal of integrated passenger transport as a part of the Hub-and-Spoke technology

Simple Point-to-Point model of delivering goods means that there are direct connections between all elements in the system. Goods are delivered from each supplier to each client therefore many vehicles are necessary so the transport costs are higher. This disadvantage is solved by the Hub-and-Spoke model, where one or more elements are transformed into hub – the assembly centre for selected area. All elements in the selected area have direct connections only with the hub and these connections are called spokes. If there is more than one hub in the system, connections between hubs represent the main transport lines in the system [10-15].

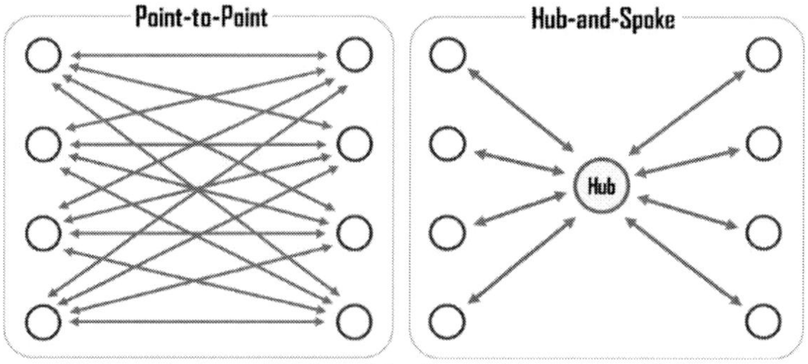

Fig. 4. Comparison of Point-to-Point and Hub-and-Spoke models. Source: [16]

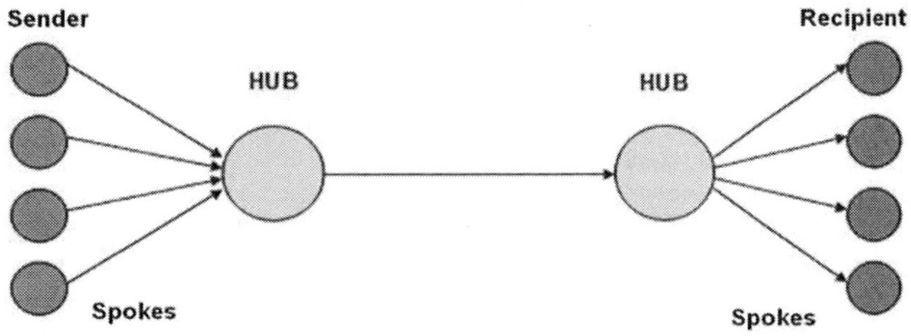

Fig. 5. Scheme of Hub-and-Spoke model with two hubs. Source: [16]

Ideas of the Hub-and-Spoke model are nowadays applied also in passenger transport. In central Europe, there is a modern trend of establishing integrated passenger transport systems in selected regions. Cores of these systems are terminals, where passengers can change vehicle and also the mode of transport, for example get off the bus and get on the train. These terminals are hubs, whence all routes and lines from some region or district are connected. Hub-and-Spoke model is also very common in air transport – airports are hubs and their importance and number of connections usually depending on location and size.

Fig. 6. Terminal of integrated passenger transport in Moldava nad Bodvou, Košice region, Slovakia. Source: [17]

Building these new terminals will improve transport accessibility in the selected region. Operators, who participate in the integrated passenger transport system, are more effective and notice increased demand for transport services. Traveling is easier and passengers can use public transport more often, when they travel regularly to work and school, or occasionally for vacation or other reasons.

4 Progress of railway passenger transport with using Just-in-Time philosophy

In general, effectiveness of passenger transport system depends on many factors, such as spatial and temporal accessibility, timetable, connectivity etc. Connectivity of passenger trains and other means of transport can be distinguished also from temporal and spatial point of view. Temporal connectivity is such sequence of arrivals and departures of different passenger trains and other means of transport, which allows changing the different passenger vehicles easily in regard to necessary time. Spatial connectivity means the distance between two passenger vehicles, among which the passenger is moving.

Passenger transport is generally considered as an effect of where people are in the exact time and their need to move. The most common reasons for moving are work, education, vacation, sport, cultural events, visiting social and health facilities, meeting family members and friends. One of the main criterions, which are related to the passenger transport, is the time length of transport, which is closely dependent on speed. It does not mean only vehicle speed, because there is also the time of going from home to the station, time for waiting, time for transport and time to reach the finish point. In some case, there is also a time for waiting to another connection, when the passenger change transport vehicles or means of transport.

Application of Just-in-Time ideas can optimize connectivity in railway passenger transport by reduction of transfer time – starting at home and finishing in the transport destination. It is important to synchronize arrivals and departures of all connected trains in all points, where passengers can get on, get off and change the vehicle. Minimizing of waiting time will increase quality of passenger transport in general. Particular emphasis must be put on reliability of all included vehicles, because delays could have serious consequences to the whole system. In freight transport, Just-in-Time technology eliminates needless warehousing, so in passenger transport this means elimination of needless waiting. This can be reached by higher number of trains, buses and other passenger vehicles, which are used in passenger transport system [15, 18-20].

Preconditions of Just-in-Time application in railway passenger transport are [13, 21]:

- dominance of customer (passenger),
- timetable dependent on passengers' needs,
- synchronized arrivals and departures in all point in the transport system,
- harmonised conditions for all operators in the transport system,
- high reliability and punctuality.

Disadvantages of Just-in-Time application in railway passenger transport are:

- overcrowding of transport vehicle may occur,
- increased number of trains could reduce the throughput of railway,
- preference of busses (more flexible).

5 Conclusion

Contemporary challenge of logistics is to ensure the connection between offer and demand, so customers would have goods and services available in exact time, exact place and exact quality. Even though logistics is mostly used in industry and freight transport, there are many principles, which can be applied in railway passenger transport and other means of passenger transport. Consecutive development of these ideas can improve quality of passenger transport system in similar way, how it improves freight transport system.

The paper was supported by the VEGA Agency, Grant No. 1/0019/17 "Evaluation of regional rail transport in the context of regional economic potential with a view to effective use of public resources and social costs of transport", at Faculty of Operations and Economics of Transport and Communication, University of Žilina, Slovakia.

This paper is supported by the research project "From horse-drawn railway to intermodal transport" within Visegrad Fund.

References

1 R. Kampf, L. Lizbetinova, K. Tislerova, Open Engineering **7**, 26-30 (2017), DOI: 10.1515/eng-2017-0006

2 J. Lizbetin, O. Stopka, Open Engineering **6**, 441-445 (2016), DOI: 10.1515/eng-2016-0066

3 J.P. Skrinjar, B. Abramovic, N. Brnjac, Technical Gazette **22**, 6, 1617-1621 (2015)

4 J. Gnap, V. Konecny, M. Poliak, Ekonomicky Casopis **54**, 668-684 (2006)

5. J. Gasparik, V. Gaborova, V. Luptak, *Transport Means – 20th International Scientific Conference on Transport Means*, 245-249 (Juodkrante, Lithuania, 2016)

6 A. Dolinayova, M. Loch, M. Kanis, Research in Transportation Economics **54,** 33-40 (2015), DOI: 10.1016/j.retrec.2015.10.023

7 V. Molnar, *CLC 2013 – Carpath. Logist. Congress*, 561-566 (Krakow, Poland, 2014)

8 L. Cerna, B. Bukova, *Transbaltica 2015*, 377-385 (Procedia Engineering **134**, Vilnius, Lithuania, 2016), DOI: 10.1016/j.proeng.2016.01.023

9 J. Lizbetin, L. Bartuska, *Transbaltica 2017*, 206-211 (Procedia Engineering **187**, Vilnius, Lithuania, 2017), DOI: 10.1016/j.proeng.2017.04.366

10 G. Fedorko, V. Molnár, J. Strohmandl, M. Vasiľ, *Transport Means – 19th International Scientific Conference on Transport Means*, 466-469 (Kaunas, Lithuania, 2015)

11 J. Gasparik, V. Luptak, P. Mesko, *ICTTE – 3rd International Conference on Traffic and Transport Engineering*, 388-392 (Belgrade, Serbia, 2016)

12 J. Majerčák, *LOGVD 99 – 2nd Scientific Conference with International Participation* (Žilina, Slovakia, 1999)

13 J. Sekulová, E. Nedeliaková, Logi – Scientific Journal on Transport and Logistics **6**, 2, 79-89 (2015)

14 L. Černá, V. Zitrický, J. Daniš, Open Engineering **7**, 1, 6-13 (2017), DOI: 10.1515/eng-2017-0002

15 L. Černá, A. Dolinayová, J. Daniš, *Transport Means – 20th International Scientific Conference on Transport Means*, 414-419 (Juodkrante, Lithuania, 2016)

16 M. Chovancova, V. Klapita, *Transport Means – 20th International Scientific Conference on Transport Means*, 112-116 (Juodkrante, Lithuania, 2016)

17 H. Böhme, *Krátka návšteva v Košiciach a v Moldave nad Bodvou*, Available online: https://www.vlaky.net/zeleznice/spravy/6026-Kratka-navsteva-v-Kosiciach-a-v-Moldave-nad-Bodvou/ available on 2017/09/25 (2016)

18. I. Kubasakova, J. Jagelcak, Communications: scientific letters of the University of Žilina **18**, 2, 109-112 (2016)

19. Z. Záhumenská, J. Gašparík, *12th International Scientific Conference of Young Scientists on Sustainable, Modern and Safe transport,* 976-981 (Procedia Engineering **192**, 2017), DOI: 10.1016/j.proeng.2017.06.168

20. J. Gašparík, V. Lendel, Logi – Scientific Journal on Transport and Logistics **1**, 1, 33-43 (2010)

21. I. Kubasakova, R. Kampf, O. Stopka, Communications: scientific letters of the University of Žilina **16**, 2, 9-13 (2014)

AUTHOR INDEX

AUTHOR INDEX